AF251486

ADHESIVE BONDING OF ALUMINUM ALLOYS

MATERIALS ENGINEERING

A Series of Reference Books and Textbooks

SERIES EDITOR

Robert S. Shane
President
Shane Associates, Inc.
Wynnewood, Pennsylvania

1. Adhesive Bonding of Aluminum Alloys, *edited by Edward W. Thrall and Raymond W. Shannon*

Additional Volumes in Preparation

2. Materials and Processes, Third Edition *(in two parts), edited by James F. Young and Robert S. Shane*

ADHESIVE BONDING OF ALUMINUM ALLOYS

edited by

Edward W. Thrall

Consultant
Aeronautical Structures
Point Arena, California

Raymond W. Shannon

Ray Shannon and Associates
Santa Ana, California

MARCEL DEKKER, INC.   **New York and Basel**

Library of Congress Cataloging in Publication Data

Main entry under title:

Adhesive bonding of aluminum alloys.

(Materials engineering ; 1)
Includes index.
1. Aluminum alloys--Bonding--Addresses, essays,
lectures. I. Thrall, Edward W., [date].
II. Shannon, Raymond W., [date]. III. Series:
Materials engineering (Marcel Dekker, Inc.) ; 1.
TS718.A34 1985   673'.72258   85-10445
ISBN 0-8247-7405-1

MARCEL DEKKER, INC.
270 Madison Avenue, New York, New York 10016

Current printing (last digit):
10 9 8 7 6 5 4 3 2 1

PRINTED IN THE UNITED STATES OF AMERICA

Preface

During the 1970s, a major effort was made by the adhesive bonding
industry to improve the reliability of structural adhesive bonding. A
lack of confidence in the reliability and durability of these adhesives
prevented their use as the primary fastening method in major structures.
It became necessary to understand the reasons why a disbond can occur in an environment that is normal for most structures. Now that
systems, such as surface treatment, adhesive selection, etc., have
been developed to prevent bonded joint failures due to corrosion, adhesive bonding of aluminum alloys can be used to advantage. It is a very
viable method of construction.

By the end of the 1970s, it was obvious that enough information
was available on all aspects of adhesive bonding to allow designers and
engineers to confidently join or fasten primary structures with adhesive
materials. Many of the ideas we believed to be true at the beginning
of the 1970s proved to be false, many things were discovered that we
had thought could not be, and there are many things that still require
a better understanding. Still, this book was compiled because of what
we do know: we can, and have proved we can, build an adhesive-
bonded primary structure that is durable and structurally sound. For
this and the following reasons, this book was written.

When planning to introduce bonded joints into any aluminum alloy
structure, the designer must have confidence in the structural reliability of this type of joint. This book is intended to furnish this confidence to the designer, the research engineer, the tool designer, the
manufacturing engineer, and the user of bonded aluminum structures.
All the latest information is presented on how to make a properly
bonded joint of aluminum alloys that will have the desired static
strength and an excellent service life. The international array of
authors in this book have devoted a major part of their lives to working

on the problems associated with the various steps or processes required
to properly join two pieces of aluminum alloy with an adhesive.

When splicing structural parts, the optimum is reached when the
joint is stronger than the parent structure being joined. The fatigue
life of the bonded joint should not be the weak link. These critical
features can be accomplished when bonding aluminum alloys if the recommendations of this book are carefully followed.

From the opening chapter, which deals with the history of aluminum alloy bonded joints and the problems and successes of the early
days, we proceed to surface preparation methods, surface analysis,
adhesive selection (including elevated-temperature-resistant systems),
mechanical properties, environmental testing, chemical analysis control,
coatings for additional environment protection, structural analysis,
tool design, nondestructive testing, and repair.

It should be apparent that the amount of information generated on
this subject is voluminous. Because of this, several chapters originally written and presented for inclusion in this book could not be included. Fortunately, some of the significant material in those chapters
has been covered here in other chapters.

We would like to acknowledge gratefully those who worked as hard
at preparing chapters that could not be included as those whose chapters are contained in the book. The following are those authors whose
chapters were not included: Frank Lennert, Douglas Aircraft Company;
Roland Cassar, Douglas Aircraft Company; Keh Wu and Robert Hocker,
Northrop Corporation; Dr. Lynn Mahony, Hysol Division/Dexter Corporation; Weldon Scardino, Wright-Patterson Air Force Base. To all, we
are deeply indebted.

Over the years there have been many successful applications of
adhesive bonding for joining aluminum alloys. We hope that those who
read this book will be able to apply the newer processes, materials,
and test inspection information presented here to produce only reliable
structures. We are confident that those of you in the research field
of structural adhesive bonding will find this volume as interesting and
rewarding as we have.

Edward W. Thrall
Raymond W. Shannon

Contents

Contributors

Walter Althof Deutsche Forschungsanstalt fur Luft-und Raumfahrt (DFVLR), Braunschweig, Federal Republic of Germany

Paul F. A. Bijlmer, Jr.* Fokker Aircraft Company, Amsterdam, The Netherlands

Mervin A. Danforth Douglas Aircraft Company of the McDonnell Douglas Corporation, Long Beach, California

Deborah K. Hadad Lockheed Missiles & Space Company, Inc., Sunnyvale, California

Donald J. Hagemaier Douglas Aircraft Company of the McDonnell Douglas Corporation, Long Beach, California

L. J. Hart-Smith Douglas Aircraft Company of the McDonnell Douglas Corporation, Long Beach, California

Ray E. Horton The Boeing Commercial Airplane Company, Seattle, Washington

Edward J. Hughes Singer Company, Kearfott Division, Little Falls, New Jersey

Raymond B. Krieger American Cyanamid Company, Havre de Grace, Maryland

---

Present affiliation: Alex Harvey Industries, Auckland, New Zealand.

Murray H. Kuperman United Airlines Maintenance Operations Center, San Franciso, California

J. Arthur Marceau The Boeing Commercial Airplane Company, Seattle, Washington

Ernest C. Millard Rohr Industries, Inc., Riverside, California

Saburo Nakahara*** Douglas Aircarft Company of the McDonnell Douglas Corporation, Long Beach, California

Narvel L. Rogers Bell Helicopter Textron, Fort Worth, Texas

Robert J. Schliekelmann[†] Fokker B. V., Schiphol-Oost, The Netherlands

Edward W. Thrall[‡] Douglas Aircraft Company of the McDonnell Douglas Corporation, Long Beach, California

---

*Present affiliation: Northrop Corporation, Pico Rivera, California
[†]Consultant, Advanced Structures Engineering, Amstelveen, The Netherlands.
[‡]Consultant, Aeronautical Structures, Point Arena, California.

ADHESIVE BONDING OF ALUMINUM ALLOYS

1

Introduction

ROBERT J. SCHLIEKELMANN* *Fokker B.V., Schiphol-Oost,
The Netherlands*

I. HISTORY OF ADHESIVE BONDING IN AIRCRAFT

Typically, adhesive-bonded joints use an intermediate layer that adheres to the faying surfaces of the parts to be joined. That layer (the adhesive) has sufficient strength of its own to transfer mechanical loads from one component into the other. Joining of material or components by means of adhesives has been done for many centuries, but always for the purpose of obtaining an improvement in the properties of the resulting materials and/or structures. There are examples dating back as far as 3000 B.C. [1]. Since then, history has shown many striking applications of adhesive bonding as a means of improving structures and materials. In the history of aircraft, adhesive bonding was used from the earliest years, as shown in Fig. 1. Large adhesive-bonded structures for wooden aircraft (Fig. 2) were used up through the last years of World War II. Those years also saw applications of structural adhesives in metal aircraft.

The first adhesive-bonded joints between metal parts emerged in two different ways. First, they were developed from vibration-damping structures which used rubber layers vulcanized between metal parts; and second, the pioneering research and development work of De Bruyne (1939) in the field of high-performance synthetic wood adhesives led to the discovery that hot curing adhesives could also adhere very well to metal surfaces. De Bruyne also contributed greatly to knowledge of the adhesion phenomena of adhesives on metal surfaces (Fig. 3).

Present affiliation: Consultant, Advanced Structures Engineering, Amstelveen, The Netherlands.

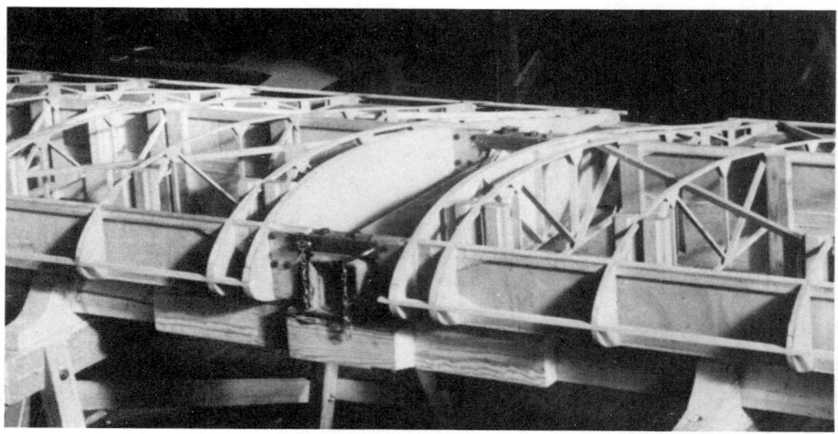

Figure 1   Wooden wing structures of early aircraft were almost entirely joined by adhesive bonding.

Figure 2   Adhesive-bonded cantilever wooden wing structure of four-engine Fokker F-36 airliner with 33-m span, built in 1934.

Figure 3   Norman A. De Bruyne, pioneer of adhesive metal bonding and composite aircraft structures. (From Ref. 2.)

The British de Havilland Aircraft Company saw the importance of the work of De Bruyne at a very early date. This allowed the British Royal Air Force during the last years of World War II to use large numbers of aircraft which had the structural efficiency of their all-wooden wings greatly improved by the application of adhesive-bonded metal reinforcements (Figs. 4 and 5). Shortly after the end of World War II the Fokker Company applied the same polyvinyl acetal/phenolformaldehyde base adhesive to entirely metal structures in both military and civil aircraft [4]. The development of rubber-based metal adhesives, which initially took place in both the United States and Germany, was continued after 1945 in the United States for structural applications in the construction of aircraft. The first large-scale application took place in the structure of the gigantic Convair B-36 strategic bomber (Fig. 6). Another milestone in the history of adhesive bonding was the publication of the results of a research and development program in 1946 by the Swiss Ciba A.G. that led to the discovery of epoxy resins. This new group of synthetic resins appeared to have properties that made them excellently suitable as a base material for structural adhesives [5].

Initially, applications of adhesive-bonded joints were limited to the solution of specific problems, such as the reinforcement of a basically wooden wing structure by menas of adhesive-bonded aluminum alloy spar caps or doubler plates. In the 1950s, the use of adhesive-bonded

Figure 4   De Havilland Hornet aircraft in which adhesive bonding was used for attachment of light alloy spar caps to plywood shear webs and skin of the wings. (From Ref. 3.)

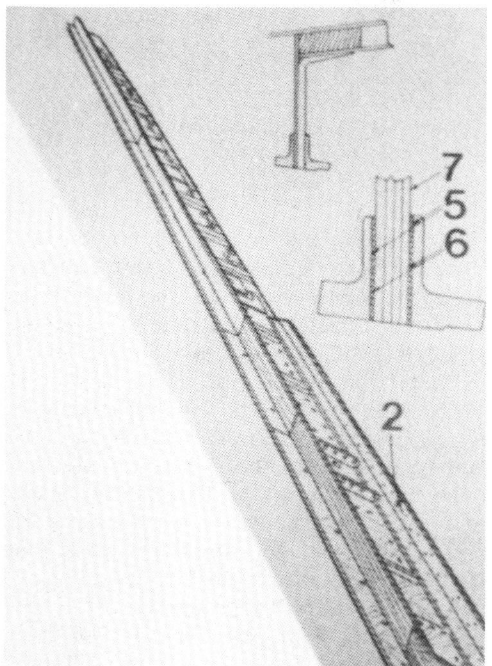

Figure 5   Details of the Hornet wing spar structure. The light alloy lower spar caps (2) are adhesively bonded to the plywood shear web (7) via a ply of wood veneer (5 and 6). (From Ref. 3.)

Figure 6 Convair B-36 bomber which had adhesive-bonded secondary structure, including magnesium alloy sheet metal.

spar caps or doubler plates. In the 1950s, the use of adhesive-bonded joints in aircraft structures increased considerably and has been growing ever since. A striking example in this respect is the structure of the Fokker F-27 Friendship, which made its first flight in 1955. Four thousand individual parts were composed into 550 bonded assemblies with an area of 370 m^2 per aircraft (Fig. 7).

Two different kinds of adhesive-bonded structures have been developed for aircraft use. First there are the metal-to-metal bonded structures, which are locally reinforced by bonded doubler plates; and second are components composed of bonded multiply laminations where each layer progressively increases the cross-sectional area, such as for stringers and spar caps (Fig. 8) [7].

Figure 7 Beginning with the prototype in 1955 and continuing to this day, the Fokker F-27 Friendship has had a large percentage of the load-carrying structure adhesively bonded. (From Ref. 6.)

Figure 8 Cutaway section of the lower wing panel of the Fokker F-27 airliner, showing the laminated metal skin and a bonded stringer.

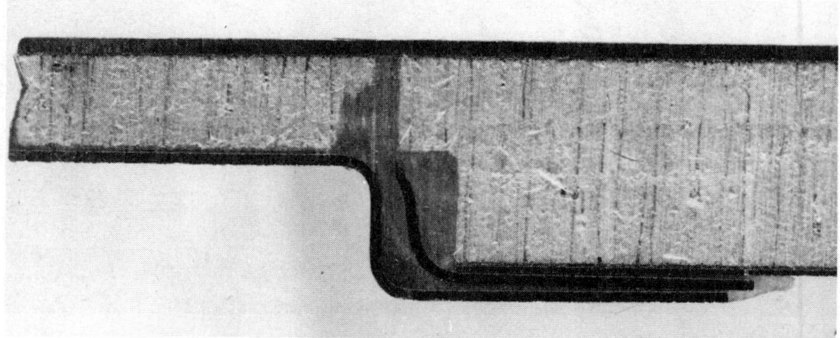

Figure 9 Cross section over a sandwich floor panel with a metal edge member and end-grain balsawood core.

Figure 10   Bonded honeycomb sandwich wing structure for the Bréguet Atlantic.   (From Ref. 8.)

The second category of adhesive-bonded structures is that which contains bonded joints between rather thin metal skin sheets and low-density core material, called "sandwich" structures (Fig. 9). Initially, plastic foam and end-grain balsa wood were used as core materials; later, honeycomb made out of paper, glass fiber, aluminum foil, and organic fiber [e.g., (Nomex) (Dupont)] was used. Later applications of sandwich structures also had bonded doubler plate reinforcements, and profile-type edge members were incorporated (Fig. 10).

II.   REASONS FOR BONDING IN AEROSPACE STRUCTURES

Reasons for the use of adhesive-bonded aircraft structures of all categories have increased considerably in recent years. Skin panels with adhesive-bonded stiffeners were originally preferred over riveted panels of similar configuration for reasons of their greatly improved local stability. This was due to the continuously bonded stiffener flanges, shown in Fig. 11. As a result of this, the load-carrying capacity under axial compression or shear loading is improved considerably. This is the case not only for static loading but also under conditions of alternating loading either at low frequency, such as that due to gusts during flight, or at high-frequency loading, such as occurs under high sound pressure in the vicinity of jet engines (Fig. 12). Only during

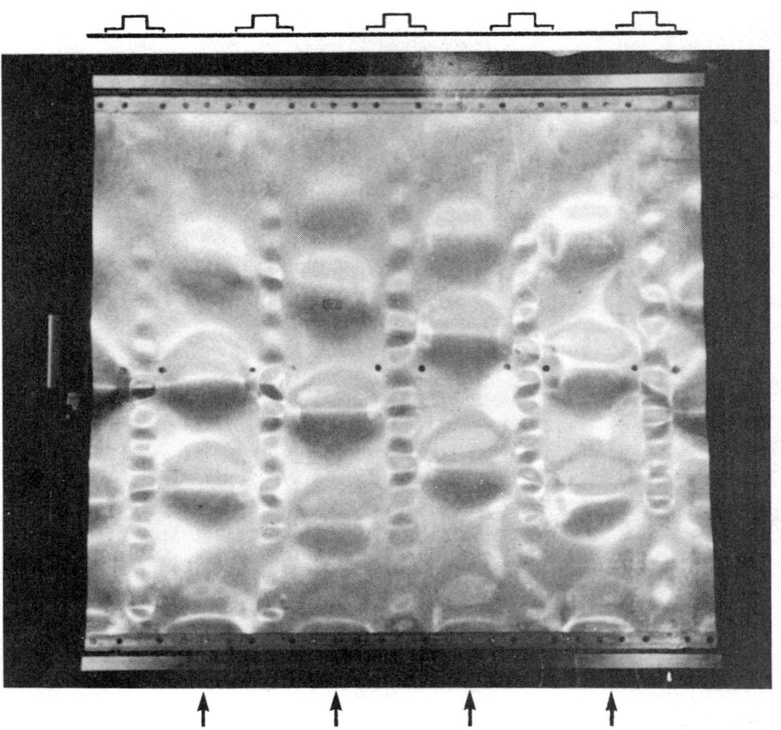

Figure 11   Skin panel with bonded tophat stringers under axial compression load shows how the stable areas of the bonded flanges define the buckling configuration.

the last decades has the background of the excellent fatigue properties of adhesive-bonded structures been fully explored. Adhesive-bonded stiffener panels, shear webs, and laminated skins have been shown to have very attractive fracture mechanical properties. These properties have been the focus of attention of modern structural designers, who now have the task of developing "damage-tolerant" structures [10]. There are structures in which not only is the operational lifetime until the occurrence of the first crack important, but also behavior in the cracked condition. Low crack propagation rates and high residual strength in the cracked state are of great interest today [11]. This will lead to an even larger application of adhesive-bonded joints in future aircraft structures. So the idea has developed not only of bonding the longitudinal stiffeners onto the fuselage skin, but also of bonding the flanges of the transverse frames, ribs, and bulkheads to skin panels.

(a)

Figure 12   (a) Riveted structure of landing gear strut fairing door, showing extensive damage due to sonic fatigue. (b) Redesigned landing gear strut fairing door with bonded honeycomb sandwich structure, showing satisfactory sonic fatigue resistance. (From Ref. 9.)

The great interest in improved fracture mechanical behavior of metal aircraft structures has derived from the rather disappointing operational results obtained with components that have been machined from solid material. The latter method had been chosen because of its high production efficiency, in particular when numerically controlled milling machines could be applied. The rather low modulus of elasticity of the cured adhesive layer compared with that of the sheet metal allows the bonded elements to deform practically independently when a fatigue crack has developed (Fig. 13). This plane stress condition is the basis for the superior fracture mechanical behavior of the metal-laminate composite. This can be attributed to the favorable properties of the thin rolled sheet, which are much better than those of thick solid plate (Fig. 14).

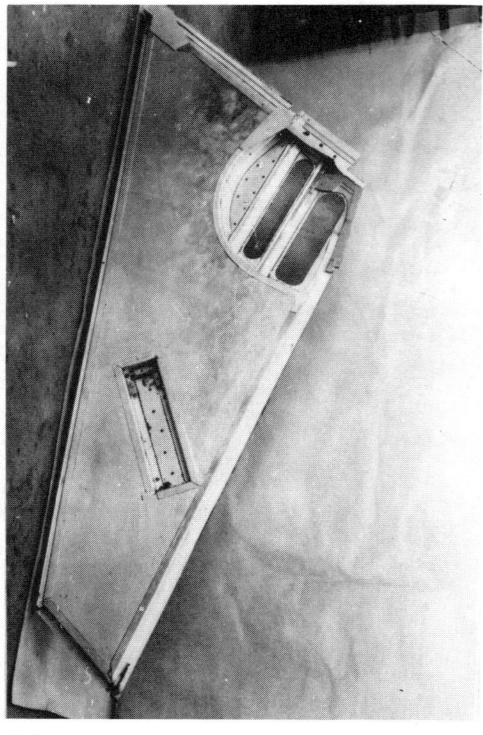

(b)

Figure 12   (continued)

    Sandwich structures, which originally were adopted because of
their attractive strength/stiffness ratio aspects, are at present also
chosen for a number of other reasons. The growing demands for close
dimensional tolerances on aerodynamic and related parts, such as the
trailing edges of fixed and movable wing and tailplane components
(e.g., ailerons, rudders, elevators, spoilers, flaps, airbrakes, slats,
tabs, and vanes), can effectively be fulfilled by the very stable and
light sandwich structures (Fig. 15). Sandwich constructions are also
used for their acoustical damping characteristics in and around engines
and their nacelles, to reduce noise for both the aircraft crew and the
passengers, and to cope with the environmental problems around run-
ways and airports from which the aircraft are operating (Fig. 16).
The same properties also lead to good resistance against high sound
pressure, which can cause serious fatigue failures in a very short time.
These favorable properties of adhesive-bonded structures guarantee
that they will be used on an even larger scale in future aircraft.

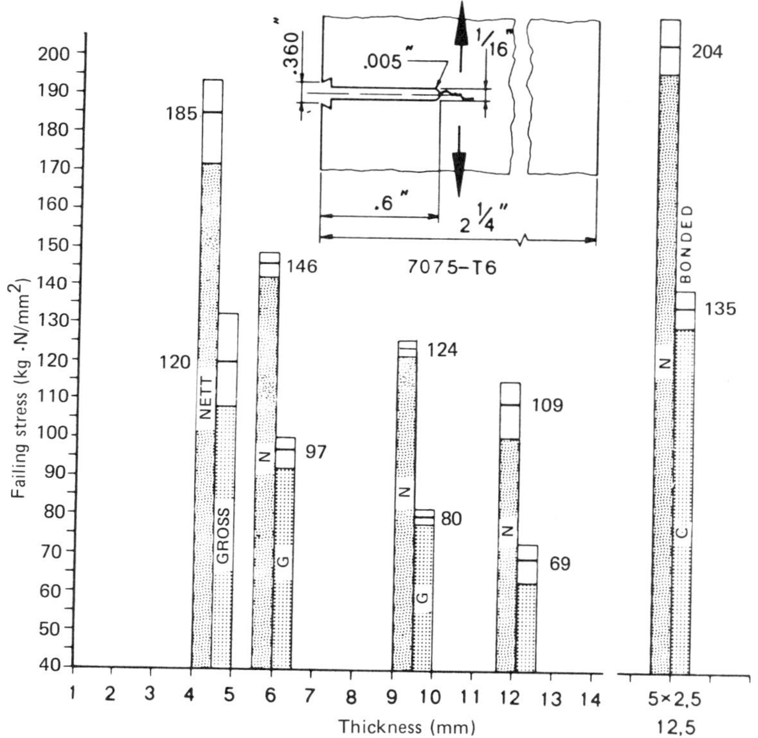

Figure 13 Residual strength of solid and laminated plate of 7075-T6
material with initial flaw. The four sets of columns on the left-hand
side show the gross and net failing stress of four thicknesses of cracked
plate. Increasing thickness shows a corresponding reduction in residual strength. The right-hand column shows the residual strength of a
bonded laminate of 5 sheets of 2.5 mm thickness each, or a total thickness of 12.5 mm. The best properties are clearly those of the individual thin sheets bonded together.

Figure 14   Specimens used in the comparative tests shown in Fig. 13. The left-hand specimen shows a brittle failure of the precracked thick plate.   The right-hand specimen shows ductile failure of the individual thin sheets bonded together.   (From Ref. 12.)

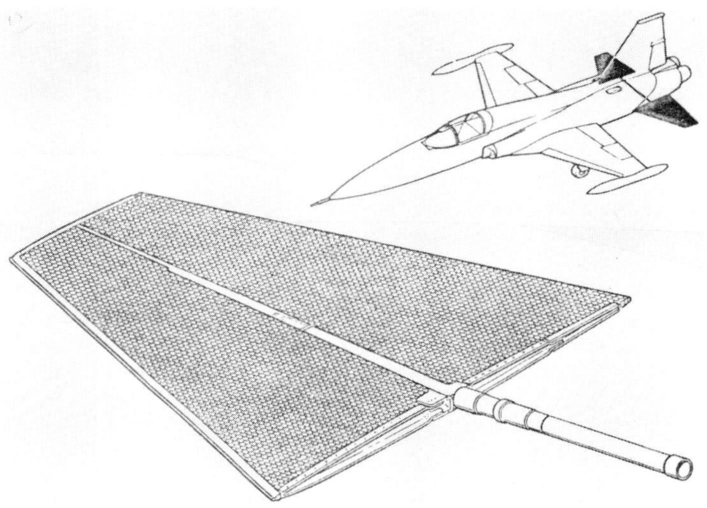

Figure 15   Close-tolerance sandwich structure for all-flying stabilizer surface for high-speed jet fighter.

Figure 16   Honeycomb sandwich structure for a fan reverser of a modern bypass jet engine which gives good sound attenuation properties. (From Ref. 13.)

III. PROBLEM AREAS

The adoption of adhesive bonding as a means for attachment of metal
components in aircraft structures reintroduced to the technology of
manufacturing aircraft an aspect that was present the early days of
aircraft manufacture. At that time it was the craftmanship of the individual worker that determined the quality and the reliability of the aircraft. With the era of adhesive bonding of metal structures this aspect
returned to aircraft manufacture (Fig. 7). It means that the mechanical properties of bonded joints are in fact created during manufacture
instead of depending on the quality of prefabricated fasteners such as
rivets and bolts. In the latter case, the aircraft manufacturer had only
to inspect the quality of the fasteners upon delivery. In case of high-
volume use of such items, the testing of statistically representative
samples would give a sufficiently reliable guarantee of the quality of
the fasteners. In the case of adhesive bonding and now also in the
manufacture of advanced composite structures, the mechanical properties of the joints and structures are the result of both the quality of
the raw materials and the quality of the execution of the manufacturing
process used.

The fact that the excellent properties of adhesive-bonded joints
are based on complex physical and chemical phenomena and that the
manufacturing processes are quite complicated makes prediction of the
long-term operational durability of these joints and structures difficult.
From the beginning aircraft manufacturers have been reluctant to adopt
the use of adhesive bonding, because they had no proof that bonded
joints would exhibit sufficient long-term durability. We have now had
over 30 years of experience in the use of adhesive-bonded joints of
various degrees of importance in aircraft and spacecraft structures.

Several years ago it became apparent that the virtues of adhesive-
bonded structures could play an important role in the solution of the
many problems that arose in connection with the new requirements for
more damage-tolerant structures. Upon reviewing the operational results obtained with adhesive-bonded structures it became apparent that
opinions were not all positive. The qualifications expressed differed
widely, from "highly favorable" through "very unfavorable," and bonded
structures were sometimes said to be "a source of continuous maintenance problems" [14].

In view of the possible increased use and more important applications of adhesive-bonded structures, it was felt necessary to make a
thorough analysis of the reasons and backgrounds for both the positive
and the negative results obtained in the past. In this respect the Primary Adhesively Bonded Structure Technology (PABST) program, initiated by the U.S. Air Force Materials Laboratory, was of primary importance. That program was carried out by the Douglas Aircraft Company as prime contractor in cooperation with many aerospace industries
and institutes in both Europe and the United States. The editors of

this book, E. W. Thrall, Jr., and R. W. Shannon, Jr., played a leading role in the execution of the PABST program, as program manager
and responsible material and process engineer, respectively. In fact,
at the time the PABST program was nearing completion, these gentle-
men took the initiative to edit a book containing all available information about the state of the art of adhesive bonding and adhesive-bonded
structures, ensuring themselves of the cooperation of experts in many
special aspects of the field. The book itself is a demonstration of the
complexity of adhesive bonding and the design of adhesive-bonded
structures [15].

IV. FAILURE MODES

An adhesive-bonded joint must be considered as a chain consisting of
a number of individual links, the weakest of which finally determines
the strength of the chain. The links themselves differ entirely from
each other, as shown in the following list (see also Fig. 17). The chain
consists of:

1. *Metal substrate* (adherend): the cohesion strength of the cured
    adhesive layer (1)

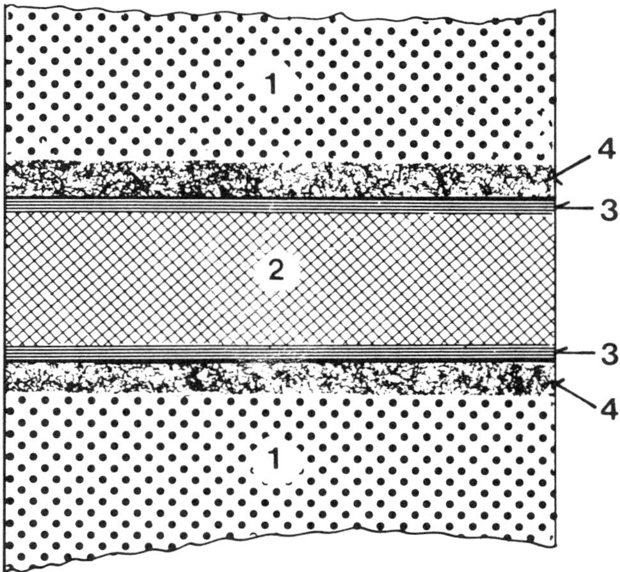

Figure 17   Metal adhesive bonding interfaces.

2. *Adhesive*: the adhesion strength between the cured adhesive layer (1) and the oxide layer (2) on top of the substrate to be bonded (3)
3. *Primer*
4. *Oxide layer* on the metal substrate

The failure of any link results in a distinct failure mode in the bonded joint. To understand the complete mechanism of the strength and durability of adhesive-bonded joints and structures it is necessary that each link be analyzed in depth. That is why so many specialists have contributed to this book. The subjects covered include the following:

Selection of the proper adhesive materials in view of the requirements of the bonded joints as dictated by the type of application that is foreseen.
Treatment of the surfaces to be bonded prior to the application of
adhesives. The adherend must not only guarantee optimum
adhesion, it must also be highly durable.

Operational experience has shown that a substrate surface that is in a condition to yield optimum adhesion is not always suitable to provide optimum durability. One of the greatest dangers to the durability of adhesive-bonded joints is the occurrence of humidity penetration

Figure 18   Seriously damaged sandwich panel which resulted from bondline corrosion.

and subsequent corrosion from the edges of the bond inward, leading finally to complete disintegration of the bonded joint, even though initially it was expected to be of high quality (Fig. 18).

There are several schools of thought on providing an adherend surface with optimum adhesion qualities combined with optimum durability. Different companies in different countries have used distinctly different approaches. These differences can be found both in the application of various surface treatments of the adherends (e.g., different etching processes whether or not followed by anodizing treatments) and in the use of adhesives based on different polymers and in the incorporation of adhesive primers prior to the application of adhesive materials. The latest views on this complicated matter, based on extensive test and operational results, are given in this book.

V. OPERATIONAL RESULTS

Inquiries directed to operators of both military and civil aircraft that utilize adhesive-bonded structures have yielded the rather disappointing result that in only a limited number of cases have the operational results been satisfactory. We discuss below the types of service problems that have been encountered. The problems can be split up into (1) general problems and (2) problems specific to metal-to-metal and honeycomb sandwich structures. A general complaint regarding the surface treatment of metal adherends has been that it was either incomplete or of too low quality. "Incomplete" refers to the lack of additional treatment to increase durability, such as anodizing and/or the use of corrosion-inhibiting primers. Insufficient etching or contamination are indicated as possible causes of poor operational performance.

Metal-to-metal bonded joints gave trouble when adhesives of insufficient water resistance and/or woven glass-cloth carriers were used. The best operational results for metal-to-metal bonded structures were obtained when chromic acid anodizing had been incorporated in the surface treatment process of the aluminum adherends and phenolic resin-based adhesives were used.

Service problems with honeycomb sandwich constructions were based on those already mentioned, but in addition, on insufficient watertightness of the sandwich structure as a whole and insufficient durability of the honeycomb core. An example of the extent of the durability problem for honeycomb sandwich components is demonstrated by the following information from one large civil aircraft operator. In 1977, in a large jet-airliner fleet that contained about 57,000 operational honeycomb sandwich components, approximately 14% had to be repaired annually, either partially or completely. Similar information has been obtained from both military and civil operating organizations [15]. In the following chapters the problem of design and manufacture of adhesive-bonded structures with the specific objective of high operational durability will be discussed.

VI. QUALITY CONTROL

Because the quality of adhesive-bonded structures is the result of an
intricate manufacturing process, tight quality control is of utmost importance. This starts with accurate control of the adhesive materials
when they arrive in the shop. Control of the surface treatment process
required is largely a chemical matter. The assembly of the components
to be bonded requires considerable skill on the part of the operator to
ensure that each part comes at the right location and in the right atti-
tude so that the adhesive-bonded joint has a good chance of being of
good quality. The actual curing process involves accurate control of
temperature, time, and pressure during the period in which the components are in the autoclave or press. In addition, many other factors,
such as the methods used for application of the adhesive materials, have
an important influence on the quality of the end product.

   This means that when the bonded structure is ready for final inspection in the bonding shop, the history of its components as to quality is of greatest importance. It is impossible to have only a final inspection for complete judgment as to the quality of bonded structures.
The final inspection therefore comprises, first of all, a review of all the
documentation to assure that each step of the manufacturing process
has been under close quality control. The actual inspection is limited
to a check on the dimensional quality of the product and of the cohesion
quality of the product, which is the result of the last manufacturing
step: the curing process. For that part of quality control advanced
nondestructive testing techniques are used, which for the most part
have been developed specifically for this purpose. Nondestructive inspection of bonded components is therefore also an important topic in
later chapters [16].

VII. CONCLUSIONS

At this moment in the history of adhesive bonding there is not only
many years of experience but also the indication that adhesive-bonded
joints will experience very strong development in aerospace and other
structures because of their unique possibilities for more reliable structures and for improved structures based on advanced fiber composites.
In the following chapters the experience of the past combined with the
knowledge of today provides an excellent basis for successful application of adhesive-bonded joints in structures of the future. This future
will be successful only if experts from many disciplines continue to co-
operate in the way they have by contributing to this book.

REFERENCES

1. Dietz, A. G. H., *Composite Engineering Laminates*, MIT Press, Cambridge, Mass., 1969.
2. De Bruyne, N. A., The Physics of Adhesion, *J. Sci. Instrum.*, Vol. 24, No. 2, Feb. 1947, pp. 29–35.
3. Golf, W. E., The De Havilland Hornet, *Aircraft Prod.*, Vol. 8, 1946, p. 211.
4. Povey, H., The Production of the Comet, *Aircraft Prod.*, Vol. 13, 1951: May, p. 134; June, p. 170.
5. Preiswerk, E., and A. von Zeerleder, Araldit, ein neues Kunstharz zum Verbinden von Leichtmetallen, *Schweiz. Arch. Angew. Wiss. Tech.*, Vol. 12, No. H 4, 1946, pp. 115–119.
6. Schliekelmann, R. J., Adhesive Bonded Metal Structures, in *Adhesion and Adhesives*, Vol. 2, R. Houwink and G. Salomon, eds., Elsevier, Amsterdam, 1967, p. 281.
7. Schliekelmann, R. J., Redux for Spars, *Aeroplane*, Vol. 8, 1952, p. 216.
8. Plantema, F. J., Sandwich Construction, in *Airplane, Missile and Spacecraft Structures*, Vol. 3, Wiley, New York, 1966.
9. Rondeel, J. H., Comparative Fatigue Tests with 2024 Clad Riveted and Bonded Stiffened Panels, Netherlands Aerospace Lab. (NLR) Rep. S411, 1952.
10. MIL-STD-1530, Aircraft Structural Integrity Program, Sept. 1972.
11. MIL-A-83444, Airplane Damage Tolerance Requirements,
12. Bijlmer, P. F. A., Fracture Toughness of Multiple Layer Adhesive Bonded Aluminum Alloy Sheet, ICAS, Lisbon, Aug. 1978.
13. Arnold, D. B., Bonding Development of Improved Adhesives for Acoustic Structures—Jet Engine Liners, *SAMPE Mater. Rev.*, Oct. 14, 1975, p. 98.
14. Schliekelmann, R. J., Operational Experience, with Adhesive Bonded Structures, in *Bonded Joints and Preparation for Bonding*, AGARD Lecture Ser. 102, 1979, pp. 1-1 to 1-30.
15. Thrall, E. W., Jr., Failures in Adhesively Bonded Structures, in *Bonded Joints and Preparation for Bonding*, AGARD Lecture Ser. 102, 1979, p. 5-15-89.
16. Schliekelmann, R. J., Non-destructive Testing of Adhesive Bonded Joints, in *Bonded Joints and Preparation for Bonding*, AGARD Lecture Ser. 102, 1979, p. 9 1-8-37.

2

Chromic Acid Anodize Process as Used in Europe

PAUL F. A. BIJLMER, Jr.* *Fokker Aircraft Company, Amsterdam, The Netherlands*

I. INTRODUCTION†

Most metal surfaces, in particular, aluminum alloys, react with their natural environment to form a complex interface. Absorption, desorption, and chemical reactions take place continously, increasing the interface thickness and decreasing the polar properties of the surface. The reactivity to oxygen is enormous because aluminum is one of the least noble materials in the galvanic series. In addition, its affinity to chlorine, fluorine, sulfates, and silicates makes the surface even more complicated. The longer a metal surface is exposed to its environment, the more likely it is that it will be contaminated and the less, likely that it will meet the requirements of an adhesive-bondable surface.

This and the next two chapters describe the importance of the aluminum (adherend) surface treatment prior to bonding to the success of that bondment in service applications over a period of many years. In general, the performance of bondments in service has been fairly good, but some bonds have been known to delaminate and cause severe corrosion problems, as well as to jeopardize the structural integrity of the parts. The location and structural configuration of these bondment failures are generally random, but there tends to be a higher frequency of delaminations when the environment around such components is severe (e.g., in the bilges of aircraft or parts exposed to warm, moist coastal environments). Figure 1 depicts typical examples of delaminated bondments showing the appearance of having failed at the metal-adhesive interfacial area and the varying degrees of corrosion that can be found. It has been concluded from many years of studying the problem at the

---

Present affiliation: Alex Harvey Industries, Auckland, New Zealand.
†*Note*: Portions of this Introduction were contributed by J. Arthur Marceau, The Boeing Commercial Airplane Company, Seattle, Washington.

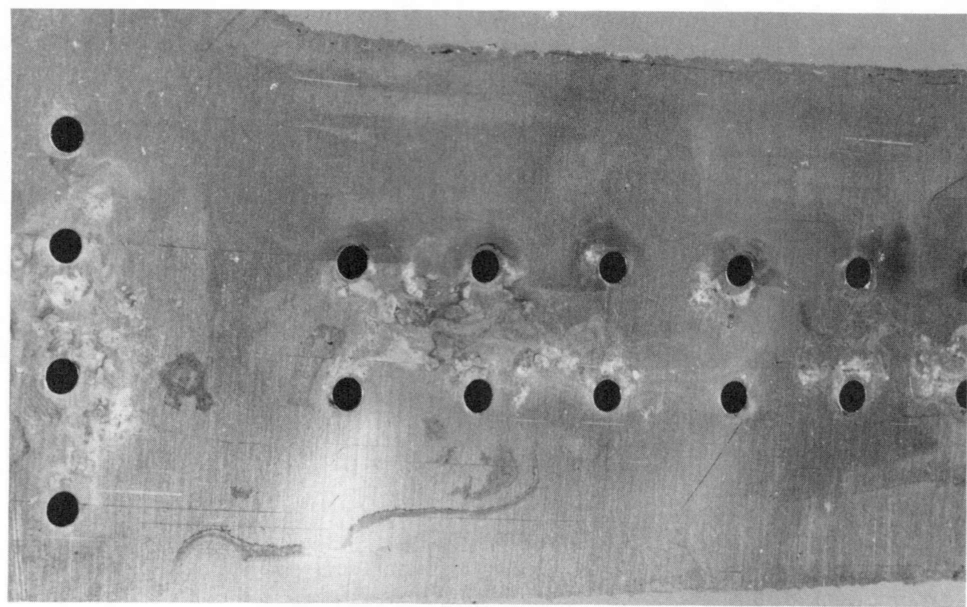

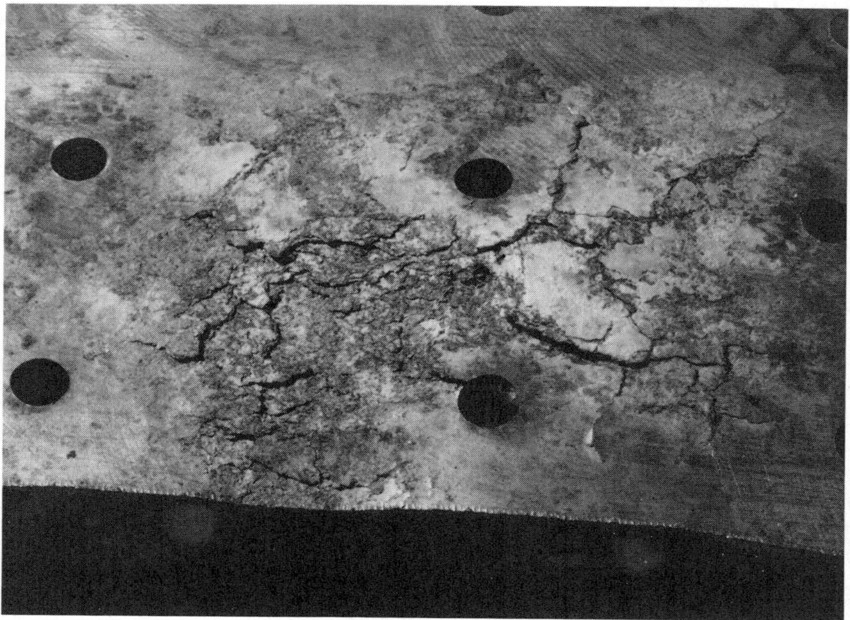

Figure 1 Typical examples of delaminated and corroded aircraft structures which had been adhesively bonded. (Courtesy The Boeing Airplane Company.)

Boeing Company, that generally, delamination occurs first and corrosion follows, its severity depending on the environment in the delamination area and the length of time the delaminated component stays in the critical environment.

To appreciate the role that oxides play in the bonding of aluminum, one must first be able to visualize the bonded system. "System" here is defined as the composite bonded joint consisting of the aluminum alloy, the oxide on that alloy, the adhesive primer applied to the oxide surface, and the bulk adhesive layer which then joins the other half of the system. Such a system is shown schematically in Figure 2.

One study determined that the locus of failure in failed bonded aircraft structures due to flight service environments is at or very near the oxide-primer interface, as illustrated in Figure 3. Therefore, it has been concluded that the nature of the oxide, physical and/or chemical, plays a very important role in the durability of bonded aluminum joints.

A surface treatment is a transformation process, changing the aluminum surface with unknown and perhaps undesirable properties to an oxide surface with known desirable properties. The quality of the oxide layer therefore defines the adhering properties of the surface. It has to be compatible with the adhesive primer so that the latter will wet the surface. But wettability alone is not a sufficient condition for a strong adhesive bond. The oxide layer must be strong enough to resist stresses, residual or applied, at the interface between oxide and adhesive. In addition, the oxide layer must resist hydration of diffused moisture to protect the base metal underneath from corrosion.

Chromic acid anodizing of aluminum alloys is by no means a new process. In the years immediately following World War I, airplane enthusiasts

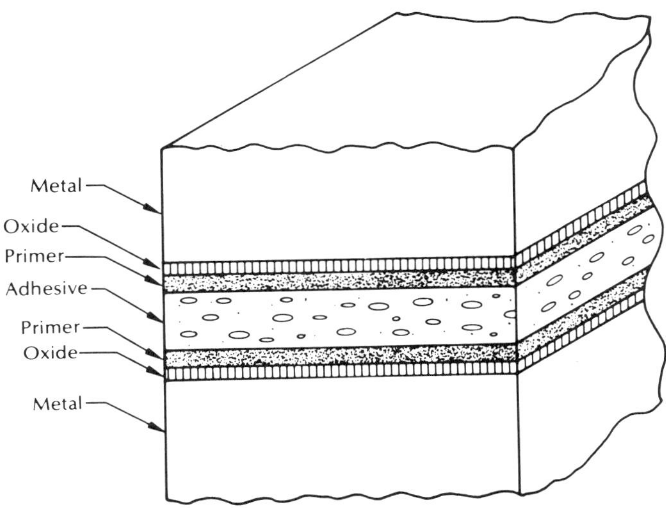

Figure 2 Composite structure of typical metal-to-metal bonded joint.

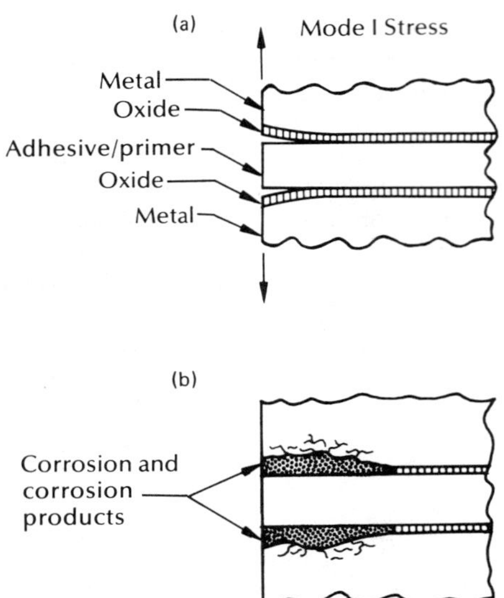

Figure 3   Bond delamination and corrosion mechanism of bonded joint: (a) the combination of stress and water weakens the oxide/primer interface, causing delamination; (b) crevice corrosion further accelerates delamination and destroys the metal.

found that the aluminum components on their craft were prone to rapid corrosive attack when exposed to salt-laden (seacoast) atmospheres [1]. It was soon discovered that aluminum could be treated electrolytically in a dilute solution of chromic acid to produce a coating that would effectively retard this corrosion. As a consequence, the aluminum alloy parts used on most early aircraft were treated by chromic acid anodize. By 1940, the anodic coating was a specification requirement for all aircraft used by the U.S. Navy. These anodic coatings were used both alone for corrosion protection and as a base for organic finishes such as zinc chromate primer and enamel.

The chromic acid anodize process found increased popularity during World War II for the treatment of aircraft parts and assemblies. Because the chromic acid electrolyte does not attack the basic aluminum alloys, it poses no problem if it becomes entrapped in joints, seams, or pores. Therefore, the process was widely used in the treatment of welded assemblies, castings, and riveted structures. Chromic acid anodizing alloys for aircraft parts prior to adhesive bonding and painting is used in Europe without postsealing treatments.

Several chromic acid anodizing processes are in use in France, Germany, and Holland. The Bengough-Stuart process, which originated in England (1925), consists of anodizing in 3% chromic acid at

40°C (104°F). The voltage is then increased to 50 V up to a total time
of 60 min. This process was modified several times and the current
process has as a minimum requirement the voltage-to-time relation shown
in Fig. 4 [1-4].

The reason for the increase from 40 V to 50 V is not clear. The
barrier layer thickness decreases as the porous layer grows under a
constant voltage. This was found by measuring the impedance at several stages of the process. Due to the increased voltage, the barrier
layer increases in thickness, improving the corrosion resistance of the
anodic layer. An improvement in bondability was never established,
but the change in anodic layer properties that occur due to this voltage
change has to be found near the barrier layer, several micrometers
away from the adhesive on top of the coating.

More important is the first stage of the anodizing process because
the oxides formed in this stage are found in the top layer of the coating
near the adhesive. No systematic investigations are known which relate
the influence of the voltage increase form 0 V to 40 V and bondability,
but it is generally accepted that a gradual increase in small steps, or
continously, is preferred. As is the case with all anodizing processes,
the properties of the chromic acid anodic coating are defined by the
anodizing solution variables, which are the composition and the temperature; the process variables, which are the pretreatment prior to anodizing, the anodizing voltage, the process time, and the posttreatment,

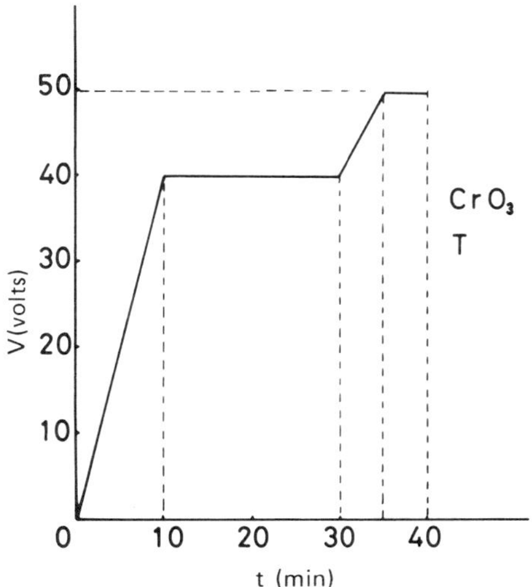

Figure 4　Chromic acid anodizing process program. Temperature =
38 to 42°C, CrO_3 = 30 to 50 g/liter.

if given; and the material properties, which are the alloy composition and the heat treatment of the alloy.

II. ANODIZING SOLUTION VARIABLES

A. Solution Composition

Normally, a concentration of 5% chromic acid in demineralized water is used and a free chromic acid content of 30 to 50 g/liter (4 to 6.7 oz/gal) is maintained during the bath life, up to a total chromic acid content of 100 g/liter (13.3 oz/gal). Interfering ions are chlorides from the rinse water and sulfates from the previous pickling solutions. Chlorides cause pitting of the anodic coating, reducing its corrosion resistance, and 200 ppm chloride in the solution should not be exceeded.

The influence of sulfate ions was investigated and these ions were found to affect anodic coating pore size and bondability. As sulfate was found to be encapsulated in the anodic layer (with the aid of Roentgen microanalysis), it might be expected that presence of sulfates decreases corrosion resistance [5]. The pore size increases with the amount of sulfates and in extreme cases weakens the anodic layer, causing low peel strength values for Redux-bonded panels. No negative influence was found up to 500 ppm, and this is an accepted limit in most specifications. The chromic acid content itself was found to influence the current density and therefore the coating weight. No influence on bondability was established up to 15% chromic acid under normal process conditions.

It was found that generally a coating weight below 20 mg/dm^2 (1.29 mg/in.^{2}) on nonclad 2024-T3 will not suffer from anodic layer fracture during peel testing, while lap joint test values scatter if alclad material with thin films are tested. In Fig. 5, peel strength values are given for anodized material with increasing coating weight.

As for bond line corrosion, a thick film is to be preferred. Values of 80 mg/dm^2 (5.16 mg/in.^{2}) for alclad materials and 40 mg/dm^2 (2.58 mg/in.^{2}) for nonclad 2024-T3 are sufficient to resist bond line corrosion. However, alclad materials are less resistant to bond line corrosion if compared with nonclad materials whatever thickness is chosen (Fig. 6).

B. Temperature Effect

The influence of temperature on current density is given roughly by the formula $i_T = 2 \times i_{(T-10)}$, which means that a 10°C (50°F) increase in temperature (T) doubles the mean current density (i), and the mean current density is about linear when related to the coating weight for a given material. The current density for alclad material is, however, much lower than for nonclad material, while the coating thickness on alclad material is much larger than on nonclad material due to the low anodizing efficiency of the latter. This result shows that a high density does not indicate a high coating thickness if different aluminum alloy materials are anodized.

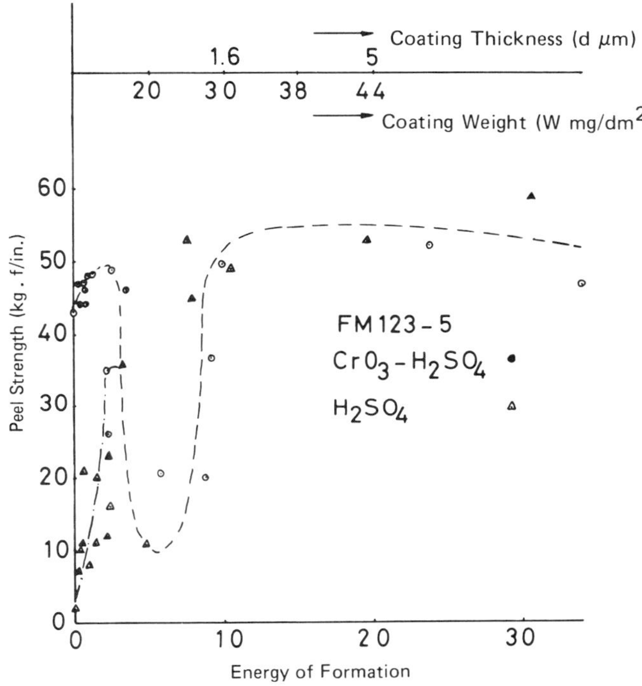

Figure 5   Peel strength values of 2024-T3 alclad panels chromic acid anodized with varying energy of formation. Note the influence of pretreatment prior to anodizing. At a bond strength of $E = 10 \text{ kJ}/\text{dm}^2$, the scatter is reduced. $E = V \times i \times t$ (V, voltage; i, current density; t, process time).

III. PROCESS VARIABLES

A. Pretreatment Prior to Anodizing

The morphology of the pickled (pretreatment) surface can be recognized on the anodized surface provided that no pores or fine pores are formed due to the anodizing process [6]. This is the case with ammonium tartarate anodizing, which completely replicates the pickled surface. To a lesser degree, sulfuric acid anodizing, with its fine porous structure, still shows pronounced evidence of the replicating effect. Chromic acid anodizing, however, with its coarse porous structure, cannot show the morphology of the pickled surface on its outer layer because the pores of the pickled surface are equal to those of the anodized surface (see Fig. 7).

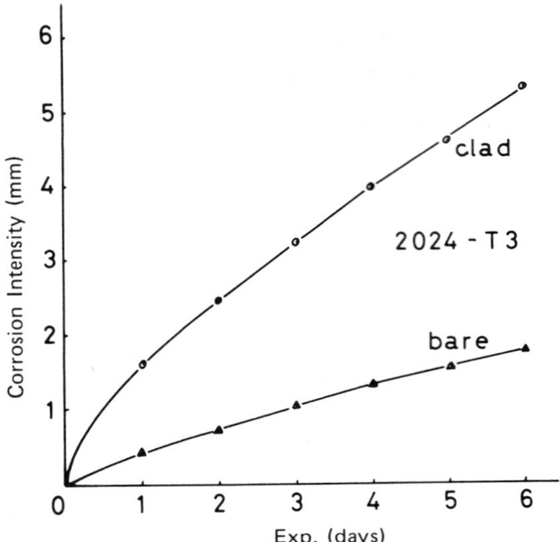

Figure 6   Filiform corrosion test results on painted panels (S15/60 primer and C21/67 top coat from Sikkens).   Each entry is the mean value of 54 observations.   (From Ref. 11.)

The dissolving power of chromic acid at 40°C (104°F) is very low except in the pores, where due to the electric current, higher local temperatures exist. It is still possible that oxides of a weak nature, formed during pickling in a depleted chromic sulfuric acid solution, exist in the top of the anodic layer. If adhesives are used that are not able to penetrate the pores of the anodized surface but only stick on the top of the anodic film, a weak bond can be formed and a low bond strength is the result.

Structural adhesives that are cured under pressure at elevated temperatures will flow and penetrate the oxide layer, leaving the weak oxides behind and finding a stronghold underneath the weak oxide. No influence of the poor pickling procedure will be evident. With the phenolic adhesive Redux, no influence of the pretreatment could be proved. Another type of adhesive, EC 2216 (3M Company), used without primer, showed some effect of the pretreatment [6]. It is, however, advised to follow the rule that an anodizing process should be preceded by a pickling process that on its own gives a sufficient initial bond strength if bonded without anodizing. Therefore, chromic acid anodizing is optimal if preceded by chromic acid sulfuric acid pickling. There are indications that the phosphoric acid process is an exception to this rule.

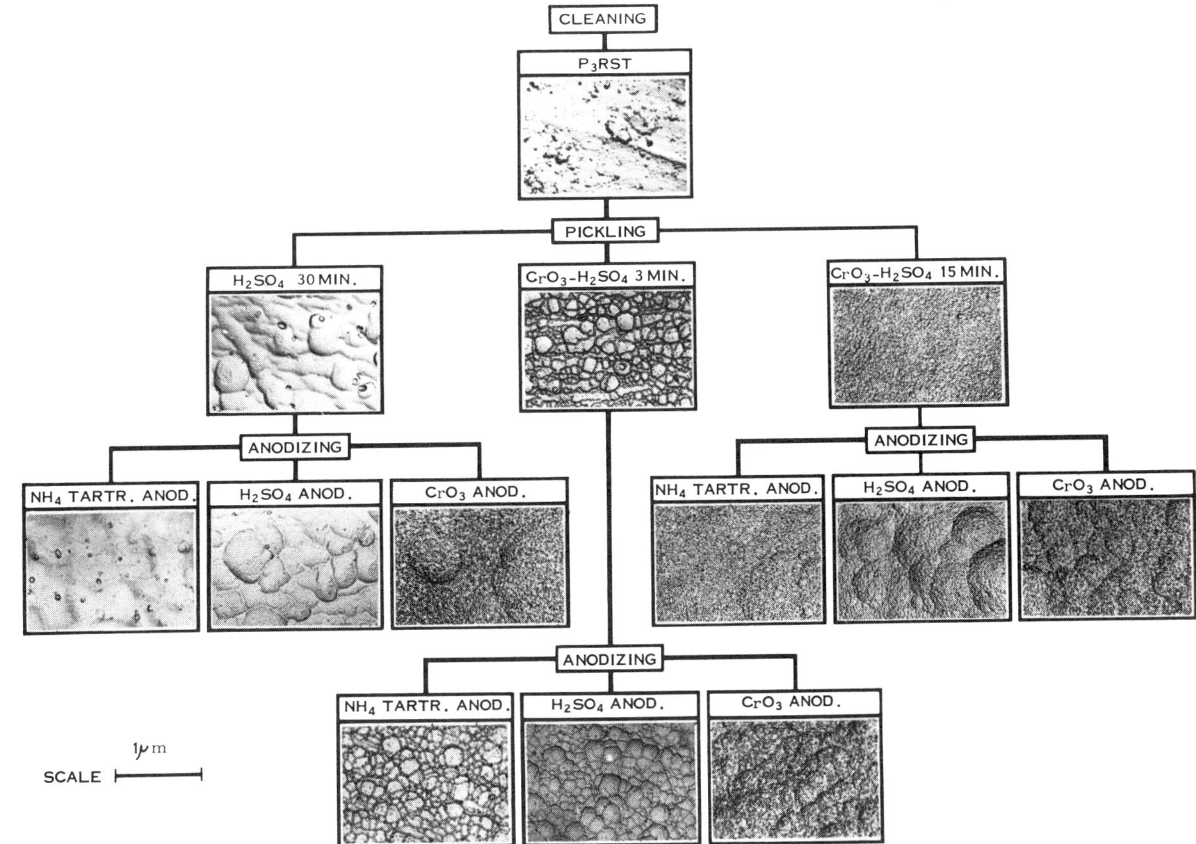

Figure 7   Influence of the pretreatment on the surface morphology after anodizing. (Ref. 6.)

B. Anodizing Voltage

The anodizing voltage has an effect on the current density in a positive
sense but is less than linearly proportional. The current density related to voltage also depends on the solution temperature and the type
of alloy (Table 1). At a high anodizing voltage, such as 40 V, the difference in coating thickness between alclad and nonclad 2024-T3 is
about 100%. At 20 V, the difference is negligible. It is perhaps for that
reason that the 20-V process is used by some manufacturers.

   If coarse grains are present on extrusions, a 40-V process will form
different coating thicknesses on the larger grains while the 20-V

Table 1   Results from Chromic Acid Anodizing on 2024-T3 Alclad

V (V)	T (min)	Variable[a]	Solution temperature [°C (°F)] 20 (68)	40 (104)	60 (140)
20	5	i	0.07	0.18	0.48
		E	0.35	0.96	2.66
		W	4.4	5.9	14.0
		P	47	48	49
20	10	i	0.05	0.19	0.52
		E	0.60	2.13	5.88
		W	5.6	11.8	26.0
		P	47	35	47
20	15	i	0.06	0.21	0.53
		E	0.96	3.52	0.27
		W	7.9	16.5	32.0
		P	44	46	37
40	5	i	0.06	0.26	0.99
		E	0.43	2.34	10.05
		W	4.9	9.5	27.0
		P	44	16	50
40	10	i	0.05	0.26	1.05
		E	0.77	5.81	23.67
		W	6.2	19.0	50.0
		P	46	21	57
40	15	i	0.04	0.26	1.11
		E	1.11	8.86	36.49
		W	8.6	28.0	67.0
		P	48	21	47

[a] i, current density (A/dm^2); W, coating thickness (mg/dm^2); E,
energy of formation (kJ/dm^2); P, lap joint strength (kgf/mm^2).

process creates no differences in thickness on the grains and thus a
more homogeneous color. In general, fewer color differences between
panels occur at the lower voltages. Also, with the lower voltages, the
average level of coating thickness values is much lower and can reach
below 20 mg/dm^2 (1.29 mg/in.^{2}) with lower bond line corrosion resistance.

C. Process Time

The process time is initially linearly proportional to the coating thickness. The rate of growth of the oxide thickness levels off as processing time increases. Other variables can also affect the growth rate.
There is an indication that the oxide growth process will reach an
equilibrium between film formation and dissolution. Powdery surfaces
are the result of extended anodizing, the time being influenced by the
temperature and voltage. At a high temperature and a high voltage,
less time will be required to form powdery surfaces. There are, however, other causes of powder formation. If a material suffers from
intergranular attack during pickling, a powdery surface will be found
after anodizing. A sensitized 5056 alloy material was found to form a
heavy powdered anodic coating. This condition was not due to an improper anodizing process but to intergranular attack during pickling
in the chromic-sulfuric acid etch. When pickled in hydrofluoric-nitric
acid, no powder was formed [7].

The anodizing time should be decreased if high temperatures and
high voltages are chosen. In principle, only 5 min of anodizing time
is needed to form a bondable surface at a solution temperature of 60°C
(140°F) and an applied voltage of 40 V. This procedure can be used
in repair of structures if care is taken to eliminate the pretreatment
pickling. Splices in the structure could cause pickling solution entrapment [8]. Mixtures of chromic acid and sulfuric acid might cause
damage, but chromic acid alone will not. Increasing the temperature
and voltage during anodizing promotes dissolution of the oxide, and
heavy etching of the aluminum might occur in extreme cases, locally
burning holes in sheet materials.

D. Posttreatments

Posttreatments are not used if adhesive bonding and painting are carried out. The rinsing after anodizing can have the posttreatment effect
of sealing the oxide if extended rinsing in hot demineralized water is
carried out to promote quick drying. It is for that reason that it is not
advised to use hot rinsing after anodizing [9].

If the material is dried after normal rinsing, a repeated rinse may
spoil the surface bondability. On the second rinse, the surface absorbs
additional water, and changes in surface potential measurements have
shown that it takes a long time to dry completely a surface that has

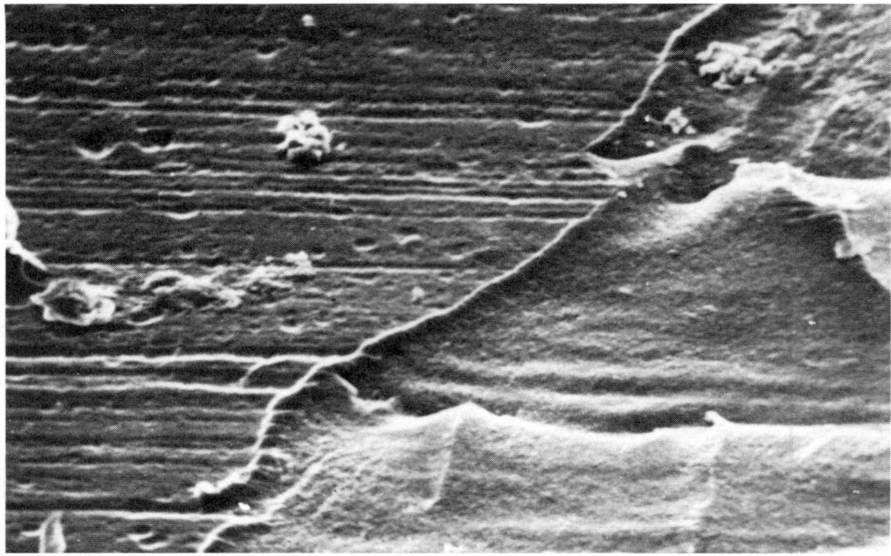

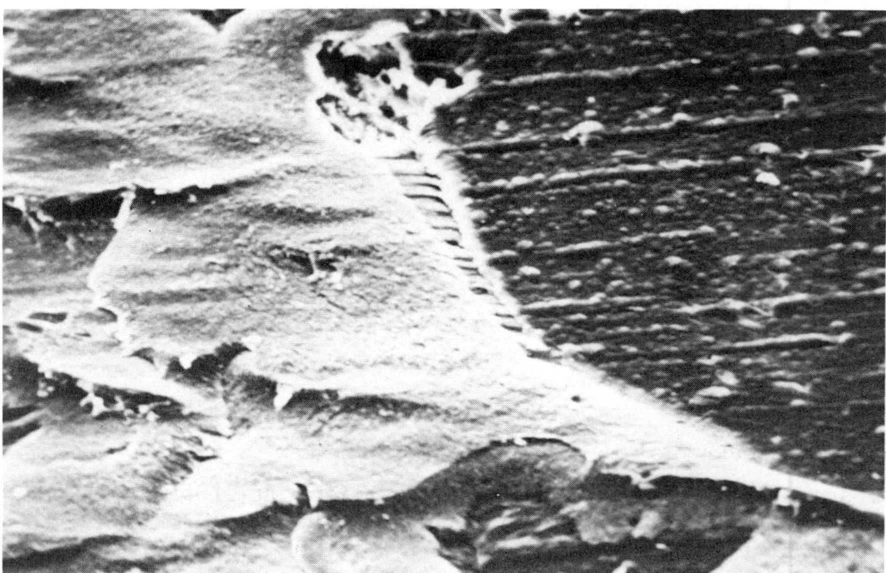

Figure 8   Scanning electron microscopic fractograph of the fracture surface after peel testing. The anodic layer is fractured off the surface. Each side of fracture is shown. (Courtesy Fokker Technological Center; Mr. A. Kwakernaak, and R. V. Veft.)

been exposed to more than one rinse. If adhesive or paint is applied to an anodized surface that has been repeatedly rinsed, poor adhesion may occur [10].

Hot-water-sealed chromic acid-anodized panels adhesive bonded with Redux show poor bond strength. Microscopic investigation of the fracture surface revealed anodic layer failure. Hot-water-sealing not only weakens an anodic layer, but the adhesive will not penetrate the pores and sticks only to the weakened cracked anodized surface layer.

The typical anodized surface morphology, with its needlelike structure, may indicate microdoublers in the layer. Due to water absorption and hydration, the oxide layer swells, causing cracks and overlaps. Another theory might describe this effect as crystallization of the amorphic oxides. It is, however, doubtful that crystallized oxide should be weaker than an amorphic one. Figure 8 shows an anodic layer fracture as it occurred after peeling the specimen.

IV. INFLUENCE OF THE MATERIAL PROPERTIES

A. Alloy Composition

The material to be anodized has a strong influence on the current density at the applied voltage. Not only does the alloy composition define the current density, but also the metallographic condition. Heat treatments have a strong influence on the behavior of material in the chromic acid solution.

The alloy composition also has a strong influence on the coating thickness after chromic acid anodizing. This influence is decreased at lower voltages and lower temperatures, tending to cause the build up of thin coatings at high efficiency. In Table 2 the coating weight, the energy of formation, and the energy efficiency are given for 2024-T3 alclad and nonclad anodized materials under various conditions [11].

The energy efficiency of an anodizing process is given as the ratio between the coating (oxide) weight and energy of formation. It indicates directly the amount of anodically formed products for the energy consumed. The energy efficiency for nonclad 2024-T3 is lower than for 2024-T3 alclad. It is clear that at low voltages and low temperatures, the energy consumption is an order of magnitude lower than at high voltages and high temperatures. The nonclad material has a tendency to etch anodically at temperatures and voltages that are too high.

The bondability of anodized material, as measured by means of climbing drum peel tests, does not correlate well with any of the anodizing parameters. There is, however, a tendency for the scatter in peel strength values to decrease at higher coating thicknesses provided that no anodic layer fractures occur during mechanical testing. The fact that no correlation can be established between sets of peel strength values obtained with Redux 775 and with EC 2216 indicate that the process variables do not influence the initial bond strength.

Table 2 Weight of the Anodic Film (W), Energy of Formation (E)$_a$, and Energy Efficiency (G) for 2024-T3 Anodized in Chromic Acid

Anodizing voltage (V)	Variable	Solution temperature [°C (°F)]					
		30 (86)		40 (104)		50 (122)	
		Al-clad	Non-clad	Al-clad	Non-clad	Al-clad	Non-clad
10	W	22	26	39	39	44	32
	E	3	6	7	12	11	22
	G	7.3	4.3	5.5	3.2	4.0	1.5
25	W	42	34	78	60	110	67
	E	13	18	26	38	53	78
	G	3.2	1.9	3.0	1.6	2.1	0.9
40	W	51	26	104	62	155	32
	E	27	22	54	64	108	125
	G	1.9	1.2	1.9	1.0	1.4	0.3

[a] Process time 45 min; W in mg/dm^2; E in kJ/dm^2; G = W/E in mg/kJ.
Source: Ref. 11.

Filiform corrosion tests were found to correlate with bond line corrosion results and, being relatively quick tests (6 days), were used to qualify the surface after anodizing. In Fig. 6, the corrosion intensity is shown progressing in time for the entire set of tested panels, both alclad and nonclad. It is obvious that nonclad 2024-T3 is less prone than alclad 2024-T3 to filiform corrosion. In Fig. 9 the weight of the oxide film is plotted against the corrosion intensity. A low coating weight results in higher corrosion intensity. The fact that the Bengough-Stuart process results in a higher coating weight for alclad materials compensates, to a great extent, for the higher sensitivity for filiform corrosion.

For a process such as the 20-V process, which tends to give low coating weight values, the use of 2024-T3 is not advantageous if filiform corrosion (or bond line corrosion, for that matter) is a concern. The minimum anodic coating weight requirement for anodized parts is less urgent for nonclad materials than it is for alclad sheets; however, most specifications ask for a minimum coating weight on nonclad materials.

B. Heat Treatment of the Alloy

Filiform corrosion resistance is highest for overaged materials, which shows the important influence of the material condition due to heat

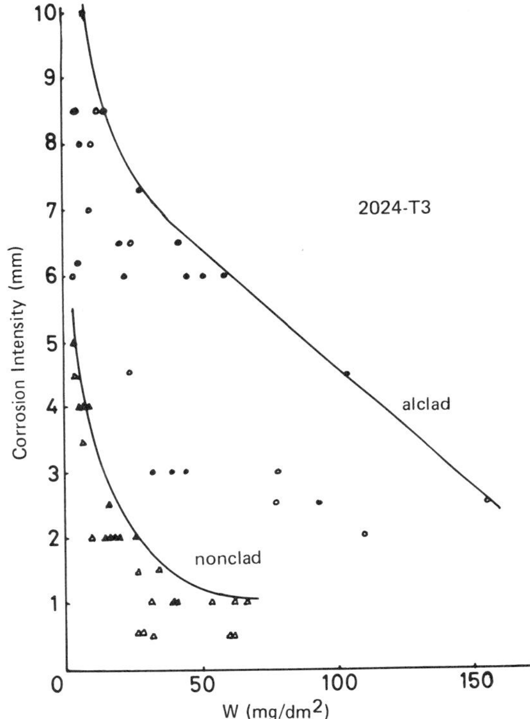

Figure 9   Corrosion intensity in relation to the anodic coating weight, formed under various anodizing conditions. (From Ref. 11.)

treatments [12]. In Fig. 10 the corrosion intensity after 20-V anodizing with different temperatures and anodizing times is given in relation to the anodic layer thickness. Three groups are found:

1.   2024-T3 alclad and 7075-T6 alclad, with the highest corrosion susceptibility, have a strong tendency for improvement (in corrosion resistance) at higher coating thickness values.
2.   The nonclad materials 2024-T3, 7075-T6, and 6061-T6, which have a much lower corrosion susceptibility, still show a tendency for improvement (in corrosion resistance) with thicker anodic layers.
3.   The nonclad overaged materials such as 2024-T8 and 7075-T73, with the lowest corrosion susceptibility, are, in this range, independent of coating thickness.

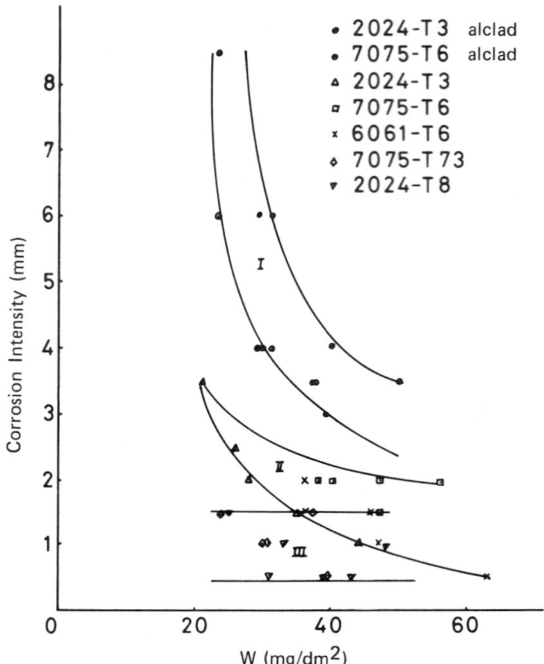

Figure 10   Corrosion intensity of painted specimens anodized at 20 V under difficult process conditions: I, alclad materials; II, nonclad materials; III, nonclad overaged materials. (From Ref. 12.)

This phenomenon leads to an important conclusion: Chromic acid anodizing, carried out at higher voltages, such as 40 V for the Bengough-Stuart process, provides a diversity of anodic coatings by supplying the thickest coatings to the materials that need it most.

The optimal 20-V chromic acid process [temperature of operation 35°C (95°F) and a process time of 30 min] and the Bengough-Stuart process are compared in Table 3. The 20-V process provides about equal anodic layers on all tested material, while the Bengough-Stuart process gives higher values for the coating weight for the alclad materials and low values for the overaged materials. The 2024-T3 and 7075-T6 nonclad specimens showed anodizing layer fracture during peel testing when anodized with the Bengough-Stuart process. The much thinner layers obtained after the 20-V process did not rupture during the peel test. The nonclad 6061-T6, with the heaviest anodic coating, did not show anodic layer fracture due to the relative low yield strength of the material. Cracking of the anodic layer took place perpendicular

Table 3 Comparison of the Optimal 20-V Process with the 40- to 50-V Anodizing Process Used at Fokker Aircraft Company[a]

	Optimal 20-V		Bengough-Stuart	
	Weight (mg/dm^2)	Corrosion intensity (mm)	Weight (mg/dm^2)	Corrosion intensity (mm)
2024-T3, alclad	29.5	3.50	71.8	2.25
7075-T6, alclad	29.8	5.50	92.4	1.75
2024-T5, nonclad	28.0	1.75	53.7	0.75
7075-T6, nonclad	37.9	1.75	76.2	1.25
6061-T6, nonclad	36.4	2.00	101.6	0.50
2024-T8, nonclad	33.1	0.75	41.8	0.50
7075-T73, nonclad	30.4	1.00	22.4	1.00
Mean value, $\bar{x}$	32.2	2.32	65.7	1.14
Standard deviation	3.8	1.66	28.1	0.66

[a]Note that the corrosion intensity is reduced to half if the coating weight doubles.
Source: Ref. 12.

to the surface, preventing unstable crack growth parallel to the surface. The specimens that showed no anodic layer fracture scored higher peel-strength values than those for the specimens treated at 20 V and showing anodic layer fracture. The mean value of the amount of corrosion for the Bengough-Stuart process after a given exposure is about half that for the 20-V process.

V. SOME OTHER PROPERTIES OF THE CHROMIC ACID ANODIZING PROCESS

One of the special advantages of the chromic acid anodizing process at high voltage is the detection of soft spots (caused by improper heat treatment) on extrusions and machined parts. The fact that at 40 V, the anodize coating thickness varies with the metallurgical conditions of the material being anodized results in different coating thickness values, and the resulting color variations make it possible to inspect the anodized material for heat treatment homogeneity [13]. At lower voltages, the coating thickness variations are less pronounced and soft spots may not show up after anodizing. Soft-spots in the alloys may occur due to a delay during the heat treating operation in the quenching of long extrusions and plates or may be caused by overheating while machining parts. The chromic acid solution must, however, be free of sulfuric acid drag-in from rinsing water or chromic-sulfuric acid pickling solutions. If it is not, the color variation between soft spots and properly heat-treated areas may not be as prominent.

Another advantage of chromic acid anodizing is its capability of detecting cracks in the metal. If the rinsing time after anodizing is reduced, some chromic acid will remain in surface cracks. After drying, this small amount of chromic acid bleeds out around the crack in the highly absorptive anodic layer, increasing the visibility of the otherwise invisible crack. Visual inspection of machined parts, extrusions, and tubes are easily carried out during the anodizing process.

Unpainted parts can be hot-water sealed after anodizing to provide adequate corrosion protection. In Europe, hot-water sealing and dichromate absorption is never carried out on the parts to be adhesive bonded and/or painted. The adhesive is applied to the parts within 16 hr after anodizing to exclude contamination and all humid environment sealing effects. The anodic coating must not be sealed with hot water before the application of adhesive, adhesive primer, or paint primer. The application of these materials performs an adequate seal.

VI. CONCLUSIONS

The unsealed chromic acid anodizing process, as a pretreatment for adhesive bonding and painting, gives satisfactory results in service. Laboratory tests for peel and shear did not always correlate with the variations in thickness of the anodic film. The scatter of the results was reduced with thicker anodic coatings.

Lowering the voltage to prevent anodic layer fractures has undesirable consequences in corrosion resistance since corrosion resistance is proportional to the coating thickness. At low voltages, the anodic layers are about equal for all materials. At high voltages, the alclad materials, nonclad materials, and overaged materials (in decreasing order) build up anodic layers, reducing their sensitivity to filiform corrosion. After low-voltage anodizing, the variation in corrosion resistance is high on the different alloys, while the difference in coating weight is low. The unsealed chromic acid anodizing process at 40 to 50 V is the best chromic acid process.

REFERENCES

1. Wernick, S. and R. Pinner, *The Surface Treatment of Aluminum and Its Alloys,* 4th ed. Robert Draper, 1972, pp. 420-435.
2. Anodizing Aluminum and Aluminum Alloys, DEF-151.
3. Chromic Acid Anodizing, Fokker Specification TH-6.7851.

4. Hartman, A., The Chromic Acid and Sulphuric Acid Anodizing, Netherlands Aerospace Lab. (NRL) Rep. M2166, 1966 (in Dutch).

5. Exalto, R., Investigations of the Presence of Sulphur in Anodic Films, Fokker Rep. R-1759, 1974.

6. Bijlmer, P. F. A., Influence of Chemical Pretreatments on Surface Morphology and Bondability of Aluminum, *J. Adhes.*, Vol. 5, 1973, pp. 319-331.

7. Zwart, R., Powder on Pickled 5456 Material, Fokker Report R-1535, 1972 (in Dutch).

8. Kwakennaak, A., Repair Method for Leading Edge Flaps, Fokker Rep. R-1535, 1972.

9. Bijlmer, P. F. A., The Influence of Hot Water Rinsing on the Surface Properties of Alcald 2024-T3, Fokker Rep. R-1050, 1969.

10. Bijlmer, P., and R. J. Schliekelmann, The Relation of Surface Condition After Pretreatment to Bondability of Aluminum Alloys, SAMPE Q., Oct. 1973.

11. Exalto, R., Determination of Optimum Chromic Acid Anodizing Conditions for Adhesive Bonding, Fokker Rep. R-1981.

12. Fokker Rep. R-2311.

13. McMechan, A., and P. Folwell, Chromic Acid Anodizing Disclosure of Soft Spots in Machined Aluminum Parts, Douglas Aircraft of Canada Ltd.

3

Sealed Chromic Acid Anodize

NARVEL L. ROGERS *Bell Helicopter Textron, Fort Worth, Texas*

I. INTRODUCTION

Early designers used the chromic acid anodic coating of aluminum in the as-formed state; that is, it was not subjected to any type of posttreatment. Later, it was learned that additional resistance to corrosion could be attained by sealing the coating in hot 93.3°C (200+°F) water [1].

Sealing can be directed toward improved adhesion or enhancement of visual appearance as well as improved corrosion resistance. Consequently, sealing and other posttreatment processing have evolved over the years into any number of particular solutions for specific applications. Most of these are still based on hot water with the addition of compounds such as dichromate, chromic acid, or acetates and in some instances, organic materials.

When designers started to look at organic adhesives as a means for fastening aluminum parts, they naturally turned to anodize as a method for preparing the adherend surface. For many years, anodize was the single most effective surface preparation process. However, as adhesive formulations were improved, anodize became less suitable because the sealing processes used to attain maximum corrosion resistance did not produce suitable bond strenghts. Therefore, it became necessary to modify existing processes to provide surfaces that would produce good adhesive-bonded joints. One such process, developed by Bell Helicopter in the early 1960s, is covered by U.S. Patent No. 3,414,489.

Chromic acid anodize is normally restricted to those alloys that contain less than 5% copper and/or 7.5% total alloying elements because of difficulties in attaining suitable coatings [2]. Those alloys that contain very high percentages of magnesium or silicon can also present extreme problems and anodize is usually not attempted.

II. ENVIRONMENTAL CONSIDERATIONS

Adhesive-bonded aluminum joints that are exposed to hostile environments such as combinations of high temperature and high relative humidity may lose much of their effective strength. If these joints are also subjected to steady or cyclic loads during exposure, the failure can be extremely rapid and often catastrophic. The usual mode of failure is moisture (water) penetration at the metal-adhesive interface. If the metal surface is very active, such as one prepared by the Forest Products Laboratory (FPL) etch process, corrosion will follow the intrusion of moisture. The consequent buildup of corrosion products can increase the stresses in the bonded joint and accentuate the rate of bond failure.

Chromic acid anodize produces an inert layer of aluminum oxide on the surface and is therefore an effective means for retarding bond line corrosion. Sealed chromic acid anodize has been used successfully for a number of years in the production of adhesive-bonded assemblies. However, each metal-surface preparation-adhesive combination must be tested in the expected operational environment or a simulated environment to determine its suitability [3].

Surfaces of 2024-T3 (nonclad) aluminum prepared by the sealed chromic acid anodize process and bonded with nitrile-epoxy adhesives are capable of withstanding 5000 hr of exposure on a special tap water test stand [4]. These joints were exposed to cyclic heating and constant moisture. Dead-weight loading was used to provide stress in the bonded joints. This is an example of specialized tests for a specific type of joint. A more conventional type of test is salt spray (fog). Anodize will produce bonded joints which will withstand exposure to salt spray (fog) testing for much longer periods than those produced with FPL etch. Test results indicate that bond line penetration on anodized surfaces in considerably less than on FPL etched surfaces, bonded with the same adhesive, after 30, 60, and 90 days of exposure to salt spray [4].

As with almost any chemical process, the impact of chromic acid anodizing on the environment must be considered. The sealed chromic acid anodize process employs solutions that contain strong acids and hexavalent chromium compounds. Adequate precautions must be taken to prevent discharge of these materials into the surrounding environment. There are no extremely volatile materials used, with the exception of the solvents used for degreasing. Therefore, air pollution is not a major area for concern. Water pollution is quite another consideration. Hexavalent chromium is reported to be toxic to most species of aquatic life as well as to human beings and animals. Therefore, it must be rendered nontoxic before disposal.

Technology is available for treating dilute solutions by either batch or continuous flow methods. These methods are generally based on chemical reduction of the hexavalent chromium to the trivalent state by

the use of sulfur dioxide. The trivalent chromium is then precipitated
as a metal hydroxide which can be disposed of in landfills, when permitted by local authority. There are also available an electrolytic process whereby the chromium can be removed and the water reclaimed.
The anodize solution, the FPL etch and/or deoxidizer solution, the final
seal solution, and the rinse waters from these operations are those areas
where hexavalent chromium are present. The size of the anodize installation will dictate somewhat the type of waste disposal system used.
In some areas it may be worthwhile to look into the possibility of re-
claiming the chemicals rather than attempting to dispose of them. Both
electrolytic and ion exchange methods may be applicable in these cases.

III. COMPATIBILITY WITH PRIMERS

Sealed chromic acid anodize may be used effectively with both corrosion-
inhibiting adhesive primers (CIAPs) and non-corrosion-inhibiting primers. Curing-type primers which are not dependent on the adhesive to
reach full cure will provide increased corrosion resistance. Primers
cured at either 250°F (121°C) or 350°F (177°C) have been used successfully. Silane (Dow-Corning) coupling agents may also provide some
added bond joint properties but should be thoroughly evaluated prior
to use.

The resistance of bonded joints to the environment, in order of increasing durability, are anodize only, anodize plus noncuring CIAP,
and anodize plus curing CIAP. The combination of sealed chromic acid
anodize and CIAP materials can greatly improve the resistance of alclad
2024 and alclad 7075 as well as the nonclad alloys in bonded joints [4].

IV. COMPATIBILITY WITH ADHESIVES

Sealed chromic acid anodize may be employed as a surface preparation
for use with any adhesive system normally used to bond aluminum.
Film-type adhesives that exhibit high peel strengths (nitrile-epoxy,
nylon-epoxy, rubber phenolic, etc.) should be used with surfaced produced by modified sealing processes. These are usually lower-temperature processes which do not impart a full "seal" to the anodize. Paste-
type adhesives and low-peel-strength films may be bonded to fully
sealed surfaces such as those produced by dichromate solutions or boiling water.

V. COMPATIBILITY WITH DIFFERENT ALLOYS

Chromic acid anodize is normally not applied to alloys that have a nominal copper content in excess of 5% or a total alloying content in excess
of 7.5%. The common alloys most often treated are 2024, 5052, 6061,

and 7075. The 2024 nonclad and alclad 5052 and 6061 alloys respond
equally to treatment at 40 V, whereas the 7075 nonclad and alclad
respond more favorably at 20 V. The alclad alloys and the 5052 alloy
will attain higher coating weights than the other alloys when processed
under the same conditions. The minimum coating weight suitable for
corrosion protection is 200 mg/ft^2 (20 mg/dm^2). This can usually
be obtained by anodizing for 30 to 60 min after reaching the specified
voltage.

   Anodize should not be applied to the heat-treatable alloys in the
"0" condition. No pronounced differences are expected when anodizing under normal heat treat conditions (i.e., T3, T4, T6, T81, T73,
etc.). However, the pretreatment for T81 and T73 may require modification. Sulfuric acid-sodium dichromate (FPL) etch solution will produce a smut (blackening) on T81 material unless time and temperature
are controlled closely. T73 material is more susceptible to pitting than
to other conditions, and the time in acid-type deoxidizer solutions should
be limited.

   Chromic acid anodize is a relatively thin coating; 200 mg/ft^2 (20 mg/
dm^2) is roughly equivalent to 0.00003 in (0.8 μm). The coating does
not have an adverse effect on the fatigue properties of the basic metal.

   The appearance of the chromic acid anodize coating will vary with
the alloy and type of surface. Forgings and extrusions will show varying amounts of grain structure when anodized. These mottled areas are
not detrimental to the effectiveness of the coating. Castings will "bleed"
after treatment; the chromic acid entrapped in porosity will exude,
causing brown stains. These are not considered to be detrimental to
the bonding properties of the surface.

VI. PROCESSING PROCEDURES

A. General Requirements

When chromic acid anodize is used as a surface preparation for adhesive
bonding, the processing facilities should be located near the area where
the adhesive priming and/or bonding is to be accomplished. The anodic
coating, like any freshly prepared surface, is susceptible to contamination by fumes, dust, or vapors. Also, the bonding quality of the surface can be degraded by improper handling. Treated parts should not
be exposed to airborne contamination nor should they be subjected to
excessive amounts of handling.

   The anodic treatment requires a source of dc (direct current) power.
Rectifiers provide the best source of power, but motor-generators can
be used successfully if the voltage application equipment is suitable.
The dc power source should be capable of producing 40 V dc at the
part, and as the name implies, the part must be anodic (+). Although
the anodic treatment operates at about 2 A/ft^2 (21 A/m^2) at 40 V, the

initial current density may be as high as 10 A/ft^2 (105 A/m^2). Therefore, this peak amperage requirement must be considered when sizing
equipment.

Racking is very important in chromic acid anodizing. Of course,
good electrical contact is an absolute necessity. Forty volts dc is capable of melting the parts being processed and will do so if loose contacts
are used. Aluminum contacts are preferred. However, aluminum racks
with titanium contact points may also be used. When aluminum racks
are used, the contact points must be stripped between loads.

The chromic acid anodize solution is tolerant toward particulate
matter and therefore filtration is seldom, if ever, required. However,
if filtration does become necessary, the solution can be handled safely
using iron or stainless steel equipment. Most plastic filtering and pump-
ing equipment is also satisfactory. Treatment with activated carbon
should not be attempted and organic filter elements should be thoroughly
tested to determine their suitability prior to use. Cloth, paper, wood,
and similar materials will reduce hexavalent chromium and should not be
allowed to contact the anodize solution.

B. Processing for Pretreatment Prior to Anodize

Preliminary cleaning is of primary importance in the anodic treatment
of aluminum. The extent of precleaning will depend on the nature of
the soils present and the history of the surfaces concerned. Scale and
oxides must be removed as well as soils and other foreign matter. This
is necessary in order that a uniform surface be available for the anodic
oxidation treatment.

The first step in cleaning is the removal of oils, greases, waxes,
and preservative compounds. This is usually accomplished by vapor
degreasing, solvent wiping, emulsion cleaning, or alkaline degreasing.
Vapor degreasing is normally accomplished with chlorinated solvents
such as trichloroethylene, perchloroethylene, or 1,1,1-trichloroethane.
Each of these has advantages and disadvantages. For instance, perchloroethylene degreasers operate at about 250°F (121°C); this makes
the parts literally too hot to handle comfortably. 1,1,1-Trichloroethane
boils at about 165°F (74°C) and provides good degreasing properties.
However, 1,1,1-trichloroethane is somewhat more sensitive to moisture
and may require addition control tests. Trichloroethylene vapors are
photosensitive and therefore are prohibited in some areas because of
air pollution requirements. In all cases, local, state, and/or federal
regulations may restrict the use of vapor degreasers.

Solvent wiping is an alternative method for removing greases and
oils. Solvents such as ketones, toluene, or naphtha can be used, but
again air pollution and fire safety regulations may restrict their application. Safety solvents such as 1,1,1-trichloroethane and/or mixtures
of solvents are probably the best materials to use.

Emulsion cleaners and alkaline degreaser/cleaners are available which may be used for tank (dip) and spray operations as well as for hand cleaning. These cleaners are usually composed of solvents, surface-active agents (surfactants), inhibitors, and pH controllers. In the past, these materials were normally used hot (140 to 180°F) (60 to 82°C), but with the energy shortage many suppliers now produce materials which are used at ambient temperatures or low temperature (110 to 120°F) (43 to 49°C). Emulsion cleaners remove greases and oils by forming oil-in-water (O/W) or water-in-oil (W/O) emulsions. The oils and soils remain emulsified until the solution is dumped and treated at the waste treatment facility. In most cases, the release of oils in the treatment facility can create problems with the equipment or it may violate pollution control regulations. It may therefore be necessary to use a displacement-type alkaline degreaser which displaces the oils and greases from a metal surface and causes these to float where they can readily be skimmed off for subsequent disposal. In each instance, the type of degreasing to be used must be consistent with the nature of the materials to be removed and local air and water pollution regulations.

Degreasing operations must be followed by alkaline cleaning to assure the removal of soils. Alkaline cleaners for aluminum are divided into two distinct types: etching and nonetching. The etching-type cleaners will remove metal and therefore this must be considered when they are used on parts with close dimensional tolerances. In most applications for aircraft assemblies, nonetch-type cleaners are preferred. These cleaners are composed of detergents, soaps, inhibitors, salts, and so on. Nonetch cleaners are further classified into silicated and nonsilicated types. Silicated cleaners contain caustic compounds together with wetting agents and a high concentration of silicate, which acts as an inhibitor to prevent the alkalies from reacting with the aluminum. Silicated cleaners are used at elevated temperatures (140 to 180°F) (60 to 82°C), and although they provide exceptionally good cleaning, they may deposit "silicate" residues on the aluminum surface. Such residues are not easily removable by subsequent rinsing and processing. Nonsilicated cleaners use chromates, nitrites, and other compounds for inhibitors. These materials will provide satisfactory cleaning at temperatures around 120°F (49°C) but are more effective at higher temperatures. To prevent possible damage to surfaces, care must be exercised to assure that alkaline cleaners are used within the concentration limits and temperature range recommended by the manufacturer. Alkaline cleaning equipment is normally made from mild steel. However, some material manufacturers recommend stainless steel tanks.

The oxide formed on an aluminum surface does not conduct electricity. Therefore, all natural oxide, heat treat scale, and other randomly formed oxide must be removed before a uniform anodize coating can be formed. The most commonly used solutions for deoxidizing aluminum are sulfuric acid or nitric acid, with additives such as chromates and fluorides. Any number of proprietary deoxidizer solutions are available and most of these can be used at ambient temperatures to prepare surfaces for anodize. The FPL etch—25 wt% sulfuric acid and 2.5 wt% sodium dichromate at 140 to 160°F (60 to 71°C) is a very effective deoxidizer and it is used in most applications for adhesive bonding.

Within the past few years, a number of nonchromated deoxidizers have been developed. These are attractive from a standpoint of pollution abatement and waste disposal. Many of these have proven to be suitable for preparing surfaces for anodize. However, each anticipated use should be evaluated. Since most deoxiders are acidic in nature, tanks and other equipment must be made from resistant materials.

C. Application of the Anodic Coating

The anodize tank should be made from mild steel and should be connected to the (−) cathode site of the voltage source. Because of the rather close operating temperature range (92 to 98°F) (33.3 to 36.7°C), the tank must be equipped with both heating and cooling coils. It is advantageous to shield the sides of the tank with glass or plastic. Shields should be placed 4 to 6 in. (10 to 15 cm) from the tank walls and should have a gap 2 to 6 in. (5 to 15 cm) about every 2 ft (61 cm) and at the bottom. The shields act as a barrier which prevents parts from contacting the tank walls during processing. They also reduce the effective cathode area. This lowers the cathode-to-anode ratio and helps reduce the consumption of chromic acid during processing.

The anodize solution is made from chromic acid (chromium trioxide) dissolved in demineralized water. The initial solution is made up to about 6 wt% as CrO_3. At this point, the total chromic acid and free chromic acid are essentially equal and this solution will not produce a coating that is suitable for adhesive bonding. The solution must be "aged" by processing scrap or dummy loads. There is some difference of opinion as to how much aging is necessary. However, as a general rule, there should be at least a 1% difference in the total and free chromic acid content. Once the bath is in operation, the free chromic acid should not be allowed to drop lower than 3%. In addition to the chromic acid content, the solution must not contain excessive amounts of chloride, sulfate, or alumina. Usually, a maximum of 0.02% chloride, 0.05% sulfate, and 0.1% alumina is permitted in an operating solution.

Voltage application is critical in the production of a good adhesive-bonding anodize. In most cases it is preferable to apply the voltage in steps of 5 to 10 V each rather than by a large number of small steps. The cleaned aluminum surface is a very good conductor when the initial voltage is applied. Consequently, the amperage will be very high—6 to 10 A/ft^2 (63 to 105 A/m^2). This amperage will decay rapidly as the anodic coating forms. There are a number of control instruments that make use of this property. They are set up to add voltage when the amperage drops to a predetermined point. This type of control does

not produce coatings which are suitable for bonding, although they are
suitable for corrosion resistance. The most effective method of voltage
application is by time delay. The initial application should be 5 to 10 V
with a delay of $1\frac{1}{2}$ to 2 min before the next applications of 5 to 10 V.
The voltage is then raised in steps of 5 V with about 1 min between
steps until 40 V is reached. This type of application will produce a
characteristic voltage and amperage chart (recorder) that may be used
for inspection purposes.

D. Treatment After Anodizing

After the anodizing voltage has been applied for the required time (30
to 60 min), the parts should be removed from the tank and rinsed in
clean water to remove the excess chromic acid. This rinsing should be
accomplished within 5 min after the voltage source has been turned off.
The rinsing should be followed by the applicable sealing operations.

The seal solution can vary from clean deionized water only, to solutions of chromic acid, sodium dichromate, or nickel acetate. The sealing temperature is a critical factor when adhesion is concerned. In
most cases, maximum corrosion resistance is attained when the surface
is sealed at 200°F (93°C) or higher. When maximum adhesion is the
primary consideration, the surface should not be sealed above 190°F
(88°C).

Drying of parts, after sealing, may be accomplished at ambient
temperature or at temperatures up to 160°F (71°C). In any case, the
surfaces must be protected against contamination by fumes, dust, and
so on. All parts must be handled using clean racks or clean gloves.
Prepared surfaces are very susceptible to damage by wiping. Care
must be exercised when handling parts to avoid wiping or rubbing the
surface.

VII. TESTS FOR DETERMINING THE BONDABILITY OF ANODIZE

Although visual inspection of an anodic coating cannot be used to determine bondability, it is an important and necessary tool for routine
process control after bondability has been established. Chromic acid
anodize produces a characteristic gray color on most aluminum alloys.
In addition, each sealing process provides a distinct visual signature.
Therefore, visual parameters for acceptable versus unacceptable surfaces can be established for each process and alloy. Some areas for
consideration are discoloration, stains, powdery coatings, and racking
marks. Evidence of burning at racking points or other areas must be
determined by close visual inspection.

The floating roller peel test has proven to be a very effective means
for testing the bondability of anodic-coated surfaces. The test may be

conducted at ambient temperature or at high or low temperatures. One
of the most sensitive areas is −54°C (−65°F) peel strength. Of course,
the peel test can be used only with adhesives that exhibit good peel
strength. When paste-type epoxies and other inherently brittle adhesives are used, the lap shear test is by far the best method for determining bondability.

There are certain test procedures, such as the ASTM Test Methods
D-1002, Shear Test; D-903, 180° Peel Test; D-1781, Climbing-Drum
Peel Test, and D-1876, T-Peel, which are used by the adhesive manufacturers to provide standardized data. These should be used together
with other test methods to establish the bondability of specific metal-
surface preparation-adhesive system combinations.

REFERENCES

1. Anodizing Aluminum by the Chromic Acid Process, Ser. 13,
   Mutual Chemical Company of America.
2. MIL-A-8625, Anodic Treatment of Aluminum Alloys.
3. Rogers, N. L., *J. Appl. Polym. Sci. Appl. Polym. Symp. 19*,
   1972, pp. 63−73.
4. Rogers, Narvel L., *J. Appl. Polym. Sci. Appl. Polym. Symp. 32*,
   1977, pp. 37−50.

4

Phosphoric Acid Anodize

J. ARTHUR MARCEAU   *The Boeing Commercial Airplane Company,*
Seattle, Washington

I. INTRODUCTION

Research has shown that an oxide characteristic common to most durable
bonded joints is that of an open, porous, columnar structure [1-8] similar to those produced by most aluminum anodizing processes used for
corrosion protection [9-11]. This should be a desirable characteristic,
since a very large and very active surface area is produced and, in
addition, adhesion can be enhanced due to keying of the primer polymers by penetration into the deep pores [4—6,12]. Evidence of polymer
penetration into these pores is shown in Fig. 1. These are results of
an electron energy loss spectroscopy (EELS) analysis of a primer ingression into the porous oxide on 2024-T3 clad alloy. The specimen was
prepared in an ultramicrotome with a diamond knife. The easiest method
of forming these porous oxides on aluminum is to immerse the aluminum
into an oxidizing electrolyte of low pH and to apply a positive dc (direct
current) potential to the aluminum with a cathode in the solution, completing the circuit as illustrated in Fig. 2. These processes and the resulting oxides are described extensively in the literature [9-11].

To orient the reader to the terminology and concept of anodizing,
the two basic types of oxides produced are illustrated in Fig. 3, and
the sequential formation of the porous oxide versus anodizing time is
illustrated in Fig. 4. Oxides produced by anodizing in phosphoric acid
differ from those produced in other common anodizing electrolytes (e.g.,
chromic acid, sulfuric acid, and oxalic acid). The major differences
are that this oxide will not hydrate or "seal," is much thinner, and has
the largest pore sizes [9-11,13]. These factors, particularly the oxide's
inability to hydrate, have limited commercial use of phosphoric acid
anodization.

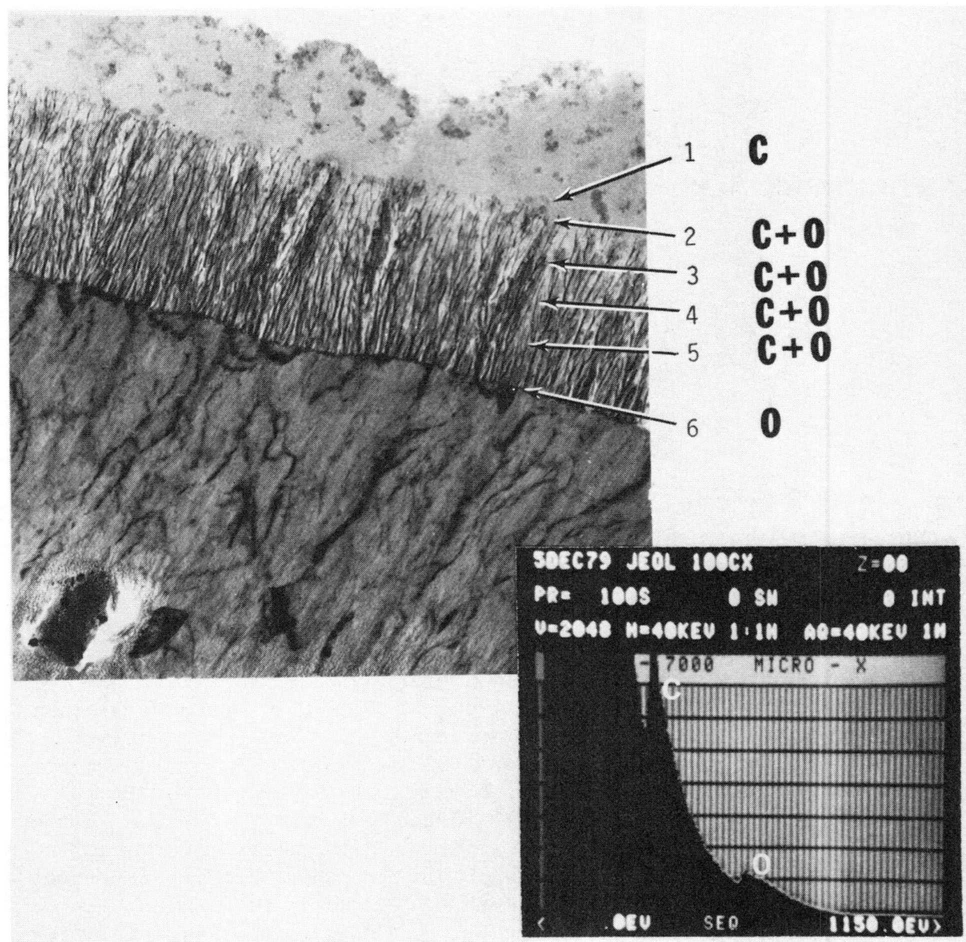

Figure 1   Results of EELS analysis of primer ingression into the porous oxide structure formed by phosphoric acid anodizing of 2024-T3 alclad alloy (C, carbon; O, oxygen). Inset is an EELS spectrum from region 3. Oxide cross section is a TEM micrograph of a thin section. Top of photo is primer, middle section is porous oxide plus primer, and bottom is alclad alloy. (Courtesy The Boeing Airplane Co.)

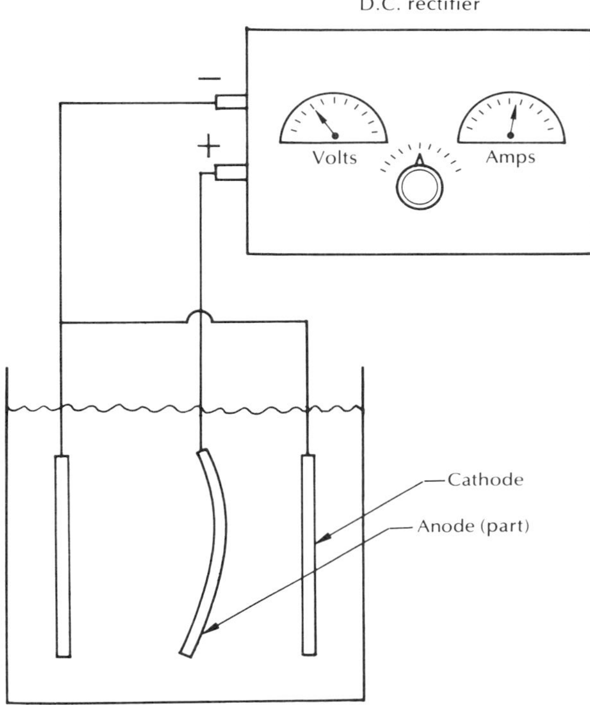

Figure 2   Schematic of a typical anodizing arrangement.

Although these oxides as such do not provide good corrosion protection, they do provide excellent adhesion of organic coatings and
adhesives. It is this combination, the oxide plus organic cover, that
provides the corrosion protection and environmental durability demonstrated in test programs [1,8,14-16].

II. COMPATIBILITY WITH PRIMERS

The following discussion is intended to provide the reader with general
information regarding the use of primers on oxide surfaces. It is not
intended to identify specific primers, but to point out what might be
expected when using various primer systems. Caution should be exercised when considering the use of a primer for which there are no
data available on those oxide surfaces, and the necessary tests should
be performed to confirm adequate or inadequate compatibility. Stressed-
durability tests should be conducted: for example, the wedge crack
test (Chapter 10) and the classic lap shear and peel tests.

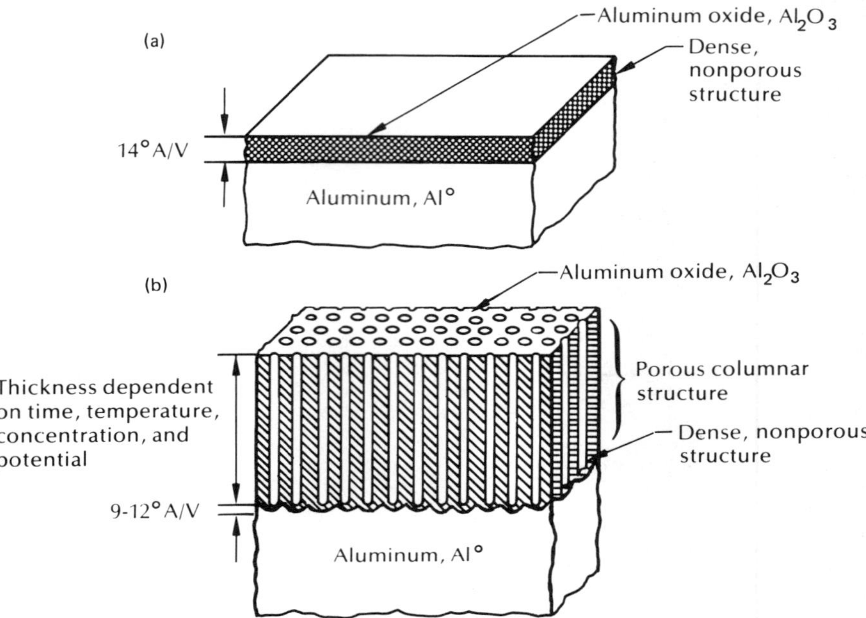

Figure 3   Oxide thickness versus forming voltage for nonetching and etching electrolytes: (a) typical anodic oxide formed in nonetching electrolyte; (b) typical anodic oxide formed in etching electrolyte.

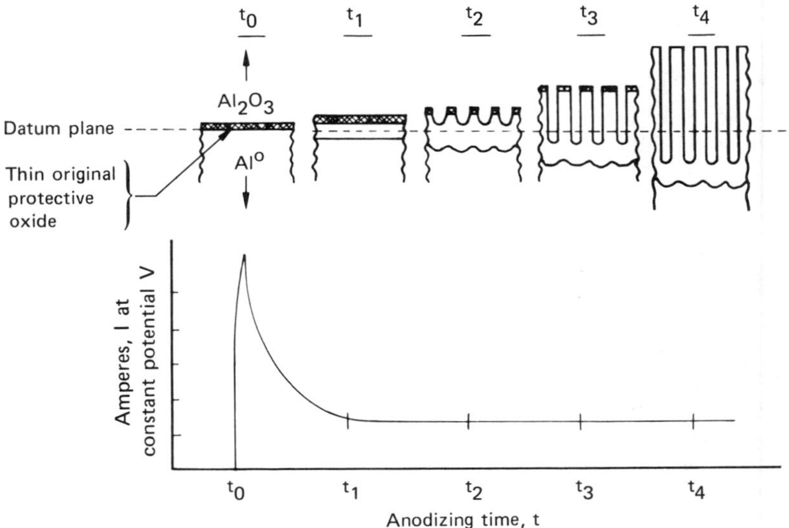

Figure 4   Mechanics of oxide formation at constant potential in an etching electrolyte.

A. 250°F (121°C) Curing Primers

The phosphoric acid anodize process described in this chapter was originally developed for use with 250°F (121°C) curing primers that are compatible with 250°F (121°C) curing adhesive films, since these were the systems around which most of commercial aircraft problems centered. Therefore, for most primers designated for use with 250°F (121°C) curing, modified epoxy adhesives have performed adequately with the subject oxides. This includes both epoxy- and phenolic-based primers. Several corrosion-inhibiting adhesive primers (CIAPs) were evaluated, with satisfactory performance indicated for all of them [16]. Not all primers will perform exactly the same; therefore, one has to determine the level of performance desired. The optimized FPL etch is used as a baseline surface treatment for performance comparisons. The generic type can influence performance.

B. 350°F (171°C) Curing Primers

Most 350°F (171°C) curing primers are compatible with these oxides; however, a few rubber-containing primers have exhibited suboptimal behavior. This behavior has been characterized by reduced lap shear ultimate strengths (approximately 30% reduction, Table 1), accompanied by a failure mode that appears to be at the primer-oxide interface when

Table 1 Room-Temperature Lap Shear Strengths of Aluminum Specimens Bonded with a 350°F (177°C) Curing Epoxy Phenolic Duplex Adhesive and Primed with Two Variations of an Epoxy Primer, One with Rubber and One Without Rubber

			Average of five lap shear strength readings (psi) for surface treatment with:		
Primer	Percent solids[a]	FPL Etch	Phosphoric acid anodize, 20-V	Phosphoric acid anodize, 5-V	Chromic acid anodize, 22-V
A with	15	3293	2317	2263	2565
rubber	30	2664	2535	2198	2348
A without	15	3577	3290	3306	3279
rubber	30	3312	3402	3405	3202

[a]Primers at 15% solids were sprayed on aluminum; primers at 30% solids were dipped on aluminum.

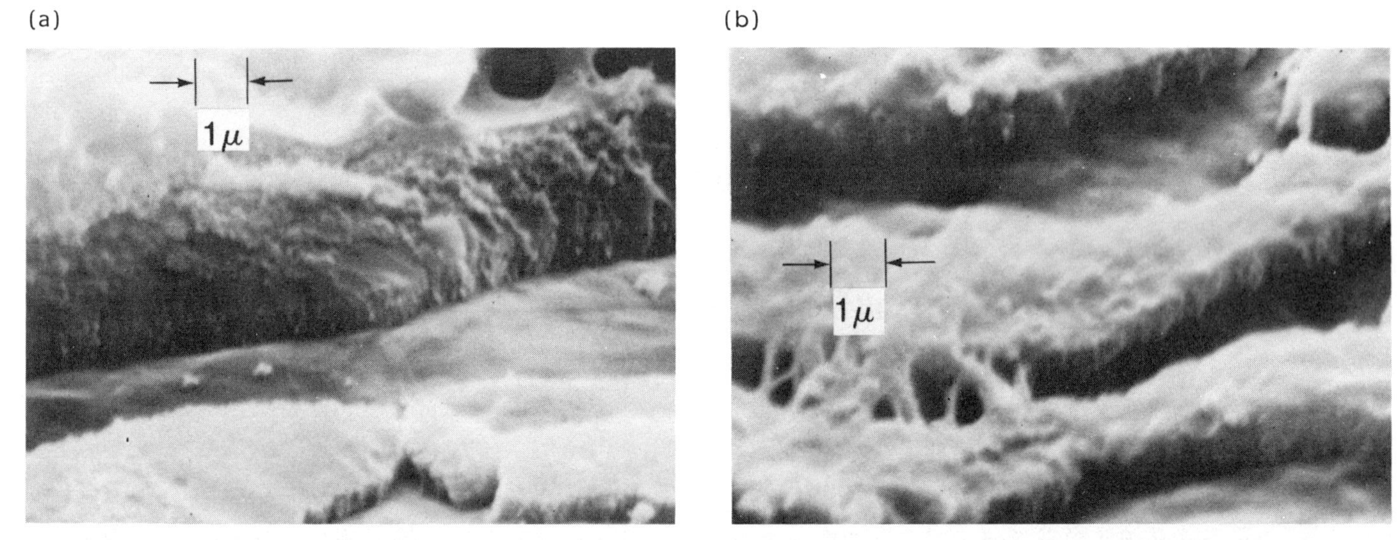

Figure 5   Two SEM (scanning electron microscope) photomicrographs of polymer-coated oxide surfaces from failed lap shear specimens. The oxide was formed by phosphoric acid anodizing of 2024-T3 alclad aluminum alloy: (a) primer A with rubber; (b) primer A without rubber. (Courtesy The Boeing Airplane Co.)

examined visually and has been reported as being an "adhesive"-appearing or "slick"-appearing failure. Further examination of these failures has revealed a thin, rubberlike layer approximately 100 nm thick on top of the oxide (Fig. 5) [17]. However, stressed-durability test performance has been excellent with lap shear specimens stressed at 1500 psi and less [15]. Poor stressed-durability performance with lap shear specimens has been reported when specimens were stressed to more than 1500 psi [18]. Ongoing studies to determine the locus of failure on adhesive-appearing failure modes have yet to reveal a failure in the oxide or at the oxide-polymer interface.

III. COMPATIBILITY WITH ADHESIVES

There has not been any evidence of adhesive incompatibility with the oxide when a primer is used, and logic says that this should be so. When the primer is omitted, a potential problem can exist. Most adhesives, particularly the film types that have been "B-staged," are of much higher molecular weight than the primers, and the chance of these larger molecules wetting into the fine porous oxide structure is much less than with the primers.

IV. COMPATIBILITY WITH DIFFERENT ALLOYS

The anodizing parameters described in this section were established initially for the anodization of four aluminum alloys commonly used in structural bonding. These are 2024, 1230 (cladding on alclad 2024), 7075, and 7072 (cladding on alclad 7075). The alloy 7075 alclad is no longer recommended for use in structural bonding applications because of the high frequency of delamination problems attributed to alclad dissolution in laboratory tests and in actual service applications.) These anodizing parameters have since proven to be applicable to other alloys [e.g., 7050, 7150, 2224, 2324, 6061, and A286 (7% Si)].

Heat treatments (e.g., T3, T81, T6, T0, etc.) do not appear to have any significant effect on the oxide properties that would affect bond integrity. The same is true for the form in which the material arrives at the anodizing process (i.e., whether the alloy is sheet, plate, extrusion, forging, or casting). The most significant effect would be the condition of the surfaces of these materials (i.e., the amount of heat treat scale and contamination). These surface conditions can ultimately influence the anodization step if not completely removed during the preanodization conditioning.

Rough surfaces formed by chemical treatments (e.g., chemical milling) or mechanical treatments (e.g., machining or shot peening) will affect the anodizing step only from the standpoint that the true surface area has been greatly increased and more current will be required. Therefore, it is very important that consideration be given to the true surface area rather than the apparent surface area when the number and type of electrical connections are being evaluated for anodization of a part.

V. PROCESSING PROCEDURES

A. General Requirements

1. Locate surface preparation and drying facilities separate from other activities and/or equipment to preclude contamination by substances detrimental to adhesion.
2. Facilities should be arranged such that parts can flow from the beginning of cleaning to the priming operation without being touched.
3. Filters and/or traps should be installed for removing airborne dust, moisture, and oil from all air lines or ducts used for solution agitation and parts drying. Check and clean or replace filters periodically to ensure proper operation.
4. Surface preparation facilities should not be used for purposes other than the preparation of parts for adhesive bonding and subsequent finishing.

Equipment requirements for this process include that equipment normally used in the chemical processing of aluminum alloys, with the following special provisions:

1. Racks and frames should be constructed of materials that will not cause corrosive reactions with the suspension wires or clips, or the parts, during the entire process. "Picture-frame" types of racks have performed satisfactorily.
2. Parts should be attached to racks or frames with conductive material that anodizes (e.g., aluminum or titanium in the form of wire, hooks, springs, clips, etc.).
3. The phosphoric acid anodizing tank lining should be of such material that interaction between the solutions, parts, frames, racks, or clips and hangers does not interfere with the anodic process and is suitable for use as a cathode. Tanks should be equipped with suitable lids to prevent contamination while not in use, and with a surface-skimming device to remove contamination, a filtering system, and an agitation system for mixing the tank contents.
4. Terminals for electrical current should be designed and positioned such that parts cannot be "burned."
5. The electrical system used during a *single*-rack anodic process should be adequate for producing 10 V dc and maintaining any set voltage within ±1 V dc.

6. The electrical system used during a *multiple*-rack process should be adequate for producing 15 V dc and maintaining any set voltage within ±1 V dc.

7. Rinse tanks should be placed adjacent to processing tanks in such a manner that parts can be moved easily and quickly from one solution to another with a minimum of transfer time, so that the parts will not become dry during successive steps of the process.

8. Incoming solution makeup water and rinse water should contain not more than the "recommended upper limit" for fluoride and not more than the "highest desirable level" of dissolved chemicals of the International Standards for Drinking Water [19, Tables 2 and 3] (1.7 ppm fluoride, 500 ppm total dissolved solids, etc.) except that (a) chloride should not exceed 25 ppm, and (b) pH should be between 5.5 and 8.0. Most regionally distributed potable water will meet these requirements. If the foregoing requirements cannot be met, the incoming water should be deionized and maintained within the following limits: (a) total dissolved solids should not exceed 150 ppm, and (b) pH should be between 5.5 and 8.0. Immersion rinse tanks should be agitated and overflowing during the entire rinsing cycle.

B. Manufacturing

Parts should be racked or suspended from frames such that they all have firm, reliable electrical contact with the anode connections, so that the parts do not contact each other and so that rinse water contacts all surfaces and drains freely. The utmost care should be taken that the parts are not touched at any time during the entire process, such as touching of adjacent parts, the rack of supporting frame, or any other item, since the surfaces and the phosphoric acid anodic coating are susceptible to contamination or physical damage prior to the application and cure of the primer. If it becomes necessary to handle or touch parts, clean white gloves should be used and contact should be limited to surfaces not to be adhesively bonded.

If parts become contaminated, corrective action should be accomplished by reprocessing, beginning with alkaline cleaning. If the anodic film has been formed prior to contamination, the film should be stripped prior to reprocessing. Agitate solutions to ensure homogeneity immediately prior to process use and after every addition of water or chemicals.

Caution: *Solutions used in this process are corrosive and toxic. Avoid breathing solution mists or vapors. Do not allow solutions to get in the eyes, on the skin, or on clothing. Consult the industrial hygiene staff for safety precautions.*

Control the time interval between withdrawal from processing solutions and rinsing so that there is no drying of the solution on the details. Parts must be water-break-free following rinsing after alkaline cleaning, after deoxidizing, and after anodizing.

C. Processing Sequence

The surface preparation process should be performed in a continuous operation as detailed on the flowchart in Fig. 6. All fabrication processes, inspections, prefit, and adjustments for individually identified assemblies should be completed before the start of the preparation cycle. All "hand work" should be completed before solvent cleaning. Once racked for alkaline cleaning, parts should not be touched by anything except the applicable processing solution and rinse waters, until the parts are dried after primer application.

The wet processing steps, from alkaline cleaning through drying after phosphoric acid anodizing, should be performed in one continuous, uninterrupted sequence of processes, with the parts not drying at any time until the oven drying after phosphoric acid anodizing.

1. Should the parts become delayed prior to anodizing during this wet processing, hold the parts in the applicable rinse tank.
2. There should be no delay in the processing sequencing after anodizing.

Should the electric current fail or otherwise be interrupted during the phosphoric acid anodizing, anodizing may be continued for an additional 20 to 25 min if the potential can be reestablished within 2 min. If the potential cannot be reestablished within 2 min, rinse and reprocess starting with the deoxidizer.

Adhesive primer should be applied within 72 hr after oven drying following the phosphoric acid anodizing. Parts should be held in a limited-contamination area prior to priming and should not be handled. However, it is most desirable to prime parts as soon as possible.

D. Rinse Requirements

Rinsing may be by spray or immersion. Single-spray or double-counter current immersion rinsing are commonly used. Feed water to rinse operations may be direct from source water or from any subsequent rinse operation, except that acid rinse waters should not be fed into alkaline rinse waters.

The final immersion rinse water after alkaline cleaning, deoxidizing, and anodizing should not exceed 100 ppm total dissolved solids above that of the incoming rinse water. When double-independent feed or

PROCESS	SOLUTION OR OPERATION	TIME (MINUTES)	TEMPERATURE (^{O}F) (^{O}C)
SOLVENT CLEAN	VAPOR DEGREASE, EMULSION DEGREASE, OR MANUALLY SOLVENT DEGREASE		
ALKALINE CLEAN RINSE - IMMERSION OR SPRAY	NONSILICITATED CLEANER (RECOMMENDED) CLEAN TAP WATER (see 2.5.h)	10-15 5 minimum	Per manufacturer's recommendation
WATER BREAKS	INSPECT FOR WATER BREAKS DO NOT DRY		
DEOXIDIZE RINSE - IMMERSION OR SPRAY	DEOXIDIZER CLEAN TAP WATER	10-15 5 minimum	110^OF (43^OC) minimum
WATER BREAKS	INSPECT FOR WATER BREAKS DO NOT DRY		
PHOSPHORIC ACID ANODIZE 1/	PHOSPHORIC ACID 9-12 wt % SINGLE RACK: APPLY DC VOLTAGE AND RAISE VOLTAGE TO 10 + 1V DC MAINTAIN 10 + 1V DC FOR TURN OFF CURRENT MULTI-RACK: APPLY DC VOLTAGE AND RAISE VOLTAGE TO 15 + 1V DC MAINTAIN 15 + 1V DC FOR TURN OFF CURRENT	 20-25 20-25	$70-82^O$F ($21-28^O$C)
RINSE - IMMERSION WITHIN 2-1/2 MIN.	AFTER CURRENT IS TURNED OFF CLEAN TAP WATER	10-15	110^OF (43^OC) maximum
	INSPECT FOR PRESENCE OF ANODIC COATING		

PRIMER APPLICATION - WITHIN 72 HOURS AFTER OVEN DRYING (PARTS MUST BE IN A LIMITED CONTAMINATION AREA)

1/ A single-rack facility is one with a single anode rack between each set of cathodes. A multi-rack has two or three anode racks between each set of electrodes.

Figure 6   Process flowchart.

double-counter current rinse is used, the first rinse after anodizing should not exceed 5000 ppm, and the time in first rinse should not exceed 2 min from time of complete immersion to complete withdrawal. Immersion tank(s) used for anodize rinse should be used for anodize rinse only.

Immersion rinse tanks should be agitated during rinsing to assure adequate rinsing and prevent stratification. Allowable concentration limits may exceed normal control limits for the first 30 sec of rinsing.

E. Quality Control Requirements

Part Inspection During Processing

Parts should be inspected during the continuous processing as noted in Fig. 6.

Water-Break Inspections

Cleaned parts should pass water-break inspection as indicated by maintenance of a continuous film of water on the surface for not less than 30 sec. Parts failing the water-break inspection should be reprocessed through the applicable cleaning operation until the surface can maintain the continuous film of water. Parts failing the water-break inspection after three cleaning cycles should be rejected.

Phosphoric Acid Anodic Coating Inspections

After anodizing and during rinsing and draining, there should be no evidence of a water break. There should be no stains, streaks, discoloration, or residue on surfaces of anodized parts after rinsing. The anodic coating should be continuous, smooth, uniform in appearance, and free from discontinuities such as scratches, breaks, burned areas, and areas that are not anodized when examined visually. Small irregularities at points of electrical contact are normally acceptable.

Note: The color change may be difficult or impossible to detect on surfaces that have been etched by marking ink, roughened by machining, sanding, deburring, and so on, or that are uninspectable because of shape. In these cases, the presence of color on tool tabs or on the opposite surface, or on undisturbed surfaces, may be evidence that the surface is acceptably anodized.

The anodized surfaces should pass the following polarized color change examination using any sufficiently intense white or near-white light (daylight is acceptable). An optical polarizing filter is available from laboratory supply houses or Polaroid, Inc., Cambridge, Massachusetts.

1. Illuminate processed aluminum part surfaces so that the incident angle of the light to the plane of the part surface

does not exceed 5°. *Note*: Verification of the presence of anodic coating may be made with the incident light at an angle greater than 5°. Before rejecting details for failure to display the color required, they should be reviewed with the incident light less than 5°.

2. Place a polarizing filter between the observer and the part surface or between the light source and the part surface and view the reflected light from the detail surface at an angle not exceeding 5°.

> Caution: *Be sure that illumination on the part being inspected has not been reflected from other anodized surfaces.*

3. Observe the detail surfaces for the appearance of "interference" colors. *Note*: Different pieces of aluminum anodized under the same conditions may show different interference colors because of differences in alloy composition and metallurgical condition. The colors most frequently seen are purple, yellow, and green hues.

4. While viewing the surfaces, rotate the polarizing filter 90°. The presence of anodic coating on a surface is verified by an observed change either from one color to the complementary color (e.g., from purple to green) or from one color to no color (e.g., from yellow green to no color). *Note*: Rotation of the polarizing filter is required because some pale shades of yellow or green are so close to white that without a color change inspection they might be considered "no color," which would falsely indicate no anodic coating.

5. Parts not passing the polarized color change test should be rejected or reprocessed starting with the alkaline cleaning step.

F. Makeup and Process Control of Processing Solutions

Deionized water: When deionized water is used in solution makeup, replenishment, and in the rinses, it should contain not more than 10 ppm total dissolved solids.

Alkaline cleaner: Makeup, replenishment, and replacement of alkaline cleaning solutions should be developed to produce effective processing.

Deoxidizer: Makeup, replenishment, and replacement of deoxidizer should be developed to produce effective processing. The sulfuric acid-sodium dichromate etch (method A for aluminum alloy in ASTM D-2651) has been an effective etch.

Phosphoric Acid Anodizing Solution

Initial tank makeup:

1. Fill the tank approximately three-fourths full of deionized
    water or clean tap water.
2. Agitate the water and slowly add 7.0 parts of phosphoric
    acid (0-0-670, class I, 85% phosphoric acid) for each 100
    parts of final solution.
3. Bring the mixture up to the operating level of the tank with
    deionized water or clean tap water not exceeding 150 ppm
    total dissolved solids and mix thoroughly.

Maintenance:

1. Maintain the operating conditions and total acid (H^+ ion) con-
    centration equivalent to the phosphoric acid concentration
    specified in Table 2. The level of solution in the tank should
    be within the effective operating level for the surface-skimming
    equipment.
2. Other selected temperatures within the range 65 to 85°F (18
    to 30°C) may be used after thorough investigation and qualifi-
    cation of a specific facility.

Qualification of phosphoric acid anodizing solution: Each
freshly mixed tank of phosphoric acid anodizing solution should be
qualified before being used for processing production parts. Qualification should be accomplished by processing and adhesive bonding no
fewer than five control specimens required by the applicable adhesive-
bonding specification. The control specimens should exceed the minimum strength requirements.

Determination of phosphoric acid concentration: Titrate a 10-ml
sample of anodizing solution to a pH of 4.2 with 1.0 N NaOH with a

Table 2 Material/Operation

Material/operation	Range
Phosphoric acid, 85%	9-12 wt%
Temperature	70-82°F (21-28°C)
Voltage, dc	9-11 (single rack) 14-16 (multiple rack)
Anodizing time	20-25 min

correction for aluminum concentration. The aluminum concentration is to be determined by atomic absorption. Use the following formula:

$$\text{oz/gal } H_3PO_4 = 1.3\,[(\text{NaOH normality})(\text{ml NaOH}) - (0.7)\,(\text{g/liter of aluminum})]$$

$$\text{gm/liter } H_3PO_4 = 9.8\,[(\text{NaOH normality})(\text{ml NaOH}) - (0.7)\,(\text{g/liter of aluminum })]$$

Replenishment and Replacement of Solutions

Tank solutions should be replenished to maintain the required concentrations and replaced when the solutions cannot be maintained or when the solutions become contaminated to the point of no longer processing parts satisfactorily.

G. Preprocess and Process Variables

The preprocess variables discussed here relate to the aluminum materials and their condition as they would arrive at the anodize process line. These variables are discussed briefly as they relate to the overall anodize process.

Materials

The different forms of aluminum that can be phosphoric acid anodized will usually play a minor role in the nature of the oxide produced during anodizing. However, the surface finishes on these materials can be a more significant factor, as will be described below. Of the different materials to be anodized, the alloys themselves and their heat treatments can play a more significant role, which will be elaborated on in the section on anodizing variables.

Surface Conditions

By the time parts are ready to be anodized for bonding, many intermediate processes usually take place that can alter the original surfaces. Generally, these alterations do not have any influence on the resulting anodize process, but some can create problems (e.g., heat treatment scales, protective oils, and lubricants, and corrosion products) that if not entirely removed in the preanodize processing can result in a poor anodic oxide to which to bond. Some surface conditions can produce aesthetic problems, for example, surface that has been sanded, or ink markings of the etching type that leave a mark on the surface, to name a few.

Rough surfaces (e.g., shot peened, sanded, and chemically milled) have larger surface areas than those of smooth surfaces. These will require more wattage per apparent square footage of area, but in theory,

the oxide formed for that particular alloy would not be any different
in structure from that formed on a smooth surface.

Preanodize

This phase of the process is very important since it paves the way
for the formation of a proper oxide in the anodizing step. Degreasing
or solvent cleaning serves the purpose of removing soluble oils,
greases, and hydrocarbons that could otherwise overtax the alkaline
cleaner. Not all parts require this step if they are relatively clean.

Alkaline cleaners should be of the nonetching type (although mild
etching, if uniform, is not detrimental). These solutions function best
at elevated temperatures [e.g., 135 to 170°F (57 to 76°C), on the
specific material]. Nonsilicated cleaners are recommended over silicated
cleaners from the standpoint that if any porous oxide (e.g., that produced by phosphoric acid anodizing which is being reprocessed) is
exposed to a silicated solution, the fine pores will fill up and retain
this solution. Rinsing, particularly short rinsing times, will not remove all of the solution. When this solution, retained in the porous
oxide, comes in contact with the acid etchant, sodium silicate is formed
and, being insoluble in the acid, quite often forms a thin film over the
surfaces. The film remains on the surface during the anodizing and
may not be removed during subsequent rinsing. If still present during the priming step, a poor bond can result.

The function of the alkaline cleaning is to remove soils and to make
the surface hydrophilic to ensure good wetting of the acid etchant for
uniform etching. The hydrophilic nature of the surface is evidenced
by a water-break-free surface after rinsing. Rinsing is necessary to
remove the alkaline solution to prevent contamination of the acid etchant.
Warm water is desirable to prevent precipitation of some constituents
of the alkaline solution onto the surfaces. The alkaline solution must
not be allowed to dry on the surfaces before rinsing, particularly if
it is a silicated solution. Once dried, many of these materials are very
difficult to remove and can subsequently affect the acid etching of these
surfaces. This could result in selective etching.

The preanodize etch or acid etch, commonly called the "deoxidizer,"
is an important and necessary step. Its function is to remove undesirable oxides (e.g., mill scales, thick weak oxides, corrosion products,
etc.) and leave a thin, uniform oxide on the surface. This oxide does
not have to be suitable for bonding by itself, but the important criterion
is for it to be completely dissolved during the anodizing process. If
any of this oxide resulting from the preetch remains on top of the
oxide produced by anodizing, this can create a weak boundary layer
in the final bonded structure. The other criterion is that the preetch
be sufficiently agressive to produce this thin oxide uniformly over the
entire surface. If any thick oxides or hydrates remain, this can result in areas of poor bond quality.

There are many acid etchants available that would be suitable for the job; however, one would have to be assured of their suitability by conducting a proper test program. Usually, an etchant should have a minimum etch rate of approximately 0.020 mil/side per hour to remove the thicker oxides effectively in a reasonable time frame (e.g., 10 to 15 min). Several acid etchants have proven to be suitable. These are FPL (Forest Product Laboratories) etch, a hot solution of sulfuric acid and sodium dichromate, and Amchem 6-16, a proprietary room-temperature chromated solution (manufactured and distributed by Amchem Products, Inc., Ambler, Pennsylvania). Table 3 shows the makeup and control of these solutions.

Nonchromated acid etchants have been tried in the laboratory with good results, provided that the etch rate was sufficient and uniformity of etching was good. However, as noted previously, one would have to evaluate the etchant prior to using it in a production situation.

Alakline etchants can also be used as preanodize etchants. Here again, they have to etch uniformly and be able to remove any heavy

Table 3   Preanodize Etchants

Material/operation	Range
FPL Etch	
$Na_2Cr_2O_7 \cdot 2H_2O$	4.1-12.0 oz/gal (30.7-89.9 g/liter) as $Na_2Cr_2O_7 \cdot 2H_2O$
H_2SO_4 (66° Bé)	38.5-41.5 oz/gal (288.5-311.0 g/liter) as H_2SO_4
Aluminum as 2024	0.2 oz/gal (1.5 g/liter) minimum
Time	5-15 min
Temperature	150-160°F (65-71°C)
Etch rate	0.00022-0.00034 in./hr per surface
Amchem 6-16	
Achem 6	4-9% wt %
HNO_3	10-20 oz/gal (74.9-150 g/liter) as HNO_3
Time	10-15 min
Temperature	65-90°F (18-32°C)
Etch rate	0.00015-0.00040 in./hr per surface

oxides that might exist on the as-received aluminum. Most alkaline etchants will leave a smut on the aluminum surface. This can be removed easily by a dip in a 20 to 50 wt % nitric acid solution (HNO_3) or it can be removed easily during the early stages of phosphoric acid anodization.

Table 4 summarizes all the variables discussed.

Table 4 Variables Related to Phosphoric Acid Anodizing

Preprocess variables	Process variables	Postprocess variables
Materials	Preanodize	Prepriming
Alloys	Degreasing	Handling
Forging	Alkaline cleaning	Airborne
Extrusions	Preanodize etch	contaminant
Castings	Rinsing	
Sheet	Anodize	Time before
Plate	Time	priming
	Temperature (solution)	Environment
Surface conditions	Voltage (part to	
Heat treatment scale	solution)	Storage
Machined surfaces	Acid concentration	Surface-
Protective oils	Solution age	primer
Corrosion products	Current density	interactions
Sanded surfaces	Racking—single,	not con-
Chemical milled	multiple	sidered
surfaces	Electrical contacts to	critical
Ink markings	parts	after
Fingerprints	Part orientation	priming
Stretch-forming	Alloy	
lubricants		
Protective maskants	Postanodize	
Shot peen	Delay—power off to	
	rinse	
	Rinsing—immersion	
	versus spray	
	Water purity	
	Rinse time	
	Rinse water temperature	
	Drying temperature and	
	time	
	Air contamination	
	Reprocessing	

H. Anodize Variables

The operating envelope described above for phosphoric acid anodizing of aluminum alloys was established such that those aluminum alloys commonly used for adhesive bonding in industry could be processed under the same conditions and still have a bondable surface that would result in an environmentally durable bondment. It was not intended that oxides produced on the different alloys be exactly the same, but that they would all contain the essential characteristics necessary to produce an environmentally durable bond with the adhesive polymer coming in contact with it.

However, the oxide characteristics necessary for durable bonds are not fully understood. One obvious characteristic of the oxide is porosity, and another essential requirement is that the oxide be strong enough to take the loads being transferred through it. The porosity of the oxide is controllable but will vary from alloy to alloy for any given anodizing condition.

Some of the most common anodize variables that influence the oxide structure are voltage, time, solution temperature, and acid concentration. Voltage will affect the barrier layer thickness, cell wall thickness, pore diameter, and the total thickness of the duplex oxide [9-11]. The relationship between voltage and barrier layer thickness is illustrated in Figs. 3 and 4.

Voltage can vary considerably at the detail part, even though the rectifier voltage is at 10 or 15 V. Assuming a single-rack arrangement with 10 V at the rectifier, the part-to-solution voltage may be as low as 9 V. A multiple-rack system presents an even different situation, since surfaces of parts not facing toward cathodes can have part-to-solution voltages as low as 4 to 5 V with 15 V at the rectifier. If racks and parts are too close together, the part-to-solution voltage can approach zero.

The reason for this is that the oxide formed on the aluminum during the anodizing process dissolves in the phosphoric acid at a significant rate, especially under the influence of the electric field present during anodizing. This is different from the situation in chromic acid anodizing, in which the rate of oxide dissolution is much less, aiding the formation of a more resistive oxide. This voltage drop through the phosphoric acid solution to part locations too remote from cathodes can be so great as to prevent development of the desired oxide. Tests in large-scale production facilities with multiple racking have shown part-to-solution voltages of 4 to 5 V to be sufficient to form the proper oxide. Less than 3 V may not be adequate.

All discussions in this book related to direct current (dc) anodizing; however, a brief discussion of alternating current (ac) anodizing is indicated. During ac anodizing, the aluminum part is alternately an anode and a cathode. During the time period in which the part functions as an anode, the oxide will form in the same manner as in dc anodizing, with similar relationships holding true (i.e., barrier layer formation, porous layer, etc.). During the time period in which the part is a cathode, the oxide formed during the anode cycle will want to dissolve. However, the dissolution rate during the cathodic cycle is much less than the formation rate during the anodic cycle, thereby creating a net growth of oxide similar to that of the duplex structure formed in dc anodizing. It is not known to the writer at this time how the oxide characteristics vary between ac and dc anodizing, but in terms of bondability and bond durability, ac anodizing appears to be equivalent to dc anodizing.

One significant advantage of ac anodizing would be the ability for one to rack parts on multiple racks in a high-density configuration with alternate electrical leads such that facing parts are alternately anode and cathode. This would eliminate the problems of voltage drops on those parts not facing directly toward a cathode during dc anodizing.

Anodizing time affects the thickness of the porous layer forming above the barrier layer. The porous layer of a duplex oxide will continue to grow until the rate of oxide formation equals the rate of oxide dissolution at or near the outer surface. The length of time needed to reach this dynamic equilibrium, sometimes referred to as the "limiting film thickness," is a function of the voltage, solution temperature, acid concentration, and aluminum alloy being anodized.

Anodize time will also affect the porosity or openness of the outer surface structure; that is, the acid will dissolve the outer cell walls because of its longer exposure to the bulk solution. The effect of time on the growth of the porous oxide layer is illustrated schematically in Fig. 4.

Solution temperature also will affect barrier layer thickness, porous layer thickness, pore diameter, and cell wall thickness, because of its effect on the solution dissolving power. High temperatures produce thinner barrier and porous layers, larger pore diameters, and thinner cell walls. Low temperatures have the opposite effect.

Phosphoric acid concentration affects the pH of the solution and, thereby, its anodizing power. This will affect the oxide structure in a manner similar to the solution temperature.

Another variable that has some influence on the process is that of solution aging or usage. As the solution sees more aluminum, phosphoric acid is consumed mainly by the dissolution of aluminum oxide, forming soluble complexes of aluminum and phosphate. In addition to the increased aluminum concentration, alloying elements (e.g., copper, magnesium, zinc, etc.) also increase. Studies at Boeing have shown no apparent effect on the bondability of oxides formed in aged solutions as long as the pH of the solution is below 1.5 and the active H_3PO_4 concentration is within the recommended range. The presence of aluminum in solution makes it difficult to analyze for H_3PO_4, but numerous procedures can be used, such as that described in Chapter 15.

At this time there has been no assessment of the life of a phosphoric acid solution used in this process in terms of usage or chronological age. One large tank (36,000 gal) at the Boeing Company has been in use for over 6 years without any change being necessary. The aluminum concentration over this period has risen to approximately 5500 ppm and appears to have reached an equilibrium concentration. This means that aluminum drag-out equals aluminum input by dissolution.

Dissolved copper in the solution will plate out on the cathode during anodization, but will redissolve slowly (e.g., over a weekend). The presence of copper does not affect the performance of the process as long as the aluminum details have proper electrical connections. If inadequate connections are made on a detail, then during anodizing a dark, sometimes reddish-brown smut will form on the detail, usually in the center or points farthest away from inadequate contacts. This situation may not show up when a new or relatively new solution is used or if only non-copper-bearing alloys are processed.

Fungus growth may occur in the phosphoric acid solution after several months of operation. It has been identified as a viable fungal mycelium. Its presence does not interfere with the anodization of aluminum, but it does create a visual problem. When allowed to proliferate, it may be difficult to rinse all of it off, creating potential adhesion problems. A simple means of controlling its growth is by filtration, either continuously or intermittently. A 5- to 10-μm filter works suitably when sized to handle a flow rate of 1 liter/min per 450 liters of solution. The most effective filtering control can be obtained when the filter pump suction is from an overflow skimmer reservoir. Another method of control is to kill the fungus by heating the solution to 176°F (80°C) for 30 min.

I. Rinsing Variables

The function of rinsing is to remove chemicals from the preceding step. As described earlier, rinsing can be accomplished satisfactorily after each chemical treatment by either spray or immersion. Each chemical step in the process must have its own rinse tanks not to be shared with other solutions. Rinsing after alkaline cleaning and anodize pre-etch are not too critical, other than that the chemicals should not be allowed to dry on the aluminum surfaces prior to rinsing and they should be adequately rinsed to prevent contamination of the following chemical solution. It is usually preferable to use a warm water rinse after the alkaline solution to aid in the rinsing.

Rinsing after phosphoric acid anodizing becomes more critical than the previous rinses. Perhaps the most critical aspect of this step is the time between termination of anodizing (turning off the anodizing power supply) to the start of rinsing. (When immersion rinsing is used, the critical time would be to the point when all the details are

completely immersed.) As soon as the anodizing power is turned off,
oxide formation ceases, but oxide dissolution continues. If too much
time elapses before the acid is removed, the oxide structure necessary
for good bonding can be destroyed or removed completely. A delay
time of up to $2\frac{1}{2}$ min is acceptable when processing 2024 nonclad alloy
with a solution temperature up to 82°F (28°C). The alloy 2024 nonclad
is more critical than others because it has the thinnest oxide with the
highest rate of oxide dissolution.

The method of rinsing is not critical as long as one stays within
the water quality guidelines. Perhaps the potentially most critical
rinsing situation is that of the first immersion tank of a mulitple-tank
immersion system. If more than 5000 ppm phosphoric acid has been
carried over into the first rinse, plus an immersion time of more than
2 min, sufficient dissolution of the oxide can occur to result in a
questionable bond surface.

J. Drying Variables

Care must be taken in drying parts after rinsing so as not to contaminate the highly active oxide surfaces. The most critical consideration
in drying is that of the incoming air quality and also the quality of the
recirculated air. Filters should be used for incoming and recirculated
air. Noncontaminating heaters and fans and covers over immersed
heated drying tanks also should be used.

Drying times and temperatures do not appear to be critical as long
as the water is completely evaporated prior to primer application. The
most common drying temperature is 140 to 160°F (60 to 82°C). Not much
has been done to investigate the effects of drying temperatures above
160°F (82°C). Room-temperature drying presents no problems other
than extending the drying time, although care must be taken to provide
a contamination-free area for this purpose.

K. Handling and Storage

Handling and storage should be considered a critical step in the preparation of aluminum surfaces for structural bonding. Until a primer is
properly applied to the oxide surface, the oxide is vulnerable to contamination and mechanical damage. How the parts are handled and
stored prior to priming should reflect all the considerations given in
preparing a bonding facility for primary structural bonding. No short-
cuts should be taken.

The ideal situation is to leave the parts on the anodizing rack or
frame to which they were attached before going into the cleaning line.
This eliminates the need to handle the parts before priming. They
can be inspected and primed directly on the rack or frame. After curing the primer, parts can be handled without concern of inflicting

damage or contamination to the oxide. The cured primer is far more tolerant of these insults than is the oxide.

Time delay before priming is not critical, provided that the parts are stored in a contamination-free area. Philosophically, parts should be primed as soon as possible after drying (e.g., within one working shift). However, this is not always possible, and one may have to wait overnight or even over a weekend before priming. Because of these considerations, a common time limit placed on industry is 72 hr, although much more time could be allowed.

Perhaps the most critical item to be considered is that of handling the parts before priming if it is not possible to leave then untouched on the racks or frames. The oxide is extremely tender and delicate, and any wiping, rubbing, or mechanically induced abrasion will cause varying degrees of damage to the oxide, which will subsequently affect bond durability. Handling parts with white gloves, wiping with cheesecloth, and wrapping in kraft paper have been reported to cause damage or contamination detrimental to bond durability [16].

Once parts have been primed and the primer cured, they can be stored indefinitely, provided that the storage environment does not deteriorate the primer. Handling of primed parts is not generally a problem since the cured primers are more tolerant of mechanical forces and contaminants. Most contaminants can be easily removed before bonding by a simple solvent wipe.

L. Reprocessing

Once a primer has been applied and cured, it is next to impossible to remove without damaging the part. The general practice at this time is not to reprocess parts that have cured primers.

Parts can be reprocessed if primer has not been applied by stripping the oxide in the preanodize deoxidizing solution and reanodizing. The best procedure is to go directly into the deoxidizer; however, it may be desirable, if not necessary, to remove contaminants (e.g., fingerprints, oils, etc.) by reprocessing through the alkaline cleaner before stripping the oxide. A silicated alkaline cleaner can be detrimental to a reprocessed part by the trapping of the silicates in the pores and the subsequent formation of silicic acid in the deoxidizer. This was discussed earlier.

REFERENCES

1. Bethune, A. W., Durability of Bonded Aluminum Structure, *SAMPLE J.* Vol. 11, No. 3, July/Aug./Sept. 1975, pp. 4-10.
2. Smith, A. W., Surface Oxide on Etched Aluminum, *J. Electrochem. Soc. Solid State Sci. Technol.*, Nov. 1973.

3. Venables, J. D., D. K. McNamara, J. M. Chen, T. S. Sun,
    and R. L. Hopping, Oxide Morphologies on Aluminum Prepared
    for Adhesive Bonding, *Appl. Surf. Sci.* Vol. 3, No. 1, July
    1979.
4. Malpass, B. W., D. E. Packham, and K. Bright, A Study of
    the Adhesion of Polyethylene to Porous Alumina Films Using
    the Scanning Electron Microscope, *J. Appl. Polym. Sci.* Vol. 18,
    1974, pp. 3249-3258.
5. Packham, D. E., K. Bright, and B. W. Malpass, Mechanical
    Factors in the Adhesion of Polyethylene to Aluminum, *J. Appl.
    Polym. Sci.* Vol. 18, 1974, pp. 3237-3247.
6. Evans, J. R. G., and D. E. Packham, Adhesion of Polyethylene
    to Copper: Importance of Substrate Topography, *J. Adhes.*,
    Vol. 10, 1979, pp. 39-47.
7. Bijlmer, P. F. A., Adhesive Bonding on Anodized Aluminum,
    Met. Finish., Apr. 1972.
8. McMillan, J. C., Surface Preparation—The Key to Bondment
    Durability, in *Bonded Joints and Preparation for Bonding*,
    AGARD Lecture Ser. 102, Mar. 1979.
9. Wernick S. and R. Pinner, *The Surface Treatment and Finishing
    of Aluminum and Its Alloys*, 4th ed. Robert Draper, 1972.
10. Diggle, J. W., T. C. Downie, and C. W. Goulding, Anodic
    Oxide Films on Aluminum, *Chem. Rev.*, Vol. 69, 1969, p. 365.
11. Diggle, John W., *Oxides and Oxide Films*, Vol. 2, Marcel
    Dekker, New York, 1973, Chap. 3.
12. Ledbury, E. A., A. G. Miller, P. D. Peters, E. E. Peterson,
    and B. W. Smith, Microstructural Characterization of Adhesively
    Bonded Joints, *Proc. 12th Nat. SAMPE Tech. Conf.*, Oct. 7-9,
    1980.
13. Hunter, M. S., P. F. Towner, and D. L. Robinson, *Tech.
    Proc. Am. Electroplaters' Soc.*, Vol. 46, 1959, p. 220.
14. Scardino, W. M. and J. A. Marceau, Comparative Stressed
    Durability of Adhesive Bonded Aluminum Alloy Joints, *J. Appl.
    Polym. Sci. Appl. Polym. Symp.* 32, 1977, pp. 51-63.
15. Marceau, J. A. and J. C. McMillan, Exploratory Development
    on Durability of Adhesive Bonded Joints, Air Force Materials
    Lab. Tech. Rep. AFML-TR-76-173, Oct. 1976.
16. Shannon, R. W. et al., Primary Adhesively Bonded Structure
    Technology (PABST), U.S. Air Forse Flight Dynamics Lab.
    Tech. Rep. AFFDL-TR-77-107, Sep. 1978.
17. Marceau, J. A., An SEM Analysis of Adhesive Primer Oriented
    Bond Failures on Anodized Aluminum, *SAMPE Q.*. Vol. 9, No. 4,
    July, 1978.
18. Schwartz, H. S., Effects of Adherend Surface Treatment on
    Stressed Durability of Adhesive Bonded Aluminum Alloys,
    SAMPE J. Vol. 13, No. 2, Mar./Apr., 1977.
19. World Health Organization, *International Standards for Drinking
    Water*, 3rd ed., WHO, Geneva, 1971.

5

Surface Analysis

MERVIN A. DANFORTH *Douglas Aircraft Company of the McDonnell Douglas Corporation, Long Beach, California*

I. PHYSICAL ANALYSIS

The transmission electron microscope (TEM) was the first instrument available for study of the physical characteristics of surface treatments for adhesive bonding [1]. The high resolution of this technique is its greatest attribute.

The scanning electron microscope (SEM), with its excellent depth of field, was used by Remmel [2] and Danforth and Sunderland [3] to investigate surface treatments. The SEM technique as developed by Remmel (Fig. 1), in which the surface treatment specimen is bent to expose the broken edge of the surface treatment, provides much information about the thickness as well as the internal structure. Venables et al. [4] have developed much information about metal adhesive bonding surface treatments using the scanning transmission electron microscope. Their work indicates that the coating used for conductivity in the SEM at high magnification obscures the fine detail at the surface of the specimen. Marceau [5] confirms the finding that a conductive coating of gold as thin as 50 to 100 Å obscures the hairlike structure on the surface of phosphoric acid anodize. Conductive coatings of gold, platinum, pladium, and their alloys applied at 20 to 30 Å by vapor deposition or sputter coating are used to provide the necessary conductivity without obscuring the surface details.

II. CHEMICAL ANALYSIS

Concurrent with the development of the instrumentation for investigation of the physical structure of surface treatment and adhesives has been the development of several techniques that can provide elemental and in some cases molecular information.

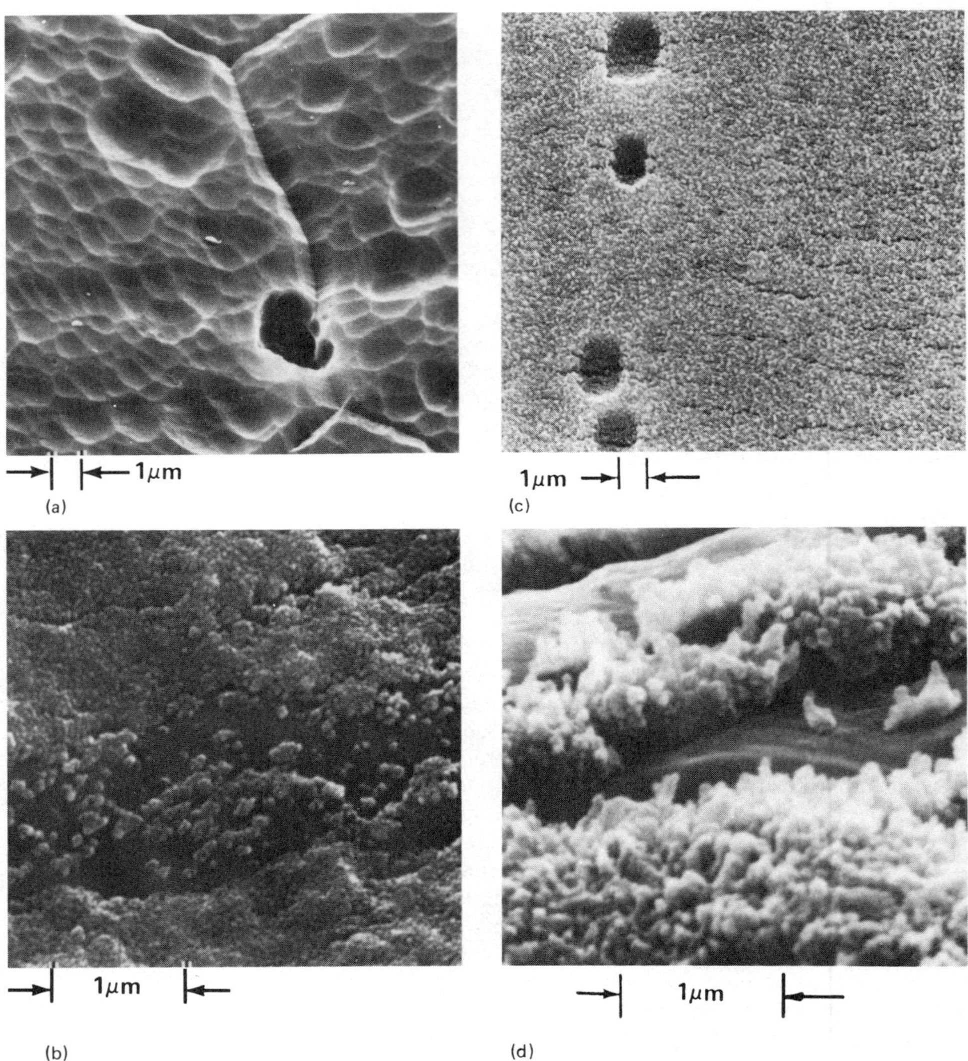

Figure 1   Scanning electron microscope technique. (a) and (b) show an FPL etch, (b) at a higher magnification. (c) and (d) show phosphoric acid anodize, (d) at a higher magnification. (Courtesy Douglas Aircraft Co.)

A. Ion Scattering Spectrometry (ISS)

When a surface is bombarded by ions and the loss of energy of the re-
bounding ions is measured, the elemental and molecular species of the
surface can be identified. This method measures the first atomic layers.

B. Secondary Ion Mass Spectrometry (SIMS)

When an ion beam bombardment results in some surface atoms being dis-
lodged, those that are ionized can be analyzed in a mass spectrometer.

C. Auger Electron Spectrometry (AES)

When a surface is bombarded by an electron beam, the resultant Auger
electron can be analyzed by a cylindrical mirror analyzer and electron
multiplier. If used in conjunction with an ion beam a chemical profile
through the surface can be obtained as it is sputtered away. A scanning Auger has been developed which offers a secondary ion image of
the surface to be analyzed as well as the capability to map an area for
distribution of an element.

D. X-ray Photoelectron Spectroscopy (XPS)
(Electron Spectroscopy for Chemical Analysis)

When a surface is irradiated with a beam of x-rays, the resulting elec-
trons are energy analyzed to identify the atoms and molecules.

E. Surface Treatment Investigation

Figure 1 shows only the cell size difference of the FPL (Forest Products
Laboratory) etch versus phosphoric acid anodize until the bend specimen is observed. Only then does the much greater film thickness together with the more columnar nature of the phosphoric acid anodize
become visible.

Figure 2 demonstrates the use of a plastic replica technique to show
the great increase in bonding surface area and the penetration of the
oxide layer by a primer. The plastic replica was made using typical
TEM methods. After stripping the plastic replica from the oxide surface it is coated with 100 Å of carbon and 100 Å of gold. The replica
is then examined in the SEM. The SEMs of the chromic/nitric/HF etch
replica is very smooth compared to that of phosphoric acid anodize.

Danforth and Sunderland [3] used SEM, Auger, XPS, ISS, and
negative SIMS to confirm surface treatment handling contamination from
cotton gloves and kraft paper. Chen et al. [6] demonstrated the value
in tracing fluorine contamination of surface treatment using XPS and
Auger. Baun [7] has demonstrated the use of ISS, SIMS, and XPS in

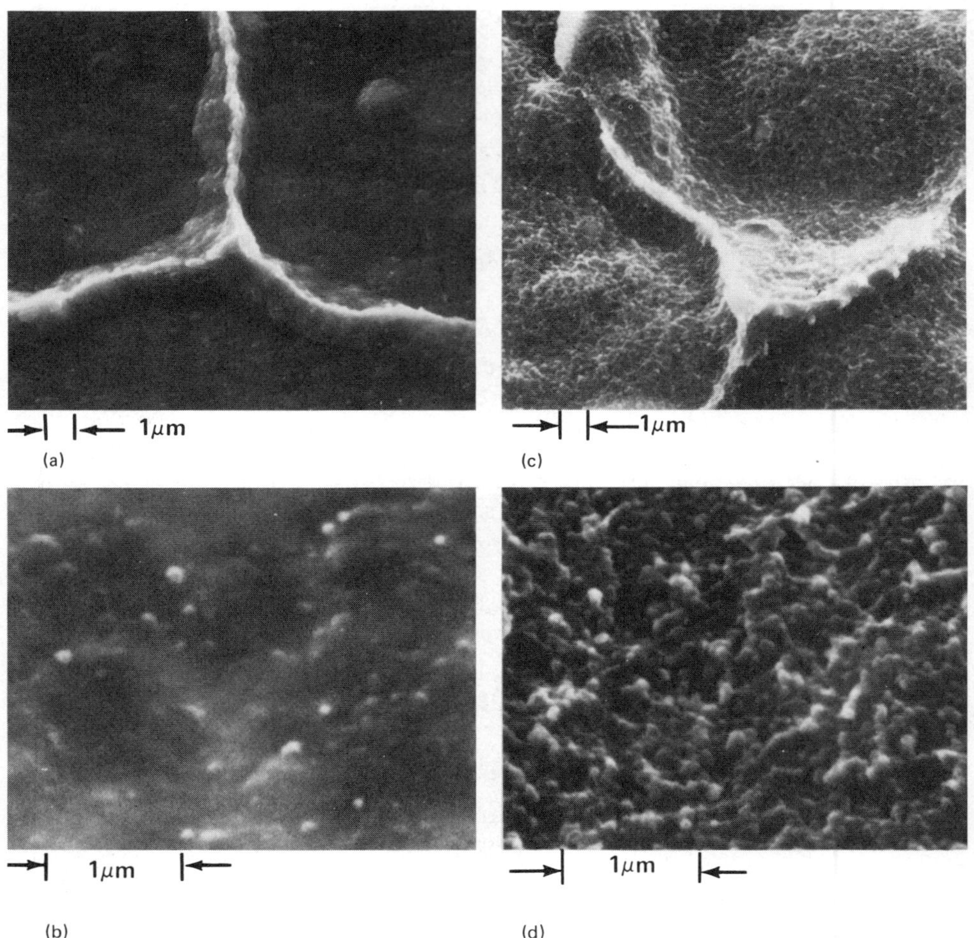

Figure 2   Plastic carbon replica. (a) and (b) show a chromic/nitric/HF etch, (b) at a higher magnification. (c) and (d) show phosphoric acid anodize, (d) at a higher magnification. (Courtesy Douglas Aircraft Co.)

determining the locus of failure in bonded joints. Through the use of combinations of these methods the true locus of failure can be determined.

When the oxide layer is coated with primer, viewing the pore structure of the oxide is not possible. Figure 3 shows a SEM of typical phosphoric acid anodize, phosphoric acid anodize exposed for 1 min to fuming nitric acid and a primed surface from which the primer was stripped using fuming nitric acid. It is evident that the acid completely removes the primer without any effect on the oxide layer. This procedure is a useful tool in investigation of surfaces which have been primed.

(a)

Figure 3 Primer removal—fuming nitric acid. (Courtesy Douglas Aircraft Co.)

(b)

Figure 3   (continued)

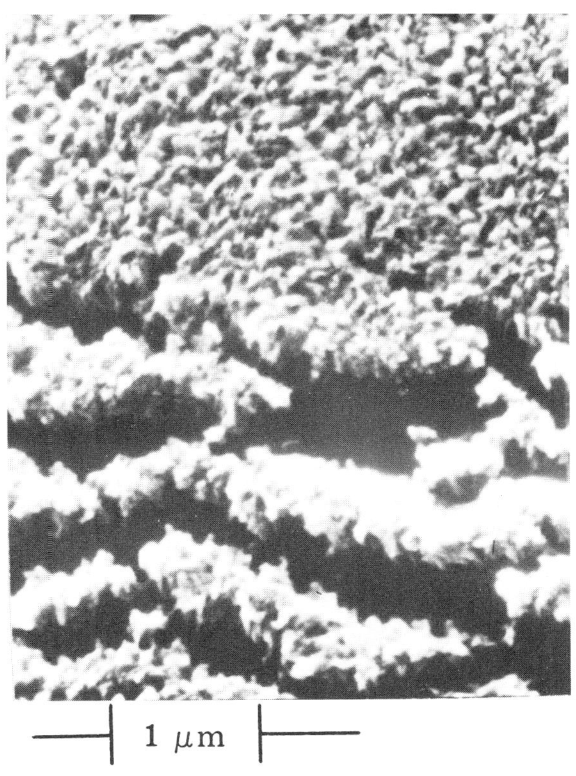

(c)

Figure 3   (continued)

(a)

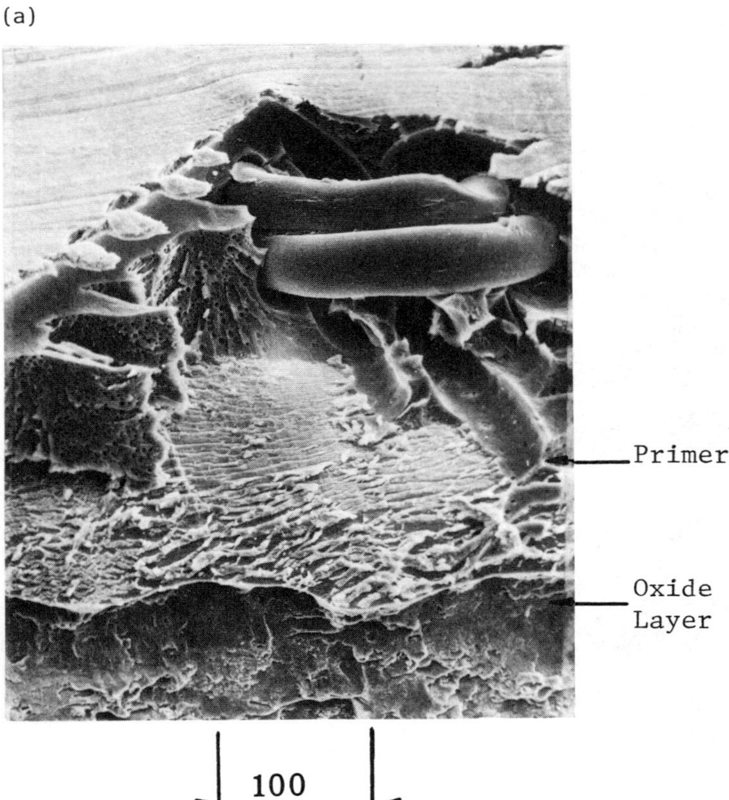

Figure 4 Metal adhesive bond failure: (a, b) adhesive side of failure; (c) metal side of failure. (d) Typical phosphoric acid anodize. (Courtesy Douglas Aircraft Co.)

(b)

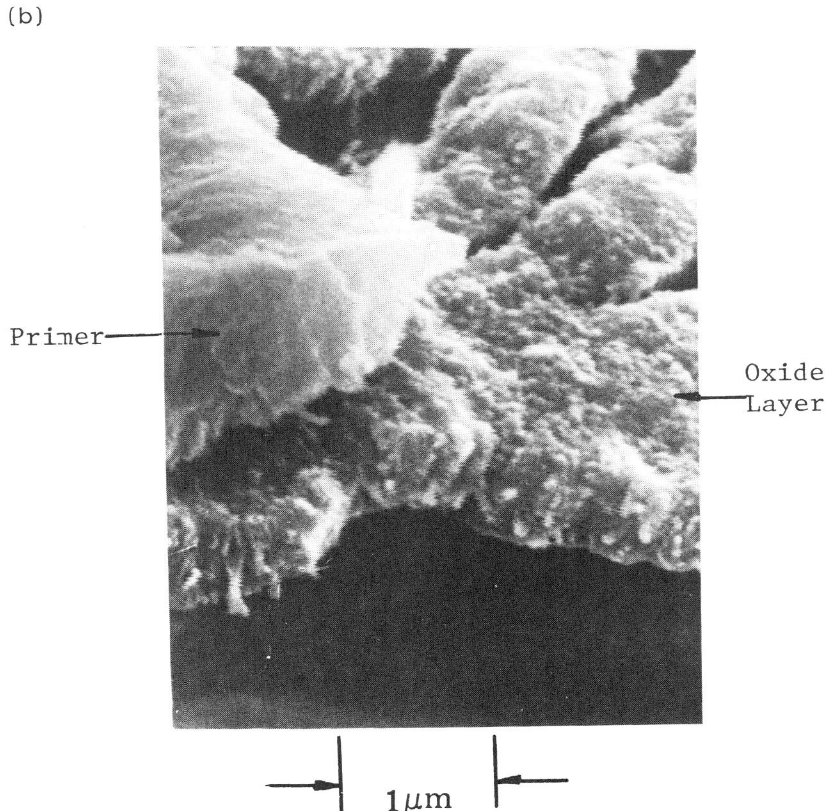

Figure 4 (continued)

(c)

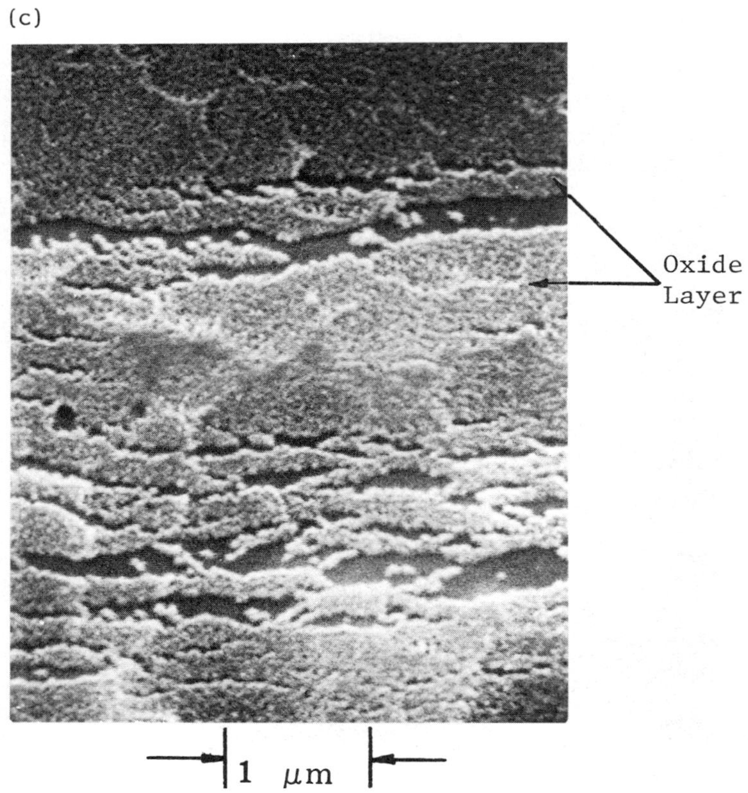

Figure 4 (continued)

(d)

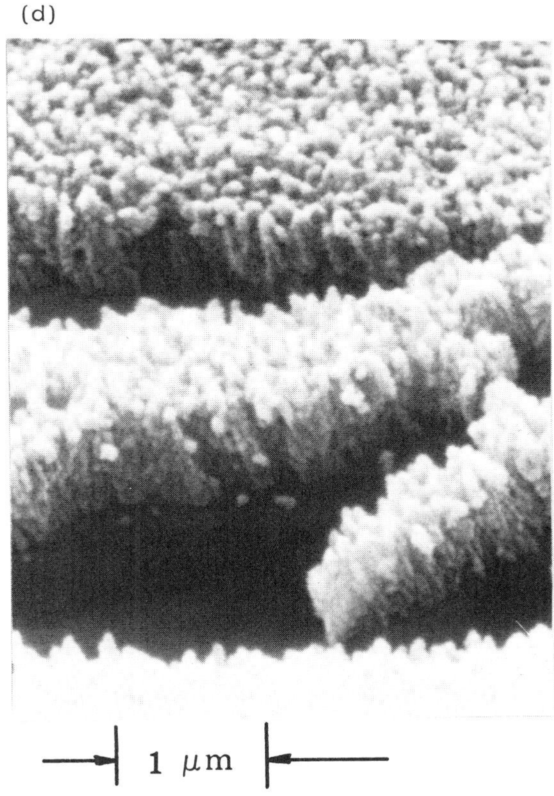

Figure 4   (continued)

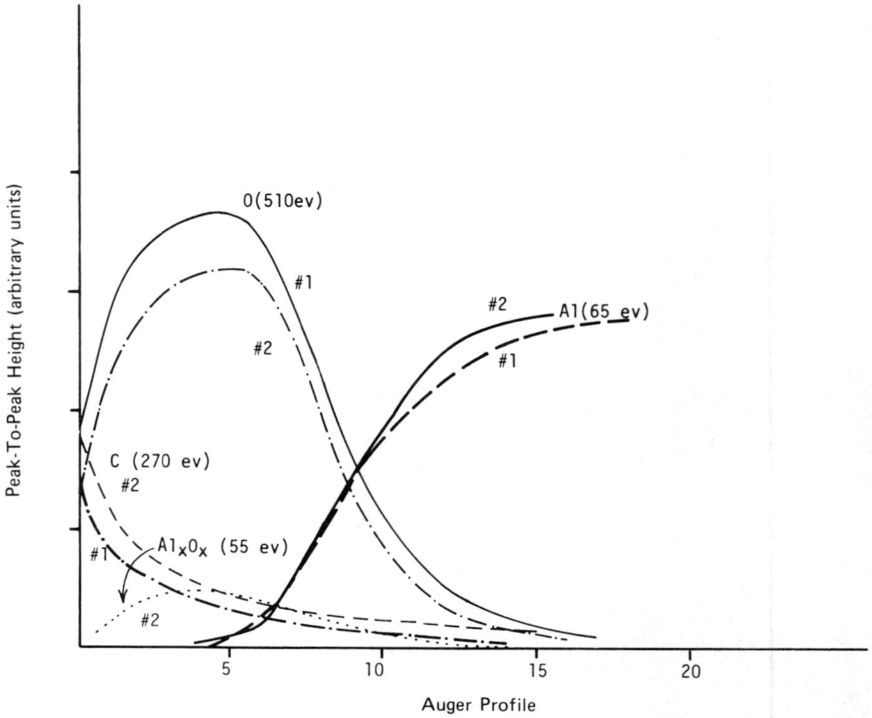

Figure 5 Failure analysis report: 2-kV argon ion beam—oxide thickness = 460 A (WRT Ta$_2$O$_5$). Samples 1 and 2 represent two different sections on the metallic side of a failed wedge crack specimen.

Figures 4 and 5 show the value of the SEM in failure analysis of an apparent adhesive failure of a 250°F cure system over phosphoric acid anodize. When the adhesive side of the failure was bent to fracture; the adhesive, primer, and a normal thickness of oxide is evident in Fig. 4. The bent specimen of the metallic side of the failure shows a thin oxide layer unlike the typical phosphoric acid anodize shown in Fig. 4. An Auger profile, Fig. 5, confirmed the thin oxide layer, >500 Å instead of 3000 to 4000 Å. The failure was attributed to improper surface treatment through loss of contact during anodizing or leaving the part in the phosphoric acid anodize for an extended period of time after the anodizing current was turned off.

REFERENCES

1. Bijlmer, P., and R. J. Schliekelmann, *SAMPE Q.*, Oct. 1973, p. 13.
2. Remmel, T. P., Characterization of Surfaces Prior to Adhesive Bonding, U.S. Air Force Materials Lab. Tech. Rep. AFML-TR-76—118, 1976.
3. Danforth, M. A., and R. J. Sunderland, *J. Appl. Sci. Appl. Polym. Symp. 32*, 1977, pp. 201—215.
4. Venables, J. D., D. K. McNamara, T. S. Sun, B. Ditchek, J. M. Chen, and R. L. Hopping, Characterization of Aluminum Surfaces Prepared for Adhesive Bonding, Struct. Adhes. Bond. Conf., El Segundo, Calif., Mar. 13-15, 1979.
5. Marceau, J. A., Relationship of Interfacial Compatability to Durability of Adhesive Bonded Joints, Interim Rep. D180-25219-1, Sept. 1978—Mar. 1979, for U.S. Air Force Materials Lab., Contract F33615-78-C-5133.
6. Chen, J. M., T. S. Sun, J. D. Venables, and R. L. Hopping, *SAMPE J.*, Vol. 14, July-Aug. 1978.
7. Baun, W. L., Experimental Methods to Determine Locus of Failure and Bond Mechanism in Adhesive Joints and Coating Substrate Combinations, ASTM STP 640, American Society for Testing and Materials, Philadelphia, 1978, pp. 41—53.

6

Adhesive Selection from the User's Viewpoint

ERNEST C. MILLARD *Rohr Industries, Inc., Riverside, California*

When designing bonded aluminum structures, one of the primary deci-
sions to be made is selection of the right adhesive. The correct choice
of adhesive is of paramount importance if the structure is to perform
the functions for which it was designed and withstand the hostile environments to which it will be subjected. The choice is the final responsibility of the user of the adhesive and no one else. The decision
must be made by utilizing all the data possible. Where data are not
available, adequate tests must be conducted to obtain it. The following
section provides a guide for the potential adhesive user on how to make
that choice and what to avoid by way of pitfalls during the selection
process.

Many factors must be evaluated, the first of which is to decide on
the basic type of adhesive system to be used. To do this, the following fundamental items must be considered. First, all adhesives cure
at certain temperatures. These may range from room temperature to
several hundred degrees Farenheit. Whatever the cure temperature
of the adhesive system selected for evaluation, the assembly to be
bonded must be capable of withstanding that temperature during the
cure cycle.

One of the problems that may be encountered is an unacceptable
change in the heat treat level of the aluminum details to be bonded.
This could, in an extreme case, result in structural deficiencies in an
assembly. The engineer must therefore make assurances that no difficulties will be encountered from this type of problem.

Second, although most modern structural adhesives are made in
films of precise thickness, there are other forms that will also provide
a good structural bond. Liquid systems are used in many applications,
and provided that good surface preparation standards are maintained,
give moderate- to high-strength bonds. An added advantage with

"

liquid systems is the ability of many of them to cure at ambient temperature. In situations where cure facilities such as autoclaves, presses, and ovens are lacking, this characteristic alone may determine the choice of whether to bond or resort to other forms of joining, such as the use of mechanical fasteners. Although liquid ambient-temperature curing adhesives have a number of limitations, they have a variety of uses and should be considered when making an adhesive selection. They are particularly helpful for use in field repairs where elevated-temperature curing is difficult to accomplish. Low-viscosity adhesives can also be used for injection into unwanted cavities or in repairing a delaminated area of a bonded assembly.

Expanding adhesives that cure at elevated temperature, the foam adhesives, are another group of materials that are used in the manufacture of bonded structures. In fact, they can really be classified as essential to the modern concept of bonding. Foam adhesives have the ability to expand and fill a gap during cure at elevated temperature, generally expanding to about two and a half times their original volume. This type of material is used principally to splice honeycomb core in sandwich structures and to provide a shear tie to edge members and machined fittings incorporated into a sandwich assembly. They should not be used for normal metal-to-metal bonding or bonding honeycomb to face sheets, except under very special circumstances.

Bulk or hot-melt systems can be another group of adhesives worthy of consideration in an evaluation program. These systems are less common but are useful where it may be desirable to coat the adhesive onto skins or honeycomb core. One particular form of construction where hot-melt systems are utilized is in sandwich assemblies where one skin contains perforations which must remain essentially unobstructed with adhesive after cure. This specialized form of structure is used for sound absorption around aircraft power plants, and maintenance of the open perforations is essential to the efficiency of the structure.

The most common form of structural adhesives, however, are films. They may be of varying thicknesses and incorporate a woven or nonwoven carrier embedded within the adhesive to improve certain mechanical properties. Films have the great advantage of providing a precise quantity of adhesive of constant thickness within the faying surface of the joint. With the application of heat and pressure, the adhesive film flows to make a homogeneous bonded joint.

All of these forms of adhesives are dealt with in more detail in Chapters 7 and 8.

The third item to be considered in determining the basic type of adhesive to be selected is the cost. Adhesives vary widely in cost, and although technical excellence is the ultimate concern when bonding structures, cost should not be left out of the selection process. Depending on the application, a less expensive adhesive will often perform as well as a high-cost material.

These three considerations—cure parameters, adhesive form, and cost—can determine the basic type of adhesive to be used. From this determination, the evaluation process can begin. In primary structures there are several important environments that must be considered, for example, temperature range to which it will be exposed, environmental resistance, mechanical properties, durability, and handleability.

The normal range of temperatures found on a typical subsonic turbine-powered aircraft are within the range of -60 to +180°F (-51 to +82°C). The subzero temperatures result while flying at high altitude, and the elevated temperatures are produced by exposure to solar radiation while on the ground. These overall temperatures, however, do not take into account those adhesive bonds exposed to local heating generated in various areas of the aircraft or other systems, such as near engines, auxiliary power units, or deicing systems. These areas must be treated separately from the rest of the aircraft structure and the temperature profile accurately determined before an adhesive selection can be made. It must be emphasized that an incorrect evaluation of temperature conditions to which the adhesive will be subjected can result in premature failure of the bonded assembly due to adhesive degradation. This could result in expensive replacement of the structure, at best, and possibly far more serious consequences.

When evaluating the temperature regime that the adhesive system will need to endure, the total time the adhesives will be exposed to normal operating temperatures must be determined. Adhesive degradation due to exposure to temperatures in excess of its normal capability has a time-temperature relationship. Many adhesive systems will withstand exposure to temperatures far higher than the long-term degradation temperature of the adhesive for short periods of time. For example, if an adhesive degrades after exposure for several hundred hours at 300°F (149°C), it may well survive 20 or 30 hr at 400°F (204°C) without permanent loss of properties. However, while the adhesive may be satisfactory for short-term exposure to high temperatures, it must be remembered that basic adhesive strength deteriorates with increase in temperature. To evaluate the performance of an adhesive system properly, data must be available on the strength retention of the material at temperatures up to the maximum likely to be encountered in service. These data should include tests made after short- and long-term exposure.

Supersonic aircraft, particularly military fighters and strike airplanes, have an additional temperature problem. The skin temperatures of these craft increase with speed due to skin air friction, resulting in kinetic heating, which at Mach 2.0 (approximately 1400 miles/hr) can be as high as 300°F (149°C) on the leading edges of the wings and empennage. Most military aircraft that experience these temperatures, experience them only for short periods of time, and in most cases the exposure can be measured in minutes. This, coupled

with the comparatively short flying life of military aircraft compared
with that of modern turbine-powered civil aircraft, results in a total
time at maximum temperature often totaling less than 100 hr. This
condition results in far less adhesive degradation due to elevated
temperatures than if it were exposed for many thousands of hours.

When supersonic civil aircraft such as the Concorde are considered,
the effects of kinetic heating are more severe. The temperature remains basically the same as for military aircraft for the same speed
regime, but the time at temperature increases dramatically due to the
greatly increased flying time at supersonic speed. Civil aircraft are
now being designed with flying lives of up to 60,000 hr, and as in the
case of a supersonic transport, a large portion of this time can be
spent at supersonic speed (Mach 2 in the case of Concorde).

The evaluating engineer must take all these factors into consideration when selecting an adhesive system and must be sure that it will
survive the projected service temperatures over the *entire* life of the
aircraft. This may require extensive long-term aging tests of the pro-
posed adhesive systems before a final selection is made. These types
of tests are expensive to conduct and very time consuming; however,
their importance to the total life of the structure cannot be overemphasized.

The second most important consideration in adhesive selection is
the expected environment to which the adhesive will be exposed. We
have already discussed temperature, which is one aspect of environment, but there are many others. Contaminants likely to come in contact with the bonded assembly must be assessed and if found to be
present at some time in the life of the structure, their effects on the
adhesive must be evaluated. Typical contaminants encountered in
modern aircraft are hydraulic fluids, engine lubricants, turbine fuels
(kerosene, etc.), and water in the form of atmospheric moisture or in
normal liquid form. Other contaminants sometimes neglected, however ,
are fluids such as aircraft cleaning liquids, normally alkaline in nature;
deicing fluids, probably with an alcohol base; and galley spills, principally confined to specific interior areas, but nevertheless problem-
makers. The latter fluids are somewhat confined to specific areas of
the aircraft, however, and are not normally included in the routine
tests to which a new candidate adhesive is subjected.

If the adhesive is to be used where any of these contaminants are
to be encountered during the life of the bonded assembly, even if the
contact is casual in nature, their effects must be evaluated. It may be
argued that atmospheric moisture or liquid water is always present
around aircraft structures and therefore should not be considered a
contaminant. This is true, but nevertheless, common water, has
probably been the most destructive contaminant to which most bonded
structures have been exposed. Many adhesives are very susceptible
to a wet environment, a situation now well recognized and understood.

In the early years of bonding, many adhesives were placed in service and subsequently suffered premature failure due to the effects of moisture. A great amount of work has been carried out in recent years on this one problem, and subsequent chapters in this book deal with it in detail.

Having determined the performance of an adhesive system at temperature and the hostile environments it may encounter in service, the evaluator must decide the type of mechanical properties most important for the particular stress the adhesive has to withstand. The normal mechanical tests carried out on an adhesive system are:

1. Tensile shear
2. Peel (metal-metal)
3. Peel (honeycomb)
4. Flatwise tension
5. Beam shear
6. Plate shear

All the tests cited above are normally carried out in the static mode before and after exposure to specific test temperatures and/or hostile environments. In addition to these static tests, it is advisable to determine the resistance of bonded joints to fatigue loading. This is typically carried out in the tensile shear mode and is conducted at stress levels determined from static loading tests. Fatigue tests are normally not conducted at stress levels more than two-thirds of the ultimate values determined from the static tests. A minimum of three stress levels will give three points for graphical presentation and the production of a curve that the stress engineer can use for determining safe loading levels in the bonded structure.

In addition to the mechanical strength evaluations noted above, there may be additional tests required by the stress or design engineer to provide further information which may be necessary for specific purposes. Typical of these rarer types of tests are:

1. Honeycomb columnar compression
2. Three-point load bending (honeycomb sandwich)
3. Four-point load bending (honeycomb sandwich)
4. Double lap shear
5. Lap shear with variable length-to-thickness (L/T) ratios

These tests are often performed as part of a design allowable test program set up for a specific project.

As part of a comprehensive adhesive evaluation program, a decision must be made whether a design allowables program should form part of the evaluation. A design allowables program is not considered essential to the basic adhesive evaluation and is not conducted to determine the characteristics of an adhesive system. Its purpose is to provide the design and stress engineer with basic information on which to base design parameters and load-carrying capabilities of the adhesive-bonded assembly. These tests will not explore the environmental resistance and durability characteristics of the adhesive, however, a caution should be exercised before committing an adhesive to production based on these data only.

Until very recently, the evaluation of a new adhesive system stopped at this point. It was considered that all the virtues and faults of a particular adhesive system were uncovered and a sound evaluation was possible based on the data generated from the tests. Founded on the knowledge available, there was no basic flaw to this philosophy. Although the tests determined the operational capability of the adhesive in terms of static conditions, they did not accurately record what happens in actual service, in other words, the way in which it performs under stress and in the environment to which it is being subjected. It has previously been mentioned that many early structural adhesives have failed prematurely due to susceptability to water contamination. In most cases these same adhesives had been subjected to water contamination tests, generally by immersing specimens in water for a month or so, and then conducting mechanical tests such as peel or shear and determining the strength retention compared to that of an unexposed control specimen. Good strength retention was obtained in most cases, and it was assumed that no problems would be encountered in service from those conditions. Yet within a few years, bonded structures were delaminating in service due to moisture or water contamination. A very great effort has been made by both suppliers and users of adhesives to resolve this problem and those efforts have resulted in a fundamental change in the methods of adhesive evaluation. As part of this change, it became apparent that the adhesive was only one factor in the delaminating mechanism. The adhesive primer and the prepared surface of the substrate were also vitally involved, and the test methods developed had to include them as an integral part of the evaluation. It was suspected for many years that the principal surface preparation used on aluminum alloys for bonding had certain shortcomings. The process, sulfuric acid/sodium dichromate deoxidizing developed by Forest Products Laboratory (FPL etch), could be susceptible to moisture penetration under certain conditions. With the widespread increase in the use of bonded assemblies it became apparent that new test methods were required to evaluate the entire bond mechanism if the problem was to be resolved. In addition to the deoxidized aluminum alloy, the primer is seen as a vital link between the substrate and the adhesive and is an integral part of the bond produced by the adhesive system. Durability must therefore be assessed with a fully representative surface preparation, primer, and adhesive if confidence is to be established from the results of the tests.

To determine the durability of an adhesive system, a range of additional tests must be added to the static and fatigue tests. These tests take the form of subjecting the adhesive to hostile environments and/or temperature while under continuous stress. The tests are generally conducted on metal-metal specimens, that is, those not containing honeycomb or other core material. The specimens are configured to evaluate the adhesive in shear or cleavage in hot, wet, and/or corrosive environments, and a number of variations in test specimen design are available for the engineer to choose from.

The tests used for evaluations in shear are conventional tensile lap shear, using the standard 1-in. × 1/2 in. bonded joint, the thick-adherend specimen, and for the most severe form of test, a specimen which is essentially the thick-adherend specimen with a hole through the bonded portion of the specimen.

A number of shear specimens are normally hung in a chain in an environmental chamber and a static load applied. These loads may be varied and are generally a fixed percentage of the ultimate failing stress achieved by the adhesive system under ambient test conditions. Depending on the adhesive and the surface preparation of the adherend, specimen failure may occur within a few hours or several weeks.

The cleavage or wedge test system imparts a more severe stress condition on the adhesive and the prepared substrate than do the shear specimens. The simple wedge specimen shown in Chapter 10 is useful for determining resistance to elevated temperature/high humidity conditions and is relatively simple and inexpensive to prepare and test. The thick-adherend specimen shown in Chapter 10 and generally referred to as the double cantilever beam specimen is more sophisticated and expensive to prepare. Both types of tests, however, have the advantage of not requiring a complex loading device for applying stress to the bond line as is required in shear testing. All that is required is a chamber capable of producing the desired hostile environment, the stress being applied by either the wedge or the loading bolts. A full description of these types of tests is given in Chapter 10.

In evaluating a new adhesive system, one of the items most likely to get the least attention is handleability. It is a sad fact that in many instances, thousands of dollars are spent on a large-scale laboratory evaluation with apparent great success, only to find that major practical problems appear when the system is placed in production. When making small laboratory specimens, the laboratory technician often cannot be aware of some of these shortcomings, and unless a thorough appraisal is made of the handling characteristics of an adhesive system in the early stages of evaluation, many problems can result.

Tack or tackiness of an adhesive film, for instance, can be very important to production operators when they have a large assembly to build with possibly compound curvature, complex cutouts, and other problem areas to negotiate. A high degree of tack can spell

disaster for this type of assembly, resulting in considerable adhesive
wastage and inflated production hours. On the other hand, very low
tack can be equally frustrating and wasteful. In the instance of very
low tack the adhesive may not remain in place during assembly of the
detail parts, causing voids or excessive mark-off in the completed
bonded assembly. Drape is another property to be considered in a
handleability evaluation. Good drapability is a desirable characteristic
in an adhesive film, particularly where assemblies containing compound
curvature are to be manufactured. Poor drape can result in time-
consuming tailoring of the film to eliminate excessive wrinkling and,
if poorly executed, can cause gaps in the coverage of the adhesive
film on the assembly. This will result in voids due to lack of adhesive,
and possible rejection of the completed assembly.

Storage requirements of an adhesive system must also be considered
as part of the overall handleability characteristics. Many modern adhesives have very limited storage life and in a large number of cases
must be stored at 0°F (-18°C) or lower. Even then they may have only
a short life at ambient temperature for assembly into a component. If
an adhesive has a maximum allowed time at ambient temperature of, for
example, 3 days, and the component being built has a total shop floor
time for assembly of 5 days, then obviously, the wrong adhesive has
been selected. This situation may not have been considered in the
laboratory evaluation.

Sub-zero storage can also be very expensive. Large refrigerators
are costly to run and maintain. Useful storage life at low temperature
should also be a consideration. An adhesive system with only a few
weeks of life, even when stored at these low temperatures, can result
in a great deal of scrap if the adhesive is to be utilized on an infre-
quent or a low rate production basis.

The application requirements for adhesive primers should also be
considered when it is planned to use them in conjuction with an adhesive. Most primers are applied by spray using either conventional
air pressure equipment or airless spray. The corrosion-inhibiting
varieties containing chromates often require special equipment to ensure adequate distribution and suspension of the chromate particles
in the resin-solvent primer solution. Inadequate chromate distribution
can result in low corrosion resistance, poor adhesion of the adhesive,
and substandard appearance. Again, this equipment can be expensive
and the possibility of its use should be considered early in the test
program. When all the testing has been completed and the report
published, the time arises when the adhesive is put into full production. If all goes well, all the characteristics of the adhesive determined by the test program will be imparted in the production assemblies.
Unfortunately, this does not always occur, and problems totally unfore-
seen in the laboratory can manifest themselves in the production effort.

Many of these problems could have been foreseen and dealt with if
a modest scale-up program had been carried out prior to full production.

Test assemblies that simulate the complexities of typical production
components can be designed with the more difficult features simulated
and possibly accentuated, which would fully tax the ability of the
selected adhesive to produce an acceptable bonded assembly. One of
the more typical problems that can be uncovered in the scale-up process is adhesive porosity in a large faying area. The typical laboratory
specimen has a small-to-modest bond area and the porosity problem
may not become apparent since it is often caused by air entrapment in
the faying surface. The larger the faying area, the greater the possibility of air entrapment. The porosity problem can be due to the
characteristics of the adhesive films and require modification of the
fibrous carrier used in the film. In severe cases a modification of the
adhesive polymer itself may be necessary. Another problem which has
been found in the transfer of adhesives from the laboratory to the production shop is volatiles. Some adhesive systems release voltailes during cure which may not cause significant problems in small-area bonds
due to their ability to escape the faying surface during cure. However,
in large-area faying surfaces, the volatiles can get trapped and cause
blisters or voids in the bonded joint. Fortunately, this particular
problem has been recognized and dealt with by the adhesive industry
and almost all the new adhesive systems have a very low volatile content, which prevents a large part of this problem. There are still
structural adhesive systems available today, however, which suffer
from this drawback and it should be considered in the adhesive selection. There are many other instances of problems arising during the
introduction of a new adhesive system to the production shop. A scale-
up program may not always reveal all the difficulties, but it probably
will prevent major problems with the material in production.

7
Epoxy Adhesives

ERNEST C. MILLARD *Rohr Industries, Inc., Riverside, California*

I. INTRODUCTION

Of all the adhesive systems developed for structural bonding of aluminum alloys, by far the most common are the epoxies. They are probably the most versatile of all the structural adhesives, and by the use of various curing systems, a wide variety of characteristics can be obtained.

There are many basic epoxy resins used in the manufacture of adhesive systems, some of which are used singularly, while others are blended to provide specific characteristics desired by the user. It is not intended to enter into the basic chemistry of epoxy adhesive systems in this chapter, since there are many reference works available that deal with the more theoretical aspects of these materials. However, some mention will be made of a few of the base resins and curing agents used in the adhesives field. Typical base resins used in adhesive manufacture are:

1. Diglycidyl ether bisphenol A
2. Tetraglycidyl methylene dianiline
3. Epoxy novalac

Curing agents used in conjunction with the base resins are the aliphatic and aromatic amines, anhydrides, and Lewis acids and bases. Typical of the amines are:

1. Dicyandiamide
2. Diaminodiphenyl sulfone

These resin and curing agents are only a very small sample of the variety of materials available to the adhesive formulator and by the addition

of various extenders, fillers, and tougheners, a tremendous variety of
systems can be manufactured to meet specific needs and requirements.
    Epoxy adhesives have a great affinity for properly prepared alumi-
num alloys, with the ability to flow and attach to the oxide layers formed
by the surface preparation processes. In the uncured state, unmodified epoxies are liquid or pastelike substances which, depending on
the formulation, can be rolled, callendered, or cast into thin films, with
or without a supporting textile carrier. They can also remain in the
liquid form, where they may be applied directly onto the surfaces to be
bonded. There are many elevated-temperature curing systems available
in liquid form which can be formulated for specific bonding purposes.
Their viscosity can be controlled by the basic resin chemistry and by
the addition of various fillers and flow control agents.
    Epoxy adhesive systems intended for structural bonding are used
primarily in film form, either with or without a supporting carrier.
The major advantages of this form are control of application and repeatability of the bonding process. By reducing the probability of production errors, which occur more frequently with liquid systems, the bonding process becomes controlled and reliable, an essential requirement
when bonding critical structures. More detailed descriptions of the
various forms of adhesive are given later in this chapter.
    In the cured condition, the unmodified epoxies become hard, horn-
like substances which, depending on the curing system chosen, can be
somewhat brittle in nature. This characteristic can be limiting in some
modes of loading, particularly the peel mode. To increase the versatility of the epoxy adhesives as a group, other polymers have been combined with them to produce a wide variety of adhesives, many of them
very tough and without the brittle character of the basic epoxy molecule.
These systems are generally referred to as "modified epoxies." By combining other substances, the scope of the original epoxy compounds can
be widened to include a much broader range of characteristics.
    Many of the formulations used in modern adhesives are highly pro-
prietory to the manufacturer and are carefully protected to prevent
duplication by competitors. Therefore, only a general description will
be given of the principal groups in universal use; specific details of
formulation will be avoided.

II. AVAILABLE EPOXY ADHESIVE TYPES

A. Nitrile Epoxy Systems

In the structural adhesive field, the most common modified epoxies in
use are probably the nitrile epoxies. In these materials, the epoxy
systems are combined with a nitrile elastomer, polybutadiene-acryloni-
trile copolymer, to produce a tough, comparatively low modulus adhesive,
offering good shear strength combined with high peel characteristics.

These systems have found widespread acceptance in the aircraft industry for bonding aluminum structures on subsonic commercial aircraft. The first generation of these adhesives were introduced in the mid-1960s and were very successful from the beginning. Before their introduction, the adhesive systems used for structural bonding on subsonic aircraft were the vinyl phenolics, nitrile phenolics, and dual-adhesive systems in which epoxies and nitrile phenolics were combined in one adhesive film, generally using the epoxy to bond to honeycomb and the nitrile phenolic to bond to the skin surface. Many of these systems had limitations such as poor flow, cure temperatures in excess of 300°F (149°C), and in the case of the nitrile phenolics, inability to form good bonds with metal honeycomb for sandwich structures. The vinyl phenolic systems, which were among the first truly structural adhesives to be developed, were also limited in peel strength. Because of these limitations, a strong demand was seen in the aviation industry for an adhesive system that was versatile enough to bond honeycomb, as well as metal to metal; which could bond large surface areas such as body skins without problems of entrapped volatiles causing voids or porosity; and which could be cured at temperatures as low as 225°F (107°C). In addition, the adhesive was required to have good shear strength, toughness, and high metal-to-metal and honeycomb peel strength. Ability to handle well and to bond large-area metal-to-metal surfaces with no undue porosity were also requirements.

From these specifications came the first-generation nitrile epoxy adhesive systems. As stated previously, they were highly successful and found immediate acceptance on such aircraft as the Boeing 727, 737, and 747 and the Douglas DC-8-60 series and DC-10. The Lockheed L1011 also made extensive use of these adhesive systems. These first-generation adhesive systems are still in volume use and continue to be used extensively throughout the world. In the mid-1970s it became apparent that although the nitrile epoxy systems performed well, there were problems related to moisture resistance and durability which had come to light following several years of service. The problems were to some degree linked to surface preparation. It was eventually determined that by a combination of circumstances, moisture and high humidity were having a deleterious effect on the durability of the nitrile epoxy systems. Several of the major aircraft manufacturers prepared new adhesive specifications which required durability testing in hostile environments as part of the qualification requirements. From these specifications came the development of the second generation of nitrile epoxy adhesive films. The new adhesives had the capability of resisting severe hostile environments while under sustained stress to a much higher degree than the earlier versions of nitrile epoxy adhesives. It must be pointed out, however, that the very great improvements in durability gained with the newer systems were achieved with a considerable upgrading in surface preparation. Details of these new techniques in surface preparation are given in Chapters 2 to 4.

Nitrile epoxy systems perform with excellent results over a service temperature range of −70 to +180°F (−56 to +82°C). The second-generation versions have a somewhat improved upper temperature limit and will give moderate to good strength up to 200°F (93°C). They are tough durable systems typified by high metal-to-metal peel strength and shear properties. They have very good resistance to all of the typical aircraft fluids, such as fuels and hydraulic and lubricating oils, and have moderate-to-good resistance to water, again the second-generation versions being more resistant than the first. Severe exposure to some hydraulic fluids, such as the phosphate ester types, can cause some softening, resulting in strength loss, and caution should be exercised when using these systems in this kind of environment. Most of the nitrile epoxy systems available cure between 225 and 250°F (107 and 121°C), generally requiring 60 to 90 min to complete the process. All of them are capable of undergoing multiple cures, where required on complex assemblies, without damage to the bond lines. It is, however, necessary to maintain pressure on the assemblies during subsequent cures. They all have excellent fatigue properties and perform well under sustained stress. Nitrile epoxy adhesives are formulated to be used with primers, many of them of the corrosion resistant type. These primers will be dealt with later in this chapter. The early primers developed for the first-generation nitrile epoxy systems were of the non-corrosion-inhibiting type.

These early primers, which are still in use, come in two basic forms: catalyzed (those containing a curing agent) and noncatalyzed (no curing agent present). The noncatalyzed versions are designed to be applied to the adhesive faying surfaces and remain in an uncured condition until the adhesive film undergoes final cure. Up to this time they can be removed from the details with solvents, enabling reprocessing to be carried out in the event of production error or necessary rework. The mechanism for final cure of the primer comes during the cure of the adhesive film. The catalyst or curing agent contained in the film for cure of the adhesive also cures the primer, the transfer being affected by the intimate contact of film adhesive and primer during cure. The catalyzed primer versions cure with the application of heat, since the curing agent is integral, either during a primer bake operation after application or during cure of the adhesive film.

The nitrile epoxy adhesives have far outstripped all other adhesive types in volume usage on current turbine-powered subsonic transports. They are used in a wide variety of applications on fuselage, empennage, and control surfaces. Any area of the aircraft that is not subjected to temperatures in excess of 180°F (82°C) can use these systems with confidence. Nitrile epoxy systems are offered by most of the principal adhesive manufacturers in the United States and Europe, the principal ones being:

American Cyanamid, Engineered Materials Dept.
Minnesota Mining & Manufacturing Company (3M Company),
  Adhesives and Sealants Division
Hysol Division, Dexter Corporation
Narmco Adhesives
B. F. Goodrich
Ciba Geigy (Reliable Manufacturing Corp., United States; Bonded
  Structure Ltd., Duxford, England)

Nitrile epoxy films, in addition to being excellent metal-to-metal adhesives, also produce good honeycomb bonds. They have the ability to
form excellent fillets to aluminum honeycomb, resulting in good flatwise
tension and honeycomb peel strengths. Sandwich structures bonded
with these adhesives perform well structurally and produce very acceptable beam shear and plate shear strengths. The high values found in
honeycomb peel can be attributed to the modifying polybutadiene-acry-
lonitrile copolymer, producing toughness, which translates into high
peel strength. Service experience with honeycomb structures bonded
with nitrile epoxies has been very good, with only sporadic problems
being apparent. Some disbonds have occurred and in most cases these
have been attributed to the ingress of moisture into the structure.
Later versions of these adhesive systems are expected to overcome this
kind of problem in the future.

B. Epoxy Polyamide Adhesive Systems

The epoxy polyamide adhesives can be obtained in both liquid and film
form. When combined with various quantities of polyamide resin, the
liquid epoxy resins produce an ambient-temperature curing adhesive
with good affinity and adhesion to prepared aluminum surfaces. This
results in a moderately tough and durable bond with fair peel strength
compared with unmodified catalyst-cured liquid epoxy systems. They
are very useful in secondary bonding operations such as applying rub
strips, local doubler reinforcements, and the application of small fittings.
They are not recommended for primary structural bonding operations.
These adhesives will provide service up to 180°F (82°C) and perform
well at subzero temperatures. They can be modified with a variety of
fillers and thixotrophy modifying compounds, which can produce a wide
range of adhesive characteristics. The addition of glass or phenolic
microballoons produces lightweight syntactic foams useful for space-
filling applications. These fillers are, of course, not confined to the
epoxy polyamide systems, and can be used in most of the liquid epoxy
adhesive formulations available.

The epoxy polyamide films were developed in the late 1950s and
early 1960s and have proved very successful over the years. They
produce excellent metal-to-metal bonds which are noted for their

toughness and high overlap shear strength. This has made them suitable for use in highly loaded dynamic applications such as helicopter
blades. Of all the structural adhesive films, the epoxy polyamides are
probably the least sensitive to surface preparation, which makes them
more forgiving than most other systems.

Epoxy polyamide films are normally used with an adhesive primer,
but these are not absolutely necessary. The development of these films
came about considerably earlier than the introduction of corrosion-
inhibiting adhesive primers and up to the present time, this class of
adhesive retains the use of non-corrosion-inhibiting types. The epoxy
polyamides are formulated to cure at 350°F (177°C); however, unlike
many other epoxy systems that cure at this temperature, they do not
perform well at temperatures above 180°F (82°C). Because of this, they
have been in strong competition with the later-developed nitrile epoxy
systems. The very high shear strength produced by the epoxy polyamide films cannot, in most cases, be usefully exploited by the design
engineer. Thus the performance characteristics of the two systems
(nitrile epoxy and epoxy polyamide) tend to be very similar, with the
exception of cure temperature. Epoxy polyamide films also perform with
excellent results when bonding aluminum honeycomb. They produce
well-formed adhesive fillets to the honeycomb cells, which provide very
high peel strengths at ambient temperature. Good flatwise tension
strengths are also achieved by these films.

Epoxy polyamide films have been in service for a great number of
years and have generally provided excellent results. There have been
problems with delaminations in service which have been attributed, in
almost all cases, to moisture ingress into the bond line. There appears
to be little doubt that these systems are sensitive to moisture after cure,
and bonds can be weakened by the continued presence of free water.
Some controversy still exists regarding this phenomenon, with some ad-
vocates maintaining the position that water presents no problem to these
adhesives. They point out the long trouble-free service achieved by
thousands of assemblies. Certainly it appears that any degradation in
bond line strength is recoverable to a large extent when moisture is
eliminated or removed by either natural or artificial means. Since there
is a possibility that a problem may exist with moisture ingress, it is
recommended that particular attention to surface preparation be maintained. Careful consideration should also be given to protection of
bond lines, such as cut or trimmed edges, and to ensuring that all
honeycomb structures are fully sealed. These precautions should prevent any serious difficulties arising due to wet environments. Epoxy
polyamide film adhesives are still in full production use by many companies. However, for the reasons given it is likely that a gradual phase-
out of these film systems will occur over the next few years in favor of
the nitrile epoxy systems. The liquid systems, however, have a very
useful place in semistructural bonding and syntactic foam formulations.

They will probably maintain their position in the marketplace for many years in the future.

C. Epoxy Phenolic Adhesive Systems

The epoxy phenolic adhesive systems are a fairly specialized group of film adhesives aimed at the high-temperature end of the service-temperature spectrum. Because of this, their use is normally confined to areas of any part exposed to high temperatures during service, such as engine nacelles, auxiliary power units, and areas subject to kinetic heating. Epoxy phenolic adhesives normally come only in film form, although there are some liquid versions available formulated for use as core splice adhesives. They cure at 350°F (177°C) and will perform well at 300 to 350°F (149 to 177°C) for several thousand hours and for short periods of time, up to 500°F (260°C).

Satisfactory metal-to-metal bonds can be achieved with epoxy phenolic adhesives, normally in conjunction with an adhesive primer. Venting of one of the adherends is necessary when using these systems over large-surface-bond areas, since they do emit volatiles during cure. Entrapment of these volatiles within a bond line can result in blistering and other forms of disbond, whereas venting will permit the volatiles to escape without harm. Good honeycomb bonds are possible with these adhesives; however, some problems caused by the volatiles have been experienced. Flatwise tension strengths are moderate. The peel strength is rather low, which can result in postbond operational problems if care is not exercised. Shear strengths generated by these adhesive systems are only moderate when compared to the nitrile epoxies. However, they are more than adequate for most structural applications and are not untypical of many high-temperature adhesive systems. Environmental resistance of these adhesives is generally very good. Service experience has been somewhat limited due to their specialized nature; however, there appears to have been no major problems revealed during the many years these adhesives have been in service. Resistance to all the fluids encountered is good and their performance in a moisture-laden environment is generally considered to be superior to that of many epoxy systems. Since the epoxy phenolic adhesives are formulated to operate at 300°F (149°C) plus, the supporting carrier fabric is normally of woven glass rather than nylon or polyester. The latter fabrics could be used if the maximum operating temperatures were limited to the capability of the fabric.

D. Unmodified Epoxy Adhesive Systems

In addition to the modified epoxy systems previously described, there are a large number of adhesives which fall in the group known as the unmodified epoxies. These adhesives can be in either liquid or film form and can utilize a number of epoxy polymers in combination with various curing agents.

The liquid systems can be formulated to cure at ambient temperature or at various elevated temperatures, depending on the processing and end-use requirements. Typical ambient temperature curing agents used for general-purpose adhesives are diethylenetriamine (DTA) and triethylenetetramine (TETA). Liquid systems using these formulations produce good shear strength on properly prepared aluminum substrates, but peel strengths are generally poor. Unmodified epoxy films will produce good shear strengths, but without the influence of modifying polymers can offer only modest-to-poor peel strengths. By the use of a variety of fillers, the peel strength of these systems can be enhanced but high strengths cannot be achieved by these methods. Elevated-temperature curing epoxy systems utilize curing agents such as *m*-phenylenediamine (MPDA) or methylenedianiline (MDA). Both of these curing agents produce good elevated-temperature performance from the adhesive. Other curing agents in common use are dicyanodiamide and diaminodiphenyl sulfone.

Unmodified epoxy adhesive films are normally formulated for elevated temperature service, generally 250°F (121°C) or above. Properties can be tailored to the adhesive end use by utilizing different starting epoxy resin polymers and the selection of one or more suitable curing agents.

Elevated temperature resistance of some adhesive formulations can be as high as 400°F (204°C) for short-term use (not more than a few hundred hours) and up to 350°F (177°C) for several thousand hours. As with the liquid systems, peel strength can present a problem, and usually the higher the temperature capabilities of the adhesive, the poorer the peel strength. There are now available adhesive films which will perform very well at the high end of the temperature spectrum, and yet provide moderate peel strengths which are adequate to impart a good degree of toughness to the bonded structure.

Most of the systems currently available are designed to be used with an adhesive primer, and the most recently developed products have primers of the corrosion-inhibiting type. The combination of the latter primers with unmodified epoxy films produce an adhesive system with excellent resistance to hostile environments.

This resistance is particularly apparent in bonded structures designed for sound suppression in the engine nacelles of modern commercial transports. These honeycomb sandwich structures have one face of the sandwich perforated with a large number of small-diameter holes which expose the internal surfaces of the honeycomb, and therefore the adhesive fillets, to the outside environment. Assemblies that have been in service for more than 8 years have shown no evidence of adhesion breakdown, bond line corrosion, or adhesive disbonds due to environmental problems.

Most of the unmodified epoxy adhesive films cure at 325 to 350°F (163 to 177°C) for 60 min. The mechanical properties produced by these adhesives are generally good. Tensile shear properties are somewhat lower than these found in some of the modified epoxy systems; however, they are more than adequate for most design purposes. Excellent bonds to honeycomb can be achieved and good flatwise tension strengths are obtained. As stated earlier, honeycomb peel strengths can be poor and caution must be taken in design to prevent any peel loading on the installed bonded assembly.

III. ADHESIVE FORMS

Epoxy adhesives are available in a number of different forms, some of which have already been discussed. Here we will deal with these forms in somewhat more detail and describe the specific uses each is designed to fulfill.

A. Liquid Adhesives

The most common form of epoxy adhesive for general-purpose use is liquid. The viscosity of liquid systems can vary from very low, for example 50 cP to pastelike materials up to 100,000 cP. Viscosity is a function of the base epoxy polymers used, together with the curing agents employed. Additional variation of viscosity can be made by the use of fillers and thixotropic control compounds, the additions depending on the end use for the adhesive product. A variety of liquid systems are used for bonding a wide range of substrates on modern aircraft structures; however, most of these end uses are confined to nonstructural or secondary bonding applications. There are exceptions to this, some elevated-temperature curing adhesive systems having been used for bonding honeycomb to face sheets and a few metal-to-metal structural bond applications. These, however, are comparatively rare, and by far the most common form of adhesive for aircraft structural bonding is film.

B. Film Adhesives

All the epoxy adhesive types discussed previously are available in film form. They are manufactured by casting or callendering the polymer into thin sheets which are precisely controlled in thickness. The adhesive is provided during the manufacturing process with protective release papers and/or polyethylene films which prevent the adhesive from adhering to itself when made into rolls of several hundred square feet. The polymers used for preparing the films incorporate the curing agents so that they are ready for use without any further treatment by the user. Adhesive films can be supplied in two basic forms: supported or unsupported. These terms are used to differentiate between an adhesive that has a supporting textile carrier or one that does not.

Supporting carriers can enhance some of the mechanical properties of the adhesive and improve the handleability of the product during lay-up of the assembly in the production shop. The carrier, depending on the type chosen, can also assist in removing trapped air or volatiles during cure in large-area metal-to-metal bonded assemblies. Normally, the carrier is situated through the center of the film; that is, an even quantity of adhesive polymer is coated on either side. There are some specialty films, however, which have the carrier close to one surface of the film, with most of the polymer on the opposite face. These are sometimes described as "one side tacky" (OST) films. They are particularly useful where the adhesive has an exceptionally high tack, making it difficult to apply to large structures, or where air entrapment is presenting a problem. In the latter case, the close proximity of the textile to the surface enables any trapped air to escape when pressure is applied to the assembly at the commencement of cure. Because of these advantages, the OST forms of film appear to be gaining in popularity, particularly with the later versions of the nitrile epoxy systems. Epoxy adhesive films can be manufactured in a range of thicknesses, generally from 0.003 to 0.015 in., although heavier versions can be made if desired. The thinner films are used for bonding metal-to-metal assemblies, while the heavier ones are generally utilized for honeycomb bonding. The thicker films may, of course, be used for metal-to-metal applications; however, the additional quantity of adhesive will increase the weight of the assembly without necessarily increasing the mechanical strength of the bond.

C. Bulk Adhesives

There are limited requirements for bulk epoxy adhesives, sometimes described as hot-melt systems. These are normally supplied as a thick, roughly callendered or cast sheet. Thickness of the sheet is not critical, since it will be cut into suitable-size pieces for placing on a hot-melt roll coater, where it will melt into a bulk form. The coater holds the hot adhesive melt and as the detail parts of the assembly are passed through, the coater deposits a precise quantity of adhesive on the part surface. This method of applying adhesives is not in general use and its principal application is for applying adhesive directly to honeycomb core. This is sometimes necessary where films cannot be applied directly to skin details, such as when the skins are perforated and the perforations must remain unblocked after cure. This is a form of construction common in engine nacelles for absorption of certain frequencies of turbine noise.

The use of thermosetting adhesive systems on a hot-melt coater must be precisely controlled since the adhesive cures by the application of heat. Lack of care in controlling time at elevated temperature of the adhesive melt can result in premature cure of the adhesive on the coater, resulting in very expensive replacement of rollers and other coater equipment.

D. Expanding Adhesives

Expanding epoxy adhesives, sometimes called foam adhesives, are a very useful form for use in splicing honeycomb core to itself or to peripheral members and internal fittings. They are normally produced in sheet form, generally from 0.025 to 0.1 in. in thickness. The sheets can be from 1 to 3 ft in length and generally around 1 ft in width. The sheets are cut into strips, which are placed between honeycomb core butt splices and around the peripheral areas, where the honeycomb meets the assembly edge members. The foam adhesive co-cures with the structural film adhesive during cure, expanding approximately two and a half times its original thickness. The expansion fills the space between the two cut rows of honeycomb cells and simultaneously bonds to the cell walls, providing a continuous path for accepting loads passing through the honeycomb structure. Foam adhesives expand during cure by the addition of blowing or expansion agents in the formulation. The agents react to the heat from the autoclave or press which, when the adhesive flows, causes it to expand, creating an open-textured, foamlike structure.

IV. CARRIERS

Carrier is the term most commonly used to describe a woven or non-woven textile material of lightweight construction, which is used to reinforce and enhance the properties of film adhesives. Carriers are generally situated in the center of adhesive films, coated evenly on either side by the adhesive polymer. Certain films, as discussed above, have the carrier strongly biased to one side and these are known as "one side tacky" (OST) adhesive films. Carriers function as a reinforcement for the adhesive film, which improves overall handleability of the material. Mechanical properties are also improved by the use of certain carriers, particularly the more brittle systems, such as the unmodified epoxies. Peel strength in particular can be strongly influenced by the use of different carrier forms and materials.

Carriers are generally made from three basic textiles; polyamides (nylon), polyesters (Dacron), and glass. The first two are the most commonly used, glass carriers being confined primarily to adhesive films formulated to perform at high temperatures. Both nylon and polyester fibers are used in a woven or knit form, the weaves being generally open in character to allow full penetration of the adhesive polymers. Both of these fibers can also be used in nonwoven form. These versions use textile fibers which are randomly assembled into a controlled scrim or mat, capable of being impregnated by the adhesive polymer. The nonwoven carriers are more common with the nitrile epoxy adhesive systems than with other modified and unmodified epoxy adhesives, and function well with these formulations.

Glass fibers have a limited use, generally in high-temperature adhesive systems where nylon and polyesters may suffer degradation after prolonged exposure to high temperature. Woven glass fabrics do not have the ability to adhere to the adhesive polymers as well as the organic materials, which can present some problems in adhesive properties. Glass carriers, for instance, will produce lower honeycomb peel strength than if other fibers were used with the same adhesive polymer. However, the high-temperature requirements may force the acceptance of lower mechanical properties, in return for long-term elevated-temperature durability.

One additional advantage, which carriers impart to adhesive films, is final bond line thickness control. The carriers, because of their inherent thickness and since the carrier does not melt or flow, are able to maintain a minimum bond line thickness during cure. Unsupported adhesive systems can, under extreme local pressure conditions, have areas where very thin bond line thicknesses are produced. This can have adverse results on the shear characteristics of the adhesive.

V. EPOXY-PRIMER COMPATIBILITY

When discussing structural epoxy adhesives, their compatibility with adhesive primers must always be included. Adhesive primers are not mandatory when using epoxy adhesives; indeed, most of the liquid epoxy adhesive systems, including the ambient-temperature during versions, require no adhesive primer whatsoever. On the other hand, most of the modern epoxy film adhesives are supplied with a recommended primer, with some requiring the primer as an essential part of the adhesive bonding mechanism.

Adhesive primers are supplied in liquid form, generally with a low percentage of resin solids dispersed or dissolved in a single or combination of solvents. The primer is applied to the adhesive faying surfaces of the assembly details by spray, roller, or flow coating, to a precisely controlled thickness, which, depending on the adhesive, ranges from 0.0001 to 0.0025 in. Thicker depositions can be applied for certain systems, but this is unusual. The primer must be compatible with the adhesive film, and desirably, aid in increasing the mechanical and/or physical properties of the adhesive. Most primers perform at least one of these functions, although there are a very few which offer little or no improvement in either. It can be argued, then, that if no advantage is gained by use of a primer, why waste time and money applying them? This would be true if primers were used only to improve properties; however, they also have one other major advantage, which can make their use more than worthwhile. This is their ability to protect freshly deoxidized or anodized surfaces from degradation due to atmospheric exposure.

By applying primers to the pretreated surfaces, the assembly details can be stored in a clean environment for an extended period of time. The actual amount of extended time depends on the adhesive primer system and the pretreatment selected. Without this extended time, the scheduling of assemblies to be bonded through a large shop would be almost impossible.

Adhesive primers, in addition to this characteristic, also have some or all of the following advantages:

1. Increase the wetting capability of the film adhesive
2. Improve the mechanical properties of the adhesive
3. Enable removal of mild accidental surface contamination without reprocessing
4. Greatly reduce bond line corrosion problems when suitably formulated
5. Give corrosion protection to the pretreated surfaces outside the bond areas

To understand these additional features, let us examine them in more detail.

A. Increase in Wetting Capability

Many adhesives benefit from primers which improve their ability to wet the faying surfaces of the details to be bonded. Adhesives with imperfect flow characteristics often partially fail to penetrate into the oxide layer formed by the surface preparation process and therefore do not adhere as tenaciously as others. The application of the primer can improve this process and adhesion is enhanced. Application of an adhesive primer to honeycomb core, particularly in peripheral areas, to improve the environmental resistance of the sandwich structure is not uncommon. This technique is employed on structural designs which do not incorporate edge members, thereby exposing the core edges to the outside environment. By improving the wetting capabilities of the adhesive and in turn the adhesion, resistance to moisture ingress into the structure is increased.

B. Improvement of Mechanical Strength

Primers can often improve the mechanical strength of epoxy adhesive systems, particularly properties such as metal-to-metal and metal-to-honeycomb climbing drum peel strength. By improving the bond between the film adhesive and substrate, failure modes of test specimens can be changed from adhesive (between film adhesive and substrate) to cohesive (within the adhesive film itself). This kind of failure mode change can also be seen at times in shear tests when comparing primed specimens with unprimed.

C. Removal of Accidental Contamination

There is an advantage with details that are primed to those which are
not. This advantage—the ability to remove accidental surface contamination that may occur during the waiting period prior to assembly lay-
up—is not often recognized. Although it may be argued that this type
of contamination should not happen, experience has shown that even in
the best-run production facilities, this kind of problem can occur. A
freshly prepared, unprimed aluminum surface is very sensitive to surface abrasion, even with soft, perfectly clean wipers. Any attempt to
remove contamination after the surface preparation operation will result
in damage to the delicate oxide layer, with subsequent substandard
bonds in the hand-cleaned area. When the pretreated surface has been
primed, the oxide layer is protected and the bond faying surface is
much less sensitive to abrasion. Because of this, mild contamination
can be removed with a suitable solvent without danger of damage to the
faying surface. It must be understood that this method of secondary
cleaning can apply only to primers which are fully cured and therefore
not susceptible to solvent attack. If a primer has only been air dried,
or received a bake at a temperature lower than the cure temperature,
it will probably be softened or partly removed by the application of a
solvent. It should be noted that many of the modern corrosion-inhibiting primers are applied and fully cured before the film adhesive is applied to the faying surfaces.

D. Inhibition of Bond Line Corrosion

The most significant event in the history of primer technology was the
introduction of corrosion-inhibiting adhesive primers. These primers
perform all of the functions described previously, with the added ability to inhibit bond line corrosion, a phenomenon that has caused a considerable number of problems in bonded structures over the years.
Corrosion-inhibiting primers are epoxy resins, which are compatible
with the film adhesive to be used, dispersed in a solvent blend. The
dispersion also contains a precise quantity of finely ground chromates;
normally strontium, barium, or zinc chromate. When the primer is
applied, the chromate-filled epoxy resin provides an even film over the
oxide surface, which after cure, makes a very receptive condition for
the adhesive film. Provided that the aluminum surface has been properly
prepared, the primer will protect the surface in warm, moist environments, and through the presence of chromates in the primer, inhibit
the formation of corrosion. The corrosion-resistant capabilities of these
primers has been well demonstrated in the laboratory and in airline
service. Tests in extremely hostile environments have shown the superiority of these primers over the more conventional non-corrosion-
inhibiting types. Corrosion-inhibiting primers were first developed
for use with the nitrile epoxy adhesive films and were placed in service

in 1969. A year or two later, primers were developed for elevated-temperature-resistant unmodified epoxy adhesives and entered service shortly after. Both classes of primer have performed extraordinarily well in service, with no significant bond line corrosion problems becoming apparent. Because of their very good corrosion-inhibiting properties, these primers will also function as a primary protective coating. In addition to applying them to the bond faying surface, the remaining surfaces of the details may be primed at the same time. After the bond cycle is completed, these nonfaying surfaces are fully protected against corrosion and require no further surface conversion treatments or organic coatings, such as conventional paint primers. The great advantage gained from using these primers as a basic corrosion protective is that they are applied on a first-class surface preparation, which ensures maximum adhesion of the coating during its service life.

VI. FILLERS

All film epoxy adhesives employ fillers of various kinds to control thixotrophy, regulate flow of the adhesive during the cure cycle, and modify the adhesive mechanical properties. By use of various fillers, the adhesive system can be tailored, by the manufacturer, to match specific requirements of the customer. For example, hard, brittle adhesive systems using unmodified epoxy resins can be modified by the use of fillers to increase toughness and thereby increase peel properties. Typical fillers used in currently available adhesive films are:

    Aluminum powder
    Calcium carbonate
    Mica
    Colloidal silica
    Asbestos fibers
    Glass fibers

Other fillers which are often used to reduce density in liquid systems are phenolic spheres and glass spheres. Fire resistance can also be improved by the addition of antimony oxide in liquid systems. An important use for fillers in film adhesives is the control of thixotrophy. Flow during cure is an important characteristic of film adhesives, and formation of good fillets to honeycomb core relates directly to adhesive flow. Control of flow can be accomplished by the use of asbestos or glass fibers in precisely controlled quantities. Colloidal silica is also a very useful thixotropic control material and is very commonly used in liquid systems to reduce the slump characteristics of the adhesive. This is an important consideration when bonding is to be carried out on vertical surfaces.

The use of aluminum powder is common in film adhesives and can enhance the elevated-temperature performance of an adhesive system. An improvement in toughness can also be achieved with the use of this filler.

Silver can be used in certain liquid epoxy adhesive systems to modify electrical conductivity. Heavy loadings of fine silver powder can create adhesives with very high conductivity, enabling them to be used for bonding details that require electrical continuity. A typical example where this type of adhesive is used is bonding of static discharge dissipators on aircraft control surfaces and empennage. The use of silver-filled adhesives is, of course, limited due to their high cost and are normally only used for special applications. (Another disadvantage of silver-filled adhesives in contact with aluminum is the corrosion susceptibility due to the difference in potential of the materials. Silver-aluminum bond joints should be protected from any type of hostile environment.)

VII. ADHESIVE PROPERTIES

Tables 1 to 11 show some typical mechanical properties for a selection of modified epoxy adhesives. The tables utilize a very limited number of available adhesive products and do not offer any recommendations.

Table 1   Tensile Shear Strength of Nitrile Epoxy Adhesive Films

Adhesive identification	Manufacturer	Test temperature [°F(°C)]	Shear strength [psi (MPa)]
FM-123-2	American Cyanamid	−67 (−55)	5180 (35.7)
		75 (24)	5140 (35.4)
		180 (82)	3300 (22.7)
FM-73M	American Cyanamid	−67 (−55)	6540 (45.1)
		75 (24)	5900 (40.7)
		180 (82)	4180 (28.2)
AF-126-2	3M Company	−67 (−55)	5070 (34.9)
		75 (24)	5200 (35.9)
		180 (82)	2890 (19.9)
AF-163-2K	3M Company	−67 (−55)	4800 (33.1)
		75 (24)	4800 (33.1)
		180 (82)	3420 (23.6)
EA-9601	Dexter Corporation	−67 (−55)	5000 (34.5)
		75 (24)	5300 (36.6)
		180 (82)	4300 (29.7)

Table 1. (continued)

Adhesive identification	Manufacturer	Test temperature [°F(°C)]	Shear strength [psi (MPa)]
EA-9628	Dexter Corporation	−67 (−55)	6100 (42.1)
		75 (24)	6000 (41.4)
		180 (82)	4000 (27.6)
Metlbond 1133	Narmco	−67 (−55)	6700 (46.2)
		75 (24)	6400 (44.1)
		180 (82)	4000 (27.6)
Plastilock 717	B.F. Goodrich	−67 (−55)	5530 (38.1)
		75 (24)	5260 (36.3)
		180 (82)	3980 (27.4)

Table 2 Metal-to-Metal Climbing Drum Peel Strength of Nitrile Epoxy Adhesive Films

Adhesive identification	Manufacturer	Test Temperature [°F(°C)]	Peel strength [in.-lb/1-in. width (N/mm)]
FM-123-2	American Cyanamid	75 (24)	70 (12.2)
FM-73M	American Cyanamid	75 (24)	85 (14.9)
AF-126-2	3M Company	75 (24)	80 (14.0)
AF-163-2K	3M Company	75 (24)	79 (13.8)
EA-9601	Dexter Corporation	75 (24)	60 (10.5)
EA-9628	Dexter Corporation	75 (24)	80 (14.0)
Metlbond 1133	Narmco	75 (24)	90 (15.7)
Plastilock 717	B.F. Goodrich	75 (24)	84 (14.7)

Table 3   Honeycomb Climbing Drum Peel Strength of Nitrile Epoxy Adhesive Films

Adhesive identification	Manufacturer	Test temperature [°F(°C)]	Peel strength [in.-lb/3-in. width (N/mm)]
FM-123-2	American Cyanamid	75 (24)	80 (14.0)
FM-73M	American Cyanamid	75 (24)	82 (14.3)
AF-126-2	3M Company	75 (24)	120[a] (21.0)
AF-163-2K	3M Company	75 (24)	70 (12.2)
EA-9601	Dexter Corporation	75 (24)	75 (13.1)
EA-9628	Dexter Corporation	75 (24)	60 (10.5)
Metlbond 1133	Narmco	75 (24)	80 (14.0)
Plastilock 717	B.F. Goodrich	75 (24)	80 (14.0)

[a]Tests were conducted using 0.085 psf (412-g/m^2) adhesive film. All remaining values are for specimens prepared using 0.060-psf (291-g/m^2) film.

Table 4   Flatwise Tension (FWT) Strength of Nitrile Epoxy Adhesive Films[a]

Adhesive identification	Manufacturer	Test temperature [°F(°C)]	FWT strength [psi (MPa)]
FM-123-2	American Cyanamid	−67 (−55)	1267 (8.7)
		75 (24	837 (5.8)
		180 (82)	420 (2.9)
FM-73	American Cyanamid	−67 (−55)	1565 (10.8)
		75 (24)	1200 (8.3)
		180 (82)	690 (4.8)
AF-126	3M Company	−67 (−55)	1427[b] (9.8)
		75 (24)	1150[b] (7.9)
		180 (82)	535[b] (3.7)
AF-163-2K	3M Company	−67 (−55)	− −
		75 (24	1025 (7.1)
		180 (82)	560 (3.9)

Table 4 (continued)

Adhesive identification	Manufacturer	Test temperature [°F(°C)]	FWT strength [psi (MPa)]
EA-9601	Dexter Corporation	−67 (−55)	− −
		75 (24)	1230 (8.5)
		180 (82)	− −
EA-9628	Dexter Corporation	−67 (−55)	− −
		75 (24)	1400 (9.7)
		180 (82)	− −
Metlbond 1133	Narmco	−67 (−55)	1400 (9.7)
		75 (24)	1050 (7.2)
		180 (82)	720 (5.0)
Plastilock 717	B.F. Goodrich	−67 (−55)	1522 (10.5)
		75 (24)	1126 (7.8)
		180 (82)	600 (4.1)

[a]All flatwise tension results are from tests conducted with $\frac{1}{4}$-in. (6.2-mm) cell honeycomb core.
[b]Tests were conducted using 0.085-psf (412-g/m^2) adhesive film. All remaining values are for specimens prepared using 0.06-psf (291-g/m^2) film.

Table 5 Tensile Shear Strength of Epoxy Polyamide Adhesive Films

Adhesive identification	Manufacturer	Test temperature [°F(°C)]	Shear strength [psi (MPa)]
FM-1000	American Cyanamid	−67 (−55)	7400 (51.0)
		75 (24)	7000 (48.3)
		180 (82)	3670 (25.3)
Metlbond 406	Narmco	−67 (−55)	5000 (34.5)
		75 (24)	5000 (34.5)
		180 (82)	3200 (22.1)

Table 6   Metal-to-Metal Peel Strength of Epoxy Polyamide
Adhesive Films

Adhesive identification	Manufacturer	Test temperature [°F(°C)]	Peel strength [in.-lb/1-in. width (N/mm)]
FM-1000	American Cyanamid	75 (24)	200[a] (35.0)
Metlbond 406	Narmco	75 (24)	70[b] (12.2)

[a]Climbing drum peel test method.
[b]T peel test method.

Table 7   Honeycomb Climbing Drum Peel Strength of Epoxy
Polyamide Adhesive Films

Adhesive identification	Manufacturer	Test temperature [°F(°C)]	Peel strength [in.-lb/3-in. width (N/mm)]
FM-1000	American Cyanamid	75 (24)	175 (30.6)
Metlbond 406	Narmco	75 (24)	150 (26.2)

Table 8   Flatwise Tension Strength of Epoxy Polyamide
Adhesive Films[a]

Adhesive identification	Manufacturer	Test temperature [°F(°C)]	FWT strength [psi (MPa)]
FM-1000	American Cyanamid	−67 (−55)	− −
		75 (24)	1200 (8.3)
		180 (82)	− −
Metlbond 406	Narmco	−67 (−55)	1200 (8.3)
		75 (24)	900 (6.2)
		180 (82)	325 (2.2)

[a]All flatwise tension results are from tests conducted with ¼-in.
(4.7-mm) cell honeycomb core.

Table 9 Tensile Shear Strength of Epoxy Phenolic Adhesive Films

Adhesive identification	Manufacturer	Test temperature [°F(°C)]	Shear strength [psi (MPa)]
HT-424	American Cyanamid	−67 (−55)	3225 (22.2)
		75 (24)	3550 (24.5)
		300 (149)	2760 (19.0)
		500 (260)	2000 (13.8)
Metlbond 302	Narmco	−100 (−73)	2300 (15.9)
		75 (24)	2300 (15.9)
		300 (149)	2150 (14.8)
		500 (260)	1900 (13.1)

Table 10 Honeycomb Climbing Drum Peel Strength of Epoxy Phenolic Adhesive Films

Adhesive identification	Manufacturer	Test temperature [°F(°C)]	Peel strength [in.-lb/3-in. width (N/mm)]
HT-424	American Cyanamid	75 (24)	33 (5.8)
Metlbond 302	Narmco	75 (24)	25 (4.4)

Table 11 Flatwise Tension Strength of Epoxy Phenolic Adhesive Films

Adhesive identification	Manufacturer	Test temperature [°F(°C)]	FWT strength [psi (MPa)]
HT-424	American Cyanamid	−67 (−55)	760 (5.2)
		75 (24)	615 (4.2)
		300 (149)	485 (3.3)
		500 (260)	290 (2.0)
Metlbond 302	Narmco	−67 (−55)	925[a] (6.4)
		75 (24)	900[a] (6.2)
		300 (149)	− −
		500 (260)	600[a] (4.1)

[a]Values obtained from specimens using 3/16-in. (4.7-mm) cell honeycomb.

VIII. CURING REQUIREMENTS

Epoxy adhesive systems have a wide range of curing parameters, depending on the curing agents selected and the epoxy resin type being used. Ambient-temperature curing formulations are very widely used for liquid systems, since their major use is in applications where elevated-temperature cures are impractical. By varying the curing agents and the quantity used, an extended range of useful application lives or "pot life" after mixing can be obtained. Pot lives ranging from 15 min up to 8 hr are quite practical. The ambient temperature curing liquid adhesives require very little pressure to effect a good bond; indeed, provided that the substrates fit well together, contact pressure is sufficient. Nevertheless, wherever practical, a light positive pressure is desirable to ensure good bond line consolidation. This can be accomplished by the use of simple clamps, spring-loaded clamps, or vacuum pressure, provided that the assembly can be placed in a sealed pressure-tight envelope. Pressure can also be applied by the use of weights directly on the assembly being bonded; however, this method can be clumsy and not too practical. A word of caution should be given on the use of vacuum to apply pressure on liquid epoxy systems. Some resins will tend to foam under high-vacuum conditions, generally in excess of 20 in. of mercury (508 mmHg), and it is recommended that vacuum levels should be kept moderately low if this presents a problem. In any event, tests should be conducted before embarking on a production program using this technique of applying bond pressure.

Film adhesives require a far different set of conditions to effect a cure and a well-consolidated bond line. They cure, depending on the type and formulation, within the temperature range 200 to 400°F (93 to 204°C); however, most of the systems in general use require a temperature range of either 225 to 250°F (107 to 121°C) or 325 to 350°F (163 to 177°C).

The nitrile epoxy adhesives cure at the lower-temperature range, while the epoxy phenolic, epoxy polyamide, and some unmodified epoxy systems cure in the upper range. Heat-up rates for all these systems are quite flexible, ranging for most purposes between 2 and 12°F (1 and 7°C) per minute. Many systems will cure satisfactorily beyond these limits; however, it is inadvisable to go lower than 1°F (.5°C) per minute, as problems may result in the flow and final cure of the adhesive. To effect full cure, a minimum time at temperature must be achieved. Most epoxy systems require 1 hr at the minimum recommended temperature, with the nitrile epoxies generally requiring 90 min. Some systems can cure with times as low as 30 min, while a very few may require up to 2 hr. The latter times are the exception rather than the rule.

All film adhesives require moderate to high positive pressures to produce good bond lines which are free from voids and other irregularities. These are normally produced by either an autoclave or a press. A minimum pressure of 25 psi (0.17 MPa) is necessary, ranging up to

100 psi (0.69 MPa). Pressure applied to an assembly can be restricted by the honeycomb density if it is a sandwich structure. Steps should be taken to ensure that the cure pressure applied to the assembly does not exceed the compressive or crush strength of the honeycomb core material. It is important also that the temperature of cure is taken into consideration, since the compressive strength of the core decreases with increase in temperature.

When curing in an autoclave, the question of whether to apply vacuum during cure must be considered. Vacuum is normally applied to an assembly prior to cure to check the integrity of the pressure diaphragm sealing. This initial vacuum should be confined to 10 in. of mercury [approximately 5 psi (34 kPa)], as some adhesive systems, particularly when bonding sandwich structures, can partially foam under high-vacuum conditions, causing porosity in the adhesive. Where possible, the vacuum applied for the diaphragm check should be maintained on the assembly until a positive pressure from the autoclave exceeds the pressure applied by the vacuum. At this point, vacuum is normally released to external atmosphere, with the pressure differential across the diaphragm maintained between normal atmospheric pressure and the autoclave pressure. There are some epoxy adhesives which cure more successfully with the continuance of vacuum pressure during the full cure cycle. The epoxy phenolic systems may require this technique to remove volatiles generated by the adhesive; however, care must be taken, particularly when bonding honeycomb sandwich structures, as under certain conditions collapse of the core can result from a possible uneven distribution of the vacuum under the pressure diaphragm. The general rule with applying vacuum is: do it only if it is necessary. Venting to the external atmosphere will present fewer problems and produce a perfectly satisfactory bond in most cases.

Cure monitoring is a very important aspect of the cure process. Instrumentation should be such that assembly temperature, autoclave pressure, and pressure under the diaphragm in the case of autoclaves is continuously monitored to ensure that all the cure parameters are being maintained. Pressure probes applied under the diaphragm will record if a rupture or leak has occurred and thermocouples attached to the assembly under cure, connected to a recorder, will properly monitor true part temperature.

IX. SHELF LIFE AND OUT-TIME CONSIDERATIONS

All the epoxy adhesive systems have a limited shelf life. Shelf life can be defined as the time from the date of manufacture to the termination of the life of the adhesive due to its deterioration from aging. This period always assumes that the particular adhesive is stored under the appropriate conditions defined by the manufacturer of the product.

Many of the room-temperature curing adhesives, which are always multiple-part systems where the curing agent(s) is added to the base resin immediately prior to application, have shelf lives in excess of 2 years. Once mixed, of course, the life is limited to a few hours at most. These resin systems are normally very stable, which results in good shelf life characteristics. Elevated-temperature curing liquid adhesives also have generally good shelf life capability. These are often one-part systems, which contain the curing agents within the base resin. Where this is the case, the adhesive will require storage at below 40°F (4°C) to further reduce the slow reaction which may take place at ambient temperature between the base resin and curing agent.

Film adhesives are, of course, all one-part materials and the curing agent(s) is an integral part of the resin system. Because of this, the shelf life of film adhesives is limited. The nitrile epoxy systems and the epoxy phenolics are particularly reactive, and therefore sensitive to storage temperature, with most requiring refrigeration at 0°F (−18°C) or below. Many of the unmodified epoxy systems also require storage at subzero temperatures; however, this is not always the case, and there are some film adhesives which cure at 350°F (177°C) that have very good storage capability at ambient temperature.

Out time is defined as the period between removal of the adhesive from refrigerated storage and the time when it is cured in the autoclave or press. While the adhesive may remain dormant at very low temperature, raising the adhesive to ambient temperature can speed up the reactions taking place between base resins and curing agents. The adhesive may become sufficiently advanced in the precure period to interfere with the flow and ultimate satisfactory cure of the system. To prevent this condition from occurring, a limitation is placed on this exposure period which is known as the adhesive out time. This out-time limitation may vary from 5 days to as much as 20 days, depending on the adhesive. These limited out times are generally conservative in the time permitted, and some safety cushion is generally included. However, recommended out times by the manufacturer should not be exceeded without extensive testing by the user to verify that there will be no problems arising during cure of the adhesive system. Adhesives which are stored at ambient temperature, due to their stability, do not have an out-time requirement but the overall shelf life of the adhesive still applies.

Many primer systems also have the same shelf life restrictions as the basic adhesives and some require storage at subzero temperatures. Out time can also apply to these materials. Some primer systems, due to their composition, may not permit the transfer of material back to refrigerated storage after removal and opening of the container.

X. DURABILITY

The durability of structural modified epoxy adhesives can generally be
described as good to excellent; however, any discussion of durability
must necessarily involve many factors other than the adhesive. A
high-quality adhesive bond can be achieved only by careful attention
to the substrate and its surface preparation, the type of primer to be
employed (e.g., corrosion or noncorrosion inhibiting), the adhesive,
and finally, the environment to which it will be subjected. All of these
items must be taken into account when describing the durability of an
adhesive and any generalization can be extremely misleading to the
reader. Of the 250°F (121°C) temperature curing adhesive systems,
principally the nitrile epoxy adhesives, the second-generation materials
have demonstrated their superiority over first-generation systems with
respect to durability. This statement assumes that all factors previously mentioned are to the best "state of the art." Typical of these
systems are:

> *FM-73:* American Cyanamid
> *EA-9628:* Hysol Division, Dexter Corporation
> *AF-163:* Minnesota Mining & Manufacturing Company (3M Company)
> *Metlbond 1133:* Narmco Adhesives

All of these systems offer excellent durability under sustained load
conditions in a hostile environment of 140°F (60°C) at 95% relative
humidity.

Although the newer 250°F (121°C) curing systems will offer superior durability over their earlier counterparts, the first-generation
systems are still in extensive use and provided that all necessary care
is taken in the bond processes (particularly surface preparation)
should continue to give excellent service for many years.

The epoxy polyamide adhesives have a more mixed history with
regard to durability. There is little doubt that these systems do have
a more than usual affinity for moisture, as do most unmodified polyamide polymers. Some are of the opinion that the moisture affinity is
very adverse to durability, whereas others discount this and describe
the excellent service history provided by these systems. These argu-
ments cannot be resolved within the confines of this chapter and the
reader must perform adequate tests to ensure that the durability of the
systems selected is adequate for the projected task.

The epoxy phenolic and unmodified epoxy adhesives both have
good durability characteristics. Evaluation of the durability of these
systems has not been as extensive as that carried out on the nitrile
epoxy adhesives; however, they do appear to exhibit good environmental durability. In particular, unmodified epoxy systems which use

corrosion-inhibiting primers have performed excellently over the past
10 years under very severe environmental conditions.

Historically, adhesive systems cured at 350°F (177°C) have generally superior performance to the lower-temperature curing adhesives
with regard to durability. However, with all the factors that must be
taken into consideration, it would be unwise to generalize on durability.
The rule of the engineer is to test, evaluate, and understand the limitations of the adhesive systems under selection. When you, the user,
can be satisfied that your final choice will give trouble-free service in
the environment to which it will be subjected, you can proceed with a
full-scale production program with confidence in the durability of the
system chosen.

XI. PROCESSABILITY

Unmodified and modified epoxy adhesive films have generally good handling characteristics. They are formulated to provide good tack properties, which enables the operator to drape and hold the film in the
correct location on the production detail part. Tack is one of the most
difficult properties of a film adhesive to define, and many attempts
have been made to measure it and put realistic values into a specification. Unfortunately, it is a property on which opinions differ widely,
and what may suit one production operator may not suit another. Because of this, no real satisfactory form of measurement has gained universal acceptance. The adhesive film manufacturers have a good feel-
ing for adhesive tack gained from experience and can generally supply
a material suitable to meet most opinions.

The epoxy adhesive films all respond to adjustments of tack by
minor modifications to the film manufacturing process. Changes can be
made to the adhesive without any drastic reformulation taking place.
This enables tack to be adjusted to meet specific requirements of the
adhesive purchaser without invalidating test data accumulated prior to
the adjustment. As a further aid to the adhesive user, adhesive films
can be colored to make them distinguishable from similar materials being
used in the production shop. Sometimes the film color is naturally produced by the particular formulation used, and sometimes as the result
of adding an inert dye to the formulation prior to manufacture of the
film. Adhesive films are frequently produced in more than one weight
or thickness. For instance, several of the nitrile epoxy systems are
produced in weights which vary from 0.030 to 0.10 psi (14.5 to 48g/m^2).
This equates to films from 0.005 to 0.015 in. (0.127 to 0.38 mm) in
thickness. To distinguish between these film weights, color is often
used. This helps to prevent errors in the production shop by enabling operators to differentiate easily between the various weights or
thicknesses.

Since epoxy adhesive films are tacky by design, they must be prevented from adhering to themselves when wound onto a roll for packaging and shipping. To accomplish this, the film is interleaved with release paper or polyethylene film or both. The paper is generally easily distinguishable from the adhesive; however, it is possible for a production operator to forget to remove the polyethylene film from the adhesive during the lay-up process. These instances are fortunately very rare; however, they have occurred and the polyethylene film remains in the bond line during cure, seriously interfering with bond strength. To help prevent this from happening, the polyethylene film is generally of a highly contrasting color to the adhesive film. This color difference reminds the assembly operator, that the release film is still in position and requires removal before closing the bond joint.

8

Elevated-Temperature-Resistant Adhesives

ERNEST C. MILLARD   *Rohr Industries, Inc., Riverside, California*

I. INTRODUCTION

When specifying an adhesive system for a particular bonding application on aircraft or other applications, one of the primary considerations
is the temperature regime to which it will be subjected. On subsonic
aircraft, the principal temperature range found in service is from -60
to 180°F (-51 to 82°C). Local areas where higher temperatures prevail
are in engine nacelles and compartments housing the auxiliary power
unit. Military aircraft capable of supersonic speed encounter higher
temperatures due to skin friction, which at Mach 2 can reach as high as
300°F (149°C). At speeds in excess of Mach 2 the temperatures generated will rise rapidly with increasing Mach number and be beyond the
capability of the more conventional adhesive systems. Areas around the
engines will also be very hot on supersonic aircraft, making the use of
adhesives very restrictive.

Until comparatively recently the very hot areas, that is, those in
excess of 400°F (204°C), could not employ the use of adhesive bonded
structure, having to rely on mechanical fasteners to assemble the structure. Adhesives are now available which can offer a good performance
at temperatures up to 600°F. Unfortunately, these are difficult to process and are limited in their field of usefulness.

To explore fully the range of adhesive systems available for use at
elevated temperatures, we first have to determine what the term "elevated temperature" means in relation to an operating environment. Elevated temperature can mean any temperature above ambient, for example,
130°F (54°C). All adhesives will perform adequately at such a temperature. We need therefore to determine the lower and upper temperature limits in which the adhesives we will be discussing will operate.
For the purpose of this chapter we will discuss adhesives designed to
perform between 300 and 650°F (149 and 343°C). To subdivide our

range of materials further we should first examine those systems capable of operating up to 400°F (short term) and then those specifically designed for the high end of the operating range, that is, between 400 and 650°F (204 and 343°C). Adhesive types capable of performing in the lower end of the high-temperature spectrum have been available for a number of years; however, the earlier materials had some limitations, such as short life at elevated temperature or difficulties in processing and handleability. During recent years, considerable work has been done to refine these adhesive systems to eliminate some of the problems. But first, in order to discuss the performance of adhesives at elevated temperature, or any other material for that matter, the relationship between time, temperature, and stress must be defined.

The relationship between time and temperature, that is the total hours the adhesive must perform at a specific temperature during the life of the aircraft, is critical when evaluating the high-temperature capabilities of an adhesive system. Many systems will give adequate performance at high temperatures for short periods of time but will degrade significantly over extended periods at the same temperature. This can be due to a variety of factors, such as polymer breakdown or oxidation. In some materials, degradation of the carrier can be a factor, as can also deterioration of one or more of the fillers in the adhesive formulation. To evaluate a system properly for its high-temperature capabilities, tests must be made at a number of different temperatures and at varying times of exposure at these temperatures. From this test matrix the performance of the adhesive can be determined and its reliability predicted with some confidence.

There is a third parameter to be considered in high-temperature adhesive evaluation. This is performance under sustained stress. Exposing an adhesive system to high temperatures by merely placing specimens in an oven will provide a profile of how it behaves under no-load conditions. When the specimen is finally tested, load is applied, the specimen fails, and the ultimate stress is calculated. Unfortunately, adhesives are rarely used in this manner. In most cases they have to perform at specific stress levels while exposed to the operating-temperature environment. To determine if an adhesive is capable of performing for extended periods of time under these conditions, sustained load or creep tests should be conducted. In these tests, the specimen is loaded to a specific percentage of its ultimate strength at the exposure temperature. The load applied is constant, with the specimen placed in an oven or chamber at the required exposure temperature. The loading system must be capable of compensating for any creep or elongation within the specimen. The tests should be continued until the required exposure time has elapsed or failure of the specimen occurs. This kind of testing is expensive to perform and will monopolize laboratory equipment for long periods of time. It is very important, therefore, that a full-scale evaluation of a high-temperature adhesive system is only carried out after a thorough preliminary investigation of basic properties and a determination that the system under test has a good chance of meeting the engineering requirements. This can be helped by working in full cooperation with the adhesive supplier and by a careful assessment of the chemical structure of the adhesive polymers and curing systems.

II. AVAILABLE ADHESIVE SYSTEMS

A. Intermediate-Temperature-Resistant Adhesives

Adhesives that fall in the operating range -60 to +400°F (-51 to 204°C) are as follows:

1. Unmodified epoxy
2. Epoxy phenolic
3. Nitrile phenolic

All these systems have the capability of curing at 350°F (176°C) at moderate pressure (up to 100 psi) and therefore fit in generally well with the cure parameters of many more general adhesive systems. None of the adhesives in this group require step cures, that is, cures where pauses in the heat-up of the adhesive are made to assist in special chemical reactions or for the removal of volatiles. This again makes them compatible with many other adhesive systems. Time of cure will vary with individual systems, but generally they fall within the range 1 to 2 hr, the nitrile phenolics sometimes requiring the longer cure period. These adhesives will cure in an autoclave, using normal vacuum bag techniques, with venting under the diaphragm to ambient atmospheric pressure. The epoxy phenolic systems can create problems, however, due to release of volatiles during cure. This can be more of a problem when curing large-area metal-to-metal surfaces, where the volatiles can be trapped and cause cracks in the bonded joint. To prevent this from occurring and to allow the volatiles to escape from the faying surface, venting holes in one of the substrates are often used. The unmodified epoxies and the nitrile phenolics do not suffer from these limitations and may be utilized in the conventional manner.

B. Very-High-Temperature-Resistant Adhesives

Adhesives that fall in the high-temperature-resistant range 400 to 650°F (204 to 343°C) are very few, and from the beginning it must be emphasied that use of these systems must be made with considerable caution, and with a thorough analysis of the adhesive properties and its end use. Information on some of the systems noted here is very limited and service experience is not available in some cases. The following adhesives have been developed, some on a very limited basis, to meet the high-temperature requirements:

1. Polyimide
2. Polyquinoxiline
3. Polybenzimidazole hectrocyclic
    polyaromatic linear polymer

Unlike the other systems, the adhesives named here mostly require curing
parameters matched to each adhesive formulation. Many require step
cures which escalate up to 650°F (343°C), with some also needing a high-
temperature postcure to develop the optimum properties of the adhesive.
These adhesives can generally be bonded to aluminum; however, because
of their high-temperature capability they were basically developed for
bonding steel or titanium, and this is their primary function. Because
of the complex curing mechanisms used in this class of adhesive, many
problems can be encountered with bonded joint configuration. Large-
area metal-to-metal faying surfaces may have voids due to entrapped
volatiles and under these conditions, venting may have to be employed.
Bonding of honeycomb can also present many problems and is generally
not recommended with the current state-of-the-art adhesives.

III. ADHESIVE CHARACTERISTICS

A. Unmodified Epoxy

Probably the best all-round-performing adhesives in the intermediate-
temperature-resistant class are the unmodified epoxy systems. The best
of these are capable of very long term durability at 300°F (149°C) (over
15000 hrs) without significant loss in mechanical properties and will perform very well for shorter periods up to 400°F (204°C). At the present
time there has not been an unmodified epoxy system developed which
can perform for 15,000+ hours at temperatures in excess of 325°F
(162°C), and it is becoming apparent that the epoxy systems have probably reached their ultimate development point with regard to long-term
temperature resistance. In recent years the major development effort
in intermediate-temperature-resistant systems has concentrated on the
unmodified epoxies, primarily due to their polymer structure and the
ease with which they are cured. They are essentially 100% solids materials, with no volatiles emitted during the cure cycle which could develop
voids and porosity in the bond faying surface. They bond metal-to-
metal and honeycomb structures equally well and have extremely good
durability in hostile environments (in addition to high temperatures),
such as high humidity and salt-laden atmospheres. They are strong,
moderately tough systems and work very well with the newer corrosion-
inhibiting primer systems. Unmodified epoxy adhesives bond very well
to honeycomb, both aluminum and nonmetallic types, and have excellent ability to form fillets at the honeycomb interface, which produces
very good flatwise tension and shear strength properties. These systems do have some limitations, however. While the shear and tension

characteristics are good, the peel strengths are generally poor. Much effort has been expended by the adhesive manufacturers to improve the peel of high-temperature-resistant unmodified epoxies. Unfortunately, a solution to this problem has not been found, although some improvements have been made during the past few years. Any form of flexibilizer or toughener that may be added to the formulation generally results in a falloff in properties at elevated temperature, thereby defeating the purpose of the original formulation. Low peel strengths need not be a problem, however, provided that the designer is aware of this particular shortcoming. Designs are perfectly feasible where peel is a nonparticipant in the stresses applied to the structure. There are several low-peel adhesives in service with 10 or more years behind them which have given no problems whatsoever. Difficulties can, however, arise in postbond operations such as drilling, trimming, and routing. Here the low peel strength can produce problems and steps must be taken, by using support and backup devices, to prevent delaminations occurring from these kind of operations.

B. Epoxy Phenolic

Epoxy phenolic adhesive systems were developed in the early 1960s and have since their introduction enjoyed moderate success as high-temperature-resistant systems. They have a capability of performing for very short periods of time at temperatures of 500°F (260°C) and for longer periods at 400°F (204°C). The epoxy phenolics, however, do not do well at temperatures in excess of 325°F (162°C) over extended periods of time (in excess of 10,000 hrs) due to degradation of the polymer-filler formulation. These characteristics have confined these systems to military or space applications due to the comparatively limited time that these vehicles spend at temperature. Civil aircraft applications are comparatively rare and are generally confined to areas subjected to intermittent temperature regimes.

Epoxy phenolic systems are not the easiest to process and cure. They have a limited time between removal from refrigerated storage and cure, which can produce shop floor problems. Their primary problem, however, is that they emit volatiles during cure. This characteristic requires special consideration by the designer of the structure to permit venting of these volatiles during cure. In spite of the difficulties, the epoxy phenolic systems have enjoyed a moderately prominent place in the high-temperature adhesives field, which is only now being taken over fully by the unmodified epoxies. Like the adhesives, the epoxy phenolic systems also have low peel characteristics. Shear and flatwise tension properties are generally good. The term "generally good" must be regarded with some reservation, and a word of caution should be interjected here, since the overall shear and tension characteristics do not approach those of the lower-temperature-resistant nitrile epoxy and epoxy polyamide systems. Shear strengths of 2000 to 3000 psi are about the

general level achieved by the epoxy phenolic systems and in many cases
with the unmodified epoxy systems also. These strengths are normally
perfectly adequate for most designs and of course the strength is maintained well into the upper-temperature spectrum. Like the unmodified
epoxy systems, the epoxy phenolics have a very good resistance to
hostile environments, particularly high humidity. They are resistant to
all the typical aircraft fluids (engine lubricants, hydraulic fluids, fuel,
etc.) and perform well at very low temperatures as well as high.

C. Nitrile Phenolics

Like the epoxy phenolic systems, the nitrile phenolic adhesives have
been in use for many years. They have a reputation for toughness and
durability and unlike the other high-temperature-resistant systems, they
have moderately good peel strength. These properties have combined to
make them a very useful system which has many contemporary uses.
Like most adhesive systems, the nitrile phenolics have some disadvantages
which limit their use for certain applications, the primary one being an
inherent low flow characteristic, which makes them unsuitable for honeycomb bonding. Their use is therefore confined to metal-to-metal applications. The low-flow characteristic does not normally present any major
problems in metal-to-metal bonding; however, cure pressures generally
need to be higher than those usually employed with other systems, particularly where large surface areas are to be bonded. In these cases,
pressures of 100 to 125 psi offer the best results. Because of their
flexibility when cured, the nitrile phenolics are often used as the adhesive for laminated sheet stock from which parts are cut and formed into
assembly details. This flexibility allows the forming operation to be
carried out without harm to the bond interface. These laminated assemblies perform very well in fatigue and in high sonic environments.

   The temperature capabilities of nitrile phenolics are good up to
temperatures of 300°F (149°C) with fair performance at temperatures
beyond this level for short periods of time. At temperatures above 350°F
(176°C) a marked falloff in properties will occur, and these systems
should not be used beyond this point. Nitrile adhesives have a good
resistance to hostile environments, particularly high-humidity conditions.
Where other systems developed difficulties in service from this condition, the nitrile phenolics performed, and are still performing, with
excellence. They have very good resistance to all normal aircraft fluids,
including phosphate ester-based hydraulic fluids. Since their introduction occurred a considerable time before corrosion-inhibiting primers
were developed, they achieve, in most cases, their excellent durability
without the benefit of these more recent primer systems. There are
some cases, however, where an epoxy-based corrosion-inhibiting primer
has been substituted for the originally developed primer and they perform equally well with this type of primer system.

D. Polyimides

The polyimide adhesive systems—not to be confused with polyamide—
are the most widely developed of the very high temperature adhesive
systems. They were introduced in the early to mid-1960s, and a
commercial material was produced and marketed. There has been considerable development effort since that time, with many systems produced on an experimental basis, but with few or no commercial
applications up to the present time. The principal development materials currently available are as follows:

> LARC-13: developed by NASA(National Aeronautics and Space
>   Administration)
> FM34: Developed by American Cyanamid

The primary polyimide adhesive system available on a commercial basis
is FM34, produced by American Cyanamid. This system has had some
success in volume-production applications and is used for bonding a
variety of substrates, including aluminum, steel, titanium, and polyimide glass and graphite composite structures. The use of aluminum
as a substrate is comparatively rare, however, since the upper operating range of the polyimide systems is beyond its capability for extended periods of time. The polyimide systems are therefore normally
reserved for use with the high-temperature-resistant substrates such
as titanium and steel. As previously stated, these systems often require a step cure, or postcure, or both, which in itself can be complex. Volatiles are released during the cure cycle and this can present
problems in large surface areas. To overcome this problem, it is often
recommended that full vacuum be maintained throughout the cure cycle
and that perforated honeycomb core (core where cells are intercon-
nected by perforations in the cell walls) be used in sandwich structure. Depending on the chemical structure of the polyimide system
used, adhesives will perform well at temperatures of 400 to 500°F (204
to 260°C). Some fall off in properties will be seen beyond 500°F
(260°C) out to 625°F (329°C), which is about their limit. Good aging
characteristics are generally apparent for several thousand hours at
500°F (260°C) and moderate strength retention after 2000 to 3000 hr
at 550°F (287°C). With the exception of American Cyanamid FM-34,
the polyimide systems must at the present time be considered in the
development stages. Because of this, care must be exercised before
any commitment is made to production use, with a full evaluation of
the problems likely to be encountered and thoroughly investigated.
Because of the experimental nature of the remaining very high temperature systems noted and the doubtful availability of them on a commercial basis, a detailed discussion of them will be omitted in this chapter.
Readers interested in their development are recommended to consult

the research literature published on them and consult with manufacturing industry and research organizations such as Air Force Materials
Laboratory at Wright-Patterson Air Force Base, and NASA Langley.

IV. ADHESIVE FORMS

All the adhesives in the high-temperature performance category are
available in film form, with and without a supporting textile carrier.
They are generally manufactured in a range of weights and thicknesses
to match specific structural and weight requirements and can be produced to match requirements for flow and tack characteristics to a
limited extent. Some of these adhesive systems can be made available
in liquid form, in particular the epoxy phenolics and the polyimides.
This form is not common, however, and is confined to very specific
applications, such as expanding foams and specialized honeycomb bonding where film adhesives are not suitable.

A third form that is available is the bulk form. This is again reserved for specialized applications, where the adhesive is to be applied
as a hot melt system using roller coating techniques. In this form, the
adhesive is supplied in sheets up to approximately 1/10 of an inch
thick. The adhesive is cut into moderate-size pieces and placed in the
coater rolls, where it melts to a moderately viscous liquid. In this way
it can be applied to honeycomb and other core materials. This form is
normally available only with the unmodified epoxy systems and only
those having good temperature stability at the temperature used for
the coating operation are suitable.

V. CARRIERS

Where the adhesives are manufactured as films, textile carriers are
most often used to enhance specific properties, for example peel. Since
this group of adhesives is formulated to operate at high temperature
the carrier also has to withstand that environment. The primary fibers
used for carriers are glass, nylon, and polyester. The latter two are
restricted to service temperatures up to 350°F (176°C); however, they
may not perform well over long periods of time at this temperature,
particularly the polyesters. Generally, these fibers are used with the
unmodified epoxy systems and are woven into open-weave fabrics of
various types. Where the adhesive system is designed to perform at
temperatures above 350°F (176°C), particularly for any extended
period of time at temperature, the carrier material must be manufactured from an inorganic material, and this role has been filled with
woven-glass fabrics. Woven glass makes an excellent carrier material
in the high-temperature regime, since it is unaffected by the service
temperatures experienced by the adhesives under discussion. The
fabrics used are generally of the square-weave type and are an open-

woven variety. Some problems exist with glass. Peel strengths are generally reduced and some difficulties with wetting and adhesion of the adhesive to the woven fibers often occurs. This will also affect the ultimate mechanical properties of the adhesive when bonded and fully cured. All adhesives are a compromise, however, and some sacrifice in overall properties must be made to gain the temperature resistance desired.

In most cases, the epoxy phenolics and the polyimides utilize woven-glass fabrics as the carrier material and the unmodified epoxies use the nylon and polyester materials. There is no hard-and-fast rule to this, however, and woven glass has been used successfully with the unmodified epoxy materials. The nitrile phenolic adhesives are generally supplied without a carrier and are used as unsupported adhesive film.

VI. COMPATIBILITY WITH PRIMERS

Performance of all the high-temperature adhesive systems is improved with the use of an adhesive primer applied to the substrate to be bonded. With some, the use of a primer is mandatory to develop the adhesive's potential fully, while with others the performance is merely enhanced with the use of primers.

Primers are formulated to match a given adhesive system, and in just about all cases the primer and the adhesive film are marketed as a complete package. In recent times, however, there have been several instances of interchanging of primers, that is, using a primer developed for a specific adhesive with that of a different adhesive. With careful development work to confirm the ultimate properties in the laboratory, this type of switch can be done successfully, resulting in a further enhancement of the basic adhesive properties. A typical example of this is the situation where the adhesive system may not have been developed with a corrosion-inhibiting primer system, that is, one containing corrosion-inhibiting material such as chromates. There have been several recent examples of substituting a corrosion-inhibiting primer for one not containing the corrosion-inhibiting substance with very considerable success. This form of substitution should not be attempted in a production mode, however, without a thorough evaluation to confirm compatibility.

Primers are prepared in many instances from the basic adhesive polymers, diluted with a suitable solvent system. With the more recently developed materials the addition of corrosion-inhibiting substances has been almost universal. These generally are in the form of chromate additives, which considerably improve the long-term corrosion resistance of the bonded joint and the overall durability of the assembly. To assist in indentifying a particular product, many manufacturers add an inert dye to the formulation to produce a distinctive

color to the product. All the primers developed for high-temperature-
resistant adhesive systems are compatible with conventional aluminum
pretreatments (sodium dichromate/sulfuric acid etch, chromic acid
anodize, and phosphoric acid anodize). They also perform satisfactorily with steel and titanium pretreatments such as nitric hydrfluoric
acid for steel, and phosphate fluoride for titanium.

VII. ADHESIVE PROPERTIES

Tables 1 to 4 show the elevated-temperature capabilities of some high-
temperature-resistant adhesive systems presently available. The results are typical of each adhesive group, but do not cover the entire
range of products in current use. These values were determined after
short periods at the test temperature (maximum 1 hr).

Table 1   Tensile Shear Strength of Unmodified Epoxy Adhesive Films

Adhesive identification	Manufacturer	Test temperature [°F (°C)]	Shear strength [psi (MPa)]
FM-150	American Cyanamid	75 (24)	2255 (15.54)
		300 (149)	2580 (17.78)
		350 (176)	2240 (15.44)
		400 (204)	—
EA9649	Dexter Corp.	75 (24)	4190 (28.88)
		300 (149)	3080 (21.23)
		350 (176)	2880 (19.85)
		400 (204)	1860 (12.82)
AF147	3M Company	75 (24)	4500 (31.02)
		300 (149)	2200 (15.16)
		350 (176)	1750 (12.06)
		400 (204)	1500 (10.34)

Table 2 Tensile Shear Strength of Epoxy Phenolic Adhesive Films

Adhesive identification	Manufacturer	Test temperature [°F (°C)]	Shear strength [psi (MPa)]
HT-424	American Cyanamid	75 (24)	3550 (24.47)
		300 (149)	2760 (19.02)
		350 (176)	2300 (15.85)
		400 (204)	2150 (14.82)
		500 (260)	2000 (13.78)
Metlbond 302	Narmco	75 (24)	2300 (15.85)
		300 (149)	2150 (14.82)
		350 (176)	2150 (14.82)
		400 (204)	2100 (14.47)
		500 (260)	1900 (13.09)

Table 3 Tensile Shear Strength of Nitrile Phenolic Adhesive Films

Adhesive identification	Manufacturer	Test temperature [°F (°C)]	Shear strength [psi (MPa)]
AF-31	3M Company	75 (24)	4430 (30.54)
		300 (149)	2550 (17.57)
		350 (176)	2350 (16.20)
		400 (204)	— —
		500 (260)	1650[a] (11.37)
FM-238	American Cyanamid	75 (24)	4300 (29.64)
		300 (149)	
		350 (176)	800 (5.51)
		400 (204)	— —
		500 (260)	— —
Metlbond 402	Narmco	75 (24)	4500 (31.02)
		300 (149)	2200 (15.16)
		350 (176)	1500 (10.34)
		400 (204)	— —
		500 (260)	— —

[a]Data developed on steel substrate after 2 hr at 500°F (260°C) using a postcure of 20 hr at 500°F (260°C).

Table 4   Tensile Shear Strength of Polyimide Adhesive Films

Adhesive identification	Manufacturer	Test temperature [°F (°C)]	Shear strength [psi (MPa)]
FM-34[a]	American Cyanamid	75 (24)	2850 (19.64)
		300 (149)	2220 (15.16)
		400 (204)	1750 (12.06)
		500 (260)	1600 (11.03)
		600 (315)	1300 (8.96)
		700 (371)	1150 (7.92)
LARC-13	NASA LaRC[b]	75 (24)	2800 (19.30) (1)
		400 (204)	2200 (15.16) (1)
		600 (315)[c]	1800 (12.40)[c] (2)

[a]Data for FM-34 developed on titanium substrates.
[b]Langley Research Center.
[c]Developed on Titanium substrate.

VIII. SHELF-LIFE AND OUT-TIME CONSIDERATIONS

The shelf life of the high-temperature-resistant adhesives varies considerably with basic polymer type. The epoxy phenolic systems are generally very sensitive to ambient-temperature conditions and for a reasonable storage life, the film adhesive and primer must be stored at 0°F (-17°C) or below. Failure to adhere to these temperatures will result in a fairly rapid deterioration of the system.

The nitrile phenolics, on the other hand, are relatively stable at ambient temperature and can be stored for as much as 6 months at temperatures up to 70°F (21°C). Most manufacturers, however, recommend that these systems be stored at 40°F (4°C). This ensures storage temperatures that are controlled and properly regulated, and prevents the possibility of materials being stored at ambient temperatures which could escalate as high as 100°F (37°C) in some areas.

Like the epoxy phenolics, polyimide systems are sensitive to ambient temperatures in the uncured state and require storage at or below 0°F (-17°C). The unmodified epoxy systems are somewhat different from their high-temperature counterparts, in that the storage temperatures can vary considerably with individual products. Some may require storage at 0°F (-17°C) or below, while others are quite stable for several months at temperatures up to 70°F (21°C). The variation depends on the particular epoxy resins and curing systems used in the formulation.

To ensure that adhesive storage is not going to be a problem, a full discussion with the manufacturer on this subject should be made. This will prevent misunderstanding and incorrect interpretation of their technical literature. Adhesive out-time, that is, the time between removal from refrigeration and final cure of the adhesive, will also depend on the type and individual formulation of the product. However, it is obvious that the more stable the system at ambient temperature and the less rigid the storage conditions, the greater out-time can be. Epoxy phenolic systems generally have very short out-times, just a few days, while the nitrile phenolics and some of the unmodified epoxy systems have out-times of 10 to 20 days. Again full consulation with the manufacturer should be made to ensure trouble-free storage.

IX. PROCESSIBILITY

Most of the high-temperature-resistant adhesive systems can be processed through cure in a conventional manner. They generally require no special treatment during layup and assembly of the structure being bonded, but there are some specific requirements necessary as part of the cure parameters that should be observed.

The nitrile phenolics are generally moderate-to-low-flow adhesive systems and as such they require higher pressures than most others, particularly where large surface areas are to be bonded. As stated earlier, they are not suitable for honeycomb applications. The epoxy phenolic systems will release volatiles during cure, and here proper venting of the faying areas is necessary to prevent entrapment. Cure cycles are conventional for the epoxy phenolics, in most cases requiring 60 to 90 min at 350°F (176°C).

Like the epoxy phenolics, the polyimides release volatiles during cure and the same precautions apply. Cure cycles, however, are not conventional and step cures are often required with high-temperature postcures as an additional requirement to complete the adhesive chemical reactions and to gain the full temperature-resistant capabilities of the system. The unmodified epoxy systems present the least problems of all the high-temperature-resistant adhesives and require little or no special treatment in the cure cycle, most of them curing at 350°F (176°C) for 60 to 90 min. All the adhesive films are supplied with separators consisting of coated papers and/or polyethylene films, which can be produced in different colors to assist in identification of the product. The separators assist in the application of the adhesive to individual detail parts, in addition to enabling the adhesive film to be wound in a roll after manufacture without adhering to itself, due to the built-in tack.

REFERENCES

1. Hendrichs, C. L., and G. H. Sylvester, Evaluation of High
    Temperature Structural Adhesives for Extended Service, SAMPE
    Symp., May 6-8, 1980.
2. Stevensen, A. A., Polyimide Adhesives for the Space Shuttle,
    Struct. Adhes. Bond. Conf., Tech. Conf. Associates, Mar.
    13-15, 1979.

9

Mechanical Properties of Adhesives

EDWARD J. HUGHES *Singer Company, Kearfott Division, Little Falls, New Jersey*

WALTER ALTHOF *Deutsche Forschungsanstalt für Luft-und Raumfahrt (DFVLR), Braunschweig, Federal Republic of Germany*

RAYMOND B. KRIEGER *American Cyanamid Company, Havre de Grace, Maryland*

I. INTRODUCTION

After selecting the surface treatment and the type or class of adhesive that a design requires, an evaluation of the mechanical properties of the adhesive must be made. As we consider the design requirements of a bonded joint and its subsequent analysis, the first thing required is an allowable. Several types of tests are available and must be considered.

The testing of bulk adhesives is a relatively straightforward procedure. For example, a test method that can contribute to the resolution of various adhesive properties is the cast (neat) tension test. This cast adhesive specimen, which may contain skrim, can be processed in the same manner as the material to be used in the adhesively bonded joint. Neat specimens are fabricated in the typical "dog-bone" shape used for metallic materials. They are tested in uniaxial tension at several strain rates and temperatures. In addition, some specimens are tested as cured, while others are conditioned for various times in a humid environment and then tested. Secant moduli and stress-strain data to failure are determined from the constant strain rate tests. From this type of test it is possible to make comparison tests between several adhesives and determine effects of load time, test temperature, and effects of humidity exposure [1]. This information, however, is not directly applicable to bonded joint analysis. The constraint that the adherend applies to the adhesive as it is strained prevents the normal

Poisson effect from taking place. Therefore, the type of test specimen must be chosen carefully.

In the design and strength analysis of the bonded joint, the major requirements are the elastic shear modulus and tensile modulus. Also helpful are Poisson's ratio, ultimate shear strain, and ultimate shear stress. The major difficulty in obtaining accurate elastic and plastic range data is making the measurements. This has required the development of an accurate strain measuring system. The magnitude of the problem can be appreciated when you consider that a typical adhesive bonded joint has an elastic range of 0.1% and a bond line thickness or gauge length of 0.005 in. (0.127 mm). Therefore, the strain measuring device must be capable of measuring displacements in the range 5×10^{-6} to 5×10^{-3} in. (12.7×10^{-6} to 12.7×10^{-3} cm) while retaining its linearity over the full range.

In addition to the measurement of strain, an important facet is the selection of a specimen configuration in which the adhesive bond is subjected to a pure shear stress without the complication of biaxial or triaxial stresses in the bond.

What follows is the presentation of three test procedures for obtaining shear data within the accuracies required.

II. TORSION SHEAR TESTING

A. Theory of the Torsion Shear Test

A napkin-ring type of specimen, consisting of two relatively thin-walled tubes bonded together end to end, was selected for test. In a test, the adhesive bond is stressed by rotating one adherend relative to the other about an axis that is coincident with the longitudinal axis of the assembly. The reasons for selecting this configuration were:

The loading mode is pure shear with no tensile or cleavage loading on the adhesive bond.

Since the specimen does not undergo dimensional changes in pure shear tests, the test data are not influenced by the effect of different Poisson contractions between the adhesive and the adherends. These effects are always present in tests such as tensile or lap shear in which the test specimens undergo dimensional changes.

In the napkin-ring specimen, the ratio of the wall thickness to the diameter of the adherends can be selected so as to minimize the variable strain in the adhesive, which arises from the fact that the strain is a function of the radial distance from the center of rotation. The tubular adherends chosen had a 3.0-in. (7.6-cm) outside diameter and a 0.125-in (0.317-cm) wall thickness.

The shear stress τ applied to the adhesive joint is given by

$$\tau = \frac{TC}{J} \tag{1}$$

where

    T = applied torque

    J = polar moment of inertia of the specimen

    C = radial distance to centerline of joint

The polar moment of inertia J is given by

$$J = \frac{\pi}{32}(d_o^4 - d_i^4) \tag{2}$$

where

    d_o = outside diameter of adherend

    d_i = inside diameter of adherend

Because of the relatively small wall-thickness-to-radius ratio, the stress across the adherend face can be considered constant. Combining Eqs. (1) and (2) provides the shear stress,

$$\tau = \frac{TC}{\pi/32(d_o^4 - d_i^4)} \tag{3}$$

The shear strain γ resulting from the applied stress is given by

$$\gamma = \frac{C}{L}\Delta\theta \tag{4}$$

where

    L = bond line thickness (gauge length)

    $\Delta\theta$ = angle of twist

From Eqs. (3) and (4), the only unknowns arising from an applied load are the torque and the resultant angle of twist, both of which must be measured during the test.

B. Test Procedure

To obtain accurate and meaningful data, there are three major considerations:

1. *Adherends*: Great care must be exercised in the specimen preparation to ensure that bond lines are uniform and the longitudinal axes of the adherends are coaxial.
2. *Torsion apparatus*: The torsion apparatus must be designed to ensure that no bending or cleavage loads are imposed on the napkin-ring specimen which must be subjected to pure shear.
3. *Strain measuring system*: The strain measuring system must be capable of detecting small displacements (ideally as low as 5×10^{-7} in. or 12.7×10^{-7} cm) and yet have a range sufficiently large to enable the measurement of strains up to failure. Finally, a reasonably wide temperature range is required if the system is to be used for a full characterization of adhesives.

Adherend Design and Manufacture

The adherends consist of two thin-walled tubes (0.125-in. or 3.17-mm wall thickness) each having a 3.0-in. (7.6-cm) outside diameter. The adherend machining specifications are as follows: (a) the bonding face to be flat within one light band of sodium and perpendicular to the surface of the inside diameter within a total indicated reading of 50×10^{-6} in. (20×10^{-6} cm), and (b) the inside and outside diameter of the adhesive face should be concentric to within 10^{-4} in. (0.0025 mm) total indicated reading. A typical adherend set is shown in Fig. 1. The grooves adjacent to the adherend faces in Fig. 1 are for mounting the strain measuring system.

Of primary importance in the specimen manufacture is the bonding procedure that must be followed to ensure axiality of the adherends and the accurate measurement of the resulting bond line. To obtain axiality, the adherends are mounted on a central arbor during the bonding procedure. The arbor (Fig. 2) consists of an aluminum rod slightly smaller than the adherend inside diameter. At the adhesive line position, the arbor is surrounded by a Teflon ring. By bonding the specimens with the arbor in place, axiality of the specimens is ensured. Joint thickness for cast adhesives is established by using gauge blocks to hold the adherends apart. The blocks are positioned between three pins positioned 120° apart on the fixed adherend and on the lip (Fig. 1) on the movable adherend. The arrangement is shown in Fig. 3. For sheet adhesives, gauge blocks are not used and the bond line thickness is determined by the sheet thickness and the bonding pressure required for the particular adhesive. Joint thickness is determined by measuring the difference in the distance between diamond-shaped indentations positioned 90° apart with dry adherends butted together before cure and again after cure. Measurement precision should be at least 2×10^{-4} in. (5.1×10^{-4} cm).

Figure 1   Typical unbonded adherend set.   (Courtesy The Singer Co.)

Torsion Apparatus

The torsion shear apparatus is designed to operate with a standard Instron machine, but is equally adaptable to any of the similar testing machines commercially available.   The loading method is shown in Fig. 4.   The load is transmitted to the torsion shear specimen through a yoke which is connected to the load cell through a universal joint.   The yoke is attached to the specimen via two chains which pass through two pulleys before fastening to a torsion sprocket.   The torsion sprocket is bolted to the movable adherend by four bolts, 90° apart.   Since the yoke is free to move, the load is distributed equally between the two chains, ensuring that the sprocket is subjected to a pure torsion. The fixed adherend is bolted to the back plate and the entire apparatus is bolted to the Instron crosshead.   The apparatus is shown removed from the tensile machine and without the extensometer mounted in Fig. 5.   The apparatus as shown can be conveniently mounted in an

Figure 2   Bonding arbor for alignment.   (Courtesy The Singer Co.)

environmental chamber to enable tests to be performed over a wide range of temperatures (−100 to 350°F or −73 to 177°C).

Referring to Eq. (3), the shear stress is proportional to the applied torque (T) on the adhesive bond. The torque is given by

$$T = \frac{P}{2}D \tag{5}$$

where

P = load cell output

D = torsion sprocket diameter

Substituting for T in Eq. (3) gives the torsion shear stress acting on the adhesive bond.

Figure 3　Gauge block arrangements for cast adhesives.　(Courtesy The Singer Co.)

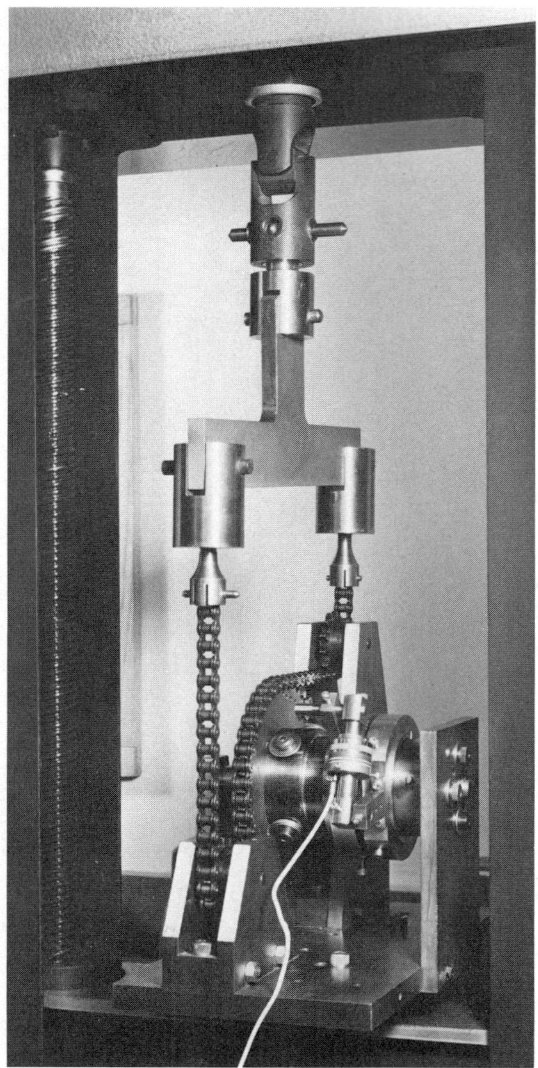

Figure 4   Torsion shear apparatus assembled in an Instron Testing machine.   (Courtesy The Singer Co.)

Figure 5   Torsion shear apparatus removed from testing machine with the capacitance gauges removed from bonded adherends. (Courtesy The Singer Co.)

Strain Measurement

When choosing a suitable strain measuring system, the factors that must be considered are

Sensitivity
Method of attachment
Linearity
Temperature capability

At Singer-Kearfott we have adopted a parallel-plate air capacitor of
the type developed by Brown and Lukens [2]. It consists of two an-
nular parallel copper plates. To calibrate the extensometer, one of
the capacitor plates is mounted on a precision micrometer thread so
that it can be moved accurately a small distance. The capacitance of
a parallel-plate condensor is given by

$$C = \frac{0.885kA}{D} \tag{6}$$

where

C = capacitance, pF

k = dielectric constant of medium between plates (for air $k = 1$)

A = effective plate area, cm^2

D = distance between plates, cm

From Eq. (6), the change in capacitance is inversely proportional to
the distance between the capacitor plates. Plate movements of 2×10^{-7}
in. (5.1×10^{-7} cm) are routinely measured with this extensometer. A
change in the plate separation produces a change in the capacitance of
the parallel-plate capacitor which is measured with a Robertshaw—
Fulton proximity meter. This proximity meter is made up of a modulated
radio-frequency (RF) oscillator, capacitance bridge circuit, amplifier,
demodulator, and output stage. The capacitance extensometer forms
one leg of the capacitance bridge. At the start of a test the bridge is
balanced by the use of two variable capacitors. As the plate separation
changes during test, the bridge is unbalanced and the out-of-balance
voltage is amplified, demodulated, and rectified. This output is fed
into the x axis of an X—Y recorder (or any other suitable recording
instrument). By feeding the output of the Instron load cell amplifier
into the y axis of the recorder, the load-extension behavior of the
material under study can be continuously recorded.

One of the most critical considerations in the strain measurement in
bonded adhesives is the method of attachment of the capacitance sensor
to the napkin-ring adherends. For the torsion shear type of test, the
best method has been found to be a pair of split rings which fit into
circumferential grooves 0.030 in. (0.76 mm) from the bonding face in
each adherend. Each ring is held in place by a collar to which one-
half of the extensometer is mounted on a radial moment arm. The arrangement of the split rings and collars is shown schematically in Fig.
6 and with one-half the extensometer attached in Fig. 7. The amount
of adherend deformation contributed by the adherends is measured by
mounting an extensometer on two grooves 0.060 in. (1.5 mm) apart on
one of the adherends. The deformation experienced by the adherends
can then be determined over the same load range used in the adhesive

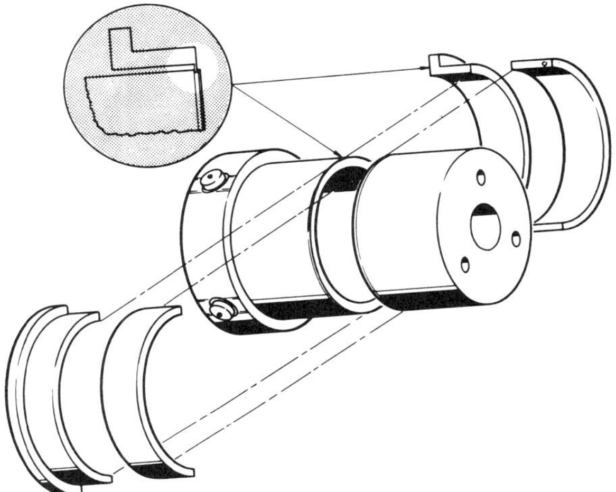

Figure 6   Schematic arrangement of split rings on the shear adherends.
(Courtesy The Singer Co.)

testing. This procedure allows the measured deformation on an adhesive bond to be corrected for the adherend deformation.
    Referring to Eq. (4), the shear strain resulting from an applied
stress is proportional to the angle of twist ($\Delta\theta$) experience by the
bond. For small displacements, $\Delta\theta$ is approximated by $\Delta s/r$, where
Δs is the displacement measured with the extensometer and r is the
radial distance to the center of the extensometer. Substituting for
$\Delta\theta$ in Eq. (4) gives

$$\gamma = \frac{C\,\Delta s}{Lr} \tag{7}$$

In Eq. (7), C and r are apparatus dimensions, L is the measured bond
line, and Δs is the measured displacement.

C. Data Generation

The torsion shear technique can be used to determine the torsion shear
modulus, generate microyield data (precision elastic limit, microyield
stress, and the irreversible work done in a load-unload cycle), and to
measure the fracture stress and strain of adhesives. Additionally, the
time-dependent properties (creep) of adhesives may be studied using
this technique.

Figure 7   Bonded adherends with one-half of the capacitor extensometer attached. (Courtesy The Singer Co.)

Shear Modulus

The shear modulus is calculated from the initial straight-line portion of the load-unload trace. The shear modulus G is given by

$$G = \frac{\tau}{\gamma}$$

(8)

Substituting for τ and γ from Eqs. (3) and (4) gives

$$G = \frac{T}{\pi/32(d_o^4 - d_i^4)} \quad \frac{L}{\Delta\theta} \tag{9}$$

From Eq. (5) the torque T is equal to $PD/2$ and $\Delta\theta$ is approximated by $\Delta s/r$, which gives

$$G = \frac{16PDLr}{\pi(d_o^4 - d_i^4)\,\Delta s} \tag{10}$$

in Eq. (10), d_o, d_i, D, and r are equipment dimensions, L is the bond line thickness, and P and Δs are, respectively, the load and corresponding displacement measured during the test.

Microstrain Data

Because of the high sensitivity of the capacitance extensometer, the technique can be used to determine the yield behavior of adhesives at extension sensitivities well below those normally used in stress-strain tests. In studying the microstrain behavior a series of load-unload cycles are made at increasing stress amplitudes (Fig. 8). At

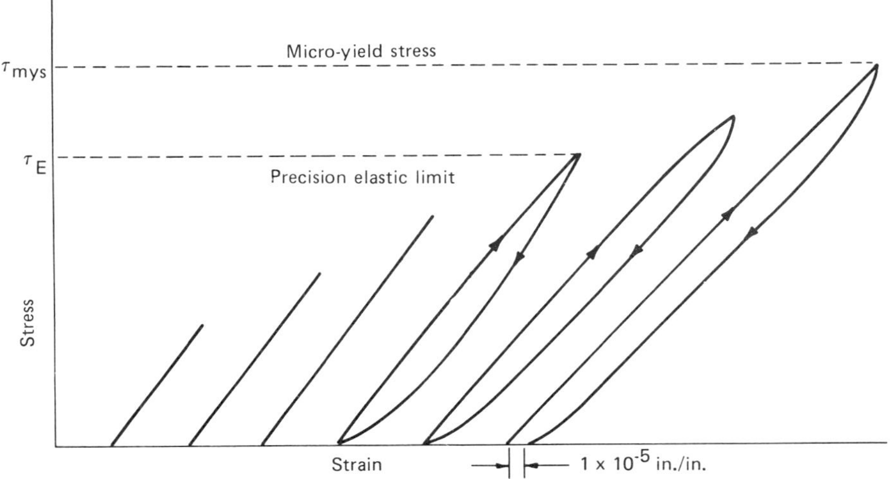

Figure 8   Microstrain behavior in load-unload tests. (Courtesy The Singer Co.)

low stresses (depending on the adhesive) the deformation is elastic; at higher stresses the precision elastic limit (τ_E) is exceeded. Deviations from elasticity may be in the form of hysteresis loops with no permanent set or by plastic strain. The stress (τ_{mys}) required for permanent strain is usually larger than τ_E. The microyield stress (τ_{mys}) is usually defined as the stress required to produce a known permanent strain; in our case 1×10^{-5} in./in. for a typical adhesive.

A knowledge of the microyield properties is particularly important when dimensional stability of the adhesive bond is a requirement. As a general rule, if the applied stress is below the microyield stress (which is usually much lower than the conventional macroyield stress) creep of the adhesive bond is not a problem.

Stress-Strain Curves

Complete stress-strain curves to failure can be recorded with the torsion shear apparatus. When ductile adhesives are being studied, the sensitivity of the test may have to be reduced to obtain the full stress-strain curve. This can be accomplished in one of three ways:

1. By increasing the capacitor plate separation
2. By decreasing the proximity meter sensitivity
3. By decreasing the voltage sensitivity of the X—Y recorder

D. Typical Data

The ability of the torsion shear technique to generate mechanical property data on a wide range of adhesive systems can best be illustrated by presenting typical data from three different types of adhesive systems.

High Strength-Low Ductility Structural Adhesives

Typical of this type of adhesive is Metlbond 329, which is a 100% solid, modified epoxy adhesive film, supported on a synthetic fiber carrier. Mechanical property values are shown in Table 1. Typical stress-strain curves to failure at 24, 66, and 75°C are shown in Fig. 9.

Ductile Structural Adhesives

Typical of this class of adhesives is FM-73. Data are presented in Table 2 and a stress-strain curve to failure in Fig. 10.

Dimensionally Stable Adhesives

These adhesives, used in the electronic and inertial guidance industry where dimensional stability is paramount, are usually highly filled two-part systems. Typical mechanical property date for four

Table 1   Typical Torsion Shear Mechanical Properties of Metlbond 329

Bond line thickness (in.)	Test temp. (°C)	Elastic shear modulus (psi)	Precision elastic limit (psi)	Microyield stress (psi)	Average fracture stress (psi)	Average fracture strain ($\times\ 10^{-3}$)
0.005	24	370,000	930	1,800	9,900	110
	191	130,000	120	450	4,600	170
	−55	472,000	—	—	10,200	72

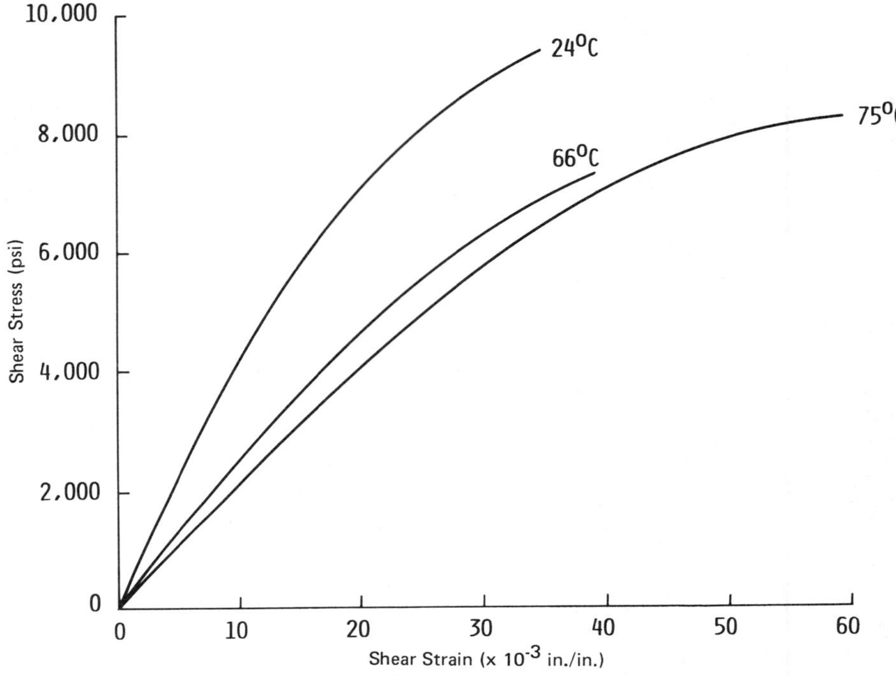

Figure 9   Typical shear stress-shear strain curves for Metlbond 329. (Courtesy The Singer Co.)

systems are given in Table 3, and a stress-strain curve to failure is presented in Fig. 11. For the design engineer, the creep behavior (strain under constant stress) is particularly important. Using the torsion shear apparatus with the capacitance strain measuring system described in Section II.B.3, creep tests can be performed for periods up to about 3 hr. A typical family of creep curves is shown in Fig. 12

Table 2   Typical Torsion Shear Properties of FM-73

Bond line (in.)	Shear modulus (psi)	Microyield stress (psi)	Fracture stress (psi)	Fracture strain ($\times 10^{-3}$)
0.0045	83,900	500	5400	615

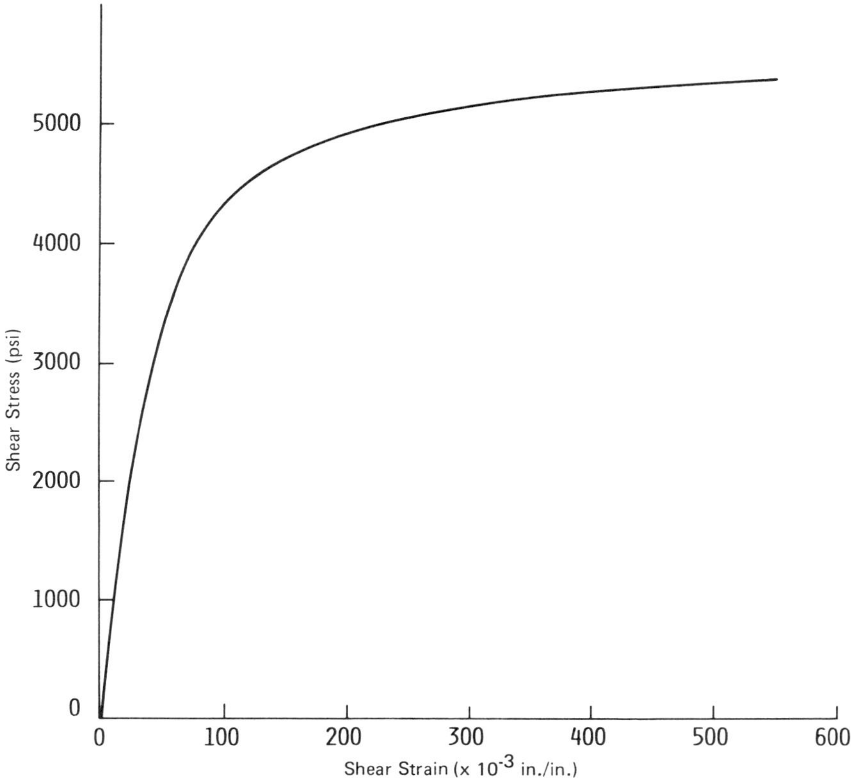

Figure 10   Typical shear stress-shear strain curve for FM-73 at room temperature.

Table 3   Static Shear Properties of Adhesives

Adhesive	Bond line thickness (in.)	Test temperature (°C)	Shear modulus prior to creep tests (psi)	Shear fracture stress after creep tests (psi)
Epotek H-74	0.0044	24	323,500	8,500
	0.0044	50	258,100	8,100
Ablebond 74-2	0.0036	24	159,200	11,200
Ablebond 72-1	0.0022	24	154,200	7,600
	0.0022	50	135,900	—
Eccobond LN-77080	0.0023	24	176,600	10,000

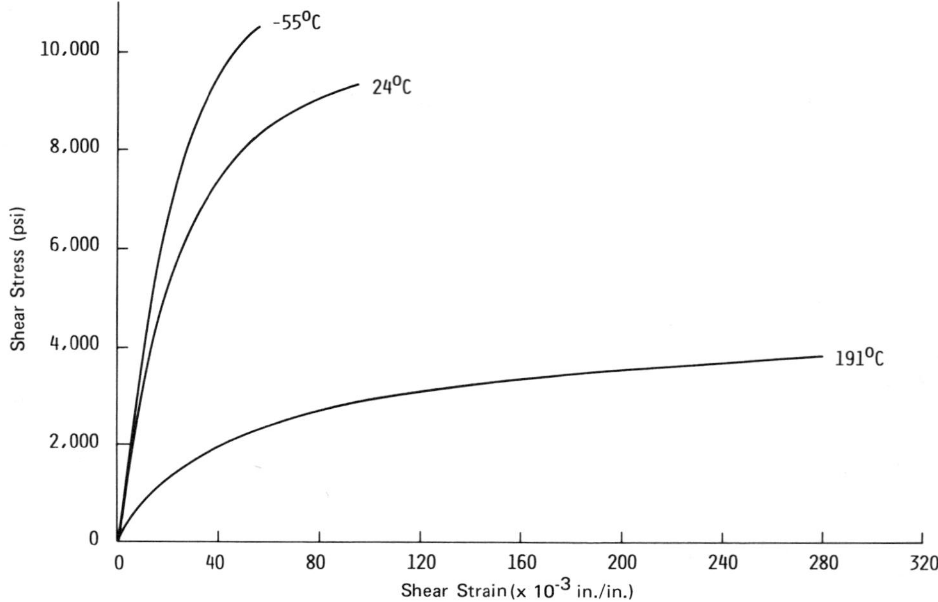

Figure 11   Typical shear stress-shear strain curves for Epotek H-74.

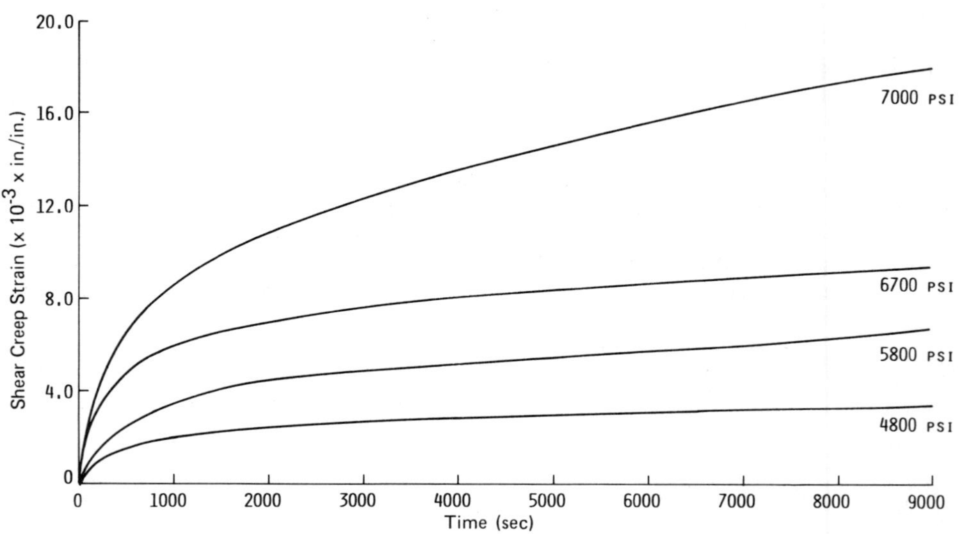

Figure 12   Room-temperature torsion shear creep curves for filled Epotek H-74.

Over longer periods of time, the data may be influenced by unpredictable instrument drift. If lower strain sensitivities can be tolerated (e.g., at high creep stress or high temperature), longer-term tests can be performed.

III. TORSION PENDULUM TEST ON CAST (NEAT) ADHESIVES

This method of testing is the one standardized in ASTM D-2236 [3] and DIN-53445 [4]. The test is aimed to determine the shear modulus G' and the logarithmic decrement Λ of an adhesive at different temperatures. For these values the following equations are valid:

$$G* = G' + G'' \qquad \frac{G''}{G'} = d$$

$$d = \frac{\Lambda/\pi}{1 + \Lambda^2 4\pi^2} \qquad d = \frac{\Lambda}{4} \text{ for } \Lambda \leqslant 2$$

where $G*$ is the complex shear modulus, G' the elastic shear modulus, G'' the loss modulus, and d the loss factor. The elastic shear modulus G' is similar to the Hooke's shear modulus, determined by static shear tests. The loss modulus and the loss factor describe the ability of an adhesive to damp mechanical vibrations.

A plot of G and Λ versus temperature indicates the thermo-mechanical characteristic of the adhesive. The transition temperature of the adhesive can also be determined. A specimen of cast adhesive (neat) is stimulated to oscillate around its longitudinal axis. The free decay oscillations are recorded. The test can be performed at high or low temperatures. The calculated values of the shear modulus G' and the logarithmic decrement Λ are plotted versus temperature in a semilogarithmic diagram and connected by curves (see Fig. 13).

The *freezing range* of an adhesive is characterized by a high and constant shear modulus and by low decrements. The adhesive in this state is hard and brittle; its deformations in a bond line are reversible. Creep is small and cracks may occur under dynamic loading conditions.

In the *softening range* the shear modulus decreases steeply, whereas the decrement increases and passes through a maximum. The softening range is also called the "dispersion range." If graduations in the shear modulus and a number of peaks occur in this range, it is divided into a "main dispersion range" and "subsidary dispersion ranges." The main dispersion range is characterized by the highest peak of decrement. The occurrence of a number of dispersion ranges is a characteristic of copolymers (i.e., polymers of a number of types of resins which have chemically reacted together). In the softening state an adhesive in a

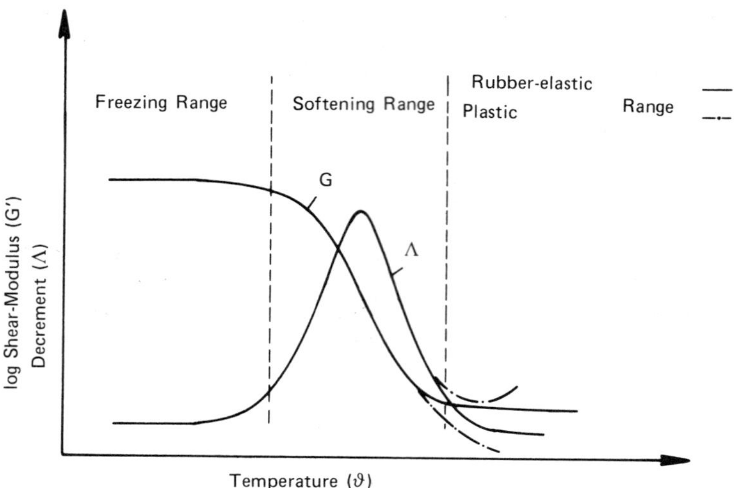

Figure 13   Scheme of a torsion pendulum diagram.

bondline deforms partly irreversibly at loading, its strength decreases,
and creep will occur.

The *rubber-elastic range* occurs in the case of cross-linked resins.
In this range the shear modulus and the decrement are low and remain
approximately constant with increasing temperature. An adhesive in a
bond line is very ductile and is creep under small applied loads.
Therefore, no mechanical loading is recommended in this case

A *plastic range* is observed in the case of resins which are not
cross-linked. In this range the shear modulus decreases and the decrement further increases with the temperature. Mechanical loading is not
recommended in this state, because the deformations of a bond line
are irreversible.

In order to manufacture the specimens, a number of layers of the
prefabricated adhesive films are compressed in a mold and cured in a
heating press under the same conditions as a bonded joint. Liquid adhesives are cast into a mold. After curing, the adhesive is cut in
bubble-free test specimens with rectangular cross section.

A flywheel mass with a known moment of inertia is fixed to the adhesive specimen and stimulated to oscillations. For recording the oscillations, for example, a mirror disposed at the axis of rotation (the
longitudinal axis of the specimen) reflects a luminous point to a moving
light-sensitive paper (Fig. 14).

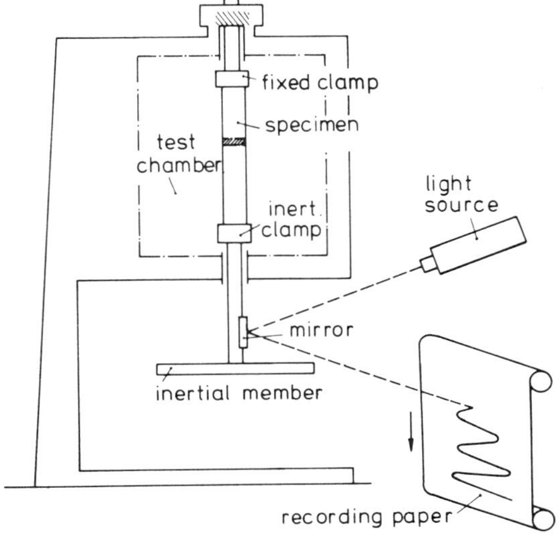

Figure 14　Schematic of the test apparatus.

From the recording, the average frequency (f) of the oscillation and the amplitudes A_n, A_{n+1} of successive cycles are determined. The *logarithmic decrement* Λ is calculated as follows:

$$\Lambda \;=\; \frac{A_n}{A_{n+1}}$$

If the oscillations are strongly damped, the equation above cannot be used; the result would be incorrect. In this case the decrement is [5]

$$\Lambda \;=\; \ln \frac{\alpha_1 - \alpha_0}{\alpha_3 - \alpha_0}$$

with

$$\alpha_0 \;=\; \frac{\alpha_1 \alpha_3 - \alpha_2^2}{\alpha_1 - 2\alpha_2 + \alpha_3}$$

and

$$\alpha_1, \; \alpha_2, \; \alpha_3, \; \alpha_0 \text{ according to Fig. 15.}$$

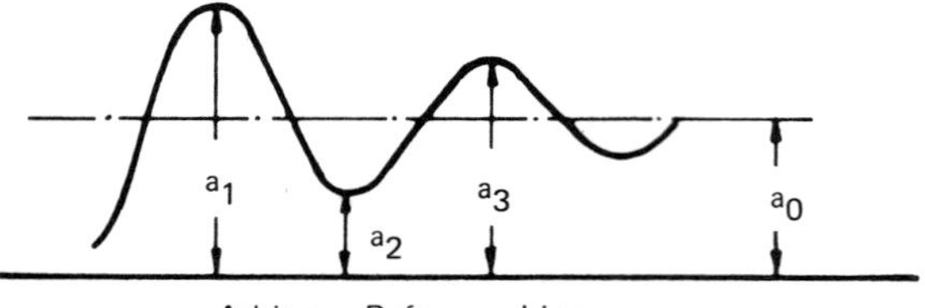

Figure 15   Scheme of strongly damped oscillations.

The *shear modulus* (the derivation of the formula is described in Refs. 6 and 7) of a specimen with a rectangular cross section is

$$G = 4\pi^2 I f^2 L \frac{3}{bt^3(1 - 0.63t/b)} F_d$$

where

$$G = \text{shear modulus dny/cm}^2$$

$$I = \text{inertia moment of flywheel mass, } g \times cm^2$$

$$f = \text{frequency of oscillation, } s^{-1}$$

$$b = \text{width of specimen, cm}$$

$$t = \text{thickness of specimen, cm}$$

$$L = \text{length of specimen between the grips, cm}$$

$$F_d = \text{damping influence on the shear modulus}$$

$$= 1 + \frac{\Lambda^2}{4\pi^2}$$

$$= 1 \text{ for } \Lambda \leqslant 1$$

Typical dimensions of specimens are b = 1 cm, t = 0.1 cm, and L = 5 cm.

The following examples will show the reliable distinction of different adhesive properties by use of a torsion pendulum diagram. The stress-strain behavior of the same tested adhesives in the bond lines of shear-loaded thick-adherend specimens will be used for a comparison [8].

A. Heat-Resistant Properties

Figure 16 shows the torsion pendulum diagram of 2 structural adhesives, an epoxy, and an epoxy nitrile adhesive. For the epoxy adhesive a first softening range begins at 130°C (266°F) with a small decrease of

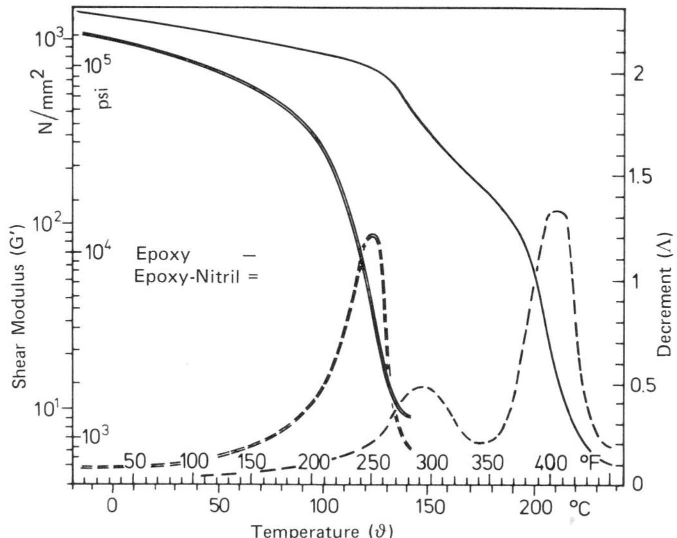

Figure 16   Torsion pendulum diagrams of two structural adhesives.

the shear modulus and a subsidary peak of the decrement. The main
dispersion range begins at 190°C (375°F); the transition temperature
is 215°C (420°F). The temperature is higher than the transition temperature of the epoxy nitrile adhesive 125°C (257°F). This adhesive
softens at 70°C (150°F).

In Fig. 17, the shear stress-strain diagrams of the adhesives are
plotted. At 100°C (210°F) the epoxy adhesive is deformed elastically
up to a shear stress half of the strength at bond line fracture. At
this temperature the epoxy nitrile adhesive is deformed in a plastic
mode. At the temperature of 70°C (160°F) elastic deformations are
limited up to stresses a quarter of the bond strength.

B. Aging in a Standard Laboratory Atmosphere

In Figs. 18 and 19 an epoxy nylon adhesive is compared with an epoxy
nitrile adhesive before and after an exposure of 8 weeks in a standard
laboratory atmosphere of 20 to 23°C (68 to 74°F) and 60 to 65% relative
humidity.

The torsion pendulum diagram shows no change of the epoxy nitrile
adhesive, and in the stress-strain behavior there is only a small decrease of the stress at which the plasticity of the adhesive begins.

For the epoxy nylon adhesive the softening range and the transition temperature change to lower temperatures. Also, the shear

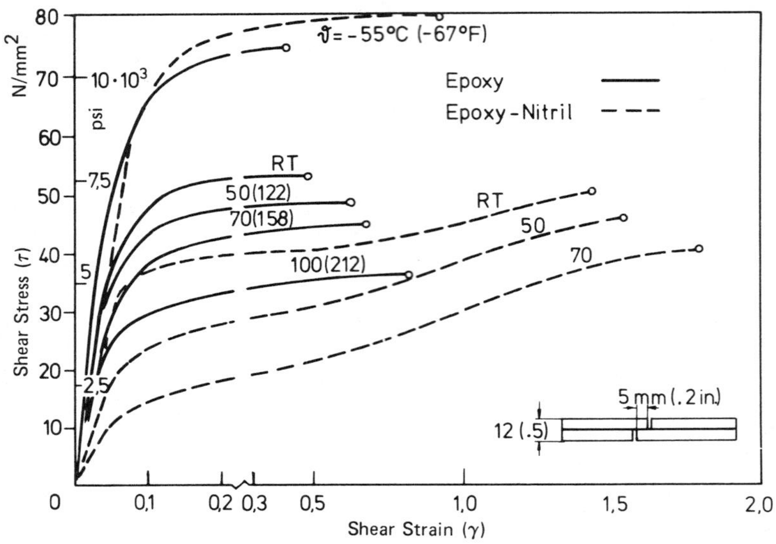

Figure 17   Shear stress-strain diagrams (thick-adherend test) of two structural adhesives at different test temperatures.

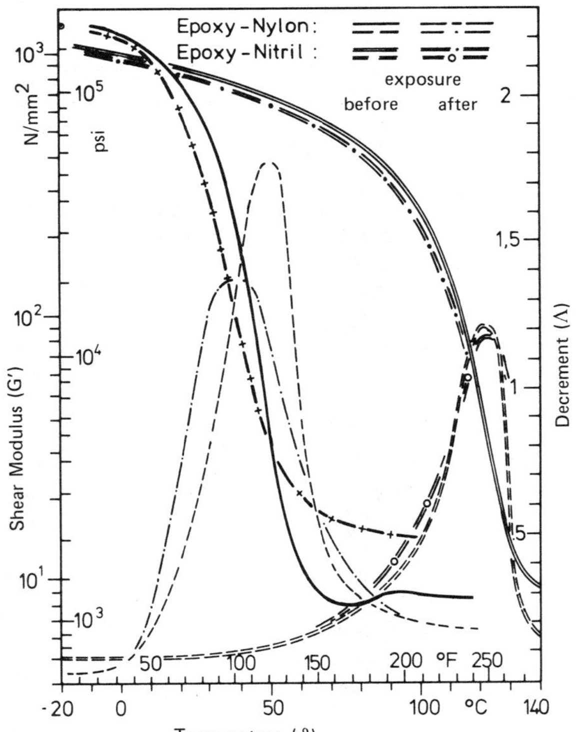

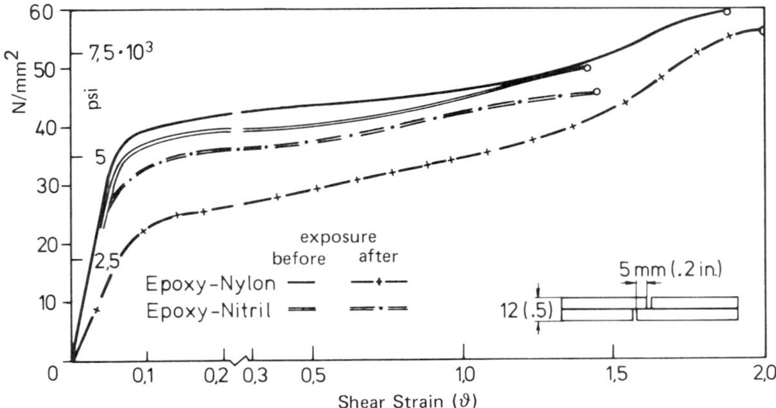

Figure 19   Shear stress-strain diagrams (thick-adherend test) of two structural adhesives before and after 8 weeks' exposure in a standard laboratory atmosphere of 20°C (68°F), 65% relative humidity.

modulus decreases at room temperature after the exposure, as shown in the torsion pendulum diagram and in the stress-strain diagram of the adhesive. A comparison of the bond strength does not explain a change of the adhesive.

C.   Aging in a Hot and Humid Environment

In Figs. 20 and 21 the test results of a modified epoxy and an epoxy nitrile adhesive are plotted before and after an exposure of 8 weeks in an environment of 70°C (160°F) and 95% relative humidity. The torsion pendulum diagram shows only a small decrease of the shear modulus of the epoxy adhesive and the transition temperature is unchanged. The same is shown in the shear stress-strain diagram.

The decrease of the shear modulus of the epoxy nitrile adhesive after the exposure is important, the transition changes from 125°C (257°F) to 95°C (203°F). Both are an effect of the adhesive softening as a result of the diffusion of water vapor into the adhesive. The stress-strain diagram also shows the great change of the adhesive in this environment.

Figure 18   Torsion pendulum diagrams of two structural adhesives before and after 8 weeks' exposure in a standard laboratory atmosphere of 20°C (68°F), 65% relative humidity.

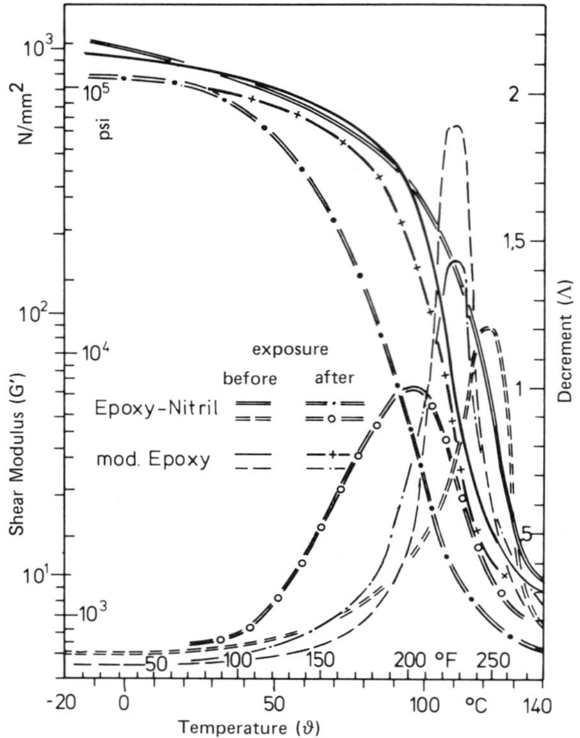

Figure 20 Torsion pendulum diagrams of two structural adhesives before and after 8 weeks' exposure in a hot and humid environment of 70°C (160°F), 95% relative humidity.

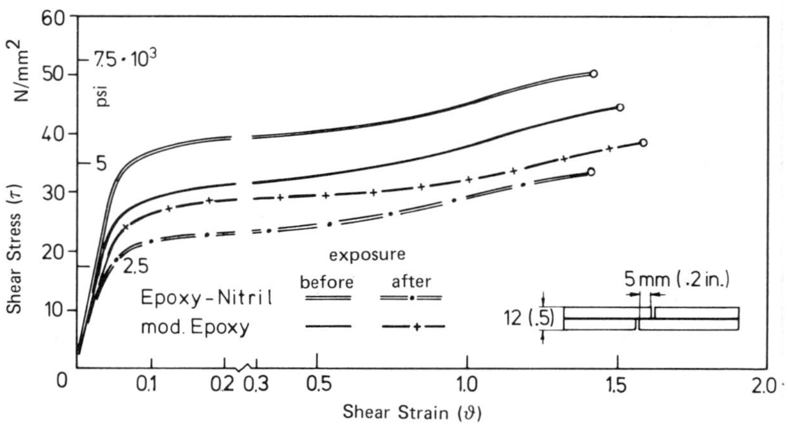

Figure 21 Shear stress-strain diagrams (thick-adherend test) of two structural adhesives before and after 8 week's exposure in a hot and humid environment of 70°C (160°F), 95% relative humidity.

As a result of the comparisons above, it may be inferred that the behavior of an adhesive with respect to a change in the molecular structure and the strain in a bond line can be predicted reliably by the simple torsion pendulum test. The torsion pendulum test gives an indication of the elastomechanical characteristics of an adhesive at different temperatures and in which manner it is changed by environments. The test has the advantage over a strength test that the results are independent of the specimen geometry and the adherend members. Therefore, the test makes possible short-time comparisons with respect to a selection of adhesives. The test does not replace strength tests with bonded joints in case the strengths of the bonds should be estimated quantitatively.

IV. THICK ADHEREND—LAP SHEAR TESTS

Considering that adhesives have been used as a fastener which joins two pieces of metal together and the load transfer is shear in the place of the glue line, it has been further determined that the shear stress distribution in the glue line is not uniform. Because of the relative stiffness of the adherends and the adhesive, there is a peak or maximum stress at each end of the splice. The stress at the center of the joint can be zero if the joint or lap is long enough.

The necessary ingredient to the determination of the stress distribution is the shear modulus, which is the ratio of shear stress to shear strain. Ideally, this could be done by measuring the shear displacement, glue line thickness, and the shear stress on a bonded specimen having uniform shear stress distribution. Two problems exist which must be solved to get these data. First, what kind of specimen should be built, and second, how do we measure the needed displacement?

A "proper" test specimen will provide stresses that are comparable to calculated stresses in a design in order to predict its services performance if possible. To begin, we consider the case of load transfer, where tension loads in a sheet are transferred in part to another sheet over a glue line bonding the two together. In the linear region, the adhesive shear stress at the beginning of the joint can be calculated using the adhesive modulus. (This is simplified if the joint overlap is infinitely long, known as the skin-doubler concept.) Two problems immediately arise. They are, first, the glue line has tension strains, in addition to shear strains, which do not match those in the stress-strain curve development. Second, the peak shear stress at the beginning of the joint may relieve itself when stiffness changes occur. These are among the reasons the stress-strain specimen is not directly suitable for allowable stresses. Nevertheless, the stress-strain curve capability has gotten us much closer to the truth than lap shear or peel. On a skin-doubler specimen, for example, we can relate adhesive stress and metal stress to test loads on the specimen. This has opened the door to the possibility of at least one useful test specimen, the skin-doubler concept.

To minimize the presence of tension in the glue line a thick-adherend short-lap specimen was designed. The metal used is 3/8 in. thick. Two plates are bonded together and cut into 1-in.-wide pieces which are 9 in. long. The center area is notched to the glue line in two places to form a lap shear specimen with a 3/8-in. overlap. A 1/2-in.-diameter hole is drilled in each end for loading pins. These holes are fitted with steel bushings 3/4 in. long, which have been turned down slightly in outside diameter for a distance of 1/4 in. from each end. This means that bearing pressure is put on the specimen over a 1/4-in.-wide band in the center, automatically preventing eccentric loading. Aluminum alloy 2024-T3 bare has become standard for several reasons. It is a representative alloy for aircraft and is without the complication of dissimilar metals of alclad (which aggravate galvanic activity under water exposure). There is no decisive problem with other alloys or metals. In fact, since metal deformation corrections must be made, it is feasible that steel be used for adhesives with extremely high modulus. There is one more important thing to remember. When machining the specimen, the cutting must be slow enough so that the adhesive will not overheat and change its mechanical properties.

The KGR-1 extensometer developed by Raymond Krieger has been designed to obtain the required data from the specimen described above. The extensometer measures the shear displacement of the bond line in a thick-adherend lap shear specimen. "Thick-adherend" means a single-overlap shear specimen with metal which is much thicker than normal. This is to minimize adherend bending and to generate a nearly uniform shear stress distribution in the test glue line. The adhesive strain signal is generated by linear variable differential transformers, whose voltage changes are used to drive a recorder. A load signal is also provided to the recorder by any of a variety of conventional means. The work can be done over a wide range of temperature, from below -67 to $+500°F$ (-55 to $260°C$). The following description sections provide the pertinent detail.

Since it is not possible to purchase an extensometer with the ability to measure the small deflections anticipated, an instrument was made, the KGR-1 (see Fig. 22). Two instruments are used; they are mirror images of each other and are mounted on opposite sides of the specimen. The signal of either individual instrument can be read, or their average signal can be obtained by combination.

The system depends on the principle of the linear variable differential transformer (LVDT). A core is made to move in coils by the same amount the glue line deforms in shear. This generates a voltage which can be recorded as the shear movement. The black cylinder contains the coils. The core rod is mounted to the front frame, a vertical bar

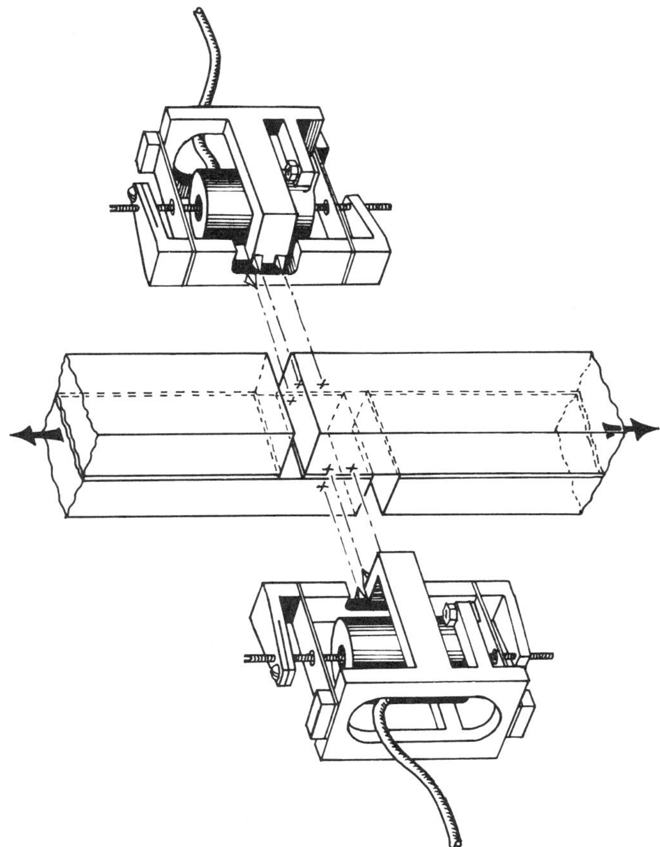

Figure 22 Functional diagram of KGR-1 extensometers and thick-adherend lap shear specimen. The exploded view is used to show precise engagement sites of extensometer gripping points. (Courtesy American Cyanamid Corp.)

carrying a hard-steel single point, which engages the specimen near the glue line. The coil is mounted to the rear frame. The rear frame has an extension arm reaching to the specimen; this arm carries hard-steel double points which engage near the glue line on the opposite side from the single point.

The front and rear frames are attached to each other by a flat blade spring at their tops and another at their bottoms. These blade springs have their ends firmly fixed (clamped) to the frames. This precludes all extraneous movement other than parallel to the direction

of shear deformation of the glue line. Also, this spring suspension precludes all errors inherent in pivoting of bearings or knife edges (i.e.,
loose fit, backlash, or dull edges).

The instruments are mounted on the specimen by being mutually
pulled together by two springs (not shown in Fig. 22), one attached
to the front frame of each instrument. These springs are connected
to each other by ball chains, which are adjustable to produce sufficient
force such that the instruments are suspended solely by their three
points pressing into the specimen.

Figure 23 shows the instruments mounted on a specimen with a view
of the mounting springs and chains. Also shown are locking devices,
which give a fixed reference setting between the front-frame single
point and the rear-frame double point. The core is properly located
relative to the coils while the lock is engaged. The instrument remains
locked until it is mounted on the specimen. Then it is unlocked and
the core is in its proper position for maximum travel with linear response.

Referring back to Fig. 22, the core is moved by turning the long
threaded rod mounted on the front frame and passing through the coil.
When correctly located, the rod is locked by tightening the screw on
the rod support arm at the top of the front frame. The technique for
proper core setting is simplified because the KGR-1 system includes an
amplifier to increase the voltage signal to the recorder. When the core
is properly located (at null), zero voltage is sent to the recorder. At
this core location, increasing the amplifier gain causes no signal and
thus the recorder will not move. If the core is not at null, a voltage
will be sent to the recorder. After the recorder has responded to this
signal, an increase in gain increases the signal and the recorder will
move. So the core rod is simply located until gain increase causes no
recorder movement; then the core is locked in its proper position.
Figure 24 gives another view of the mounting springs, chains, and
locks. The single and double points are also shown.

The KGR-1 system includes a tool which allows the points to be
moved for any predetermined amount. This allows the amplifier gain
to be accurately set for a useful, convenient scale on the recorder
chart. This tool consists of two thick aluminum plates separated by
an artificial glue line of Teflon film. One plate is fixed. The other
plate is driven by a large micrometer to simulate the shear displacement
of the adhesive in the actual test specimen. The KGR-1 is mounted on
this tool exactly as on the actual specimen. This tool is used to calibrate and functionally check the KGR-1, including positioning the core.

Figure 25 shows (1) the KGR-1 mounted on a calibrating tool (2)
a calibrating tool mounted on a micrometer, and (3) an amplifier, with
its power switch and gain controls for each side of the KGR-1.

The prime reason for a preamplifier is that the movement to be
measured is far less than the capability of conventional strain recorders

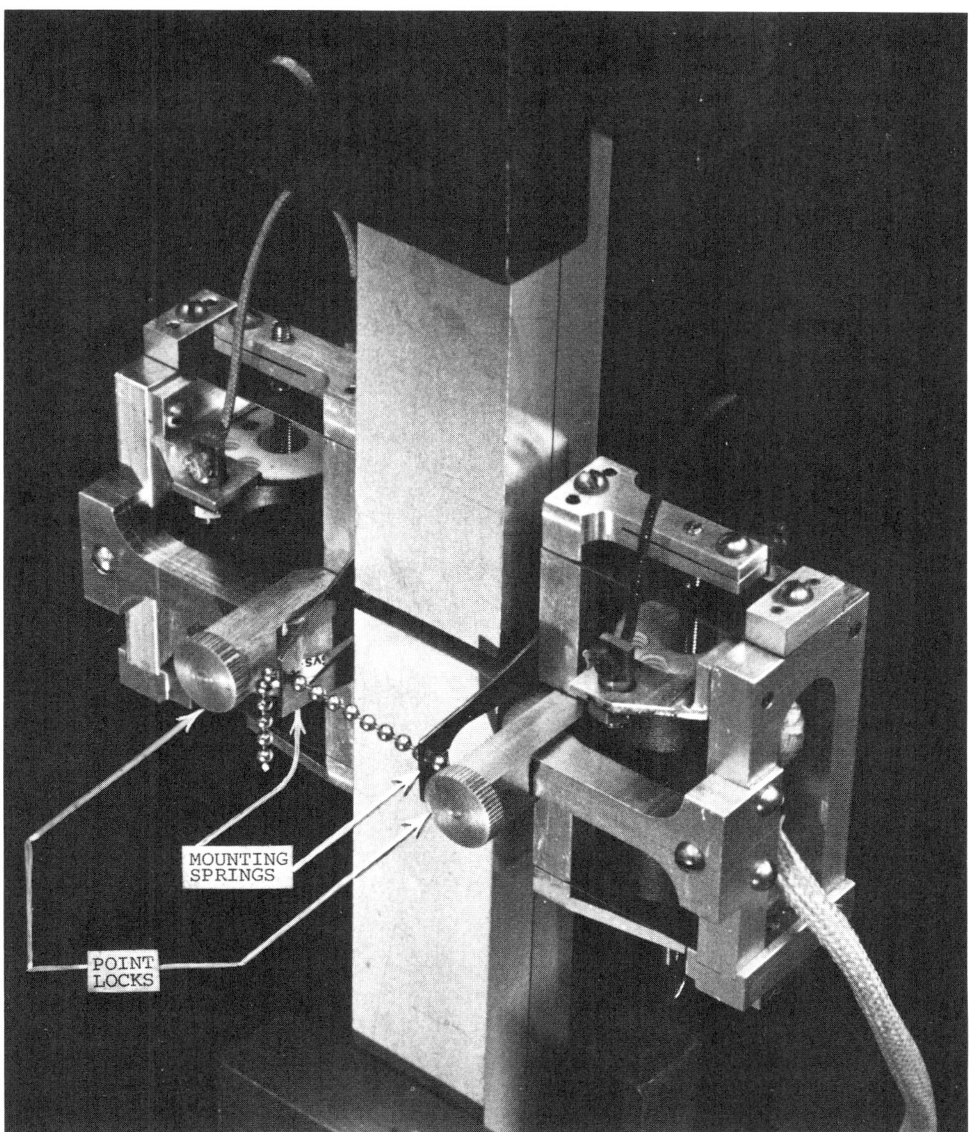

Figure 23   KGR-1 extensometers mounted on the thick-adherend lap
shear test specimen. (Courtesy American Cyanamid Corp.)

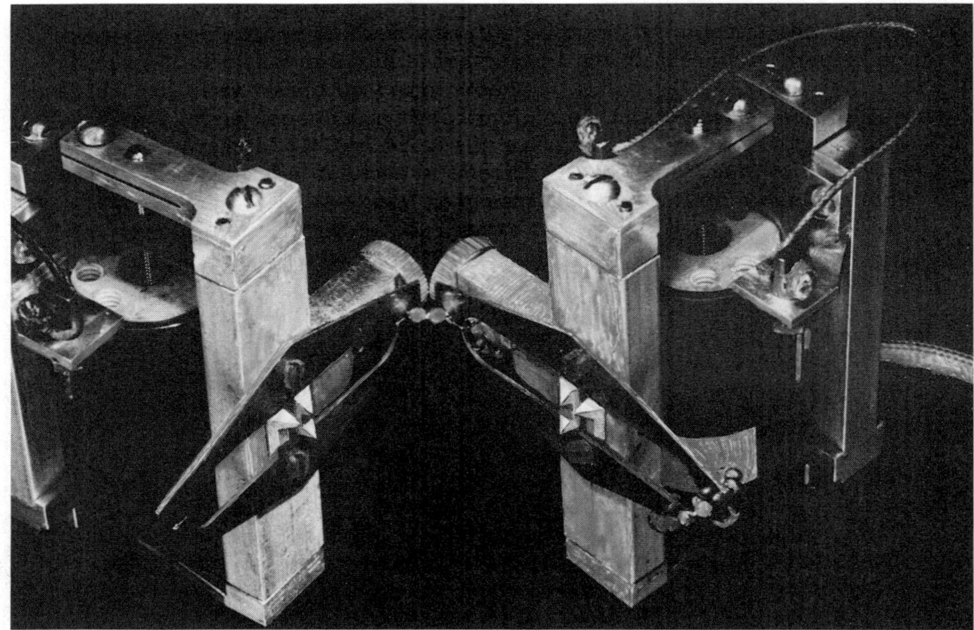

Figure 24   KGR-1 extensometers showing shape and arrangement of gripping points. (Courtesy American Cyanamid Corp.)

found in company laboratories. For economy, it is necessary to use these recorders, and this is readily done by feeding them amplified signals. Experience indicates that recorder charts should produce a 2-in. movement for each one thousandth of an inch movement of the specimen.

The second reason for a preamplifier is that it represents a very convenient technique for feeding the voltage from a specific LVDT to many different recorders set up for other LVDT characteristics.

The third reason for a preamplifier is that it can, and does, present gain controls for "fine-tuning" the accuracy of the strain signal. This is important because experience has shown that a very worthwhile reduction in standard deviation is achieved if this gain setting is calibrated fresh for each day's testing.

The calibration tool has been described and is pictured in Fig. 25. In essence the extensometer is driven by a large micrometer. The smallest division of the micrometer is 0.0001 in., making it quite feasible to obtain an adjusting of the preamplifier gain until a 1-mil movement on a micrometer will provide a repetitive 2-in. pen movement on

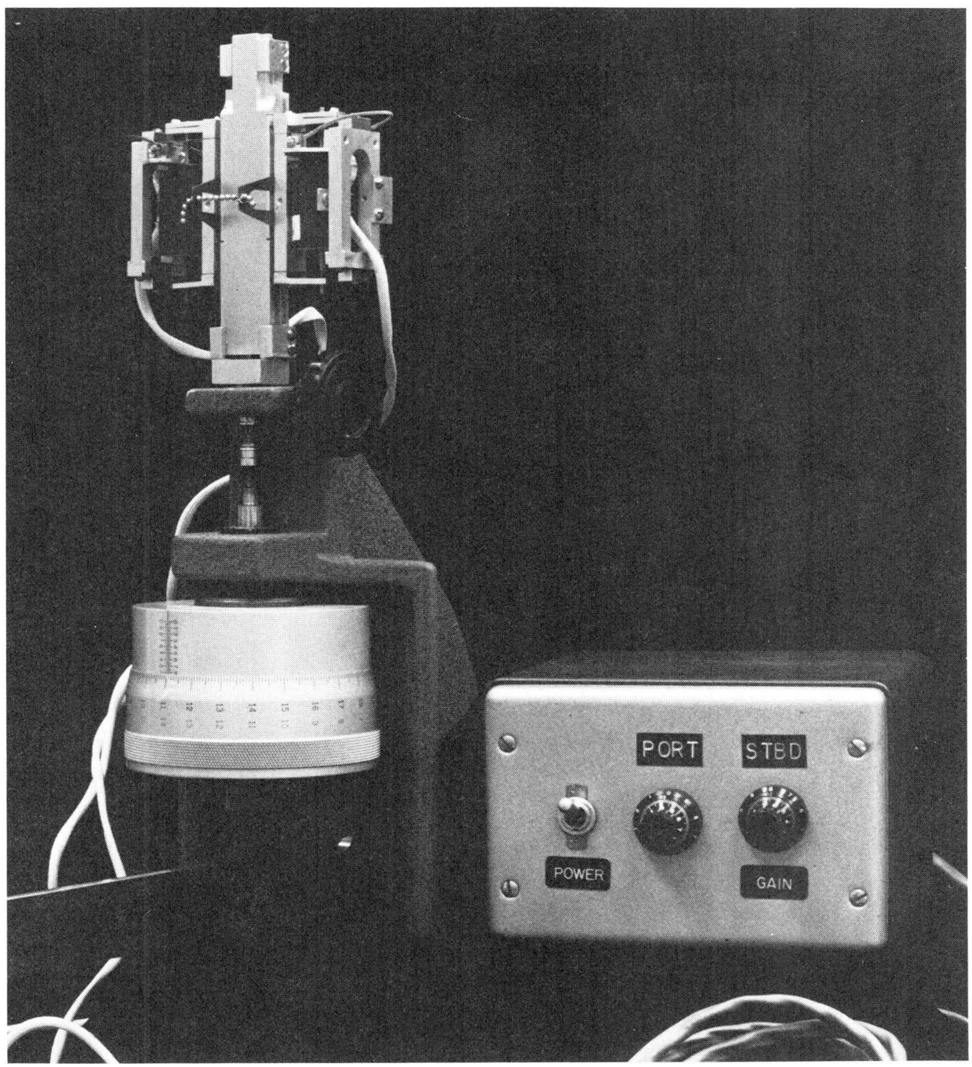

Figure 25 KGR-1 extensometers mounted on the calibration fixture together with the preamplifier. (Courtesy American Cyanamid Corp.)

the recording chart. The LVDT signal has been shown acceptably linear over the test range, particularly when overheating or shock damage to the glue line. The surface must be reasonably smooth for proper mounting point engagement.

As in all test specimens, elimination of signal errors is desired. To eliminate the adherend deflections a specimen is made by notching a solid 3/4-in.-thick bar to represent two 3/8-in. plates bonded together. The KGR-1 instrument is attached to this specimen and a deflection curve made for the test load cycle. The resulting deflection can be subtracted from the results of the bonded specimen test.

Test data for the same adhesive have been obtained from many specimens where adherend thicknesses and overlap distances were varied. In many designs, the adherends are relatively thin (0.063 in.). Lap shear fracture test with this thickness yield an ultimate stress of 4480 psi when using 0.5-in. overlap. Althof used 0.125-in.-thick adherends with 0.2 in. overlap and the ultimate fracture stress was 6400 psi. Krieger's tests with 0.376 in. thick adherends with 0.375 in. overlap gave virtually the same fracture stress levels (6400 psi) as those from Althof's tests but the shear strain was greater. The napkin-ring tests with an adherend thickness of 0.125 in. gave a fracture stress

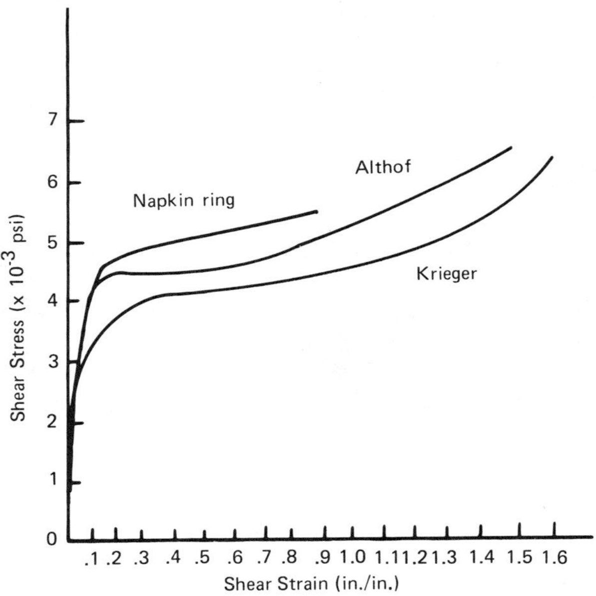

Figure 26   Comparison of stress-strain data.

level of 5350 psi, but the stress strain value at failure was about one-half that obtained from the Krieger tests. Figure 26 is a plot of the foregoing data. Chapter 13 will show how to incorporate these data into the strength analysis.

REFERENCES

1. Hughes, E. J. and J. L. Rutherford, Study of Micromechanical Properties of Adhesive Bonded Joints, Contract DAAA21-67-C-0500, Picatinny Arsenal, Aug. 1968.
2. Brown, N. and K. F. Lukens, *Acta Metall.*, Vol. 9, 1961, p. 106.
3. ASTM Designation D 2236-70, Standard Method of Test for Dynamic Mechanical Properties of Plastics, American Society for Testing and Materials, Philadelphia, 1973.
4. DIN 53 445, Prufung von Kunststoffen, Torsionsschwingungsversuch; Beuth Vertrieb, Cologne, West Germany, 1965.
5. Riekmann, in *Praktische Physik*, F. Kohlrausch ed., Teubner, Stuttgart, West Germany, 1955.
6. Timoshenko, S., *Vibration Problems in Engineering*, D. Van Nostrand, New York, 1955.
7. Timoshenko, S. and J. N. Goodier, *Theory of Elasticity*, MacGraw-Hill, New York, 1951.
8. Althof, W., et al., DLR FB 77/63, Environmental Effects on the Elastic-Plastic Properties of Adhesives (in German), Wissenschaftliches Berichtswesen der DFVLR, Cologne, West Germany, 1977.

10

Environmental-Durability Testing

J. ARTHUR MARCEAU *The Boeing Commercial Airplane Company, Seattle, Washington*

EDWARD W. THRALL* *Douglas Aircraft Company of the McDonnell Douglas Corporation, Long Beach, California*

I. INTRODUCTION

Every manufacturer of bonded assemblies has generated its own experience with surface treatments described in commercial or governmental specifications and selected an adhesive from lists of available materials that gave a successful assembly. When a new generation of adhesives appears or a new surface process is recommended, what is the manufacturer to do? The making of a strong durable bond will depend on the proper combination of surface treatment on the desired alloy, the adhesive primer to be applied to the metal surface oxide, and the adhesive. It is important to answer the question "is it suitable for practical use?" rather than the often used question "does it meet the requirements of the specification?"

Destructive testing of specimens must be carried out to indicate the quality of adhesive and pretreatment combination used to bond the test specimens, not the surface pretreatment alone. The type of test specimen can affect the results. A large number of test specimens (T-peel and lap shear) with the same pretreatment were tested after different environmental baths. The results showed the uniformity of peel strength values for a given adhesive in contrast to lap shear tests where shear results can vary over a large range for the same adhesive system. This says that peeling tests are most selective for measuring surface treatment quality differences. A series of Boeing tests has also confirmed that wedge crack tests, which also put adhesive in tension as do the peeling tests, will show up small differences in surface treatment quality.

Present affiliation: Consultant, Aeronautical Structures, Point Arena, California.

Comparative tests using alclad and nonclad aluminum alloy test specimens have shown interesting results. Using the same surface treatments but different adhesives for the T-peel specimens gave conflicting results. One adhesive test suggested that surface quality was good, while the other test suggested that it failed. This suggests that a material that is difficult to pretreat (2024-T3 alclad) should be tested with an adhesive that gives high peel strength. This type of adhesive will be most sensitive to surface defects.

II. WEDGE CRACK TEST

The wedge crack test method simulates in a qualitative manner the forces and effects on an adhesively bonded joint at the adhesive-adherend interface. It has proven to be highly reliable in determining and predicting the environmental durability of adherend surface preparations. The test has been used in controlling surface preparation operations and in screening surface preparations and primer and adhesive systems for durability. In addition to determining and comparing crack growth rates, the failure modes can be evaluated for more in-depth information. The method has proven to be correlatable with service performance as will be shown later in this chapter. The test specimen is a self-contained stressed specimen which can be placed in any environment desired without the need of sophisticated and/or expensive devices to maintain a stress on the specimen (Fig. 1).

The test method is based on the concept of fracture mechanics, but in a qualitative sense. Figure 2 shows the loading modes for bonded joints. The mode I component, tension in bond, is always at the load transfer edge. Also, the maximum shear load mode II is near the load transfer edge, but always preceded by mode I component [1—3], suggesting that crack propagation will be controlled by mode I fracture as in metals [4]. Therefore, a mode I only specimen was considered to be a practical approach. Much of the development work used thick-adherend double cantilever beam (DCB) specimens to study the behavior of bonded systems having a history of poor service performance. The specimen configurations are shown in Fig. 3.

Figure 1   Wedge crack test specimen. (Courtesy The Boeing Airplane Co.)

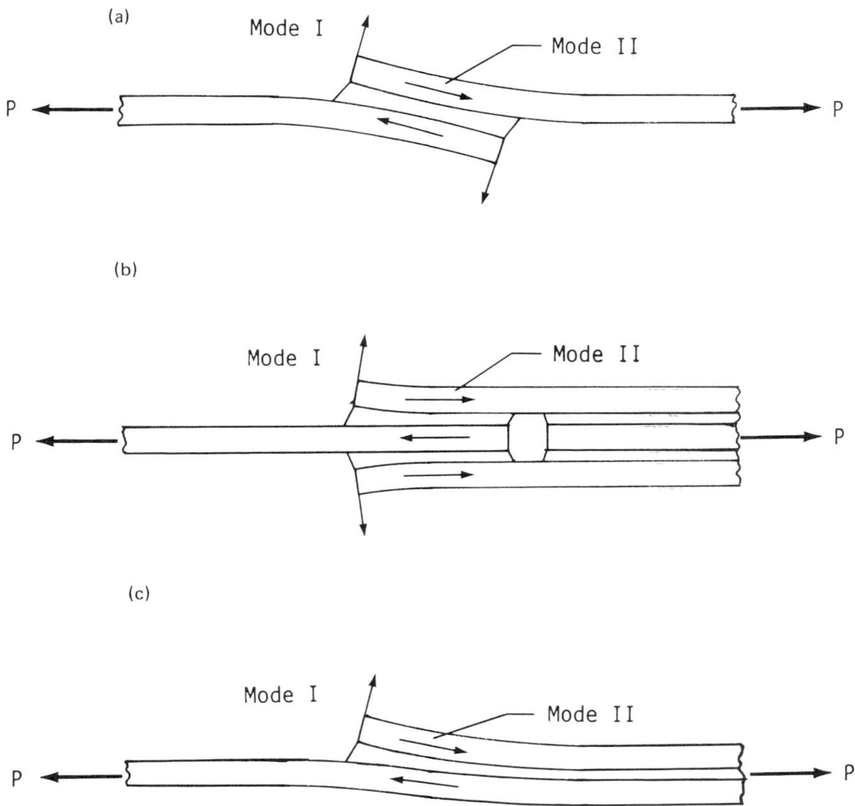

Figure 2   Examples of combined modes I and II loading: (a) simple lap shear; (b) double lap shear; (c) skin-doubler.

Studies using DCB specimens showed that all the characteristics of in-service bond failures were duplicated when poor or marginal aluminum surface preparations were used. From these studies, the simplified less expensive wedge crack test specimen configuration was developed. The basic purpose of the wedge crack test is to determine the quality of adherend surface preparation methods as related to in-service durability performance of adhesively bonded joints.

A minimum of five 25.4 mm × 152 mm (1 in. × 6 in.) specimens from a single assembly is recommended for each test (Fig. 4). The stress level at the crack tip must be the same for all specimens with the same adhesive system to justify comparison of results. This stress is controlled by the stiffness of the adherends, the thickness of the wedge, the length of the precrack, the distance the crack travels into the test

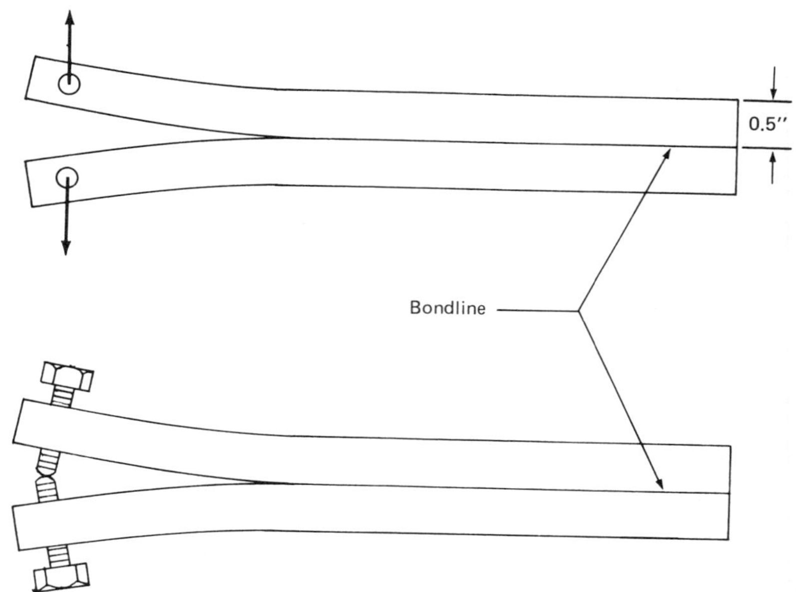

Figure 3   Examples of thick-adherend double cantilever beam (DCB) specimens: (top) specimen stressed externally; (bottom) specimen stressed with bolts (self-contained).

specimen, and the toughness of the adhesive. It is preferred that the test specimen adherends (beams) do not plastically deform during insertion of the wedge. Plastic deformation will reduce the stress at the crack tip.

The specimens may be exposed to any appropriate environment (e.g., humidity, heat, thermal shock, salt spray, etc.). However, for most applications water is the most deleterious environment for the polymer-adherend interface. A typical accelerated aging environment commonly used is 50°C (122°F) and condensing humidity. The time of exposure may be varied. Significant crack growth will usually occur within 1 hr if the joint is affected by the chosen environment.

The test procedure is as follows:

1. Prepare the surfaces of two pieces of aluminum sheet 152 mm × 152 mm × 3.2 mm (6 in. × 6 in. × 0.125 in) using the surface treatment process of interest for the test requirements.
2. Prime the faying surface of each panel (if a primer is required), apply the adhesive, assemble the panels, and cure the adhesive as required by the appropriate specification or manufacturer's recommendation. (For later convenience in inserting a wedge,

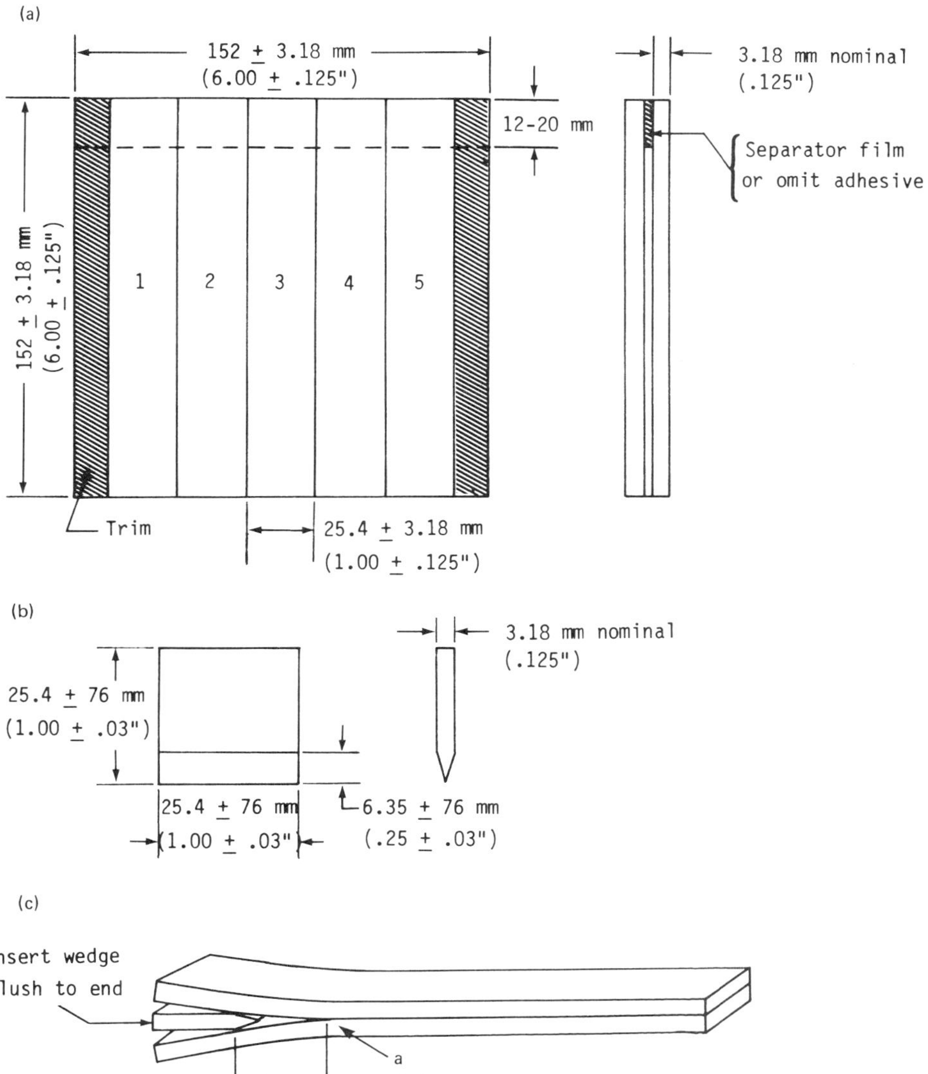

Figure 4 (a) Wedge crack test specimen configuration; (b) aluminum or stainless steel wedge; and (c) bonded assembly with wedge inserted.

a separation film may be inserted along one edge of the assembly shown in Fig. 4 or the adhesive may be omitted along one edge.)

3. Cut the test assembly into five 1-in.-wide test specimens as shown in Fig. 4. At least one cut edge of each specimen should have a smooth surface finish such that there will be no burrs or smearing of adherend material onto the bond line. Do not overheat or damage the bond when cutting or finishing the edge.

4. Mark all test specimens to identify them appropriately with respect to their processing in a manner such that testing in a high-humidity environment will not destroy the identification markings.

5. Crack one end of each test specimen by inserting a wedge as shown in Fig. 4. Insert the wedge by using a pushing force or several relatively light taps with a hammering device. When inserting the wedge, do not hold the specimen in any way that will restrain initial cracking. Position the wedge so that the blunt end and sides are approximately flush with those of the specimen. (in any use of an auxiliary tool to precrack a test specimen to facilitate insertion of a wedge, the extent of pre-cracking must be less than that which is caused by insertion of the wedge.)

6. After 1 hr at room temperature and using 5 to 30 power magnification, locate the tip of the crack, a, on the smooth edge of each specimen. This is the point farthest from the contact point of the wedge where the specimen (the adhesive, primer, and/or adherends) has separated. Mark the location on both sides of the adherend edge using a fine stylus or scribe. If the specimen is to be used in salt spray or other environment expected to be corrosive and liable to obliterate the mark, the scribe mark can be deepened with a triangular file.

7. Expose the wedged specimens to the environment of interest.

8. After 1 hr of test environment exposure, remove the specimens from the environment and within 15 min mark the location of the tip of the crack, Δa, in the same manner as above. Tests to date indicate that primary crack growth occurs within the first hour. Tests can be run for longer times as desired.

9. Measure the initial crack length a_0 and the crack extension Δa of each specimen to 0.254 mm (0.01 in).

10. At the conclusion of the test, forceably open the specimen and note the failure mode in the test section.

The crack extension Δa and the crack extension failure mode, that is, adhesive failure at the interface or cohesive within the adhesive, is reported. The initial crack length a_0, the crack extension Δa, and the crack extension failure mode are all a function of the adherend, surface treatment, and the adhesive-primer systems being considered. Because of these variables, the acceptance criteria for a bonded system of interest will have to be established. Examples of acceptable and unacceptable criterion for Forest Products Laboratory (FPL) etched aluminum adherends bonded with 121°C (250°F) curing high-peel modified epoxy structural adhesives are shown in Fig. 5.

The test method has had many practical uses over the years and a few examples will be described to give the reader an idea of how else the test can be used to solve a particular surface interaction problem.

One very important use the test had was to show a correlation between laboratory tests and in-service performance of adhesive-bonded aluminum joints. Whenever bonded aircraft components were returned to the manufacturer for evaluation of disbonds, test specimens were prepared from those areas which still had a sound bond.

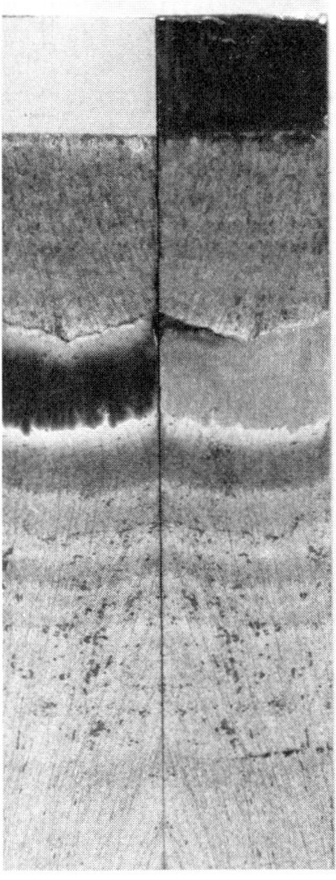 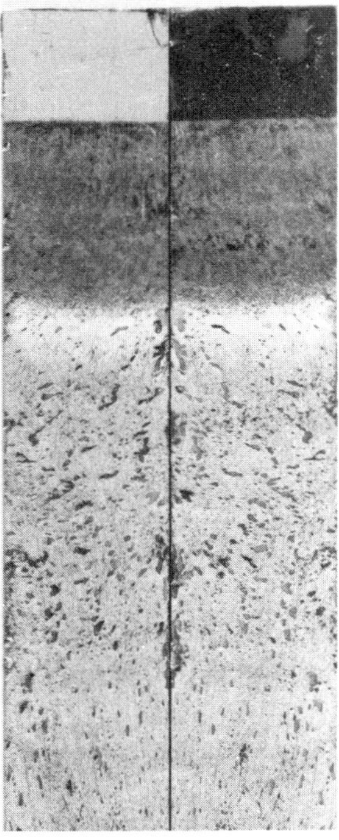

Figure 5   Examples of (left) acceptable and (right) unacceptable crack growth. Wedge crack test specimens opened after exposure to test environment.

Usually, the tests conducted were the traditional lap shear and peel tests, and sometimes an "on-assembly" shear test [5]. Invariably, these tests would demonstrate perfect bonds, whereas areas adjacent to them had already been delaminated under service conditions, proving that traditional tests could not predict in-service durability. When wedge crack specimens were prepared alongside the traditional specimens, they would always exhibit long crack growths and adhesive failure in the crack extension zone, duplicating the in-service failure modes. Table 1 presents lap shear, on-assembly shear, and wedge crack data for four aircraft components, two of which maintained good bonds in service and two of which exhibited disbonds.

One of the most extensive uses for the test method has been as a processes control test for the optimized FPL etch described in an earlier section. The Boeing Company has required the use of the test for all its subcontractors throughout the world to control prebond surface treatment. This has resulted in tens of thousands of data points demonstrating the improvements of the optimized FPL etch.

Perhaps the most value of the wedge crack tests is its use as a research and development (R&D) tool. It has been the primary tool used in defining parameters of the optimized FPL etch. phosphoric acid anodize, field repair surface treatments, titanium surface treatments, and stainless steel surface treatments, to name a few applications. When materials other than aluminum are used (e.g., other metals and nonmetals), consideration must be given to their thickness, modulus, and strength.

Table 1   Correlation of Laboratory Testing with Service Performance

Bonded assembly	Service disbond	Lap-Shear avg. (psi)	Porta shear[e] avg. (psi)	Modified wedge test a_0	Modified wedge test Δa[a]
A	Yes	4828[b]	4198[b]	1.66[b]	1.84[c]
B	Yes	4795[b]	4428[b]	1.10[b]	1.16[c]
C	No	4415[b]	4118[b]	0.48[d]	0
D	No	4425[b]	4762[b]	0.33[d]	0

[a] 1 hr exposure at 140°F (60°C)/100% RH.
[b] Mixed adhesive-cohesive failure modes.
[c] 100% adhesive failure mode.
[d] 100% cohesive failure mode.
[e] Porta Shear are on-assembly shear results.

III. TESTS FOR ADHESIVE SELECTION

When an adhesive system must be selected for a specific design, a large
battery of tests must be considered. The tests to be performed should
include the mechanical properties (Chapter 9), evaluation of environ-
mental-durability resistance, and verification of the processing tolerance allowables put forth by the manufacturer. Also, there is a need
to verify the adhesive performance, strength, and durability when exposed to other variables that could be expected while fabricating the
production article. These variables include excess atmospheric humidity during lay-up, time out of refrigeration, nonstandard heating rates
during cure cycle, and so on. Many new adhesives recommend the application of a primer over the anodized surface before the parts are
bonded. Several variables can occur here and must be tested to establish manufacturing tolerances such as primer thickness, contamination
limits, and solvent tolerances in the primer mix. Detrimental results
have arisen when a manufacturer was forced to purchase an alternate
solvent because of a labor strike. The substitute solvent was purchased
to the same specification and new batches of primer were manufactured
with resulting adhesive failures of the bonded joint in service. Extreme
caution must be the byword of the team building bonded components.

There are many tests to be considered for the metal-to-metal and
honeycomb bonded assemblies. These include lap shear, double lap
shear, T-peel, creep, climbing drum, flatwise tension, beam shear,
thick adherend (fatigue and shear modulus), napkin ring (shear and
tension modulus), torsion pendulum (shear modulus), wedge crack,
and double cantilever beam (see Figs. 3, 4, and 6). Most of these tests
should be conducted for the temperature range that bonded part will
see in service, for example, the maximum low temperature where the
adhesive is brittle, a test point at room temperature, and then the highest temperature plus a margin where the adhesive is soft and may creep.

As stated elsewhere, it has been found that most adhesives are
degraded by 100% relative humidity at an elevated temperature around
140°F (60°C). Another devastating condition is salt-laden sea air.
This condition can be obtained in an environmental test box or by having a test site on an ocean beach where condensing moisture conditions
can exist almost once a day. A partial list of environments includes
fuel, hydraulic fluid, temperature variations, humidity, salts, cleaning
compounds, fire extinguisher compounds, and deicer compounds. Further consideration must be given to the loading cycles that will be seen
by the structure during its lifetime. The loading can be shock, high
cycle rate, low cycle rate, or steady.

Bonded joints in service may be subjected to cyclic loads which can
vary from zero and up, or from a compression state up to a maximum
tension stress. Since bonds are primarily in shear, these loads will
determine if a load reversal exists or just one that goes from zero up to
a maximum and then back to zero.

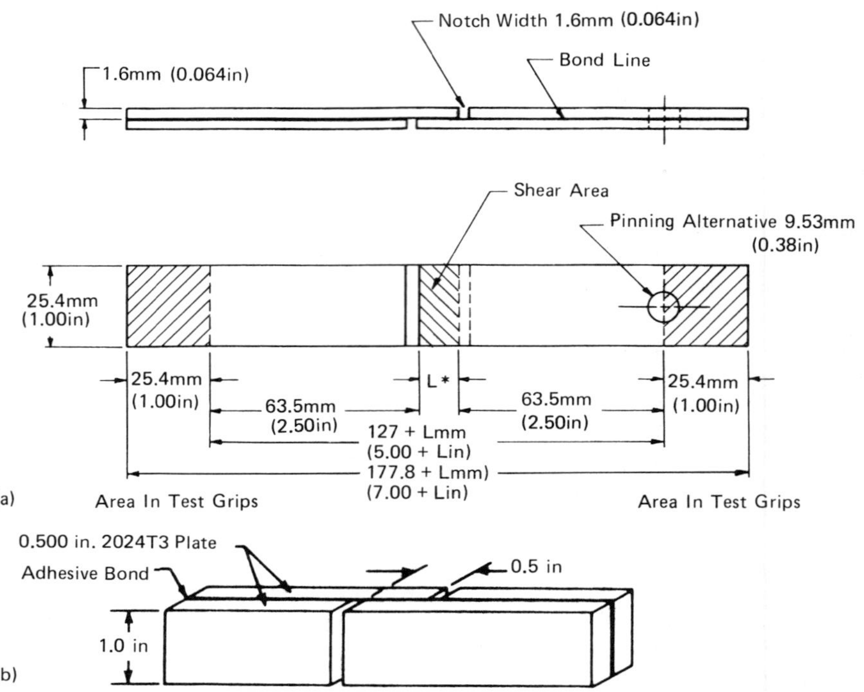

Figure 6   Test specimens. (a) Lap shear specimen. L* = length of test area. Recommended lap length is 0.50 ± 0.01 in. (12.1 ± 0.3 mm). (b) Thick-adherend specimen. (c) RAAB specimen. (d) Standard peel test for metal-to-metal bonding adhesives: (top) peel specimen; (bottom) results of the test, showing the average peel strength. (e) Sandwich beam shear test. S = specimen width, T = specimen thickness, T_F = facing thickness. End support plates are 1 × 3 × 0.250, with grooves for alignment. Loaded edges are rounded to 0.05 radius. Load bars are 0.500 round. End support plates should be used under round load bars. Tolerances as indicated, or ±0.020. (f) Creep test specimen. All dimensions in inches. (g) Bell peel specimen. Specimen width may be 1 or 1½. Test value is expressed in pounds per 1-in. width. All dimensions in inches. (h) Flatwise tension specimen. All dimensions in inches. (i) Sandwich peel specimen. All dimensions in inches. (j) Sandwich shear specimen. All dimensions in inches.

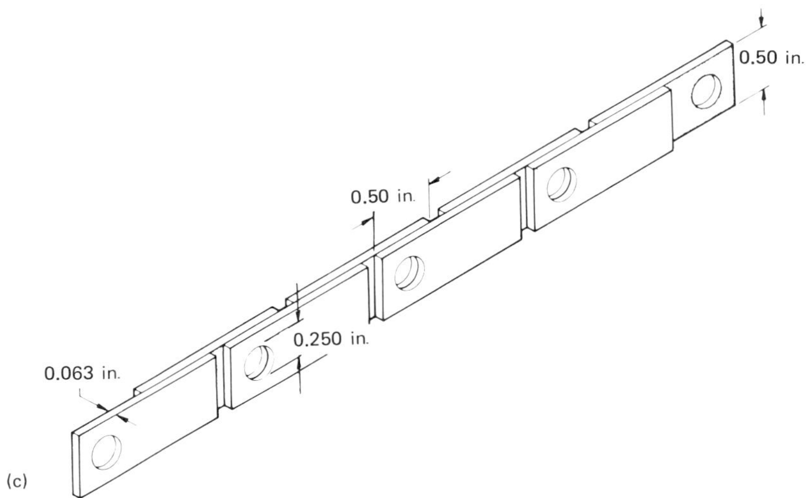

Figure 6   (continued)

Simple lap shear specimens have long been used for these environmental tests because they are easy to make and load. Weights can be used for constant loads, for example. Also, it has been a practice to soak the specimen in the test environment for long periods (up to 120 days) so that the adhesive would get wet throughout the joint. Then the specimen can be static loaded to failure, held at a constant load less than failure, or put in a cycle loading environment. It should be noted that the "lap" specimens usually have only a 1/2-in. overlap of the metal at the joint. This is much less than the minimum that would be used with even the thinnest metals being joined in a production article. For these tests the failure should be in the adhesive and not the metal, while in the production part, the adhesive will not fail and the metal is the limiting strength member. Chapter 13 will explain this in great detail.

Modern-day adhesives have shown a greater resistance to water intrusion, so it has been nearly impossible to get the adhesive in a lap shear bonded joint (0.5 × 1.0 in.) wet. Therefore, a special lap specimen was developed. The RAAB specimen (Fig. 6) was developed by the American Cyanamid Company. It consists of a blister detection type of lap shear joint of 0.5-in. overlap and 0.5 in. wide. A 0.25-in.-diameter hole is drilled into the center of the overlap. This provides a specimen with approximately 0.2 in.^{2} of bond area and 2.75 linear inches of exposed glue line and a very short penetration distance. A standard lap specimen 0.5 × 1.0 has 0.50 in.^{2} of bond area and 3 linear

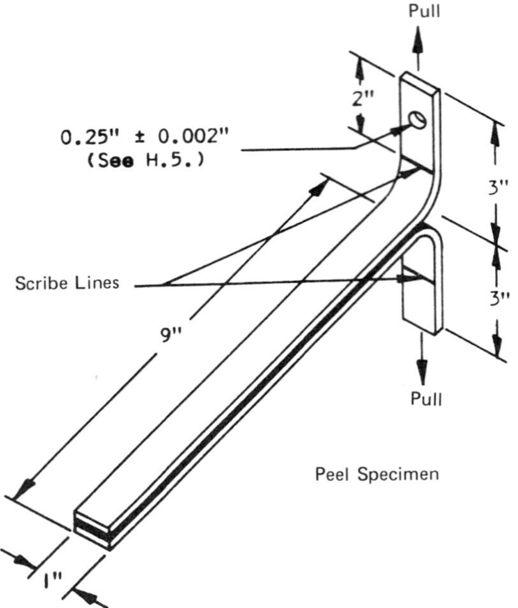

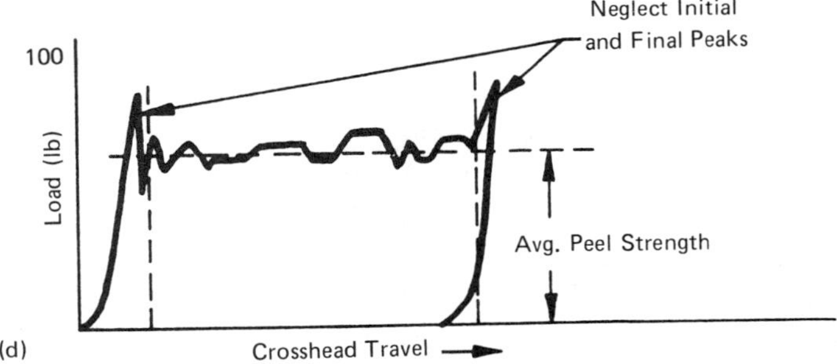

Figure 6   (continued)

inches of exposed glue line and more than four times the penetration
distance. These RAAB specimens can give good comparative data for
evaluating adhesives. A string of these specimens are usually made
(see Fig. 6) and when a failure occurs at one joint a bolt is inserted in
the 0.25-in.[2] hole and testing continued until all joints have failed.

Cycle load tests are very important. Also, the cycle rate has an
effect on cycle life. On one program, a series of cycle tests showed
some astonishing results. Thick-adherend (1.0 in. with 0.5 in. overlap)

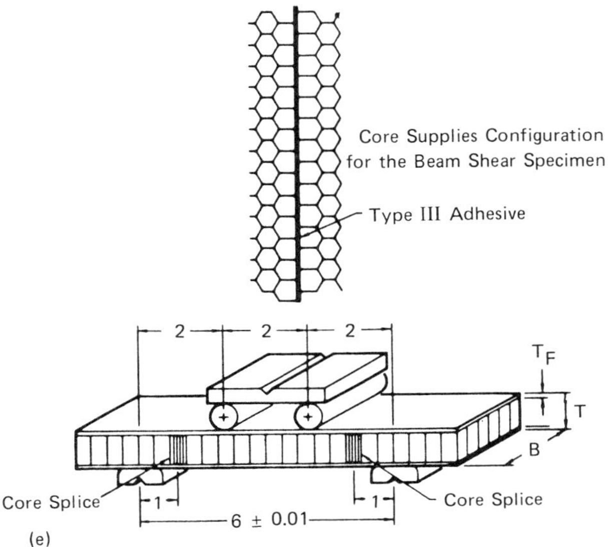

Figure 6 (continued)

lapshear specimens were bonded and placed in 140°F and 100% relative humidity environment for 69 days. They were then cycle tested at 30 Hz at room temperature and −50°F. Tests were stopped after 10^6 cycles and no failures. Tests were run at stress ratios of +0.2 and −0.2 and a peak stress of 30% ultimate. An alarming thing happened when the same specimens were run at a load cycle of 1 hr load on and 15 min at no load. The specimens were subjected to the same environment as the fast-cycle specimens, but the endurance was only 800 cycles. It should also be noted that these same specimens were loaded with a constant load and were still intact after 3000 hr.

Another example of how cycle tests can show weaknesses in adhesives is that during the course of a year's program, some RAAB specimens in a cycle test suddenly gave short lives when compared with the other tests. An investigation showed that moisture may have gotten into the adhesive during layup before the cure cycle. Strangely, though, the same moist adhesive showed no loss in strength for the standard lap shear specimen. These lap shear specimens are often cured along side production parts and tested afterward to give assurance that the cure cycle was correct and proper shear strength is assumed.

It has been determined that loading rate and test temperatures can affect the shear strain and shear strength. Figure 7 shows the shear

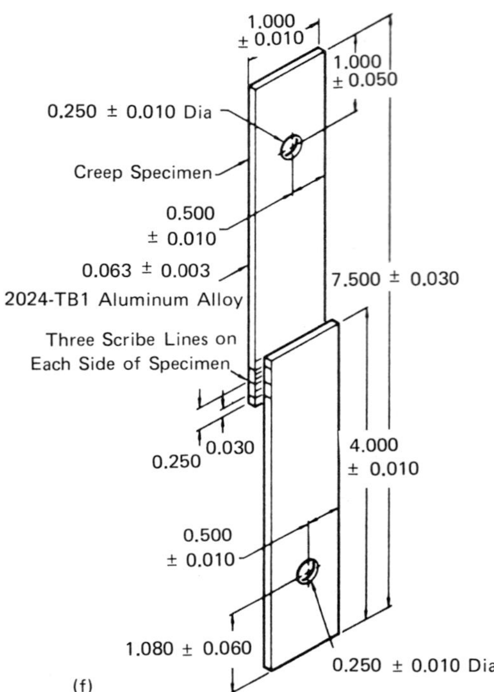

Figure 6   (continued)

stress-strain curves for a thick-adherend specimen run at three loading rates and two temperatures. The fast rate gives a higher failing
stress and a lower strain when compared to the slower-rate test specimens. Therefore, it is important to know the end use of the bonded
assembly so that the test allowables are arrived at with suitable loading
rates.

There is another interesting failure phenomenon that can exist in a
test specimen. If the lap joint is long enough so that the adhesive joint
will not fail first, it has been observed that when slowly loaded, say 10
min to go from zero to ultimate, the adherend will yield at the top of
the bond and as the load increases, the Poisson effect on the metal will
cause the adhesive to fail slowly at a lower strength than would be the
case where no metal yielding occurred.

Seldom are bonded joints in tension. That is, the primary load
across the joint is in tension. Figure 8 is an example where such a

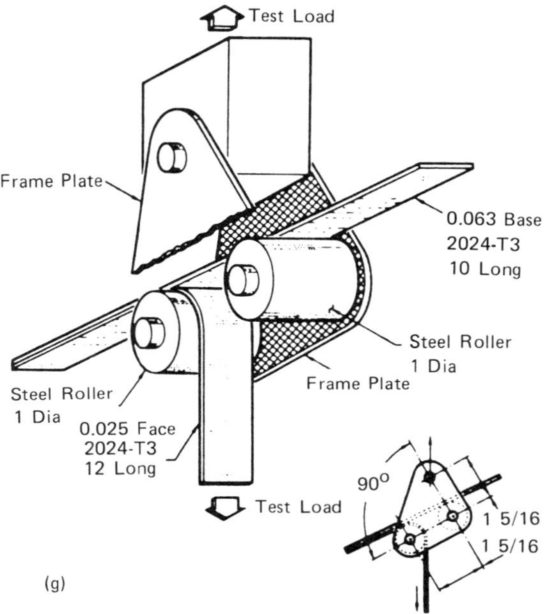

Figure 6 (continued)

joint results from the structural arrangement. The structure pictured
is a fusealge frame-to-skin joint of a pressurized aircraft. Cabin pressure is resisted by the fuselage skin, which resists the load by hoop
tension. Since the fuselage frame resists very little of the pressure
load, the skin wants to pull away from the frame and puts tension on
the bonded joint. Since adhesives are known to be weak under this
loading, a special test must be developed to assure adequate strength
exists. If a special test is developed, the deflections of the joined
production parts must be duplicated in test as close as possible. Also,
T-peel and flatwise tension tests can give clues when comparative tests
on different adhesives are tested.

Test matrix are very helpful when trying to evaluate several variables at a time. Consider, for example, adhesive out time and heat-up
rates. Three or more out times can be tested for two or more heat-up
rates. Tests have shown that some adhesives really deteriorate in
strength after a total of 2 months out time, while others will bubble if
the heat-up temperature rise exceeds 7°F per minute.

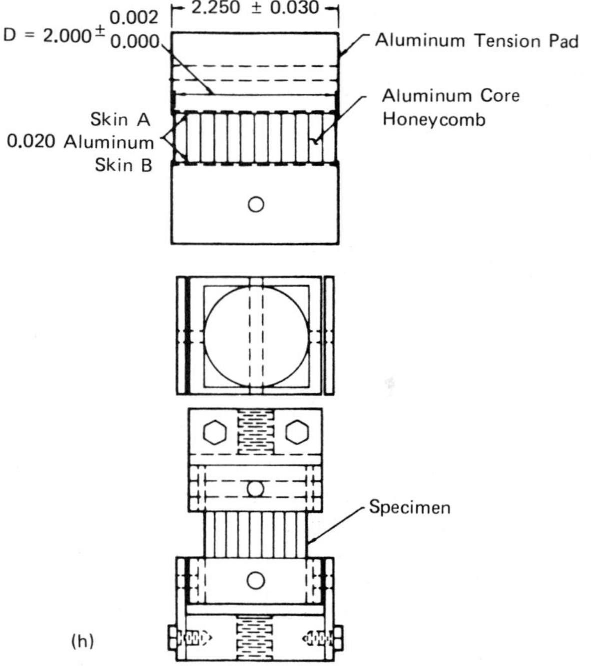

Figure 6   (continued)

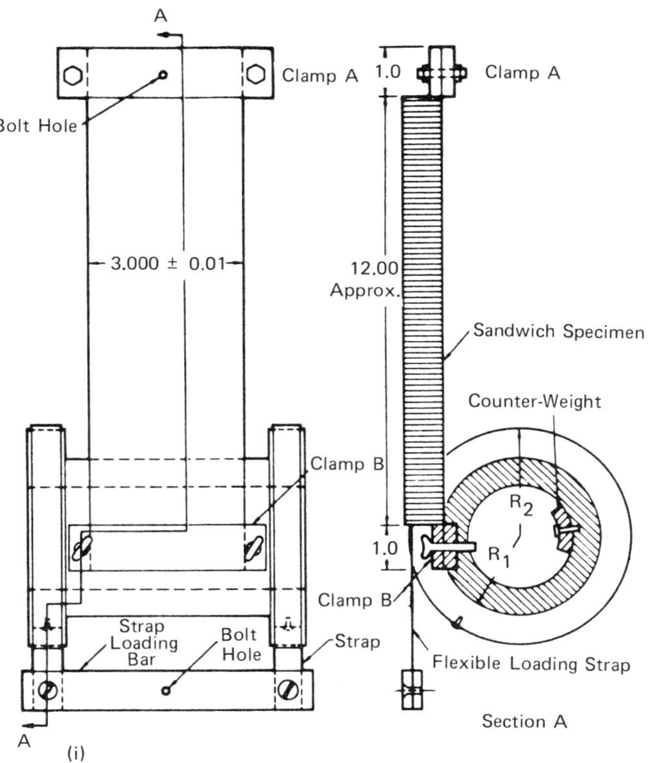

Figure 6 (continued)

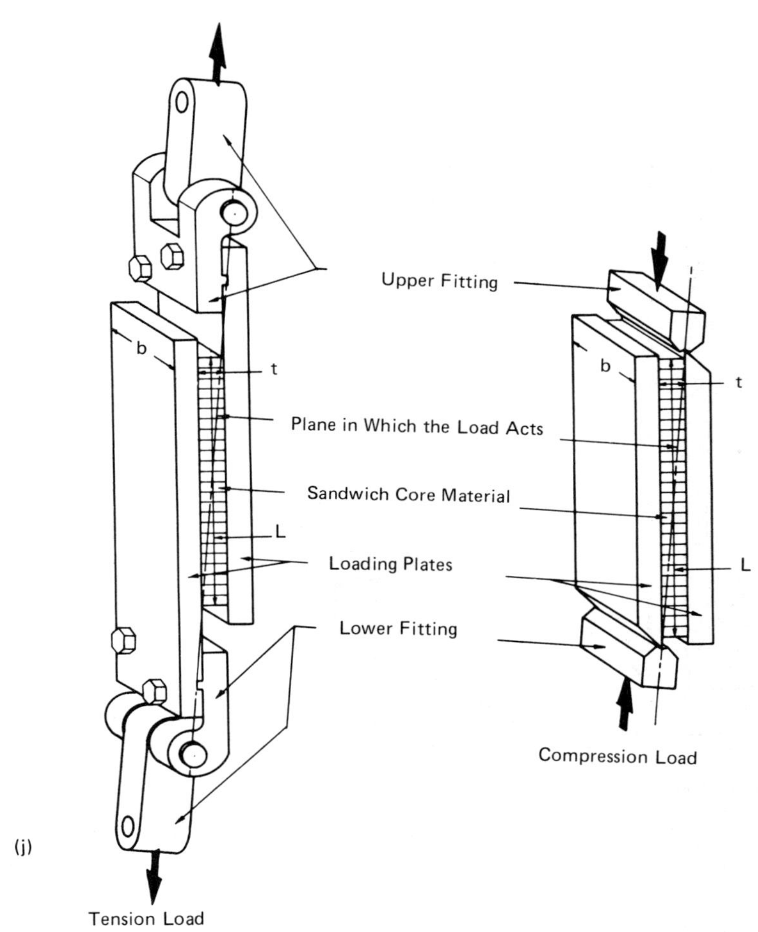

Figure 6 (continued)

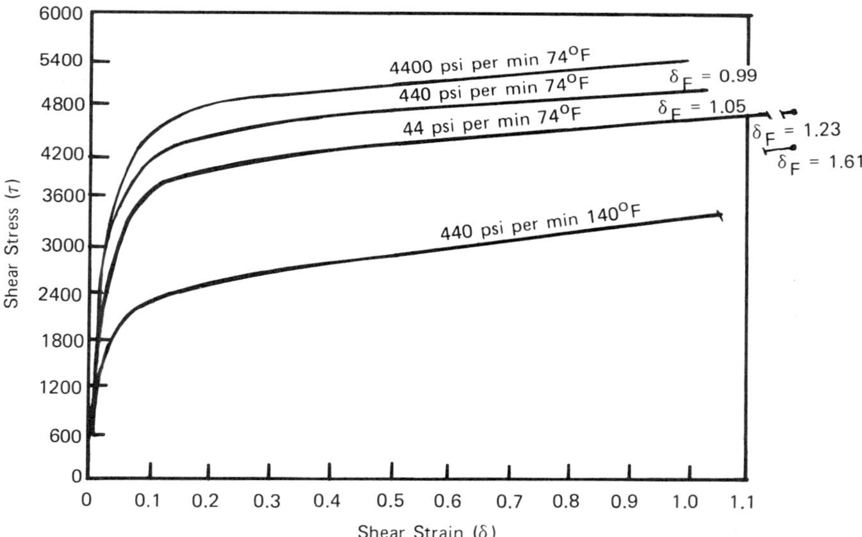

Figure 7   Effect of loading rate on stress-strain values.

Marceau and Thrall

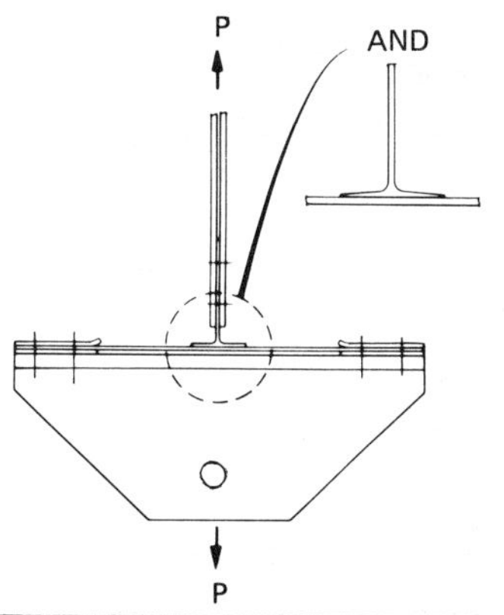

| Skin | Adhesive | Primer | Failure load test temp | | | Design load |
			−50 ± 5°F	R.T.	140 ± 5°F	−50°F
0.090 7075-T6	FM 73	BR 127	1740 lb			249 lb
0.040 7075-T6	FM 73	BR 127	1595 lb			389 lb
0.090 7075-T6	AF 55	XA 3950			4000 lb	249 lb
0.090 7075-T6	M 1133	BR 127	2170 lb			249 lb
0.090 7075-T6	AF 55	XA3950	5910 lb	5050 lb	5057 lb	249 lb
0.090 7075-T6	M 1133	BR 127	2640 lb	4650 lb		249 lb
0.040 7075-T6	AF 55	XA 3950	1670 lb	3700 lb	4220 lb	389 lb
0.040 7075-T6	M 1133	BR 127	2105 lb	3275 lb	3358 lb	389 lb

Figure 8   Tension tee test.

IV. CONCLUSIONS

It is possible to perform a series of tests on a new surface treatment and/or a new adhesive to determine the in-service durability and strength. Specific test specimens must be used to verify that the production processing is giving the proper durability and strength results for each production run. An excellent reference that summarizes environmental testing is Shannon et al. [6].

REFERENCES

1. Jemian, W. A., and M. B. Ventrice, The Fracture Toughness of Adhesive-Bonded Joints, *J. Adhes.*, Vol. 1, July 1969, pp. 190–207.
2. Jemian, W. A., and R. C. Wilcox, Study of the Onset of Permanent Deformation in Structural Bonded Joints, AD736875, Jan. 1972, Contract DAAH01-70-C-1425.
3. Krieger, R. B., Stress Analysis of Metal-to-Metal Bonds in Hostile Environment, *Adhes. Age*, Vol. 21, No. 6, June 1978.
4. Wilson, W. K., W. G. Clark, and E. T. Wessel, Fracture Mechanics Technology for Combined Loading and Low-to-Intermediate Strength Metals, U.S. Army Contract DAAE 07-67-C-4021, Nov. Nov. 18, 1968.
5. Marceau, J. A., Y. Moji, and J. C. McMillan, A Wedge Test for Evaluating Adhesive-Bonded Surface Durability, *Adhes. Age*, Vol. 20, No. 10, Oct. 1977.
6. Shannon, R. W. et al., General Material Property Data, U.S. Air Force Flight Dynamics Lab. Tech. Rep. AFFDL-TR-77-107, Sept. 1978.

11

Chemical Analysis for Control

DEBORAH K. HADAD *Lockheed Missiles & Space Company, Inc.,*
Sunnyvale, California

I. INTRODUCTION

A. Background

Adhesives are complex mixtures of resins, curing agents, accelerators, various modifiers, and some form of carrier material. Until recent years the actual chemical composition of these materials has been unknown not only due to the proprietary nature imposed by manufacturers but also because complete analysis of a newly developed or completely unknown material is difficult and requires the successful use and data correlation of many chemical and physiochemical techniques.

Advances in our knowledge of the chemical composition of polymer systems used in the aerospace and aircraft industries are largely due to new or improved instrumental testing methods. These methods are fast, highly sensitive, accurate, often allow the sample from one test to be recovered for other studies, can be applied directly to difficult samples (films, dispersions, rubbers), and also provide information on the physical behavior of a material which is not obtainable by purely chemical methods.

The most basic requirement for producing high-quality, reliable finished parts is to start with controlled and consistent materials. Years of qualification tests are run on adhesives as well as other resin matrices to establish a statistical base for design analysis and long-term environmental stability. However, most material acceptance requirements have been based on simple but time-consuming performance tests such as mechanical properties of test parts, gel times, and flow tests for lot-to-lot comparisons. Although these methods were generally satisfactory for the early users of advanced composite materials, they cannot always detect chemical changes which can affect long-term

performance. With the current widespread use of adhesives in critical
applications such as primary aircraft structures where failure can lead
to loss of huge investments and, more important, loss of human lives,
it becomes imperative to establish rigid test criteria specific to the
actual chemical composition and reaction condition of the resin system.
It is this resin chemistry that ultimately determines the processability
of the material and long-term durability of the finished parts.

B. Technical Approach

To establish physiochemical quality assurance procedures for a material,
the chemical composition of the formulation must be known. As previously mentioned, these data are not always readily available. Several
approaches for establishing quality control on advanced composite resin
systems are available in the literature [1–3]. Briefly summarized, this
involves first an initial qualitative chemical analysis followed by an extensive quantitative analysis for each component in the resin matrix.
Obviously, it is not feasible to perform a complete quantitative analysis
on every lot of incoming material. This type of information obtained on
an acceptable and successful material forms the base on which the selection of test methods suitable for routine, cost-effective quality control
is made.

For most adhesives the chemical starting material have been partially reacted to improve their handleability and processability in part manufacture. Since this reaction advancement, or B staging, affects the
quality of a finished part, quality control tests must also be developed
which define the degree of resin reaction.

Once a baseline has been established, those techniques chosen for
quality control can be used to assure the fabricator that no significant
chemical changes have been made on purchased materials over the lifetime of that product. The expenditures spent on establishing this data
base will be more than recovered by eliminating bad material before it
is used to produce expensive parts.

The cooperation of individual suppliers with the fabricator in establishing chemical controls is the ultimate goal in establishing any
quality control requirements. Not only will the expertise and knowledge provided by the supplier's resin chemists assist the fabricator's
chemists in determining meaningful tests, but it is to the vendor's advantage to participate in test development which affects their proprietary resin formulation as well as being the basis for acceptance or
rejection of their material.

It is also desirable that physiochemical test procedures become
standardized similar to those ASTM methods already available for
mechanical tests. An example of a cooperative effort by composite
users and suppliers to develop meaningful standard test methods is a
compilation of chemical quality assurance test procedures developed for
materials in advanced composite preimpregnated (prepreg) materials [4].

C. Evaluating Techniques

Instrumental techniques usually associated with physiochemical characterization are infrared spectroscopy, liquid chromatography, thin-layer chromatography, thermal analysis, and dynamic dielectric analysis. Carbon-13 and proton nuclear magnetic resonance (NMR) spectroscopy as well as some wet chemical test methods are also used. These techniques offer the capability of rapid qualitative fingerprinting as well as quantitative information on complex formulations. Some characteristics obtainable by these methods include identification of chemical functional groups, presence and amounts of individual components, degree of cure advancement, aging of a material, and the affect or processing on the chemistry of these materials.

The quality assurance techniques ultimately chosen for a particular resin matrix will vary and it is possible that several of the foregoing procedures will not be used. However, all the methods have value either singularly or coupled with another method. Certainly, no one technique will solve the quality assurance problem of the complex materials, but by utilizing well-established analytical methods and carefully evaluating their applicability to a specific material and its end use, cost- and time-effective quality assurance programs can be established, resulting in higher-reliability products.

D. Determining Limits

To set quality acceptance limits on resins utilizing physiochemical test procedures involves investigating the effect of variations in resin chemistry on the matrix-critical mechanical properties of cured parts as well as effects on long-term durability. Multiple-batch testing must be conducted to provide a statistically sound basis for setting compositional limits to control chemical uniformity.

Selected quality control methods should be used for testing on several batches of materials representing the suppliers' standard production material to obtain an allowable "spread." Special batches with known compositional variations should then be tested to help establish the sensitivity and accuracy of each method. An excellent study on one commercial product illustrated the successful use of chemical characterization techniques for just such an evaluation [5].

II. INFRARED SPECTROSCOPY

A. General Method

The infrared absorption spectrum of an organic material represents one of its truly unique chemical properties. Except for the case of optical isomers, no two compounds have identical spectra. Because of this characteristic infrared spectroscopy is used most widely for identification of compounds and as a guide for determining molecular structure.

The infrared region ranges from approximately 13,000 to 33 wave numbers (cm^{-1}). However, the majority of applications are limited to the region from about 4000 to 600 cm^{-1}. Briefly, the technique involves passing a continuous beam of electromagnetic radiation through a sample. The samples may be solid, liquid, or gas. The chemical composition of the sample determines how much radiation is absorbed and how much is passed through the sample unchanged. After the beam leaves the sample it is dispersed (i.e., separated into its spectrum) by means of a prism or grating. This device is rotated so that each individual wavelength is focused on a detector. The energy of the sample beam is simultaneously measured and electronically compared at each individual wavelength with the energy of an identical beam that has passed through a reference. Usually, the reference is air, however, for specific applications it can be a variety of materials. If the energy of the sample and reference beams are the same, no absorption occurs and 100% of the radiation is transmitted. If specific wavelength energy is absorbed by the sample, an energy difference exists between the two beams. This phenomenon leads to the sometimes complex curve known as the infrared spectrum illustrated in Fig. 1.

It is possible to identify a material by its infrared spectrum because each molecule has characteristic absorption bands. In addition, certain groupings of atoms within a molecule also give rise to particular absorptions which are essentially independent of the rest of the molecule. These groupings of atoms include all the common organic functional groups (C=O, C=C, C—H, CH_3, C≡C) and many inorganic ones (—OH, N—H). Therefore, the procedure used to determine an unknown is to obtain its spectrum, identify the specific functional groups present, determine what types of molecules contain these groups, and

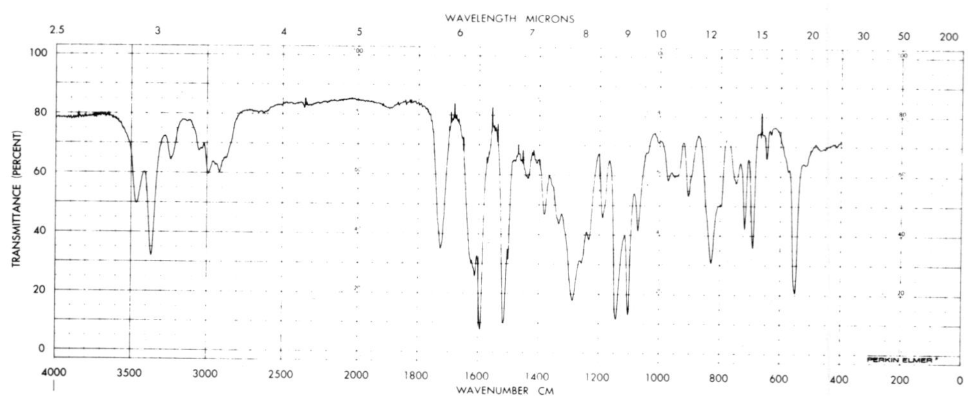

Figure 1   Typical infrared spectrum.

match the total spectrum of the unknown material with a reference spectrum of a known material. This sounds simple and straightforward. However, a multitude of molecular and atomic interactions lead to uncertainties as to the true identification of chemical groups, but it is just this effect that provides the unique features of the infrared spectrum that make it such an excellent "fingerprint" for materials.

B. Prisms and Gratings

Both prisms and gratings are used for dispersing infrared radiation, although the general use of gratings predominates. Several materials have been used to make prisms and a list of these materials together with some of their characteristics is given in Table 1. All the prism materials are easily scratched and many are water soluble, thus requiring protection from moisture condensation with desiccants or with heat. For this reason grating spectrophotometers are more popular. Reflection gratings are advantageous for dispersing in the infrared region because they possess better resolving power due to minimal loss of radiant energy. They also have a more linear dispersion than prisms and are water resistant. An infrared grating is usually made from aluminum-coated glass or a plastic replica.

C. Sample Handling

Several modes of analysis are available using an infrared spectrophotometer. One of the most common approaches to analyzing a material involves placing a sample on the surface of a crystal (NaCl, KBr, KRS-5) and directing a beam of radiation through it and perpendicular to the

Table 1 Prism Materials

Material	Use range	Characteristic
Quartz	Near IR; 12,500–3300 cm^{-1}	Absorbs strongly past 2500 cm^{-1}
Sodium chloride	5000–667 cm^{-1}	Most common material used; affected by H_2O
Potassium bromide	Far IR; 675–250 cm^{-1}	Affected by H_2O
Lithium fluoride	Near IR; 10,000–2000 cm^{-1}	Not affected by H_2O
Cessium bromide	Far IR; 675–250 cm^{-1}	

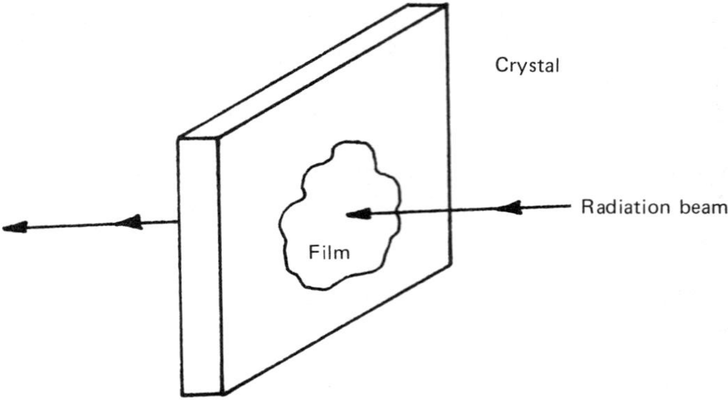

Figure 2   Schematic of infrared transmission mode.

surface (see Fig. 2). This is called the transmission mode and is the
simplest method for neat resins and films. Transmission can also be
used for liquid solutions by utilizing sealed liquid cells made from similar crystalline materials.

    Another useful technique, although one not as widely used, involves
multiple internal reflection (MIR), where the radiation beam is reflected
many times within the crystal (see Fig. 3). A sample placed on the
crystal surface will absorb energy at its characteristic wavelengths.
Polished or flat surfaces on rigid samples may be analyzed as well as
materials that may contain undissolved particles, such as fillers or
catalysts which will scatter radiation on conventional transmission cells.
Also, when only a very small amount of sample is available, as in the
case of liquid chromatographic fractions, MIR multiplies the absorbances
so that an acceptable spectrum can be obtained.

D. Fourier Transform Infrared (FTIR) Spectroscopy

One of the most significant technological advancements in the field of
infrared spectroscopy has been the introduction of interferometer-
equipped and digitally computerized infrared spectrophotometers.

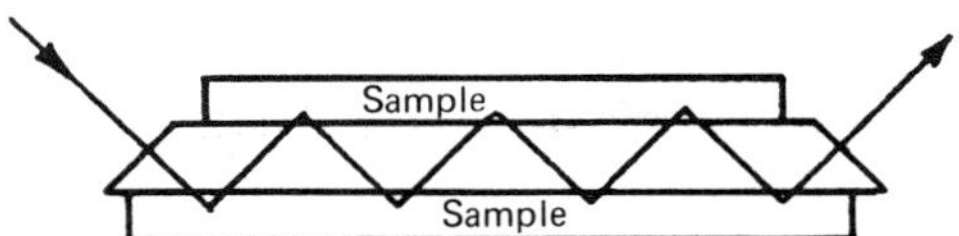

Figure 3   Schematic of infrared multiple internal reflectance mode.

Rapid analysis times coupled with high resolution are two of the advantages of FTIR over conventional infrared analysis. FTIR instruments do not require energy-wasting slits to isolate individual frequencies, thus improving the sample signal. All radiation frequencies are incident to the sample throughout the scan and the resulting signal, called an interferogram, is a plot of intensity as a function of time. The final infrared spectrum is obtained by completing a Fourier transformation of the interferogram by a digital computer. The presence of a computer has added advantages in that it automatically provides the capability for data handling, such as scale expansion, spectral comparisons, and spectral additions and subtractions.

The FTIR is ideally suited for MIR spectroscopy. The high-energy throughput and signal-averaging advantages have made MIR a valuable everyday tool where it once was used only occasionally. FTIR is being used today for studying epoxy composite weathering as well as following the cure and postcure kinetics of epoxy and polyimide materials [6,7].

E. Qualitative and Quantitative Uses for Quality Control

The power of IR analysis for identification of materials has been discussed. Because of the unique characteristics of a material's infrared spectrum, it can be used as a standard for comparison with other samples. Qualitatively, a visual or computer comparison of spectral curves between a sample and an approved "standard" is the basis for many quality control test requirements. By simple overlaying it can be determined whether two materials are similar or dissimilar. An example of this type of comparison is shown in Fig. 4. Here two materials having basically the same formulation are compared. It is quite evident where an unapproved additive has been introduced. This material would subsequently be rejected following incoming purchase specification tests.

Up until now our concern has been "what" our sample has been comprised of, not "how much" of each ingredient is present. Qualitative identification depends on the position of absorption bands, whereas quantitative analysis depends on the intensity of these bands. Since concentration is directly proportional to absorbance, with the proper choice of absorption bands and instrument parameters accurate quantitative analyses can be performed. For example, infrared analysis has proven successful as a means for determining the concentration of dicyandiamide (Dicy), a common curing agent used in adhesives. Known amounts of Dicy are dissolved in a mixture of 80:20 tetrahydrofuran:methanol and a calibration curve obtained using a liquid cell technique by measuring the change in transmittance of the 2190-cm^{-1} band due to the $C \equiv N$ stretching of Dicy (Fig. 5). A calibration curve is shown in Fig. 6. An actual adhesive sample would be prepared in a

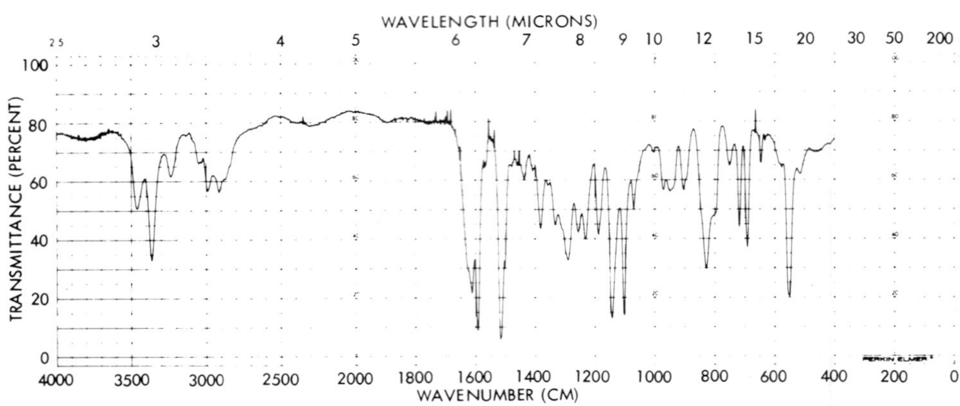

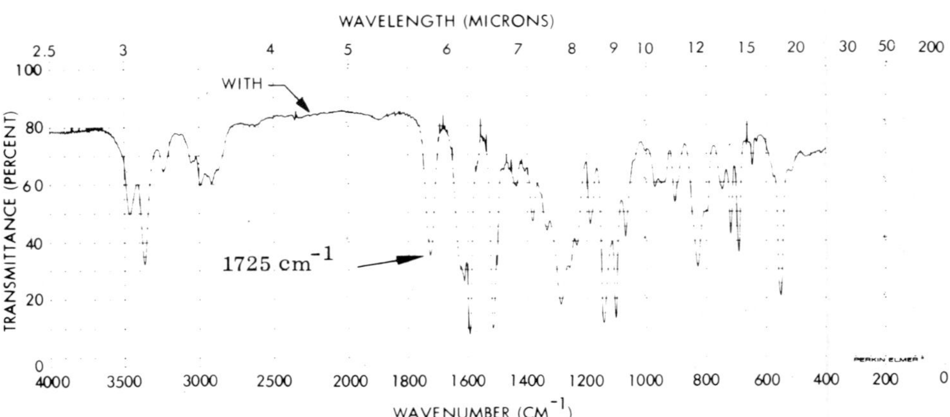

Figure 4 IR spectra comparison of resin matrix (a) with and (b) without additive.

similar manner, an infrared scan obtained, and the %T determined. With this value and the calibration curve it is possible to calculate the percent curing agent present in the sample by the equation

$$\% \text{ Dicy adhesive} = \frac{\text{Dicy concentration (mg/ml)}}{\text{sample concentration (mg/ml)}} \times 100$$

The same type of procedure is applicable to many other absorption bands. The sulfone, $-SO_2$, group and the epoxy, $-C_2O-$, group

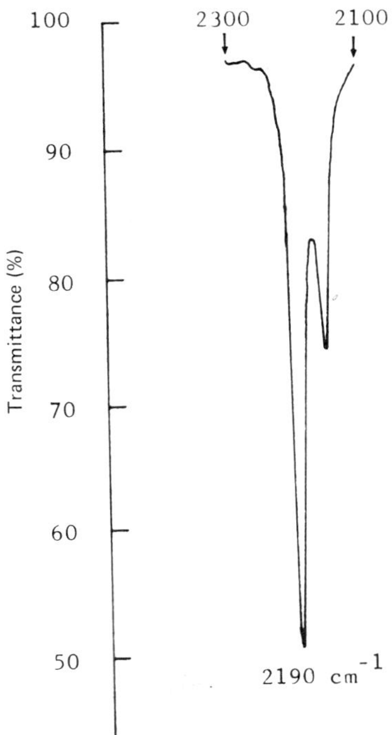

Figure 5   Infrared absorbance band for dicyandiamide quantitative determination.

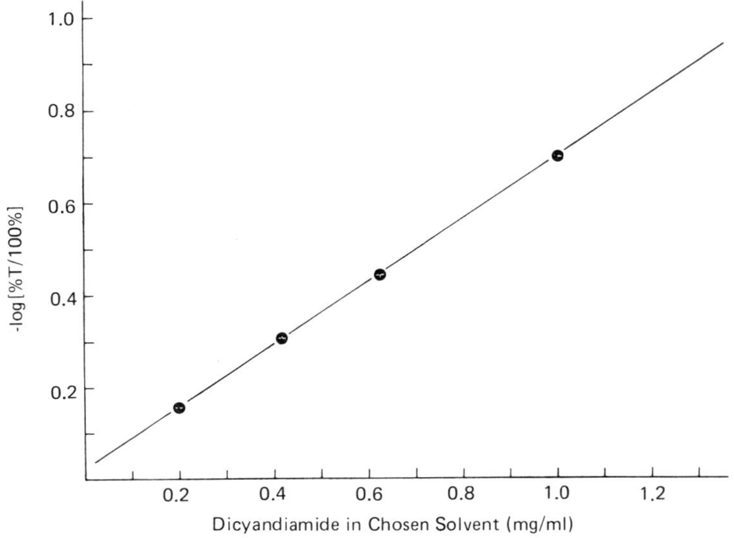

Figure 6   Dicyandiamide curing agent calibration curve at 2190 cm⁻¹.

are associated with other common materials used in adhesives and lend themselves to quantitative analysis.

F. Evaluating Techniques

As with any test procedure, the analysis chosen for quality control must be tailored to the specific material to be purchased. If the purchased item is an unfilled material, a transmission mode may be adequate. However, if particles are present, such as metal powders, carbon, or silica fillers, a reflectance mode may be the best approach. The ultimate end use of a product together with the materials' chemical behavior will determine if a simple qualitative test is adequate or if a more extensive quantitative test is required. Much hard work goes into the development of any chemical characterization test procedure to determine these parameters.

III. CHROMATOGRAPHY

A. High-Performance Liquid Chromatography

General Method

Although volumes have been written on the subject [8—10], chromatography applies basically to a wide variety of techniques where separation of a sample into its individual components is accomplished by their interaction between a moving or mobile phase and a stationary phase. Liquid chromatography utilizes a liquid mobile phase and a solid stationary phase and the separation mechanism involves how sample molecules distribute themselves between each phase and the time spent in each. A simple schematic of a liquid chromatograph is shown in Fig. 7 and

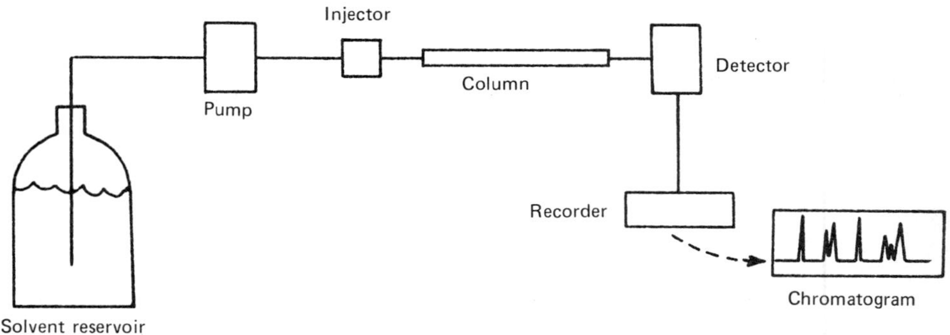

Figure 7   Schematic of basic liquid chromatograph.

consists of a solvent delivery system, an injection system, a column (holds stationary phase), and a detector.

The mobile phase in polymer analyses is an organic solvent. Common materials are tetrahydrofuran, chloroform, and acetonitrile. This solvent is driven through the system by means of a pump that produces constant liquid flow. The sample (previously dissolved in solvent) is injected and swept through the column and detectors by the mobile phase. The detectors sense the presence of each molecular fraction as well as its quantity and the total separation is displayed on a strip chart as a chromatogram. An example of an epoxy adhesive resin matrix is shown in Fig. 8.

Today, advances in both instrumentation and column packing materials are occurring at such a rapid rate that it is a full-time job to remain cognizant of the latest innovations. The label "high-performance" added to the rather innocuous term "liquid chromatography" indicates the emergence of this method from the tedious, time-consuming procedures of the past, involving large-diameter glass columns and gravity-flow mobile phase. Although HPLC has several subbranches, only two will be discussed as applied to the analysis of adhesive polymeric materials: size exclusion or gel permeation chromatography and liquid-solid chromatography.

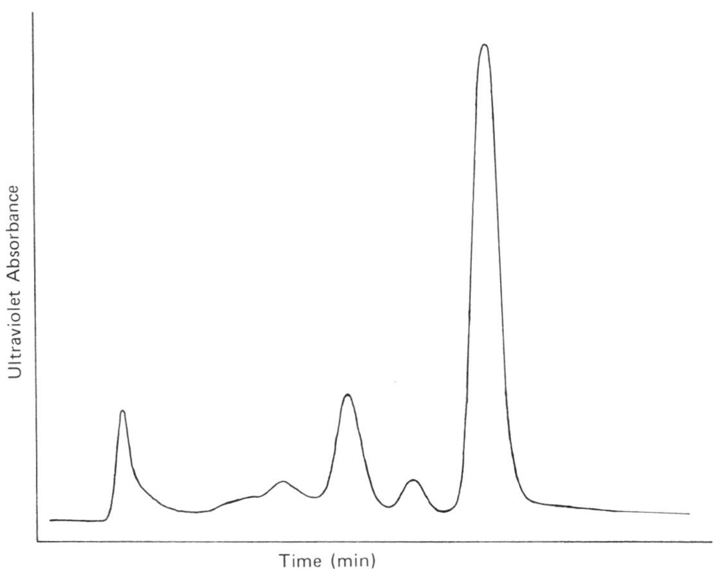

Figure 8   Typical liquid chromatogram.

Gel Permeation Chromatography (GPC)

Gel permeation or size separation is a type of liquid chromatography in which molecules in solution are separated by their permeation or lack of permeation into a porous packing gel. Large molecules are excluded from some or all of the pores, therefore eluting more quickly from the column than smaller molecules, which can permeate into a greater percentage of the pores. In a resin system there are many molecular weight species present resulting from resin advancement, or B staging, the interaction of the resin and curing agent, and the accelerator. Much information may be obtained about the composition and aging history from the ratios of the various weight fractions. For example, as a resin system cures or ages, polymerization and crosslinking take place, thus increasing the length and molecular weight of the polymer chains. Therefore, the concentration of the low-molecular-weight species decreases while the high-molecular-weight fractions increase. Accordingly, gel permeation chromatography which monitors these changes is a valuable tool for studying thermoset resins.

Liquid-Solid Chromatography (LSC)

Liquid-solid chromatography is based on affinity separation. Separations by affinity normally involve a stationary phase which is more polar than the mobile phase but not always. When the adsorbent is less polar the technique is called "reverse-phase" partition chromatography. The determining factor in the relative adsorption of a sample is its functional groups, and this adsorption increases as the polarity and number of functional groups increase. The uniqueness of affinity chromatography results from a competition between sample and solvent molecules for a place on the adsorbent surface and the multiple interactions between functional groups on the same molecule and the rigidly fixed sites on the adsorbent surface.

Since the separating power of liquid chromatographic method depends on interactions between sample molecules and the stationary and mobile phases, it follows that the choice of these two elements has an extreme effect on the success or failure of an analysis. Usually, chemists are not confronted with a completely "unknown" material and their knowledge of the basic material characteristics obtained from other sources (manufacturer's data, other chemical analyses, previous experience) helps in determining the most likely separation mode to be used.

Isocratic and Gradient Elution Separations

The manipulation of the mobile phase composition presents a unique opportunity for very carefully controlling and maximizing resolution. Many solvents are used in the various HPLC modes and two common methods used to introduce these liquids into the LC are isocratic and gradient elution.

Isocratic separation involves using a mobile phase of constant solvent strength. The mobile phase can be comprised of 100% of one solvent or a mixture of two or more different solvents. All size exclusion separations are carried out isocratically and this mode offers definite advantages over other methods, as will be discussed later.

Gradient elution increases the solvent strength or polarity of the mobile phase during the course of the separation. This is accomplished by carefully mixing two (or sometimes three) solvents at a programmed rate which tends to shorten the retention time of the molecules strongly retained on the stationary adsorbent. The gradient profile is variable from a straight linear curve to several variations in between (see Fig. 9) and can be varied even more complexly on some sophisticated solvent programmers. However, even with the most advanced instruments, the analyst must select the optimum gradient conditions empirically.

Gradient elution is usually used on complex samples where optimum separation and reduced analysis times are desired. There is improved and consistent peak shape in this mode, which increases the effective sensitivity. The resolution of these peaks is improved as well. Because of this sensitivity, however, gradient separations are very dependent on the purity of the solvents used as mobile phases. If strongly adsorbed impurities are present in bulk solvents, they will be concentrated at the head of the column during equilibration, and as the solvent

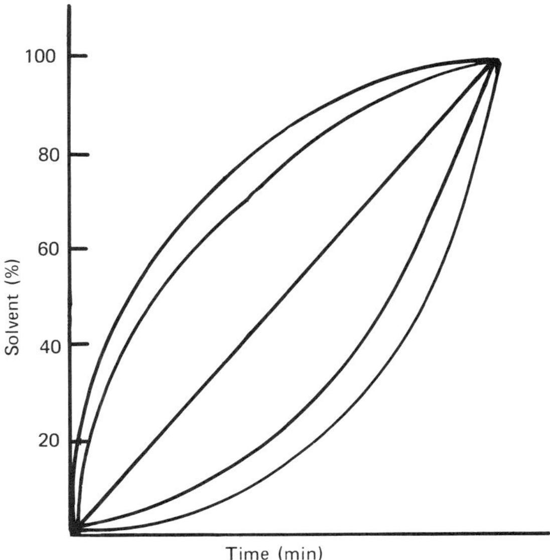

Figure 9   Various gradient profiles.

strength increases during a run these impurities will be eluted and separated along with the sample mixture. The already complex chromatograms are then compounded by a myriad of extraneous and often confusing additional peaks. When working in the isocratic mode solvent integrity is not as critical a parameter as in gradient work. Since only one mobile phase of constant composition is used, the detector can be electronically balanced or a blank (containing the same mobile phase) can be used in a reference cell. Thus isocratic separations are simpler to run since there are fewer variables to be concerned with: no dynamic mixing of solvents is necessary requiring additional complex equipment, ultrapure solvents are not needed, and baseline shifts, reequilibrations, and impurity peaks are not a problem. Gradient elution is best when flow control is required for consistent peak area calculations. Isocratic methods give better precision of retention times and are easier to automate using less complex and expensive equipment.

Detectors

The most commonly used detector in HPLC is the ultraviolet (UV) type. Based on principles similar to those discussed under infrared spectroscopy, the UV detector separates molecular constituents from a mixture by virtue of their absorbing characteristics in the UV at a discreet wavelength. In polymer analysis, specifically involving epoxies, most separations are performed at wavelength settings of 230 or 280 nm. Other compounds require varying wavelengths in order to obtain optimum separation. Because of this, variable UV-visible detectors are becoming more popular than fixed-wavelength ones since they are more versatile in detecting a large range of materials. The UV detector is extremely sensitive and separations of very dilute solutions are possible.

The refractive index (RI) detector is used for materials that do not absorb well in the UV as well as being an all-around multipurpose detector. This type of detector responds to the change in refractive index of the mobile phase caused by the presence of eluting sample molecules. It is not as sensitive as the UV detector and is strongly affected by solvent and ambient temperature fluctuations. The RI is used extensively in GPC separations and, with proper care and understanding of its capability, it is a very acceptable device. Other types of analyses are the infrared, fluorescence, flame ionization, and mass spectrometer.

Qualitative and Quantitative Uses for Quality Control

HPLC has been proven to be a powerful technique for the separation of complex adhesive and advanced composite resin matrices into their individual components [5,11—13]. This analytical method is thus ideally suited for quality control applications. Qualitatively, data as to the composition and to some extent the amount of chemical reaction of a resin matrix are readily available using basic liquid chromatographic procedures. However, to obtain repeatable quantitative results the analytical test procedures must be carefully controlled. The subject has been discussed fully in the literature [4], but a summary of the important parameters in developing an accurate quantitative procedure is given in Table 2.

As in any other analytical procedure, quantitative HPLC methods require knowledge as to their accuracy and precision (reproducibility). With HPLC being used more routinely for quality assurance purposes, the emphasis on reproducibility and consistency is essential. Statistical treatment of data may include standard deviation, variance, and confidence level.

Evaluating Techniques

All HPLC tests must be tailored to the specific type of material to be analyzed. As already discussed, true unknown materials are rare and experienced analysts will have the necessary knowledge to make an educated guess as to the type of HPLC mode best suited for their needs. The refinement of techniques for quantitative purposes requires only time and care in setting test parameters and statistically evaluating the data. For novices, all good texts on chromatography include guides to HPLC mode selection, but a literature search of past research dealing with the general type of material to be analyzed will probably yield much more informative and specific information.

B. Thin-Layer Chromatography

General Method

Thin-layer chromatography (TLC) is an extremely simple form of chromatography. A thin sorbent layer is applied to a support material and serves in place of a packed column. A drop of solution containing the sample is applied to the sorbent and placed in a closed container holding the developing solvent. This mobile phase migrates upward by capillary action, carrying the individual matrix components with it. After a preselected time the support medium is removed and the solvent allowed to evaporate. The locations of the various components can be determined by several methods.

Compounds advance various distances up the sorbent depending on their adsorption coefficients, R_f. This is the ratio of the amount of material adsorbed to the amount of material remaining in solution at equilibrium. As the solvent moves over an adsorbed spot, the equilibrium is shifted and the compounds are desorbed, the more tightly held materials to a lesser extent than the loosely adsorbed ones. Because of this mechanism, TLC can be used to assure that the chemical ingredients in adhesive resin matrix have not been altered. An example of this is shown in Fig. 10.

Table 2 Steps for Developing Quantitative LC Procedures

Test parameter	Comments and effects
Sample preparation	Highly purified solvents. Physical removal of fillers and scrim. Typical concentration range: 15–60 μg of resin injected Storage of prepared resin solution not advisable over 12–24 hr (must be determined for each individual material).
Column equilibration	For chromatogram reproducibility and analytical precision. Temperature control to eliminate solvent compressibility, flow, and solubility fluctuations.
Analytical sequence	For autosampling of a large number of samples, precise time sequencing for equilibration, test run, and reequilibration of the column produces a consistent test environment. Run blank prior to sample series to verify purity of mobile phase(s). Run samples in replicate. For a large number of samples run blanks during the course of the series to check for contamination buildup and to clean column.
Detector linear concentration response	Determine linear operating range of detector specifically when using ultraviolet. Sample concentration vs. peak area.
Standard calibration solution	No specified set of instrument parameters will yield identical sample chromatograms between different instruments. Absolute quantitative results are not possible strictly by instrumental electronic response. This parameter as it relates to quantitative evaluation is determined by an operator. Every polymer mixture has its own response characteristic; thus a generalized method for many materials is not possible.

Table 2   (continued)

Test parameter	Comments and effects
Standard calibration solution (continued)	Standard solution containing the components of a mixture is used to determine response factors which remain constant regardless of the instrument used. For a large number of samples, the standard solution should be run during the course of the series.
Integration techniques	Single largest error-producing parameter in HPLC analysis Must evaluate computer parameters and specify conditions which produce consistent integration for each fraction in a mixture.

The analyses are simple to run, inexpensive, labor cost effective, and ideal for repetitive quality control testing. TLC uses less solvent than is needed for liquid chromatography, and solvents and solvent mixtures can be changed rapidly with no equilibration time required between runs.

Sorbents

As in liquid chromatography, TLC can be performed using various separation modes: adsorption, ion exchange, partition, or reverse phase. The choice of media is limited only by its ability to be coated onto a support.

The most popular and perhaps versatile sorbent material, silica gel, offers diverse modes of separation, some completely opposite from one

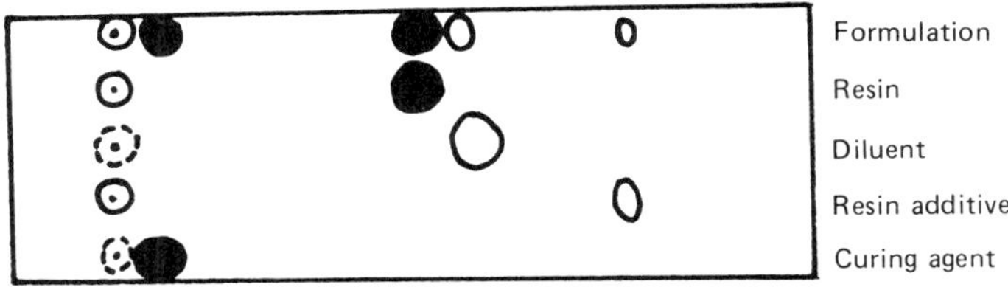

Figure 10   TLC of resin + constituents.

another. Hydrophilic silica gel having a strong affinity for water
separates compounds on the basis of polarity (i.e., polar materials
are held more tightly than nonpolar materials). The opposite type of
silica gel, hydrophobic, has a porous nature which performs in the
reverse-phase mode. That is, nonpolar materials are retained more
strongly than are polar ones. Because of this porosity molecular size
separations can also be made using reverse-phase TLC.

Aluminum oxide or alumina is also a widely used TLC sorbent.
Since alumina is crystalline in nature, its fixed pore size allows the
aluminum ion to participate in adsorption separations. Other sorbents
are available. However, the trend today is to avoid an infinite selection of specialized materials in favor of a few, very versatile sorbents.
Surface area, porosity, ambient conditioning, and type of binders are
all factors affecting capacity, efficiency, reproducibility, and selectivity
of TLC sorbents. Once a sorbent has been chosen for a particular
separation, the remaining tasks involve solvent selection and the mode
of development.

Developing Modes

Selectivity or how effectively compounds are resolved is obtained
through the action of the mobile phase on the material and sorbent.
The choice of a solvent or solvent mixture is dependent on the nature
of the compounds to be analyzed as well as on the type of separation
mode to be used. A systematic study of how to select mobile phases
in TLC has been offered in place of the historic trial-and-error procedure of the past as well as an extensive discussion of developing
techniques [14].

The way a TLC development is carried out determines how effective
the ultimate resolution will be. Aside from the standard ascending
development method where solvent moves upward carrying sample components with it, there have been advances in obtaining increased resolution and sensitivity by more complex developing modes.

Multiple development involves several redevelopments of the same
support until a satisfactory separation is obtained. This type of procedure essentially increases the length of the layer. Continuous development involves a constant flow of solvent along the sorbent and
excellent work has been done using this technique [15]. X—Y or two-
dimensional development, programmed multiple development [16],
gradient elution, and high-performance radial chromatography [17] are
other available techniques.

Visualization and Detection

There are several methods available for locating the positions of
compounds following a thin-layer separation. These include detection
by ultraviolet light, charring, color reaction, densitometry, and flame

ionization (FID). With the exception of FID, these forms of visualization have been covered extensively in the literature [14]. The flame ionization detector is relatively new in its application to TLC and only minimal work has been done on polymers [18]. This system combines conventional TLC with a hydrogen flame ionization detector. Not only is the detection device unusual for TLC but the support medium is also unusual. Separation is achieved on a thin-layer rod rather than a plate, which offers the distinct advantage of being reusable up to 100 times. Development is performed conventionally on the rod but the detection technique is automatic, requires no visualization reagents, and the separation is quantitative. This technique has been used successfully on a simulated resin matrix. An example chromatogram is shown in Fig. 11 and resembles the curve obtained in liquid chromatography. These areas under the peaks are proportional to the concentration of sample components, and thus quantitative analysis becomes even more rapid than with quantitative visual techniques.

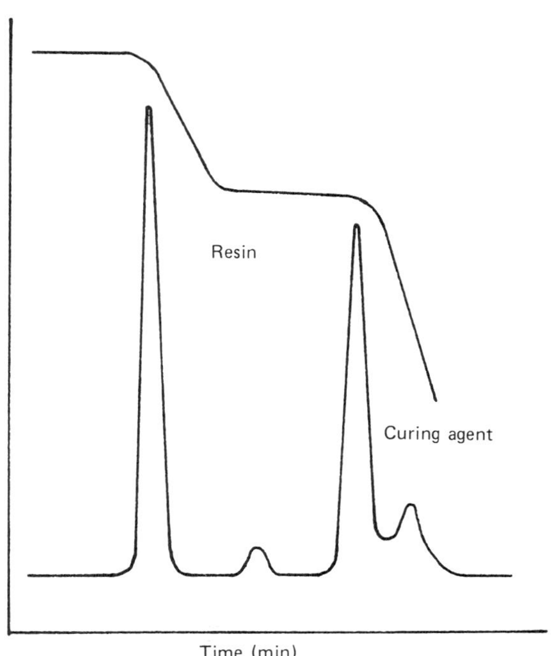

Figure 11   Typical TLC/FID chromatogram.

Qualitative and Quantitative Uses

Although TLC has been used successfully in qualitatively separating resin matrices [11], not much has been done for its quantitative applications. However, with emphasis now shifting from research to routine testing, where time and cost are important, TLC is becoming more appealing over the more established but expensive methods of liquid and gas chromatography. The sensitivity and quantitative attributes of TLC in the biological field are well known, and only time and effort are required to adapt these principles to the analysis of polymers.

IV. THERMAL ANALYSIS

A. General Method

Thermal analysis is the measurement of changes in the chemical or physical state of a material as a function of temperature. This simple dependency allows the resin chemist access to a wide variety of information relating to polymer performance and processing with minimum effort. Thermal analysis can be used for material quality assurance and process control as well as being invaluable in new process development and problem solving. Heat capacity, weight loss, compression and expansion properties, and kinetic properties are some of the measurements that can be obtained using thermal analysis. All of these relate to the ultimate performance of a finished part, and these performance-related properties provide for better materials testing and evaluation.

The four thermal analysis techniques used most commonly today are differential scanning calorimetry (DSC), thermogravimetric analysis (TGA), thermomechanical analysis (TMA), and dynamic mechanical analysis (DMA). A brief description of each follows. More detailed information is available in the literature [19,20].

B. Differential Scanning Calorimetry (DSC)

The DSC is designed to measure the amount of energy absorbed (endotherm) or given off (exotherm) by a material as a function of temperature or time. Temperature differences between a sample and an inert reference material are recorded as a function of the sample temperature, with the area under the output curve being directly proportional to the total energy (q) transferred in or out of the sample. These differences are recorded as heat in millicalories per second. The temperature in the calorimeter housing is controlled by a programmable heater which can be made to simulate press or autoclave heating cycles. The DSC may also be programmed to cycle, hold at temperature, or cool at a specified rate. The range of the DSC is from liquid-nitrogen temperatures, $-190°C$, to $600°C$ and can be heated from 0.5 to $100°C/min$.

A DSC curve is shown typifying the resin matrices used in some advanced composite and adhesive materials. The abrupt drop (heat

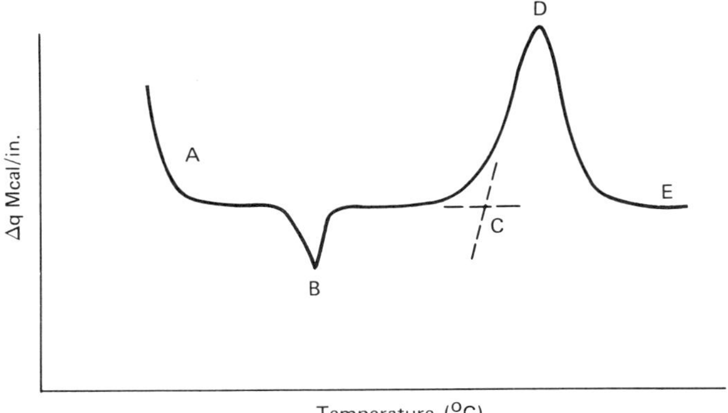

Figure 12   Typical DSC thermogram.

endotherm) from the beginning to point A is due to the heat capacity of
the sample. At B, a more pronounced endotherm is shown and is indicative of phenomena such as evaporation of a liquid, melting of a solid, or
another change of state requiring energy. In lower-temperature regions,
it is associated with true phase changes such as boiling or melting. In
adhesives this could indicate solvent loss due to residue remaining after
manufacturing.

Point C is the onset of the main reaction or exotherm (heat evolution). This is an important area since it is where the reaction rate
between resins and curing agents increases rapidly. From the area
under the exotherm peak (from C to E) it is possible to calculate the
heat of reaction. The exotherm peak, D, denotes the temperature at
which the curing reactions are proceeding at a maximum rate for the
particular cure cycle involved. Point E is the end of the exotherm,
and where the cure levels off indicates complete cure under the given
conditions.

A few applications of DSC in the study of polymers include measurements of glass-transition temperatures (T_g), evaluation for degree of
cure of a fabricated part, analyzing polymer mixtures from melting or
crystallization data, and developing cure schedules by exotherm onset
temperatures. Also available for DSC studies is the ability to control
the sample environment, from inert to reactive gases and pressure from
the micrometer range to 1000 psi. This adds a new dimension to the
variables that can be studied.

C. Thermogravimetric Analysis (TGA)

TGA is a method for measuring the weight gain or loss of a material
either as a function of increasing temperature or at a set temperature

(isothermal) over a period of time. Basically, a TGA is an extremely
sensitive microbalance utilizing a photosensitive null detector. The
temperature range for this technique is from ambient to 1200°C, where
a purge gas (inert or reactive) flows over the sample during the test.
This effluent may be collected for additional studies by gas chromatography, mass spectrometry, or infrared spectroscopy.

Composition and thermal stability can be determined for polymeric
systems. TGA can also be used to study the effect of additives such
as stabilizers, antioxidants, and flame retardants, as well as controlling
or maximizing the amount of fillers and extenders.

A typical curve for epoxy adhesive is shown in Fig. 13. This type
of curve can be used effectively for "fingerprinting" a group of similar
materials by comparing their thermal degradation.

D. Thermomechanical Analysis (TMA)

Thermomechanical analysis measures vertical dimensional changes of a
material with respect to temperature and time. The technique involves
a sample resting on a quartz stage surrounded by a heater with a quartz
probe contacting the top of the sample. Any movement of the sample is
transmitted to the probe, which results in a proportional displacement
on a chart.

With the appropriate probes any of six measurements can be made.
These include compression properties (softening or penetration under
load), expansion properties, tension properties (which includes the
expansion or shrinkage of materials under tension), single-fiber properties, dilatometry involving volumetric expansion of a material within a
confining medium, and kinetic measurements using various modes at a
fixed temperature as a function of time.

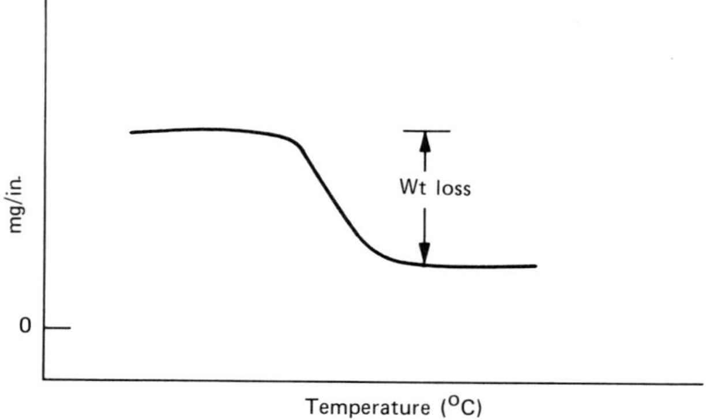

Figure 13 Typical TGA thermogram.

Glass transition temperatures (T_g), melting or softening points, coefficients of thermal expansion (CTE), elastic or single-fiber modulus, and specific volume are some of the data obtainable using TMA. Both the CTE and T_g values of a thermoset resin are closely related to the extent of cure and hence are invaluable in the development of cure cycles. A typical TMA curve for an epoxy adhesive is shown in Fig. 14.

E. Dynamic Mechanical Analysis (DMA)

DMA involves the flexing of a specimen at constant amplitude as a function of temperature. This technique is becoming more popular for characterizing polymer systems since it is related more directly to end-product usage than are other thermal analysis methods (i.e., it measures mechanical properties dynamically and therefore yields rapid information on future part performance. DMA measures both energy dissipation and resonant frequency and because of its sensitivity can detect lower-order molecular transitions due to the dramatic change in damping and modulus.

Some applications of DMA include evaluating degree of matrix cure, effectiveness of reinforcement and coupling agents on the resin, presence of moisture, degree of crystallinity and branching of a base polymer, transition temperatures, the effect of additives in processing, and sample toughness (presence of secondary transitions at low temperatures). A typical DMA scan is shown in Fig. 15.

F. Uses for Quality Control

The value of thermal analysis in the characterization and control of adhesives is obvious in both quantitative and qualitative uses. Each

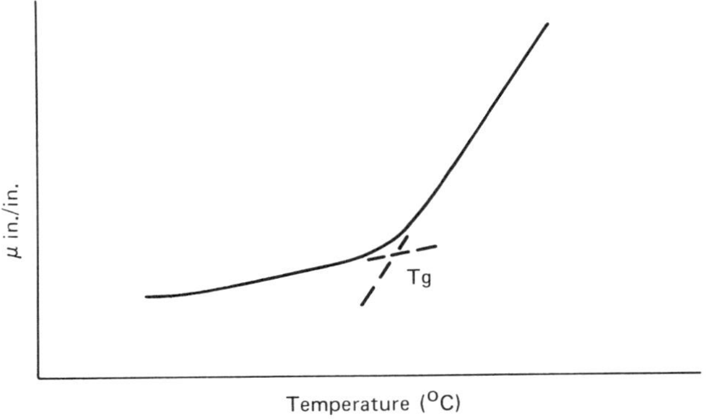

Figure 14   Typical TMA thermogram.

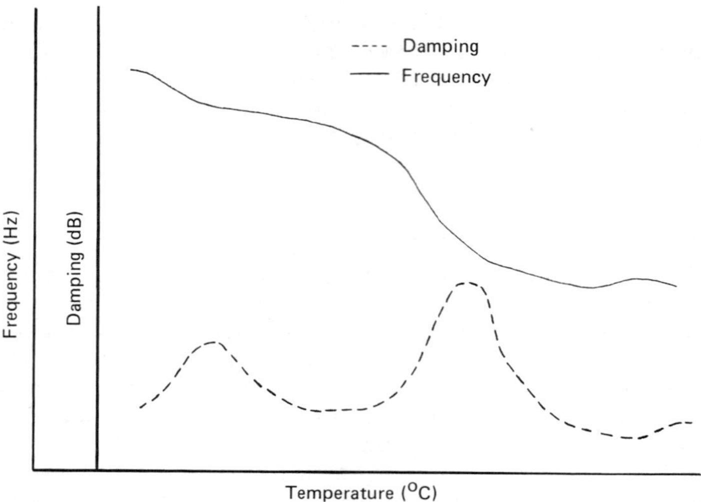

Figure 15   Typical DMA thermogram.

technique offers solutions to specific temperature-related problems,
providing information on processing and performance that other analytical techniques cannot. The ultimate method chosen for quality control
must be tailored to that specific material and its end use, as must the
specification limits.

V. DYNAMIC DIELECTRIC ANALYSIS (DDA)

The fabrication of high-quality aircraft and aerospace structures requires well-designed cure cycles on chemically understood materials.
The strength of any adhesively bonded joint is affected by many factors, including a specific cure cycle based on a thorough knowledge of
what is happening both chemically and physically during cure. The
curing mechanism of any type of adhesive is a constantly changing
process of altering molecular structure. Dielectric analysis is a method
that follows the changing dielectric state of a resin matrix as a function
of temperature and time and can be related to rheological events.
    Cure monitoring is dealt with in detail in the literature [21,22].
Briefly, DDA is a measurement of dielectric changes as related to the
molecular mobility of a resin matrix. Most organic resins are polar in
nature, thus behaving as dipoles that will orient in an alternating
electrical field. The movement of these dipoles is restricted by the
physical state of the material. Therefore, the degree of dipole orientation relates to the rheology of the resin. When a resin is in the liquid

state, dipoles move quite easily. However, as the matrix cures and
the molecules begin to polymerize it becomes increasingly difficult for
them to align in the field. When final cure and maximum cross-linking
(rigidity) is reached, the dipoles are no longer free to move. One
method for tracking these phenomena is by recording the dissipation
factor or loss tangent. This is a measure of the loss of energy due to
increased difficulty of dipole rotation.

A typical curve is shown in Fig. 16. For epoxy-type matrices two
peaks are normally observed. The first of these maxima is related to
the flow of the resin, while the second results from gellation and cure.
The valley between the two peaks is a low-viscosity region. It is at
this portion of the curve where processing changes are made (i.e.,
pressure application and hold temperature). The DDA signature varies
as the chemical structure of a material changes. As resins age or enter
the B stage, the two peaks move toward each other and eventually tend
to merge. An example of this type of trend is shown in Fig. 17.

Since DDA relates to the physical state of a material relative to
temperature, this technique can be used both as a quality control tool
and to define processing parameters. Incoming quality assurance of a
material is based on several properties necessary to produce good products. Traditionally, tests on adhesives have been things like gel time
and resin flow. Although these techniques are relatively fast and inexpensive, they do not give a complete picture of the actual condition
of the resin itself. Thousands of dollars worth of material is discarded
yearly because, based on the foregoing tests, the gel or flow no longer
falls within the specification limits. Yet it is still possible to make good
parts with this same material. By utilizing DDA, which gives information

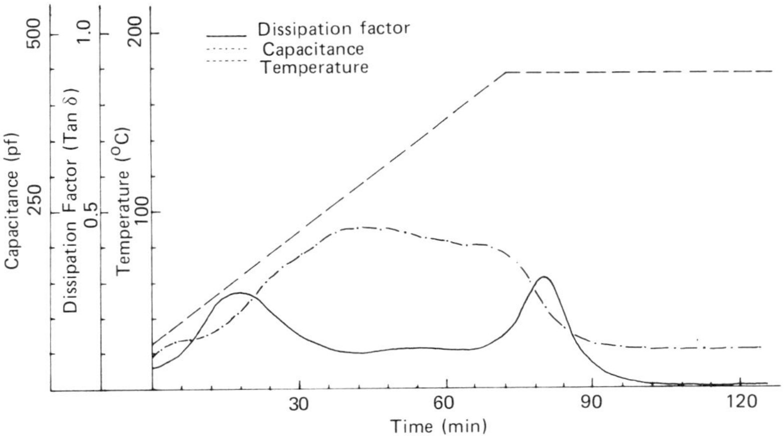

Figure 16 Typical DDA curve.

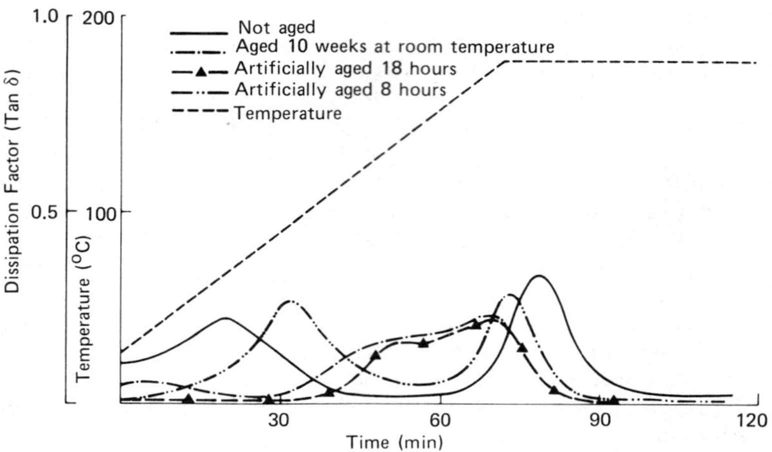

Figure 17   Effects of aging on DDA curve.

on the properties of the actual resin state, the degree of chemical advancement can be followed accurately and easily on a day-to-day basis. In this way part fabricators may base cure schedules on a sound, scientific basis rather than by trial and error, thus eliminating the chance of scraped parts due to guessing the state of resin reactivity. DDA offers a convenient and rapid test responsive to the dynamic state of resin curing systems.

VI. OTHER TECHNIQUES

Many other chemical techniques are available and are being used for quality control of polymeric materials. These include carbon-13 and proton nuclear magnetic resonance (NMR), spectroscopy [23], atomic absorption and emission spectroscopy, as well as a variety of wet chemical tests for determining pertinent functional groups (epoxies, amines) and elemental analysis (C, H, N, S).

VII. CONCLUSIONS

The goal of any quality control scheme for adhesives and advanced composite matrix materials is to assure that purchased material will produce high-quality and reliable manufactured parts. To achieve this, it becomes mandatory to start with well-characterized and consistent raw materials. Thus the material purchase specification must ensure that incoming material has the same chemical formulation and has been processed correctly. It is the user's responsibility to choose the necessary tests, determine the acceptable variation range, and relate these to changes in mechanical properties as well as long-term durability and environmental stability. Mechanical property testing may still be required, however, if preceded by chemical quality control testing, the elimination of much unnecessary labor can be realized. This time savings together with the need for very little material leads to extremely cost effective techniques for the performance of a very vital function.

REFERENCES

1. May, C. A., D. K. Hadad, and C. E. Browning, Physiochemical Quality Assurance Methods for Composite Matrix Resins, 33rd Ann. Conf. Reinforced Plastics/Composites Institute, Sec. 15D, Washington, D.C., Feb. 1978.
2. May, C. A., Composite Matrix Quality Assurance—An Art Becomes a Science, *SAMPE*, San Francisco, Vol. 24, No. 1, May 1979, pp. 390–403.
3. May, C. A., T. E. Helminiak, and H. A. Newey, Chemical Characterization Plan for Advanced Composite Prepregs, *SAMPE*, Seattle, Wash., Vol. 8, Oct. 1976, pp. 274–294.
4. Hadad, D. K., *Chemical Quality Assurance Test Procedures for Advanced Composite Resin Matrices*, Lockheed Missiles & Space Company, Inc., Sunnyvale, Calif., May 1980.
5. Carpenter, J. F., Test Program Evaluation of 3501-6 Resin, U.S. Naval Air Systems Command Rep. N00019-77-C-0155, Final Report, May 1978.
6. Sprouse, J. F., Fourier Transform Infrared Spectroscopy as a Method for Studying Weathering of Glass-Fiber Epoxy Composites, *Proc. TTCP-3 Crit. Rev.;* Techniques for the Characterization of Polymeric Materials (AMMRC MS 77-2), Jan. 1977, pp. 43–54.
7. Hagnauer, G. L., et al., Evaluation of New Techniques for the Quality Control of Epoxy Resin Formulations, U.S. Army Materials and Mechanics Research Center Rep. TR 78-8, Jan. 1978.
8. Heftmann, E., *Chromatography*, 3rd ed., Van Nostrand Reinhold, New York, 1975.
9. *A User's Guide to Chromatography. Gas, Liquid, TLC*, Regis Chemical Company, Morton Grove, Ill., 1976.
10. Johnson, E. L., and R. Stevenston, *Basic Liquid Chromatography*, Varian Associates, Palo Alto, Calif., 1978.
11. Hadad, D. K., J. S. Fritzen, and C. A. May, Exploratory Development of Chemical Quality Assurance and Composition of Epoxy Formulations, U.S. Air Force Materials Lab. Tech. Reps. AFML-TR-T6-112 and AFML-TR-77-217, Final Reports, June 1976 and Jan. 1978.

12. Hagnauer, G. L., and I. Setton, Compositional Analysis of
     Epoxy Resin Formulations, *J. Liquid Chromatogr.* Vol. 1, No. 1,
     1978, pp. 55—73.
13. Hindrichs, R., and J. Thuen, Advanced Chemical Characteriza-
     tion Techniques Applied to Manufacturing Process Control,
     SAMPE, Vol. 24, May 1979, pp. 404—421.
14. Touchstone, J. C., and F. D. Murrell, *Practice of Thin Layer
     Chromatography*, Wiley, New York, 1978.
15. Thin-Layer System Raises Chromatography Capabilities, *Ind.
     Res./Dev.*, Aug. 1978.
16. Perry, J. A., and Louis J. Glunz, Programmed Multiple Develop-
     ment: The Intrinsic Grid, *Ind. Res.*, Oct. 1977, pp. 117—120.
17. Kent, D. M. and R. Vitek, Quantitation and Correlation, *Ind.
     Res./Dev.*, May 1978, pp. 99—102.
18. Hadad, D. K., New Developments in Chromatographic Character-
     ization and Quality Assurance of Composite Materials, 72nd Annu.
     AIChE Mett. San Francisco, Nov. 27, 1979.
19. Levy, Paul F., Thermal Analysis: An Overview, *Am. Lab.*,
     Jan. 1970.
20. Wehrenberg, Robert H., II, Thermal Analysis: The Hot New
     Technique for Testing Plastics, *Mater. Eng.*, Sep. 1979.
21. May, C. A., Dielectric Measurements for Composite Cure Con-
     trol—Two Case Studies, Nat. SAMPE Symp. Exhib. San Diego,
     Vol. 20, Apr. 1975, pp. 108—116.
22. May, C. A., J. S. Fritzen, and A. Wereta, Jr., Cure Monitor-
     ing Techniques for Adhesive Bonding Processes, U.S. Air Force
     Materials Lab. Tech. Rep. AFML TR-77-58.
23. Poranski, C. E., Jr., W. B. Moniz, and D. L. Birkle, Carbon-
     13 and Proton NMR Spectra for Characterizing Thermosetting
     Polymer Systems, U.S. Naval Research Lab. (NRL) Rep. 8092,
     June 1977.

12
Coatings

SABURO NAKAHARA* *Douglas Aircraft Company of the McDonnell Douglas Corporation, Long Beach, California*

I. INTRODUCTION

Corrosion control of adhesively bonded assemblies is important for durable bond strength. Delamination of bond joints with the destruction of bond integrity will eventually result if corrosion in any form is present. Aluminum and its alloys all respond differently to the influences of the environment and to the many factors that contribute to corrosion [1]. Degradation of material systems, applied for protection in these environments, will result in subsequent corrosion of the underlying substrate. Corrosion of aluminum occurs in various forms, with most being electrochemical in nature. Corrosion characteristics are generally well known and recognized. The forms of corrosion usually found in bonded joints, trimmed edges, and at attachment areas can be identified as surface, crevice, pitting, exfoliation, and/or filiform-type corrosion [1, 2]. During fabrication of a bonded assembly, numerous scratches and gouges can occur. To protect these bared areas from subsequent corrosion and to maintain integrity of the bond joint and adjacent surfaces, the use of a protective coating system is necessary. The interior and exterior areas requiring corrosion protection are the fuselage shell, skins, stringers, frames, and dissimilar metal joint areas. Design considerations for the interior of the fuselage with respect to minimizing corrosion problems should include good drainage and the elimination of crevices. The ease of cleaning and inspection should also be considered [1]. If these design features are lacking, the effectiveness of the protective system will ultimately be affected. Organic coatings are normally semipermeable membranes that allow moisture to diffuse

---

**Present affiliation*: Northrop Corporation, Pico Rivera, California.

"

from one side of the membrane to the other. Constant saturation of
these membranes for extended periods with moisture, toilet chemicals,
galley liquids, and hydraulic and cleaning fluids would result in organic-
coating film failure and provide sites for corrosion [3]. Coatings to be
used should be selected with these factors in mind.

II. ADHESIVE PRIMER

Corrosion-inhibiting adhesive primers are, at times, applied on the
bonding surface soon after surface treatment. Their primary function
is to provide a stable surface for bonding. They also perform three
other functions essential to successful bonding: (1) to preclude contamination of the surface that have been freshly chemically etched or
anodically processed, (2) to provide the basic corrosion inhibition to
the nonbonded and bonded areas of the assembly, and (3) to serve as
an excellent paint base for application of subsequent finishes.

III. FINISH SYSTEMS

A. Primers

Normal finish systems for aircraft are composed of more than one coating type to give the structure the protection needed for its expected
environment. Primers are applied over the adhesive bonding primer to
ensure corrosion protection to any bare metal resulting from scratches
and as a primer coat for adhesion of the top coat. These primer systems are all normally pigmented with chromates, such as calcium, strontium, zinc, or a combination of these materials. The mechanism of corrosion inhibition provided by the chromates in the presence of moisture
can briefly be described as the passivation of the aluminum surface and
preventing or reducing the cathodic evolution of hydrogen by the reduction of the hexavalent chromium to the trivalent state [4].

Types

There are several primer systems that can be used to provide corrosion inhibition. They are listed in Table 1 and are used with or
without a top coat, depending on the area and the anticipated environment. Figure 1 shows a typical application for aircraft use.

Properties

Epoxy: The epoxies are usually resins that become hard and
interlocking when curing agents are reacted with them. The resulting
film has a high degree of cross-linkage. Aliphatic amine-cured epoxies
are very inert chemically and can be subjected to prolonged exposure
at 121°C (250°F) without affecting their fluid and chemical resistance.

Table 1 Types of Primer and Top Coat

Coating	Generic type	Usage	Film weight $(lb/ft^2/mil)$
Primer	Epoxy (amine-cured)	Interior and exterior	9.7×10^{-3}
	Epoxy (polyamide-cured)	Interior and exterior	8.8×10^{-3}
	Polyvinyl butyral	Exterior	6.3×10^{-3}
	Urethane	Exterior	9.1×10^{-3}
Top coat	Epoxy (amine-cured)	Interior	8.0×10^{-3}
	Urethane	Interior and exterior	8.5×10^{-3}

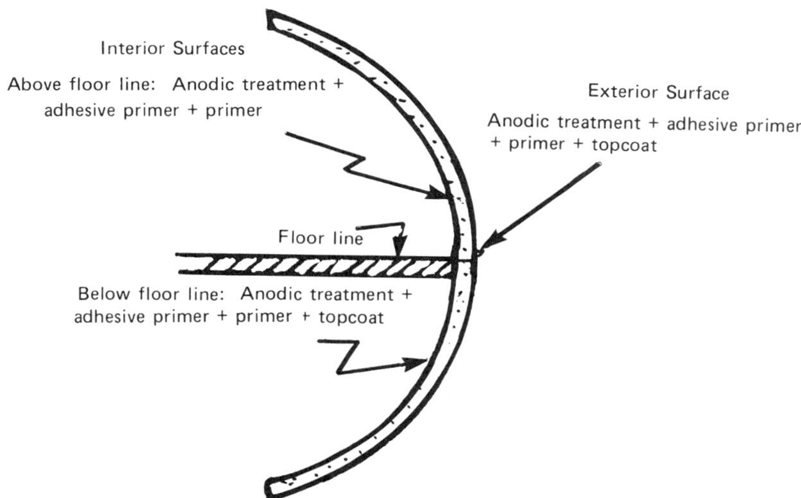

Figure 1 Protective finish system.

They have excellent moisture and corrosion resistance, are very hard, are compatible with other coatings, and adhere well to pretreated metal and previously coated surfaces. As the film progressively cures it exhibits increased tensile strength, while showing a reduction in impact resistance and elasticity. Stripping of aged primer film is usually difficult [5].

In comparison to the amine-cured epoxies, the polyamide-cured epoxies have slightly lower chemical resistance, tensile strength, and hardness. Their impact resistance is higher and they have excellent moisture and corrosion resistance [5]. Strippability of aged primer from pretreated metal surfaces is much easier than with amine-cured primer.

Polyvinyl butyrl/urethane: Primer formulated with vinyl butyral resin is commonly known as wash primer (pretreatment) and catalyzed with an acid component. This coating is applied as a very thin film over treated or non-surface-treated metal. Its purpose is to etch the metal surface to provide adhesion for subsequent coatings. It is sensitive to moisture during application and cure and has low chemical resistance. Excessive humidity may cause "blushing," thereby affecting the adhesion of subsequent coatings.

The urethane primer is applied over the wash primer as an intermediate prime coat prior to top coating. It provides the corrosion resistance of the system. The urethane primer properties will vary depending on the components of its formulation, but normally it has good fluid and chemical resistance and flexibility.

B. Top Coat

Types

As noted in Table 1, there are two generic-type coating systems that can be used on interior and/or exterior surfaces. They are the amine-cured epoxy and the urethane. Of the two, the one that provides the best protection and durability when used with a chromated primer is the two-component aliphatic urethane. This coating exhibits good versatility compared to many of the other coatings in today's state-of-the-art coating technology. Urethane coatings can generally be described as coatings that are a product of a reaction between a polyisocyanate polymer and a resin containing active hydroxyl groups [6]. The performance of these urethanes will vary largely depending on the selection of the reacting components. There are two classes of two component urethanes; the aromatic and the aliphatic. Use of the latter type will provide a durable system especially when applied to surfaces where exposure to ultraviolet radiation is continuous. The use of the epoxy top coat would mainly have application on nonultraviolet exposure areas.

Properties

Urethane: The aromatic urethane has good fluid and chemical resistance but has poor weathering and color retention properties. It exhibits severe yellowing and premature chalking when exposed to heat and/or ultraviolet radiation for extended periods. Degradation of this nature will eventually affect such properties as elongation, adhesion, gloss, and flexibility [6].

Aliphatic urethane is more stable and resistant to discoloration and chalking and is designated as nonyellowing [6]. In comparison to the epoxy top coat, they exhibit comparable or better performance in the areas of fluid and moisture resistance, corrosion resistance, flexibility, elongation, tensile strength, impact resistance, and chemical resistance, as noted in Table 2 [4, 6].

Epoxy: The properties are basically the same as those stated in the epoxy primer section. In addition, the pigmented top coats discolor and chalk rapidly on exposure to ultraviolet and on weathering in general. Use of the epoxy top coat over fasteners and attachments is not recommended, due to the loss of elongation of the film upon aging. Removal of the epoxy top coat from the epoxy primer using phenolic-type strippers is usually not difficult.

Table 2   Properties of Top Coats

	Coating	
Properties	Urethane (aliphatic)	Epoxy (amine-cured)
Flexibility	Good to excellent	Poor
Adhesion	Good to excellent	Excellent
Elongation	Excellent	Poor
Tensile strength	Good	Excellent
Hardness	Medium to hard	Very hard
Weatherability	Excellent	Poor
Gloss retention	Excellent	Poor
Impact resistance	Excellent	Poor
Solvent resistance	Good to excellent	Excellent
Chemical resistance	Good to excellent	Excellent
Corrosion resistance	Excellent	Excellent
Abrasion resistance	Excellent	Excellent
Moisture resistance	Excellent	Excellent
Skydrol resistance	Excellent	Excellent

C. Environmental Resistance

Environmental testing falls into two categories: laboratory and service. Laboratory tests are to accelerate testing to determine in a short period whether the finish system conforms to the performance required in any given exposure [1]. These tests, together with data from outdoor weathering and an industrial-marine environment, provide an assessment for the selection of durable finish systems. Service data determine the durability, the reliability, the ability to minimize corrosion, and the maintainability of the system in a real-life environment.

Laboratory

Resistance of finish systems: Finish systems subjected to environmental resistance tests are compared in Table 3. Systems were applied on 7075-T6 alclad and/or non-alclad aluminum which were surface treated with chromic acid anodize and/or a conversion coating such as Alodine. No adhesive bond primer was applied prior to application of the protective finish system.

Corrosion inhibition of adhesive primer: Aluminum surfaces used in bonding are normally given an anodic or etch treatment. If left uncoated and exposed to a corrosive media (5% salt spray), the treatments exhibit moderate-to-severe corrosion, as shown in Fig. 2. When these same treatments are coated with 0.0001 to 0.0003 in. of adhesive primer the panels exhibit no corrosion or minimal corrosion, as indicated in Fig. 3 and Table 4.

Of the corrosion tests, the acidified salt spray induces the most severe environment. The use of dissimilar metal specimens accelerates the corrosive reaction and provides an assessment of the effectiveness of corrosion inhibition. This inhibition is shown in the specimens of Fig. 4. All the adhesive primers by themselves exhibit substantial film failure and do not inhibit corrosion, especially in the nonfastener areas. Overcoating each of the adhesive primers with the epoxy polyamide primer greatly reduces coating failure and corrosion [7].

Figure 5 shows the 18-month beach (industrial-marine) exposure panels for the four adhesive primers overcoated with the epoxy polyamide primer/aliphatic urethane top coat system. Adhesive primer (a) shows a correlation with the laboratory test of Fig. 4, primer A. When overcoated it exhibits better corrosion resistance than do the other adhesive primers. There is minimal corrosion around the fasteners without exfoliation, whereas exfoliation and delamination of the bond joint are quite evident on the other three panels [7].

Figure 6 shows the effects of an acidified salt environment on the various aluminum alloys and surface treatments overcoated with an adhesive primer and the protective finish system. The use of the epoxy polyamide primer shows its effectiveness in providing corrosion inhibition [7].

Table 3 Accelerated Environmental Tests

Test[a]	Protective finish system				
	Epoxy amine primer	Epoxy amine primer + epoxy top coat	Epoxy polyamide primer	Epoxy polyamide primer + urethane top coat	Wash primer + urethane primer + urethane top coat
Adhesion	Pass	Pass	Pass	Pass	Pass
Humidity (1 month)	Pass	Pass	Pass	Pass	Pass
5% salt spray (2000 hr)	No corrosion	No corrosion	No corrosion	No corrosion	No corrosion
Acidified salt spray (1 month)	—	Fair	—	Fair	Poor
Beach site (12 months)					
Chalk and gloss retention	—	Poor	—	Excellent	Excellent
Filiform corrosion resistance	—	Poor	—	Good	Poor to good
Exfoliation corrosion resistance	—	Fair	—	Good	Poor

[a]Dissimilar metal panels used in acidified salt apray and beach exfoliation corrosion tests.

(a)   (b)   (c)

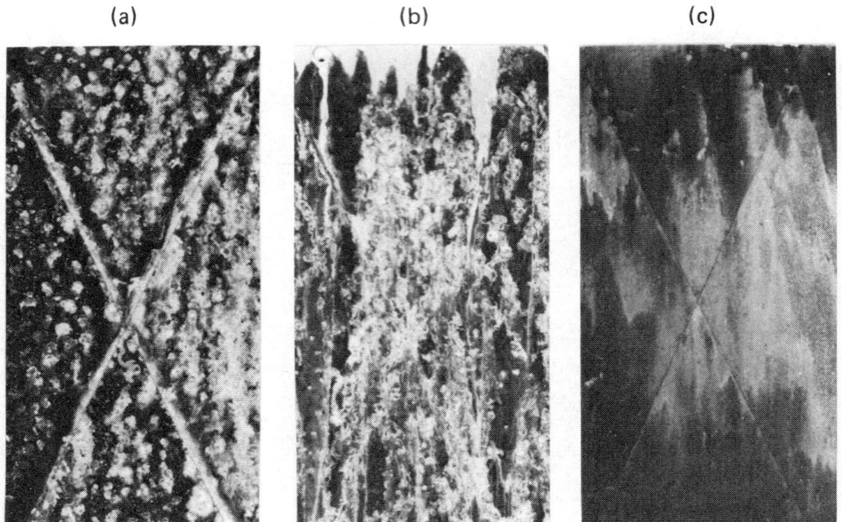

Figure 2 Corrosion resistance of uncoated surface treatments to salt spray (5%): (a) FPL etch (exposure time: 21 hr); (b) phosphoric acid anodize (exposure time: 141 hr); (c) chrome acid anodize (exposure time: 336 hr). (Courtesy Douglas Aircraft Co.)

Figure 3 Corrosion inhibition of adhesive primer to salt spray (5%) for 1000 hr: (left) coated; (right) uncoated. (Courtesy Douglas Aircraft Co.)

Table 4 Corrosion Resistance of Surface Treatments

Surface treatment	Adhesive primer (mils)	5% salt spray (hr)	Observation
FPL etch	None	21	Heavy corrosion with pits
	0.13	3000	Scattered blisters with corrosion
Phosphoric acid	None	141	Heavy corrosion with pits
anodize	0.15	3000	No corrosion
Chromic acid	None	336	Scattered minute corrosion pits
anodize	0.18	3000	No corrosion

Service

Use of finish systems in normal and adverse environment conditions has been going on for a number of years. The amine-cured epoxy primer and top coat have been used for both interior and exterior application on commercial and certain military aircraft. The epoxy amine primer has been proven to be a very adequate corrosion-inhibiting coating, in addition to its economical cost. The use of pigmented epoxy top coats for exterior aircraft use results in discoloration and severe chalking. In addition, cracking of the film around fasteners and rivets results in filiform corrosion on some fuselage skins. Since the implementation of the urethane top coat system the primary application for the epoxy has been for use on interior aircraft surfaces. The epoxy polyamide/aliphatic urethane system is currently being used on interior and exterior aircraft surfaces. The service performance on commercial aircraft has shown it to be a durable and reliable corrosion-inhibiting system. The Navy and the Air Force have also used this system with success [8, 9]. This system can generally be removed without difficulty.

A modified epoxy primer/aliphatic urethane top coat system is currently used on some commercial aircraft. It provides a more flexible primer and top coat over fastener areas, which enhances the inhibition of filiform corrosion. Its overall corrosion resistance is comparable to the epoxy polyamide/urethane top coat system. Complete stripping of the system is difficult; usually, only partial removal of the primer can be accomplished.

A two-primer coating system consisting of the wash primer and urethane primer overcoated with an urethane top coat is used by some American and European airlines. Some systems presently used are reported to provide improved filiform corrosion inhibition. Historically,

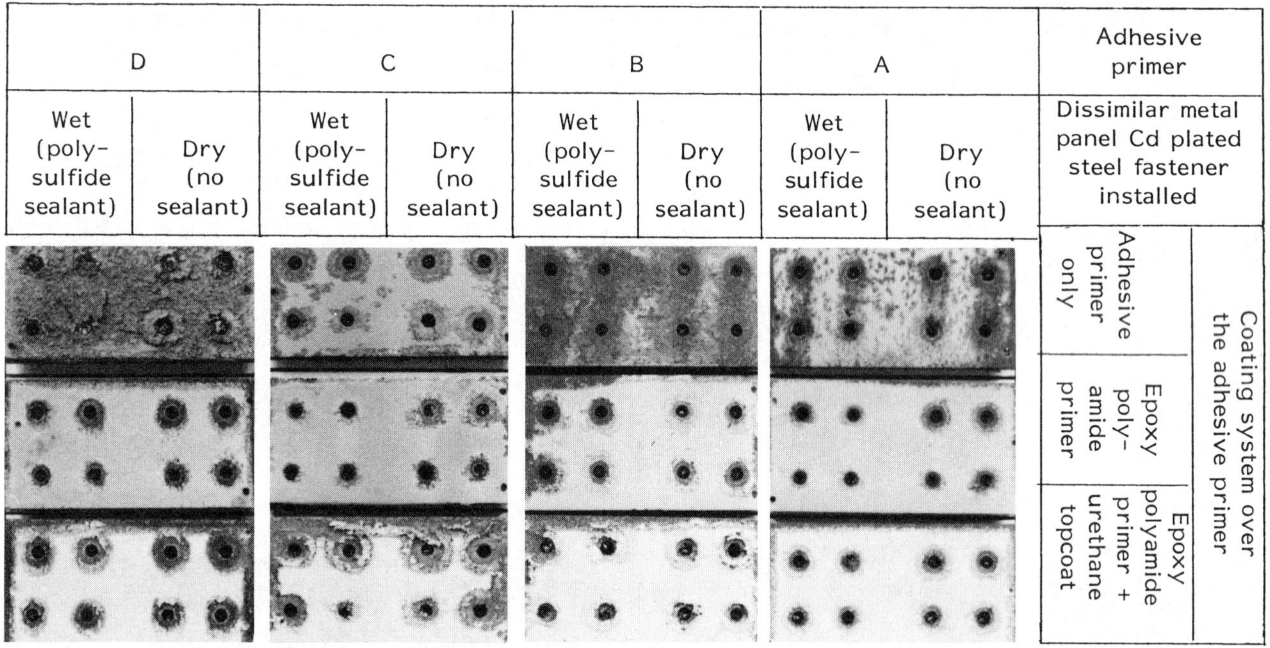

Adhesive primer	Dissimilar metal panel Cd plated steel fastener installed	Coating system over the adhesive primer		
		Adhesive primer only	Epoxy poly-amide primer	Epoxy polyamide primer + urethane topcoat
A	Dry (no sealant)			
A	Wet (poly-sulfide sealant)			
B	Dry (no sealant)			
B	Wet (poly-sulfide sealant)			
C	Dry (no sealant)			
C	Wet (poly-sulfide sealant)			
D	Dry (no sealant)			
D	Wet (poly-sulfide sealant)			

Figure 4 Acidified salt spray (1 month)—comparison of adhesive primers. The aluminum alloy is 7075-T6, and the surface treatment is phosphoric acid anodize. (Courtesy Douglas Aircraft Co.)

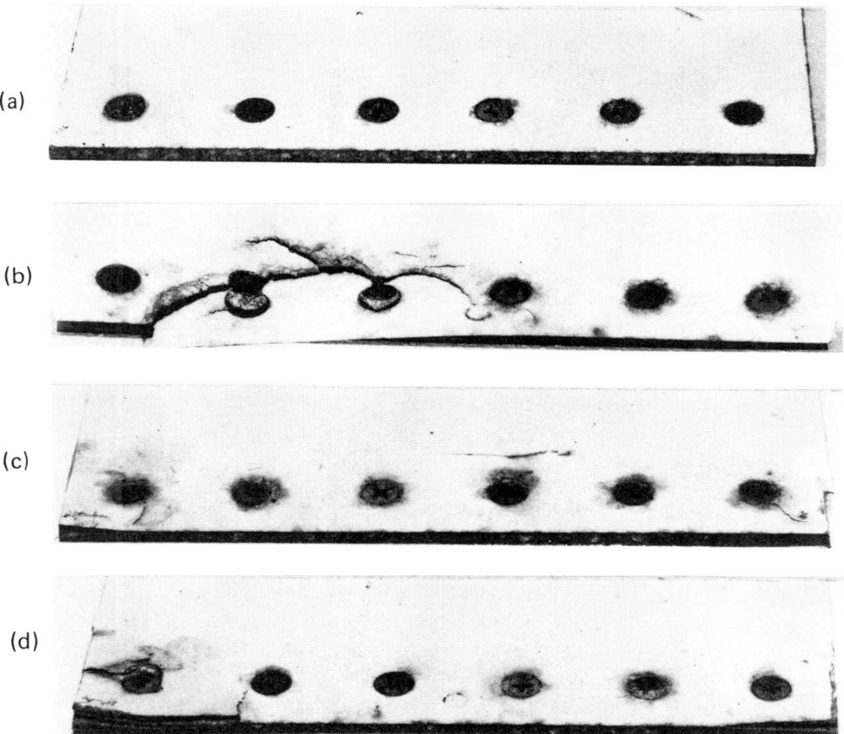

Figure 5 Beach exposure (18 months). Finish system (epoxy polyamide primer and urethane top coat) applied over adhesive primer.
(a)—(d) = different adhesive primers. (Courtesy Douglas Aircraft Co.)

this has been the weakness of this type of system in addition to its low
resistance to phosphate ester hydraulic fluid. Removal of this system
has always been accomplished with ease due to the wash primer.

D. Repair of Damaged Finish

Removal

The finish system on bonded assemblies consists of the adhesive
primer overcoated with a primer or with a primer and a top coat system.
In the event that any one or all of the coatings or systems require removal, the following methods can be used.

Adhesive primer: The standard phenolic acid strippers normally
are ineffective. The adhesive primer can be removed from detail

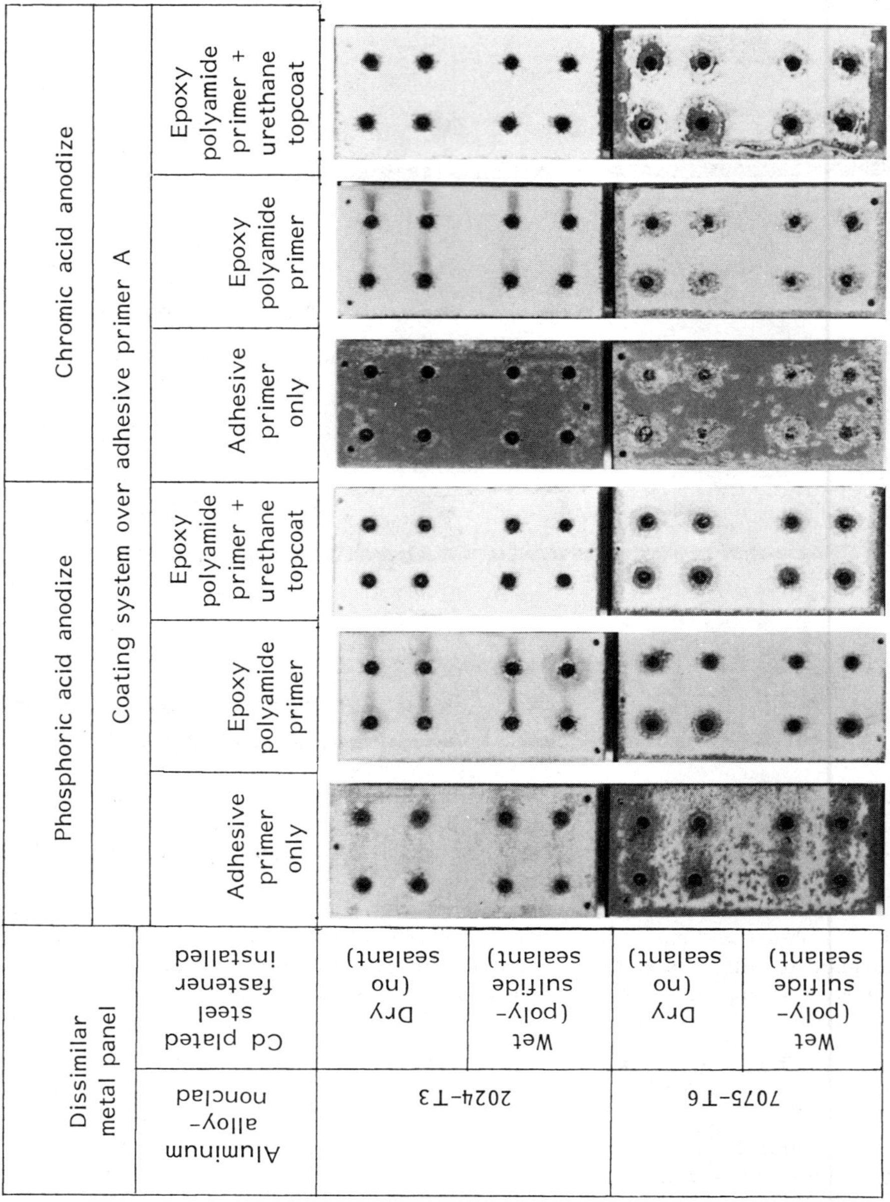

Figure 6 Acidified salt spray (1 month) of coated anodized aluminum. (Courtesy Douglas Aircraft Co.)

parts using hot chromic acid. Bonded parts or assemblies should not be immersed in this solution [7].

Finish system over adhesive primer: The finish can be removed with phenolic or acid stripper; the adhesive primer is not affected. Therefore, if removal of the primer is necessary, mechanical or abrasive removal methods must be employed [7].

Corrosion Protection

To assure corrosion protection, proper surface treatment should be provided prior to application of the primer and top coat. Surfaces which have the coating removed down to the metal substrate should be abraded and surface treated with an Alodine-type conversion coat. Those surfaces where the adhesive primer is intact and no metal is bared require only reactivation by lightly abrading and solvent cleaning.

IV. CONCLUSIONS

1. In addition to bonding properties, the adhesive primer provides corrosion inhibition to the underlying surface treatment [7].
2. The epoxy polyamide primer or similar chromated primers applied over the adhesive primer reduce coating failure by greatly improving corrosion inhibition [7].
3. The adhesive primer is not strippable with either a phenolic or an acid brush-on stripper. The removal of primer from detail parts can be done with hot chromic acid [7].
4. Based on laboratory and historic in-service data the following systems applied over the adhesive primer will be most effective in providing corrosion inhibition.
   a. *Interior surfaces*: Above the floor line, use amine-cured epoxy primer. Below the floor line, use amine-cured epoxy primer and top coat.
   b. *Exterior surfaces*: Use the epoxy polyamide-cured primer and aliphatic urethane top coat [7] or the modified epoxy primer and flexible aliphatic urethane top coat. The primary advantage of the two-primer (wash primer and urethane primer) coat and aliphatic urethane to coat system is that it provides easy strippability.

REFERENCES

1. Perry, R. H., C. H. Chilton, and S. D. Kirkpatrick, eds., *Chemical Engineering Handbook*, 4th ed., McGraw-Hill, New York, 1963.

2. Slabaugh, W. H., W. Dejager, S. E. Hoover, and L. L. Hutch-
    inson, Filiform Corrosion of Aluminum, *J. Paint Technol.*, Vol.
    44, No. 566, 1972.
3. Tomashov, N. D., *Theory of Corrosion and Protection of Metal*,
    Macmillian, New York, 1966.
4. Boises, D. B., B. J. Northan, and W. P. McDonald, Corrosion
    Inhibitors in Primers for Aluminum, NACE Paper 31, 1969.
5. Lee, H., and K. Neville, *Epoxy Resins: Their Application and
    Technology*, McGraw-Hill, New York, 1967.
6. Rosenberg, P., *Urethane Paint Technology*, Society of Automo-
    tive Engineers, 1969.
7. Shannon, R. W. et al., General Material Property Data, Primary
    Adhesively Bonded Structures Technology (PABST), U.S. Air
    Force Flight Dynamics Lab. Tech. AFFDL-TR-77-107, Sept. 1979.
8. Miller, R. N., and G. G. Seeliger, Full Temperature Range Pro-
    tective Finish for Fastener Areas of Carrier Based Naval Air-
    craft, Rep. LG 76ER0020, 1976.
9. U.S. Department of the Air Force, Polysulfide Primer, Final
    Report, MEP (AFR 66-8), 1976.

13

Structural Analysis of Adhesive-Bonded Joints

L. J. HART-SMITH *Douglas Aircraft Company of the McDonnell Douglas Corporation, Long Beach, California*

EDWARD W. THRALL *Consultant, Aeronautical Structures, Point Arena, California*

I. INTRODUCTION

The preceding chapters have presented the importance of surface preparations that yield desired oxides, the adhesive selection, the mechanical properties, and necessary durability testing. Now we shall examine
the end product of the whole process—the bonded joint. A basic
criterion that must be accepted by designers is the fact that the weak
link in a bonded joint should never be the adhesive. The overlap length
and other metal details, such as tapered edges, will prevent adhesive
failure from occurring before the metal parts fail. It has also been
shown in the preceding chapters that adhesives are relatively weak in
tension, so designers must make primary load transfers with the adhesive in shear. If no alternative method is possible except with the
adhesive in tension, careful attention to the metal details is required
to eliminate "hard" spots in the joint. This will give better load distribution and allow more of the adhesive material to carry the load.

II. DOUBLE-STRAP BONDED JOINTS

The double-strap bonded joint is probably the most desirable joint transferring loads between two members. The loads in the metal parts can
be in tension, compression, or in-plane shear. Ideally, each splice strap
should be precisely one-half as thick as the basic parts being joined,
so that the adhesive is equally effective at each end of the overlap (see
Fig. 1). An important factor has been obtained from fatigue tests. It

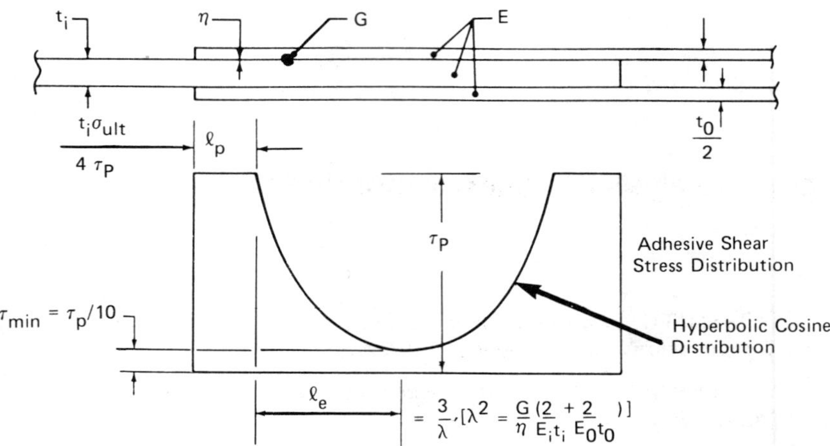

Figure 1 Graphic representation of the adhesive shear stress distribution in a double-lap bonded joint. The length of the plastic zone, ℓ_p, is limited by the ultimate strength of the inner or outer adherend. $\ell_p = t_i \sigma_{ult}/4\tau_p$. The plastic end zones are equal in length only when joint is balanced or $t_i = t_o$. The center elastic trough, $2\ell_e$, shall be long enough to prevent creep at the middle, $\ell_e = 3/\lambda$ and $\lambda^2 = (G/\eta)[(2/E_i t_i) + (2/E_o t_o)]$.

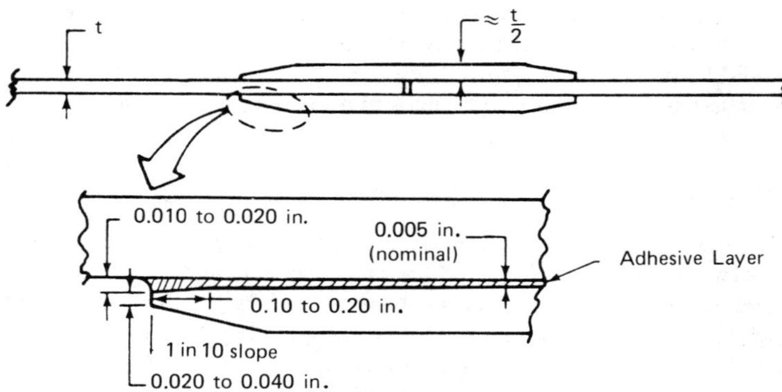

Figure 2 Tapering of edges of splice plates to relieve adhesive peel stresses. All dimensions in inches.

was found that the failures consistently occurred in the splice plates
of the double-strap bonded joints. Therefore, it seems prudent for
designers to increase the splice strap thickness by 25%. When splicing
sheet or member thicknesses of 0.032 to 0.090 in., the straps can be
of uniform thickness. If the primary member thickness is greater than
0.090, tapering of the ends of the splice plates must be given serious
consideration. This is shown in Fig. 2. The prime purpose of this
tapering is to reduce the tension or peel load or stress on the adhesive
produced by the joint eccentricities, which would cause a joint failure
at a lower load than the adhesive could sustain under pure shear loading. Figure 3 is presented to show graphically how the shear stress
distribution changes with length of the bonded overlap length, ℓ.
Figure 4 shows the detail of a skin splice for a pressurized fuselage.

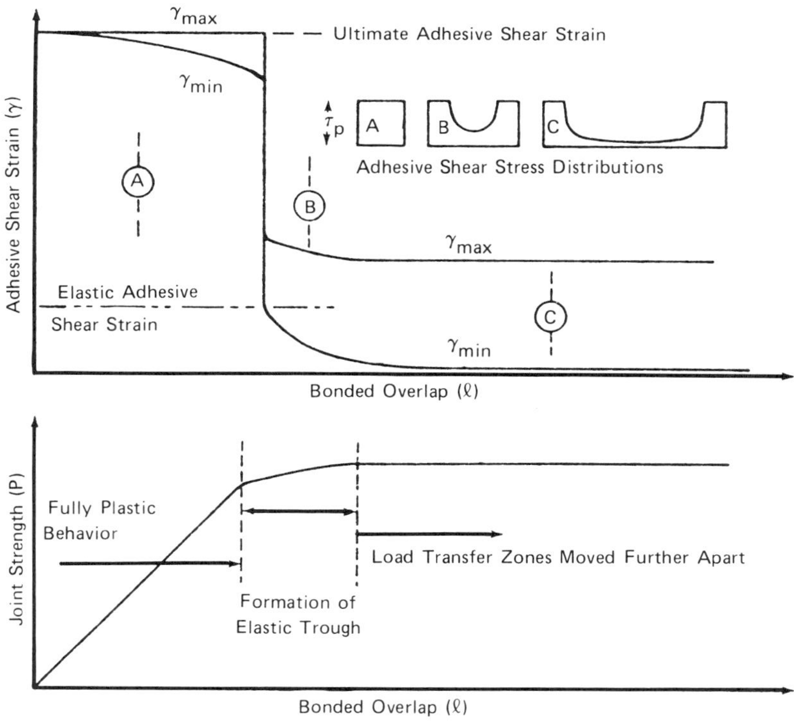

Figure 3   Influence of overlap on maximum and minimum adhesive shear
strains in bonded joints.

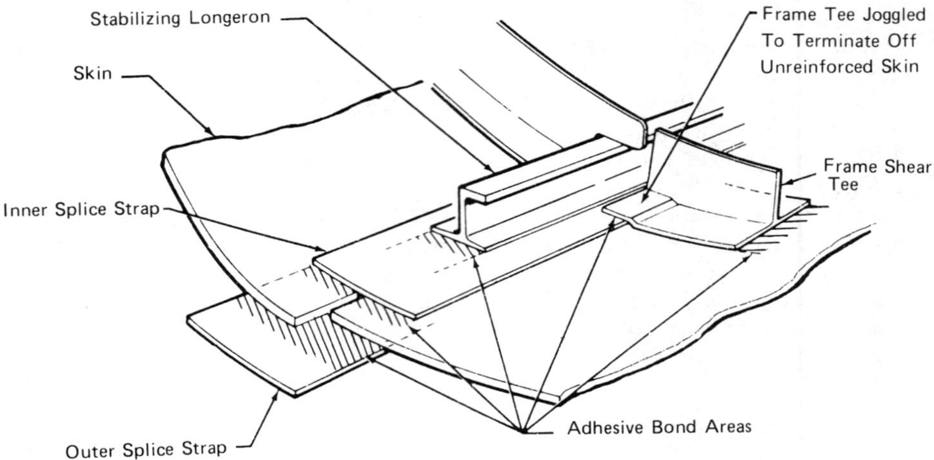

Figure 4   Typical longitudinal double-strap bonded splice for pressurized fuselage.

III. SINGLE-LAP BONDED SKIN SPLICES

Single-lap joints have less bond strength than double-strap bonded
splices, but that does not matter for gauges so thin that the bond is
still stronger than the metal. It simply means that single-lap joints are
much easier and less expensive to manufacture, because there are no
splice details to make or to locate in the bonding fixture. The single-
lap joint is also considerably easier to inspect ultrasonically, because
a complete inspection can be made from only one side of the joint. The
prime disadvantage of the single-lap joint is that it imposes an eccentricity in load path which, in turn, induces bending moments in the
adherends and much higher peel stresses in the adhesive than for
double-strap joints. However, those problems can be alleviated substantially by using a sufficiently generous overlap to smooth out the
eccentricity gradually. The overlap, of course, adds weight to the
splice—but surprisingly little. At an overlap thickness (ℓ/t) ratio of
75:1, the splice weight is roughly equal to that of 30:1 overlap double-
strap splice with the splice straps one gauge thicker than half the skin
thickness.

Single-lap bonded joints for longitudinal fuselage skin splices in
pressured fuselages need not be quite as structurally efficient as
double-strap bonded splices. As long as the splices represent less of
a stress concentration than exists in the skin alongside the longerons
as a result of cabin pressure, they do not impose any structural inef-
ficiency.

The precise analysis of single-lap joints is very complicated because of the interactions between each of the three potential failure modes: adherend yielding, adhesive peel, and adhesive shear. See Ref. 1 for further information.

For design purposes, the following simplified procedures are recommended for the ductile adhesive:

1. Set the length-to-eccentricity ratio (length-to-thickness ratio for uniform adherends) in the range 50 to 100, depending on the criticality of the application. At maximum possible loads for 2024-T3 and 7075-T6 aluminum alloys, the ratio $\ell/t = 100$ corresponds to bending stresses that are only 10% of the membrane stresses, while $\ell/t = 50$ gives a corresponding joint efficiency of 80%. In contrast, $\ell/t = 10$ for aluminum produces an efficiency of only 35% (i.e., the bending stresses are 65/35 times as high as the membrane stresses that induced them). Figure 5 presents a chart with which to calculate the joint efficiency factor for both balanced and unbalanced joints. The joint efficiency is a function of the applied stress level, with lower efficiencies for lower stresses.
2. To relieve peel stresses, taper the edges of the adherends down to 0.030 ± 0.010 in., if the basic thickness exceeds 0.050 in. The taper angle should be about 0.020 in. per 1/4 in.
3. Make an approximate check on the shear strength of the adhesive by assuming that the joint is one-half of a double-lap joint which has twice as thick a midskin. Effectively, this means that single-lap joints should not be used with aluminum gauges in excess of about 0.063 for ductile adhesives and subsonic aircraft.

The effect of using brittle, high-temperature adhesives in single-lap joints is to restrict still further the applicable range of adherend thicknesses that could be bonded effectively into a single-lap joint.

The simple rules stated above can be used for design because the correct answers are so strongly dependent on the ℓ/t ratio. It is important to stress that the design condition is one of ℓ/t for the bending of the adherends and not one of bond area for the shear strength of the joint.

The key difference between double-lap and single-lap joints is that bending moments are set up in the metal due to the single-lap load path eccentricity. A simple linear analysis considerably overestimates the correct bending stresses, typically by a factor of 2 to 3. A nonlinear analysis must be used to account for structural distortions under load to relieve the eccentricity in load path. Such refined analyses indicate that the combination of direct and bending stresses can be only about twice as high as the normal stress instead of the fourfold increase that the linear analysis would have predicted. Refer to Ref. 2 for a more

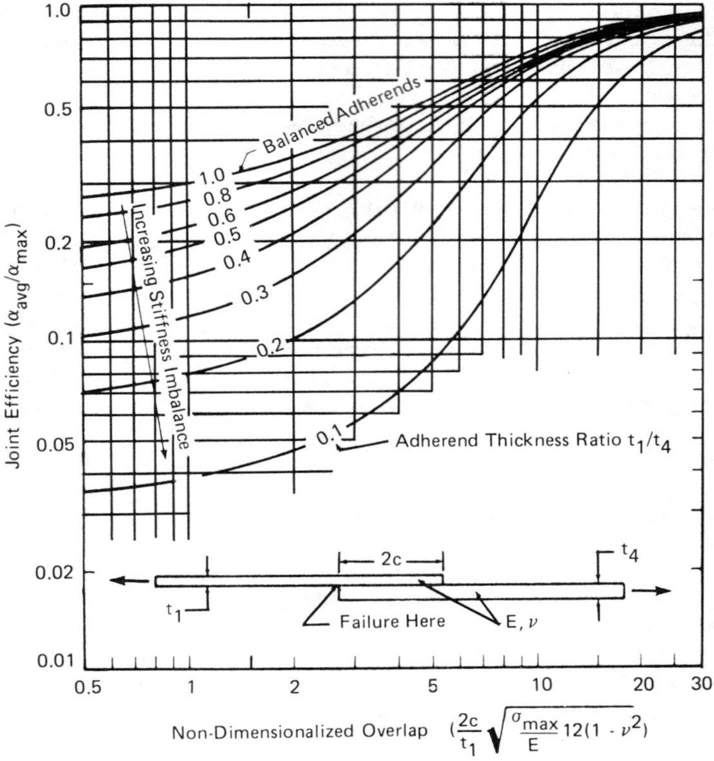

Figure 5 The curves show the effect of the ratio of bond overlap length 2c, to the thinner adherend thickness, t_1, versus the joint efficiency, $\sigma_{avg}/\sigma_{max}$. The thinner adherend, t_1, is always critical in combined bending and axial load at the end of the splice area or overlap.

complete discussion. The key to this bending-moment relief is a long overlap to absorb the eccentricity smoothly. Figure 6 shows that, for aluminum adherends, an average stress of 20 ksi will induce a maximum sheet stress of 55 ksi if the overlap/thickness ratio is 10, but only 30 ksi if the ratio is 50. There is usually no point in increasing the ℓ/t ratio beyond 100, but an ℓ/t ratio of 50 should be regarded as an absolute minimum for unsupported single-lap joints. PABST program testing of riveted single-lap joints with sealant having an ℓ/t ratio of 80 showed that the joints usually attained an adequate fatigue life unless there was countersinking into sheet 0.50 in. thick or less. This is where hot bonding in combination with fasteners would be valuable.

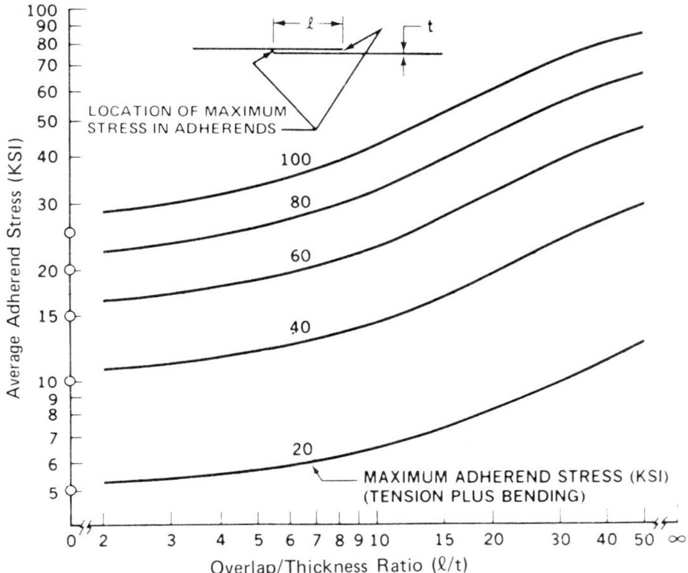

Figure 6   Stress amplification of single-lap joints due to eccentricity
in load path.

IV. SINGLE-STRAP (FLUSH) BONDED SPLICES

Figure 7 shows two examples of single-strap bonded joints that must
be flush around the circumference of a fuselage or a spanwise wing
skin splice to minimize aerodynamic drag.   The splice in Fig. 7a joins
relatively thin (0.050 in.) sheets together, and the joint is supported
at right angles to the load by a substantial member, the frame.   The
splice in Fig. 7b joins two fairly thick (0.090 in.) skins with a thick
doubler that would be tapered as shown in the figure.   It would also
have support normal to the load.   The splices shown in Fig. 7 are re-
finements of the basic splice shown in Fig. 8.   This basic splice failed
after one lifetime (the goal was four).   Figure 8a shows where crack
initiation was and because that spot is next to impossible to find during
inspection it could lead to a damaging failure.

The basic difficulty with this kind of joint is that it is like two
single-lap joints butted together, so that in the middle there is no
length over which to alleviate the eccentricity in load path.   There
is thus a high bending moment in the middle of the splice plate, with
very high adhesive peel stresses where the skins are butted together
and sometimes even high bending moments in the skin at the edges of
the splice strap(s).

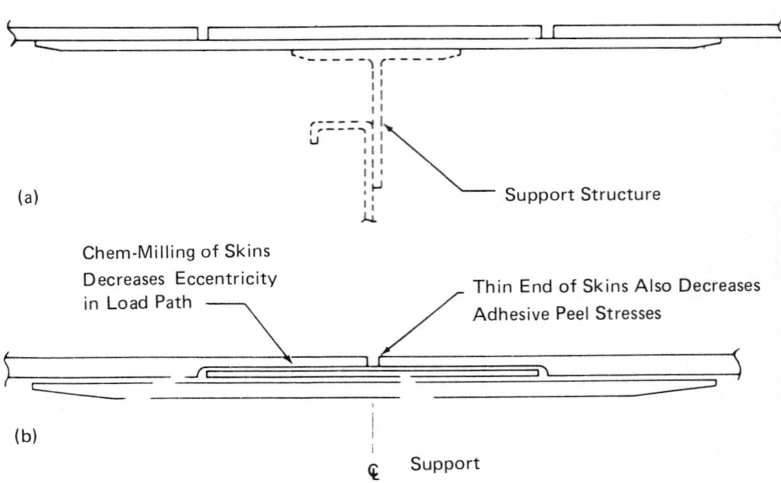

Figure 7   Two design concepts for structurally improving flush bonding splices. Design (a) shows how the doubler area with maximum bending moment is helped with the addition of a bonded doubler. The inner doubler overlap on the skin is sufficient to transfer load. The ends of the inner doubler can be tapered to reduce peel stresses. Design (b) shows how the skin, being spliced, can be chemically-milled at the ends to decrease the eccentricity of the load path. The position of the chemically-milled step occurs at an area of reduced axial stress. The thin ends of the outside members decrease the adhesive peel stresses. The inner splice plate can be laminated as shown or made from a single member. The increased thickness at the center increases the bending strength.

The problem with the single-strap joint is that reinforcing the splice plate(s) to increase the bending strength also increases the eccentricity in load path, and hence the bending moment. For other designs, one should also account for the bending restraint afforded by any curvature of the skin being spliced. The analysis of this kind of flat joint is now documented in Ref. 3. The real concern about this kind of joint is that it is inherently weaker than the skins being bonded together. It therefore lacks the excellent damage tolerance of double-strap splices and has the potential for complete failure, either in the splice plate as observed under test, or in the adhesive. A narrow strip of adhesive peeled apart immediately adjacent to where the skins butted together on the test panel (Fig. 8a). This occurred over most of the width of the panel but did not propagate more than about 0.25 in. across the bond because the peel stresses decrease as the disbond grows.

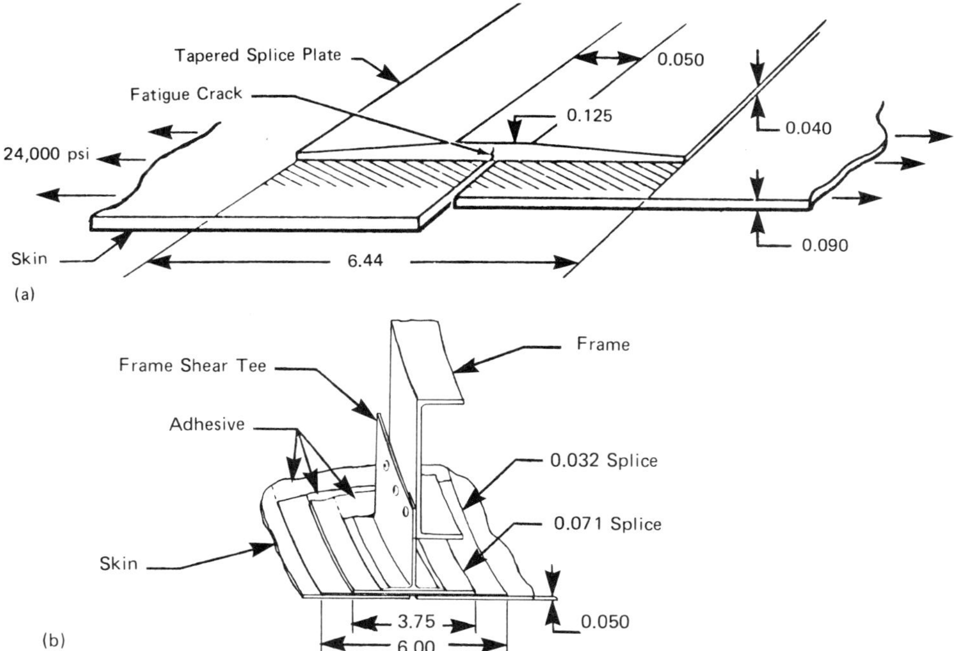

Figure 8   Two examples of single-strip (flush) splices that have been used in design.   The tapered splice plate in (a) was used on a test panel that was fatigue tested.   Failure initiated where shown.   Z stiffeners crossed the splice at 12 in. on-center.   The splice shown in (b) is a circumferential bonded splice for an aircraft fuselage.   All dimensions shown are in inches.

If the splice in question is for a circumferential splice on a pressurized fuselage, the pressure-induced axial load is only half as high as the corresponding hoop load.   This is why there should be no problem with the bonded flush circumferential splice.   Bonded single-strap splices would not be appropriate for longitudinal joints in pressurized fuselages.

Study of single-strap bonded joints indicated that increasing the overlap was a powerful technique for alleviating the effects of the eccentricity in load path, but that such joints would be much heavier than equivalent mechanical splices.   The simple design rules of thumb which are apparent are that the total splice width should be about 150 to 200 times the thickness of the sheet being spliced and that the thickness of the tip of the splice should not exceed about 0.04 in.   It has also

been confirmed that a linear analysis of the bending stresses is far too
conservative. No analysis exists for the shear stresses in the bond, so
a simulation of the shear transfer in the joint as part of a double-strap
splice is all that can be used at present.

Experimental evidence on this subject for the tapered single-strap
splice can be summarized as follows. Fatigue tests were run on a panel
of 0.090-in. 7075-76 aluminum alloy 24 in. wide and 48 in. long. The
splice plate was 0.125 in. thick, tapered down to a 0.040 in. at the
edges, and was a total of 7 in. wide. Two Z longerons were bonded
longitudinally (at right angles to the splice doubler) to restrict the
bending deflections of the panel. The test results are recorded in
Ref. 4. While the panel survived a considerable number of cycles prior
to failure, the final failure was a crack running straight across the middle of the splice, from one side to the other, with both stiffeners completely disbonding. The initial fatigue damage prior to fast fracture
had not been detected and consisted of a crack in the splice only 1.4
in. long on the buried surface and 0.5 in. long on the visible side. The
average skin stress at the time of failure was only 24 ksi, but analysis
predicted a stress of 50,400 psi at the origin of the crack. In other
words, this type of bonded joint has poor fail-safety and is difficult to
inspect. The analysis also predicted disbonding throughout a band
across the middle of the splice. Testing confirmed the presence of such
a band, of width 0.3 to 0.4 in. on each side of the skin joint. The peel
stresses decrease as that disbond grows because the splice plate is able
to deflect more easily, just as in an equivalent riveted joint (which has
a finite distance between the inner rows of rivets instead of just a narrow gap over which to deflect).

V. SKIN-TO-STIFFENER BONDED JOINTS

In thin-sheet metal structures, shear webs are light and carry the loads
with diagonal tension, and this factor causes wrinkles to form at approximately 45°. To increase the load capacity of a thin shear web, vertical
stiffeners are spaced at various distances extending from the upper cap
to the lower cap. The stiffeners consequently pass over the wrinkles,
flattening them out. The resulting loads from wrinkles tend to push
the stiffener away from the shear web or put a tension load in the at-
taching rivets or adhesive.

Perhaps the best testimonal to the success of the bonding of the
stiffening tees to the skin is given in Fig. 9. Despite quite deep skin
wrinkles in this shear panel test, the skin flange of the stiffener stayed
bonded to the skin while the inner leg was ripped off, as shown. This
happened not just once but several times on test panels. The selection
of the stiffener proportions for bonding was more a matter of skilled
judgment than precise analysis, and is likely to remain so because all
the necessary analyses have not yet been developed. Actually, the

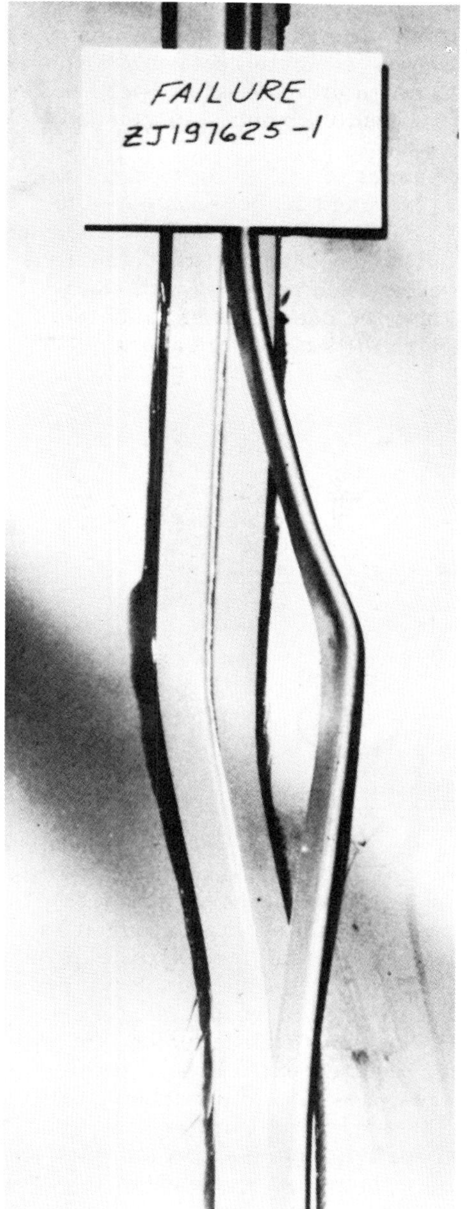

Figure 9   Shear test failure detail.   (Courtesy Douglas Aircraft Co.)

number of constraints on the stiffener proportions are such that they
almost design themselves. The stiffener area and moment of inertia
were sized for a rivet attachment design. It would be easy to argue
that an adhesive connection could handle a stiffener of greater area
and stiffness and thereby increase the load-carrying ability of the
shear web.

The process by which the skin flanges of stiffening tees are designed is explained in Fig. 10. The extruded sections have a wide
symmetric bonding flange, with a T or J section, rather than an asym-
metric angle or Z section. The latter have a hard heel that would disbond easily, unless the structure were buckle resistant with close-
spaced stiffeners. Analysis shows that the peel stresses induced
by cabin pressure are very sensitive to the base-to-thickness (b/t)

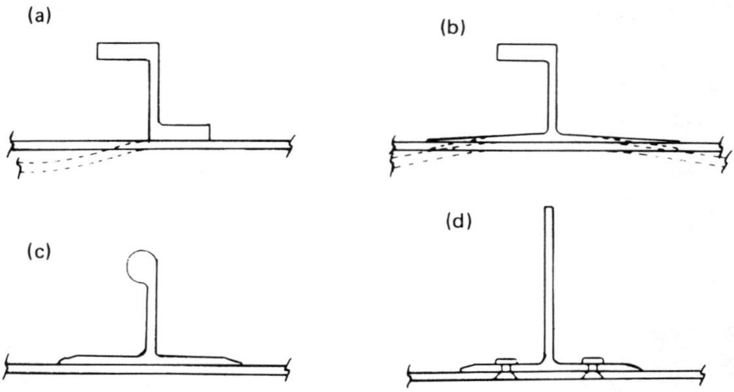

Figure 10   Design considerations for the proportions of stiffeners to
be used in bonded stiffeners. (a) a Z section can have problems.
Bond width may be too narrow. Excessive peel stresses occur at the
heel of the Z due to skin shear wrinkles. Bonding pressures on over-
hang may make the stiffener roll over. (b) The thin flexible tip promotes low adhesive peel stresses because the stiffener and skin can deflect together under low loads. (c) The thin flexible tips of skin legs
promote low adhesive peel stresses. Easier to extrude than taper
shown in (b). Constant thickness over most of width makes bonding of
matching details easier. Also, leg thickness makes stiffener more effective in stopping skin cracks. Less handling damage than stiffener shown
in (b). (d) Outer flanges of stiffener should be wide enough to permit
riveted repairs when necessary. In any design it must be remembered
that damage tolerance (crack stopping) ability requires the bond width
to be large enough when considering stiffener area and spacing. This
will prevent the unzipping of the bond.

ratio (Fig. 11). The peel stresses decrease in the ratio $(t/b)^2$. The most important upper limit on the b/t ratio is the ability to extrude such sections without gaps and tears in the metal. Most of the bonded stiffeners used in tests had a flange thickness of 0.050 in. and 0.090 in. and a total base width of 1.75 in., the taper over the last 0.25 in. reducing the tip to a nomimal thickness of 0.020 to 0.030 in.

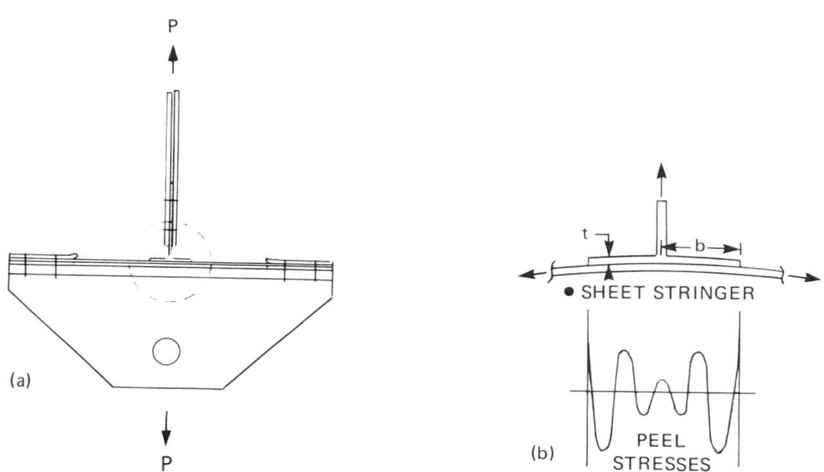

Figure 11   (a) Tension-tee tests were conducted to see the effects of adhesive selection, adherend primer, and test temperature. The test results are shown in the chart. (b) Possible peel stress distribution when skin wrinkles from shear loads or tension pulling skin away from stiffener. High b/t ratios reduce peak stresses.

			FAILURE LOAD TEST TEMP			DESIGN LOAD
SKIN	ADHESIVE	PRIMER	-50 5°F	R.T.	140 5°F	-50°F
0.090 7075-T6	FM 73	BR 127	1740 LB			249 LB
0.040 7075-T6	FM 73	BR 127	1595 LB			389 LB
0.090 7075-T6	AF 55	XA 3950			4000 LB	249 LB
0.090 7075-T6	M 1133	BR 127	2170 LB			249 LB
0.090 7075-T6	AF 55	XA 3950	5910 LB	5050 LB	5075 LB	249 LB
0.090 7075-T6	M 1133	BR 127	2640 LB	4650 LB		249 LB
0.040 7075-T6	AF 55	XA 3950	1670 LB	3700 LB	4220 LB	389 LB
0.040 7075-T6	M 1133	BR 127	2105 LB	3275 LB	3358 LB	389 LB

Two analyses that should be performed to aid in proportioning
stiffeners for adhesive bonding relate to the peel stresses induced by
cabin pressure and to damage tolerance. More analysis of the peel
stress is needed before testing can become unnecessary, but the analysis to date has identified the dominant parameters and guided the test
program. Figure 11 shows the typical distribution of adhesive peel
stresses between the skin and frame tee or longeron, as well as the
results from a test program. It is clear that the bonded joints tested
had more than adequate strength for this load condition. Figure 9 indicates that these bonded joints are also strong enough to resist shear
wrinkles, for which analysis is not yet available. The most important
element of this investigation is the need to represent the skin membrane
stresses as well as the bending stresses. Otherwise, the test deflections are much greater than occur in practice, and premature failure
results. For this reason the test specimens shown are bolted tightly
to a very stiff, thick steel base, or spreader bar. These, in turn, restrict the deflection to close to what is observed on the real structure,
about 1/16 of an inch, rather than the more than 1/4 of an inch which
was observed when the bolt holes were drilled with a slight clearance.
Tests such as these are wasted if the structural deformations under
load are not represented properly. Note that the conventional angle
shear clip used to tie the skin to the frames in riveted construction
would not be adequate for adhesive bonding because of the hard heel
under the shear web. The flexibility of the wide symmetric base on the
tee extrusion is most important for bonding. It is interesting that,
even so, there is tremendous amplification of the adhesive peel stresses
with respect to the nominal P/A stress; so much of the bonded area is
in compression rather than tension that it is fair to say that nearly 49%
of the bonded area is trying to push apart the other 51%.

VI. LOAD TRANSFER IN ADHESIVE-BONDED JOINTS

The key to understanding the transfer of load from one member to another through an adhesive-bonded joint is that the load transfer is inevitably nonuniform. The great bulk of the load transfer is effected
in narrow zones in the immediate vicinity of the ends of the bonded
overlap. It is important that the peak strains in the adhesive not exceed the material capabilities either for ultimate strength or for the
regularly recurring fatigue loads. It is equally important that the minimum adhesive shear strain, near the middle of the overlap, should not
exceed some critical value if the joint is to resist creep accumulation to
ensure a long service life. Both of these criteria apply to all types of
bonded joints, from simple configurations to the more complex stepped-
lap joints used to bond fibrous composites to titanium edge members on
modern fighter aircraft.

It is most important that the adhesive strength must be greater than that of the members being joined, to provide for manufacturing imperfections such as flaws, porosity, and variable-thickness bond lines, as well as for the processing variations that occur at any one stage from the manufacture of the adhesive film to its final cure. Adhesively bonded structure performed well on the PABST program [5], just as have the environmentally resistant adhesive systems (such as Redux and chromic acid anodize) used in service. However, treating the adhesive-like metal or composite structural elements and designing to minimum margins would invite a disaster more serious that that which resulted from past use of adhesive systems and surface preparations with inadequate resistance to moisture.

VII. ADHESIVE SHEAR CHARACTERISTICS

To design adhesive-bonded joints, one must have some characterization of the adhesive mechanical properties (see Chapter 9). Even today, the most widely measured adhesive property is the average failing shear stress on the standard single-lap test coupon (ASTM specification D-1002), which has an overlap of 0.5 in. and adherend thicknesses of 0.063 in. Yet the test values so obtained have never been usable directly as design allowables. The typical test results from these tests for tough adhesives suitable for use on subsonic transport aircraft are in the range 4000 to 6000 psi, yet in the past design allowables used were in the range 400 to 600 psi. Even with those values, one could still design a weak joint because the adhesive was never actually uniformly stressed at 400 to 600 psi. What made early applications of adhesive bonding effective despite such crude analytical tools was that the applications were confined to thin, mainly secondary structure in which the metal was inevitably weaker than the bond, even if it had not been analyzed rigorously.

The application of adhesive bonding to much more heavily loaded primary structure necessitated a far better characterization of the adhesive properties. The first major improvement came with the introduction of the napkin-ring torsion test, in which the adhesive was loaded in almost uniform shear throughout, with negligible peel stresses. Rutherford and his associates at Singer have done valuable work in this field, as documented in Chapter 9 and Ref. 6. These test specimens, however, are relatively expensive to manufacture because of the precise alignment needed. The test specimen used more widely today to generate close-to-true adhesive shear stress-strain characteristics is the thick-adherend short-overlap coupon described in Ref. 7. Frazier at Bell Helicopter and Krieger at American Cyanamid pioneered much of the work with this coupon. The end product of such tests, described in Fig. 12 as the true characteristic, is the basis for the analyses used

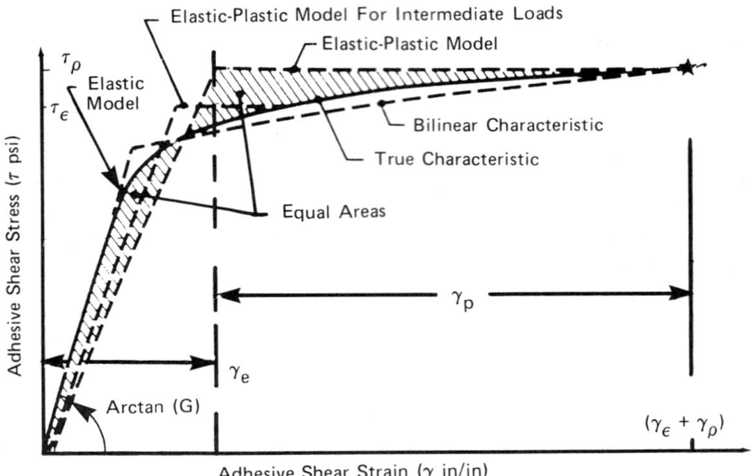

Figure 12   Representations of adhesive nonlinear shear behavior.

today throughout the industry for the analysis of structural bonded
joints.

Figure 12 also shows some of the various adhesive models used in
analyzing bonded joints. Not shown are the even more precise charac-
terizations, such as the Ramberg–Osgood model, which have led to such
mathematical complexity that very few such solutions have been or can
ever be derived. Reference 1 shows explicitly that closed-form analysis of the double-lap joint predicts precisely the same joint strength
for any two-straight-line characteristic having the same failure stress
and strain, and strain energy to failure. Thus the precise shape of
the stress-strain curve is not all that important. The elastic-plastic
model in Fig. 12 is a particular case of the bilinear model and is the
simplest possible model that can account for adhesive nonlinear behavior.
This has permitted more solutions [1–4,8,9] to be calculated on that
basis than for any other model.

Once the knee of the stress-strain curve has been exceeded, permanent damage is being done to the adhesive. Consequently, repeated
loading far beyond the knee must accumulate fatigue damage in the adhesive. It is possible to restrict all frequently recurring loads to no
more than 50% of the adhesive elastic capability, leaving the remainder
of the adhesive strain energy for accommodating bond line flaws and
irregularities. Experience with brittle adhesives, in other more highly
loaded structures, indicates that adhesives can be operated safely
slightly beyond the knee in the stress-strain curve. There are no data
yet to indicate just how far past the knee is acceptable, but the message
is clear—only slightly past, not all the way to the end.

Figure 13 shows how the adhesive properties change with temperature. That is why strength checks must be made at the extremes of the operating environment rather than just in the middle. Actually, the ultimate joint strengths will not vary as much with temperature as any one property would suggest, because the strain energy to failure is nearly constant. The table in Fig. 13 gives properties that can be used as a first approximation for many ductile adhesives cured at 250°F (epoxy nitriles and epoxy nylons).

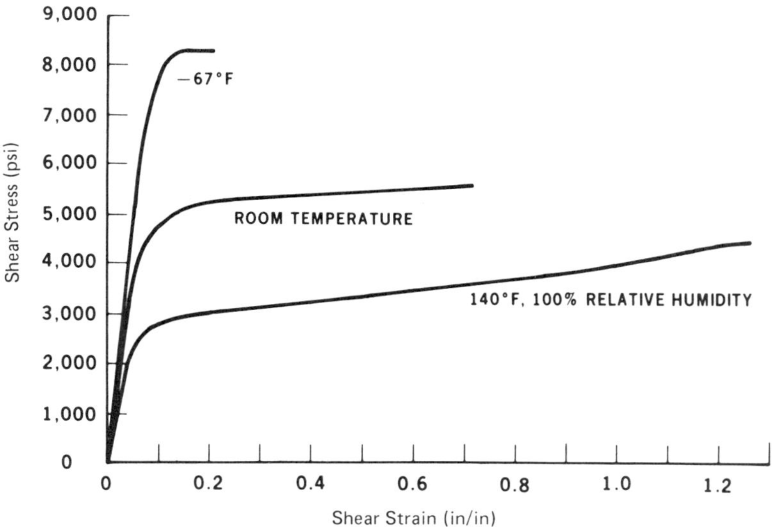

Figure 13   Stress-strain curves for the adhesive FM-73 in shear. Results were obtained from a thick-adherend specimen, bond thickness 0.0056 in., and measurements made with a KGR-1 extensometer. Typical design characteristics for 250°F curing ductile adhesives are given in the following chart:

Temperature (°F)	P (psi)	G (psi)	γ_e	γ_p
R.T. (70)	5,000	50,000	0.1	1.0
−50	7,000	60,000	0.117	0.5
+160	2,500	40,000	0.063	1.5

VIII. ELASTIC–PLASTIC ANALYSIS OF BONDED DISCONTINUOUS MEMBERS

The elastic-plastic adhesive analysis methods developed by the first author in Ref. 10 and 11 for adhesive-bonded joints form the basis of the analyses presented here. Peel stresses in the adhesive due to eccentricities in the load paths are also important, whether or not associated with terminated or broken elements. Limited analyses predict stresses for the peel aspects of bonded joints.

Example 1

Figure 14 depicts the geometry and nomenclature for the analysis of bonded laminate with one member (identified by subscript 1) cut. The analysis applies equally if the other member is severed, by interchanging the subscripts 1 and 2. Out-of-plane displacements are

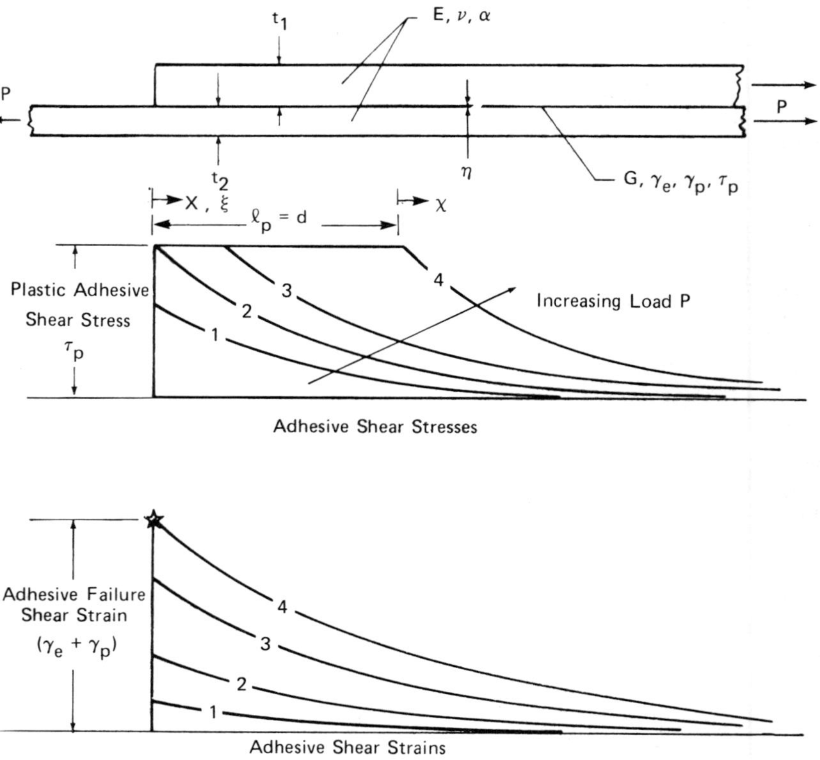

Figure 14   Typical adhesive shear stress and strain distributions.

considered to be constrained to eliminate peel stresses in the adhesive and bending stresses in the adherends. The analysis is performed in terms of an equivalent thickness of each member per unit width of bond. The conditions of longitudinal force-equilibrium for a differential element dx within the joint are

$$\frac{dT_1}{dx} - \tau = 0, \quad \frac{dT_2}{dx} + \tau = 0 \tag{1}$$

where the subscripts 1 and 2 refer, respectively, to the cut (or terminated) and intact member. The stress-strain relations for the assumed elastic adherend yield

$$\frac{d\delta_1}{dx} = \frac{T_1}{E_1 t_1} + \alpha_1 \, \Delta T \,, \quad \frac{d\delta_2}{dx} = \frac{T_2}{E_2 t_2} + \alpha_2 \, \Delta T \tag{2}$$

As a first approximation, the adhesive shear strain is taken to be

$$\gamma = \frac{\delta_1 - \delta_2}{\eta} \tag{3}$$

Within the elastic adhesive region, more than d from the end of the discontinuous member, the adhesive shear stress is assumed to be

$$\tau = G\gamma = \frac{G}{\eta}(\delta_1 - \delta_2) = f(x) \tag{4}$$

while throughout the plastic region adjacent to the end of the discontinuous member.

$$\tau = \tau_p = \text{constant} \tag{5}$$

This analysis covers both elastic and elastic-plastic situations, just by adjusting the adhesive characteristics. Eliminating δ_1 and δ_2 between Eqs. (3) and (2) produces

$$\frac{d\gamma}{dx} = \frac{1}{\eta}\left[\frac{T_1}{E_1 t_1} - \frac{T_2}{E_2 t_2}\right] + \frac{(\alpha_1 - \alpha_2) \, \Delta T}{\eta} \tag{6}$$

The use of Eq. (1) to eliminate T_1 and T_2 then yields the governing differential equation

$$\frac{d^2\gamma}{dx^2} = \frac{1}{\eta}\left[\frac{1}{E_1 t_1} + \frac{1}{E_2 t_2}\right]\tau \tag{7}$$

Within the elastic region, this equation becomes

$$\frac{d^2\tau}{dx^2} - \lambda^2\tau = 0 \tag{8}$$

in which

$$\lambda^2 = \frac{G}{\eta}\left[\frac{1}{E_1 t_1} + \frac{1}{E_2 t_2}\right] \tag{9}$$

Within the elastic zone, the solution of the differential Eq. (8) is

$$\tau = \tau_p e^{-\lambda X}, \qquad \gamma = \gamma_e e^{-\lambda X} \tag{10}$$

where the origin for the X coordinate is at the edge of the plastic adhesive zone, of width d. Within the plastic zone,

$$\gamma = \left[\frac{\lambda^2 \tau_p}{2G}\right]\xi^2 + C\xi + (\gamma_e + \gamma_p) \tag{11}$$

At $\xi = 0$ (at the discontinuity in member 1), from Eqs. (11) and (6),

$$\frac{d\gamma}{d\xi} = C = -\frac{T_2^0}{E_2 t_2 \eta} + \frac{(\alpha_1 - \alpha_2)\,\Delta T}{\eta} \tag{12}$$

in which the superscript 0 denotes the value at the discontinuity
($\xi = 0$), so that, in the plastic zone,

$$\gamma = \left[\frac{\lambda^2 \tau_p}{2G}\right]\xi^2 - \left[\frac{T_2^0}{E_2 t_2 \eta} - \frac{(\alpha_1 - \alpha_2)\,\Delta T}{\eta}\right]\xi + (\gamma_e + \gamma_p) \tag{13}$$

Now the integral of the bond stresses over the entire bond area is
obviously equal to the load in the broken member at infinity
($x = X = \infty$). Thus, from Eqs. (5) and (10),

$$T_1^\infty = \tau_p\left(d + \frac{1}{\lambda}\right) \tag{14}$$

in which d is the extent of the plastic zone. It is assumed that at station $x = \infty$ far removed from the discontinuity in member 1, both members are strained identically so that, from Eq. (2),

$$\frac{T_1^\infty}{E_1 t_1} + \alpha_1 \, \Delta T = \frac{T_2^\infty}{E_2 t_2} + \alpha_2 \, \Delta T \tag{15}$$

Overall force equilibrium requires that

$$T_2^0 = T_1^\infty + T_2^\infty \tag{16}$$

$$T_2^0 = T_1^\infty \left[1 + \frac{E_2 t_2}{E_1 t_1} \right] + E_2 t_2 (\alpha_1 - \alpha_2) \, \Delta T \tag{17}$$

The expression (13) at $\xi = d$ has the form

$$\left[\frac{\lambda^2 \tau_p}{2G} \right] d^2 - \left[\frac{T_2^0}{E_2 t_2 \eta} - \frac{(\alpha_1 - \alpha_2) \, \Delta T}{\eta} \right] d + \gamma_p = 0 \tag{18}$$

since

$$\gamma = \gamma_e \qquad \text{at } \xi = d \tag{19}$$

Now, from Eq. (14),

$$d = \frac{T_1^\infty}{\tau_p} - \frac{1}{\lambda} \tag{20}$$

so that the unknown extent of plastic adhesive zone can be eliminated between Eqs. (18), (20), and (17). A quadratic expression in T_1^∞ results; this is

$$\frac{\lambda^2 \tau_p}{2G} \left[\frac{T_1^\infty}{\tau_p} - \frac{1}{\lambda} \right]^2 - \left[\frac{T_1^\infty}{E_2 t_2 \eta} \left(1 + \frac{E_2 t_2}{E_1 t_1} \right) \right] \left[\frac{T_1^\infty}{\tau_p} - \frac{1}{\lambda} \right] + \gamma_p = 0 \tag{21}$$

whence

$$T_1^\infty = \sqrt{\frac{(\gamma_e/2 + \gamma_p) 2G \tau_p}{\lambda^2}} = \frac{\tau_p}{\lambda} \sqrt{1 + 2 \left(\frac{\gamma_p}{\gamma_e} \right)} \tag{22}$$

In terms of the load, at infinity, in the continuous member, Eq. (22) is modified by Eqs. (19) and (20) to become

$$T_2^\infty = \frac{E_2 t_2}{E_1 t_1} \frac{\tau_p}{\lambda} \sqrt{1 + 2\left(\frac{\gamma_e}{\gamma_p}\right)} + E_2 t_2 (\alpha_1 - \alpha_2)\, \Delta T \tag{23}$$

so that the total load at infinity is given by

$$T_2^o = T_1^\infty + T_2^\infty = \left[1 + \frac{E_2 t_2}{E_1 t_1}\right] \frac{\tau_p}{\lambda} \sqrt{1 + 2\left(\frac{\gamma_p}{\gamma_e}\right)}$$
$$+ E_2 t_2 (\alpha_1 - \alpha_2)\, \Delta T \tag{24}$$

The adhesive stress and strain distributions can be expressed once the extent d of the plastic zone is established. This is accompanied by eliminating T_1^∞ between Eqs. (14) and (22), leading to

$$d = \frac{1}{\lambda}\left[\sqrt{1 + 2\left(\frac{\gamma_p}{\gamma_e}\right)} - 1\right] \tag{25}$$

Typical adhesive stress and strain distributions are illustrated in Fig. 14. For load levels sufficiently low not to load the adhesive into the plastic state (i.e., $\gamma_p \equiv 0$ and $\gamma_{max} \leqslant \gamma_e$), Eq. (25) predicts correctly that d = 0. In the expressions for elastic loads, such as (22) and (24), it is necessary to set $\gamma_p \equiv 0$ and replace τ_p by $\tau_{max} = G\gamma_{max} \leqslant \tau_p$.
Just as in the case of adhesive-bonded double-lap joints, Eqs. (22) to (24) indicate that the strain energy in shear is the single necessary and sufficient characterization of the adhesive for this problem (provided that peel stresses and eccentricity effects are eliminated).

Example 2

Consider the case of a doubler bonded on a skin and terminating as is shown in the upper sketch of Fig. 14. Using a unit width and 2024-T3 nonclad sheet:

$t_1 = t_2 = 0.063$ in. $\gamma_e = 0.117$

$p = 1$ in. $\gamma_p = 0.5$

$\eta = 0.002$ in. $G = 60,000$ psi

$E = 10.5 \times 10^6$ psi $F_{TY} = 42,000$ psi

$\nu = 0.3$ $F_{TU} = 63,000$ psi

$\tau_p = 7000$ psi

(adhesive properties are for a $-50°F$ condition). Adhesive properties are given in the table in Fig. 13.

Evaluate the residual strength for Example 1 using the parameters of the table in addition to looking at the effects of a thicker glue line ($\eta = 0.005$) and then the effects of using $160°$ adhesive properties ($G = 40,000$).

$\eta = 0.002$ in.	$\eta = 0.005$	$G = 40,000,$ $\eta = 0.005$

Using Eq. (9), we have

$$\lambda^2 = \frac{60,000}{0.002}\left(\frac{1}{10.5 \times 10^6 \times 0.63} + \frac{1}{10.5 \times 10^6 \times 0.63}\right)$$

$\lambda^2 = 90.7$	36.38	24.19
$\lambda = 9.2$	6.02	4.92

Using Eq. (22), we have

$$T_1^\infty = T_2^\infty = \frac{7000}{9.52}\sqrt{1 + 2\,\frac{0.5}{0.117}}$$

$T_1^\infty = 2272$ lb	3593	3547
$T_2^\infty = T_1^\infty + T_2^\infty = 4544$ lb	7186	7094

Stress in T_2^o (per unit width),

$\sigma_2^o = \dfrac{T_2^o}{A_s} = \dfrac{4544}{0.063} = 72,127$ psi	114,000	112,6000

Using the maximum available shear strength of the adhesive with thin (0.002) and thick (0.005) bond lines, the doublers will be loaded by values shown as T_1^∞. However, the stresses in the single sheet, ahead of the bonded doubler, are high enough to fail the material. Allowable $F_{T_U} = 63,000$ psi. The answers above also show that very thin bond lines (0.002) are most critical for the adhesive (lowest load transfer) and that there is really no difference in the load transfer between the $-50°F$ adhesive properties and that for the $160°F$ properties.

Next, it is appropriate to see what length of bond is required to transfer the loads calculated above. The length of the plastic zone is determined from Eq. (20).

		$G = 40,000$
$\eta = 0.02$	$\eta = 0.005$	$\eta = 0.005$

$$d = \frac{T_1^\infty}{T_p} - \frac{1}{\lambda} = \frac{2272}{7000} - \frac{1}{9.52} = 0.221 \text{ in.}$$

0.347	1.215

The length of the elastic zone is determined
from the equation $\ell_e = 3/\lambda$

$$\ell_e = 0.315 \text{ in.}$$

0.489	0.610

Therefore, the total length of bond
necessary to transfer the calculated
load is $\ell = d + \ell_e$:

$$\ell = 0.221 + 0.315 = 0.546 \text{ in.}$$

0.836	1.825

If the material in Example 1 were 7075-T6 nonclad and the allowable F_{T_U} is 78,000 psi, the thin adhesive, $\eta = 0.002$, would not have developed loads high enough to break the single sheet and the bond would fail. The room temperature and hot adhesive can support a bonded doubler up to the failing stress of the 7075 nonclad material with a thin 0.002 in. glue line.

Since the strength of the adhesive will break the metal, it is appropriate to calculate the length of the bonded splice based on metal strength, thick bond line (0.005), and 160°F adhesive properties.

$$T_1^\infty = \frac{63,000}{2} \times 0.063 = 1985 \text{ lb}$$

$$= \frac{T_1^\infty}{T_p} - \frac{1}{\lambda} = \frac{1985}{2500} - \frac{1}{4.92} = 0.80 - 0.52 = 0.28 \text{ in. (plastic)}$$

$$\ell_e = 0.61 \text{ in. (elastic)}$$

$$\ell = 0.28 + 0.61 = 0.89 \text{ in.}$$

This shorter overlap of 0.89 in. is equivalent to the design overlap in Fig. 1. The value 1.825 in. above corresponds with the greater load that would have been needed for the adhesive bond to fail were the metal not to fail first.

Example 3

Next, examine a double-strap bonded joint. A prime consideration in the design of the bonded splice joint shown in Fig. 15 is that sufficient

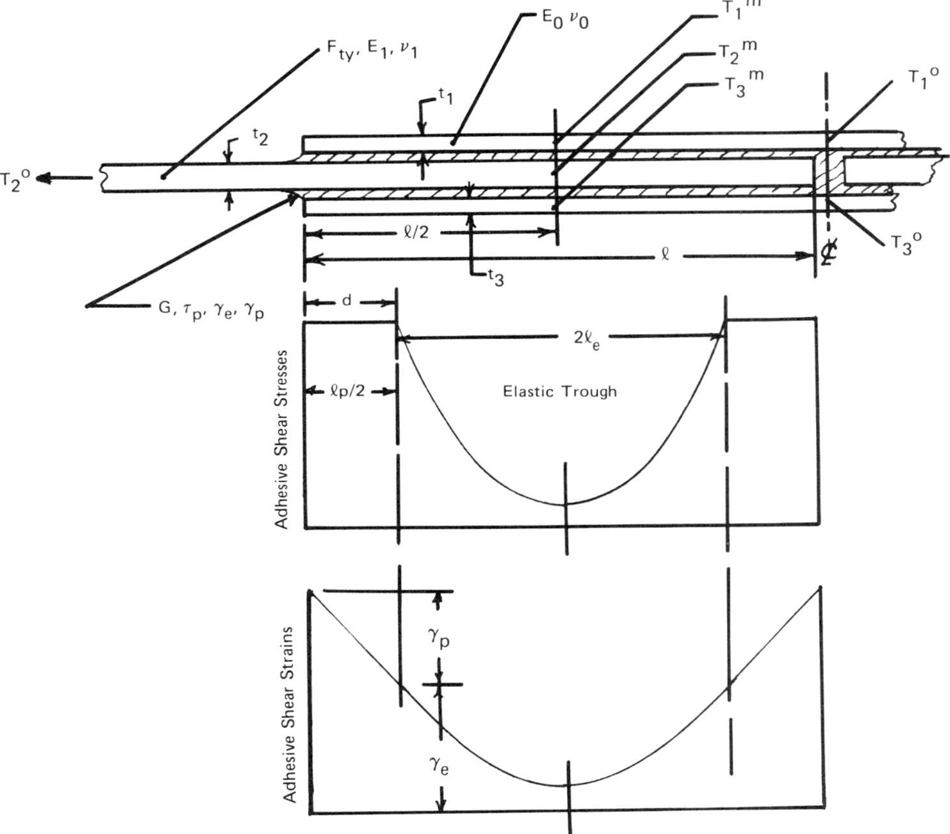

Figure 15   Design for double-strap bonded joints.

total overlap length ℓ each side of the joint centerline exist to restrict adhesive stress in the middle of the overlap length to no more
than 10% of the maximum under the most severe environment. Therefore, let us examine the double-strap splice joint to see what variations
are required in Eqs. (9), (22), and (25) to solve for the minimum safe
overlap. At distance $\ell/2$ the total load T_2^o is carried by the three adherends in proportion to their thickness. If it is a balanced joint where
external adherands are one-half the thickness of the internal member,
half the load is transferred out of central member at $\ell/2$. This is the
same as that for the example in Fig. 14 if t_1 equals t_2. Let us assume
that the metal will be stressed to F_{TU} so that the maximum overlap will
be determined.

   If we use Eq. (9) to solve for λ^2, it will have to be multiplied by

2 because there are two bond lines transferring the load; $t_1 = t_3 = 0.032$ in., $t_2 = 0.063$ in., $E = 10.5 \times 10^6$ constant. Using 160°F adhesive properties, $F_{T_U} = 63{,}000$ psi (2024-T3).

$$\lambda^2 = \frac{G}{\eta}\left[\frac{2}{Et_2} + \frac{2}{E(t_1 + t_2)}\right]$$

$$= \frac{40{,}000 \times 2}{0.005 \times 10.5 \times 10^6}\left(\frac{1}{0.063} + \frac{1}{0.032 + 0.032}\right)$$

$$= 48$$

$$\lambda = 6.928$$

$$T_1^m = T_3^m = \frac{63{,}000}{2} \times 0.032 = 1008 \text{ lb}$$

From Eq. (20),

$$\frac{\ell_p}{2} = d = \frac{1008}{2500} - \frac{1}{6928} = 0.403 - 0.144 = 0.259 \text{ in.}$$

$$\ell_e = \frac{3}{\lambda} = \frac{3}{6.928} = 0.433$$

$$\ell = 2\left(\frac{\ell_p}{2}\right) + 2\ell_e = 2(0.259 + 0.433) = 1.384 \text{ in.}$$

If 7075-T6 material had been used, then $F_{T_U} = 78{,}000$ psi.

λ^2 will remain the same

$$\lambda = 6.928$$

$$T_1^m = \frac{78{,}000}{2} \times 0.032 = 1248 \text{ lb/in. (one-half of load transferred)}$$

$$\frac{\ell_p}{2} = d = \frac{1248}{2500} - \frac{1}{6.928} = 0.499 - 0.144$$

$$d = 0.355 \text{ in.}$$

ℓ_e remains unchanged (0.433)

$$\ell = 2(0.355 + 0.433) = 1.576 \text{ in.}$$

The following formulas for the potential shear strength P of the bond should be used to verify the adequacy of the joints as designed above. The bond shear strength per unit width (for long overlap joints) is given by relationships of the form

$$P = \text{lesser of } \sqrt{2\tau_p \eta\left(\frac{\gamma_e}{2} + \gamma_p\right) 2E_i t_i \left[1 + \frac{E_i t_i}{E_o(t_{o_1} + t_{o_2})}\right]} \qquad (26)$$

and

$$P = \sqrt{2\tau_p \eta\left(\frac{\gamma_e}{2} + \gamma_p\right) 2E_o(t_{o_1} + t_{o_2}) \left[1 + \frac{E_o(t_{o_1} + t_{o_2})}{E_i t_i}\right]} \qquad (27)$$

where the equations examine the criticality of the adhesive at each end of the joint in turn. Both equations give the same answer when the two splice plates together equal the thickness of the center member; t_i = center member thickness and $t_{o_1} = t_{o_2}$ = splice plate thickness.

$$P = \sqrt{2 \times 2500 \times 0.005\left(\frac{0.063}{2} + 1.5\right) \times 2 \times 10.5 \times 10^6 \times 0.063\left(1 + \frac{0.063}{0.064}\right)}$$

$$= 10.026 \text{ lb/in.}$$

and

$$P = \sqrt{2 \times 2500 \times 0.005 \times 1.5315 \times 2 \times 10.5 \times 10^6 \times 0.064\left(1 + \frac{0.063}{0.064}\right)}$$

$$= 11,004 \text{ lb/in.}$$

These values are for the harshest environment (160°F, 100% relative humidity). Again, checks at room temperature and $-50°$F show more strength for the adhesive shear strength. For brittle adhesives and for any adhesive bonding to thermally dissimilar materials, it is the lowest temperature that is most frequently the worst.

Considering the maximum load that has to be transferred (i.e., the ultimate strength of the metal outside the bonded joint), it is $F_{TU} \times t_2$ or $78,000 \times 0.063 = 4914$ lb/in. Both of the values above exceed this value, so there is considerable margin or safety. If pinch-off is considered, the bond line thickness of 0.005 in. used above could be reduced to 0.002. This reduces the allowable shear strength by a ratio of $\sqrt{0.002/0.005}$, so the 10,026 value becomes 6341 lb/in. This will yield an adequate margin of safety at the ultimate strength of the metal (7075-T6):

$$\text{M.S.} = \frac{6341}{4914} - 1 = 0.29$$

If the assembly is subjected to a repeated (fatigue) loading it is advisable to determine the maximum load that will not exceed the elastic limit of the adhesive. For this evaluation, use of $-50°F$ adhesive properties. Using Eq. (26), solve for P when γ_p is put at zero and $-50°F$ properties are used.

$$P = \sqrt{2 \times 7000 \times 0.005\left(\frac{.117}{2}\right) 2 \times 10.5 \times 10^6 \times 0.063 \left(1 + \frac{0.063}{0.064}\right)}$$

$$= 3279$$

Again, examining what happens when the glue line is squeezed down to .002 inches, the allowable load would be reduced by the square root of the thickness change.

$$P_{thin} = 3279 \sqrt{\frac{0.002}{0.005}} = 2074 \text{ lb/in.}$$

This reduces the maximum metal stress to 32,900 psi, for a safe fatigue loading of the adhesive.

The final check on the joint strength is to examine the peel stress. The peak peel stress is given by the equation

$$\sigma_{peel} = \tau_p \left[\frac{3E_c'(1 - \nu^2)t_1}{E_1 \eta}\right]^{1/4} \tag{28}$$

where

    t_1 = splice plate thickness

    E_c' = effective peel modules of the adhesive film as constrained by the metal adherends

500,000 is an estimated value based on the biaxial constraint of the aluminum, which makes the adhesive nearly infinitely stiff because it is close to incompressible. There are very few test data in this area.

    $(1 - \nu^2) = 0.91$ using $\nu = 0.3$ for aluminum

$$\sigma_{peel} = 5000 \left(\frac{3 \times 500,000 \times 0.91 \times 0.032}{10.5 \times 10^6 \times 0.005}\right)^{1/4} = 4775 \text{ psi}$$

This stress level is about one-half the typical peel strength of ductile adhesive at room temperature. It should be noted that some adhesives have poor low-temperature peel strengths, so appropriate elastic properties need to be obtained when selecting an adhesive.

The previous two analyses have examined discontinuous structures in the form of skin or sheet splices. For built-up skin-stiffened structures, which are common in aircraft fuselage designs, it is important to examine the consequences of a failed sheet and an intact stiffener and vice versa. In both cases, the intact member can hold the total structure together if no fast fracture of the adhesive occurs adjacent to the fractured or discontinuous member.

Example 4

Consider the case of a bonded stiffener and sheet combination in which the stiffener is cracked through and the sheet is intact. By neglecting the shear distortion in the sheet, the analysis presented can be used to approximate this condition. Evaluate the residual strength of such a bonded panel made from 2024-T3 nonclad sheet, in which

$$A_{str} = 0.2 \text{ in.}^2 \qquad\qquad \tau_p = 7000 \text{ psi}$$

$$w = 1.5 \text{ in.} \qquad\qquad \gamma_e = 0.117$$

$$p = 12.0 \text{ in.} \qquad\qquad \gamma_p = 0.5$$

$$t_{skn} = 0.063 \text{ in.} \qquad A_{skn} = pt_{skn} = 0.756 \text{ in.}^2$$

$$\eta = 0.002 \text{ in.} \qquad\qquad G = 60{,}000 \text{ psi}$$

$$E = 10.5 \times 10^6 \text{ psi} \qquad F_{T_Y} = 42{,}000 \text{ psi}$$

$$\nu = 0.3 \qquad\qquad F_{T_U} = 63{,}000 \text{ psi}$$

$$\sigma^\infty = \text{disbonding stress away from crack}$$

and identify the weak link, be it adhesive or sheet. The terminology for this problem is identified in Fig. 16. The equivalent thicknesses are

$$t_1 = \frac{A_{str}}{w} \quad \text{and} \quad t_2 = \frac{t_{skn}}{w} \tag{29}$$

$$= 0.133 \text{ in.} \qquad\qquad = 0.504 \text{ in.}$$

$$\lambda^2 = \frac{G}{\eta}\left(\frac{1}{E_i t_i} + \frac{1}{E_2 t_2}\right) = 27.15 \qquad \text{[from Equation (9)]}$$

$$\lambda = 5.21$$

Equations (22) (9) and (29) yield, for the cut stiffener,

$$\sigma^\infty = \frac{\sqrt{E}\,\sqrt{\tau_p(\gamma_e + 2\gamma_p)}}{\sqrt{A_{str}/\eta w}\,\sqrt{1 + A_{str}/A_{skn}}} \qquad \text{away from the cut} \tag{30}$$

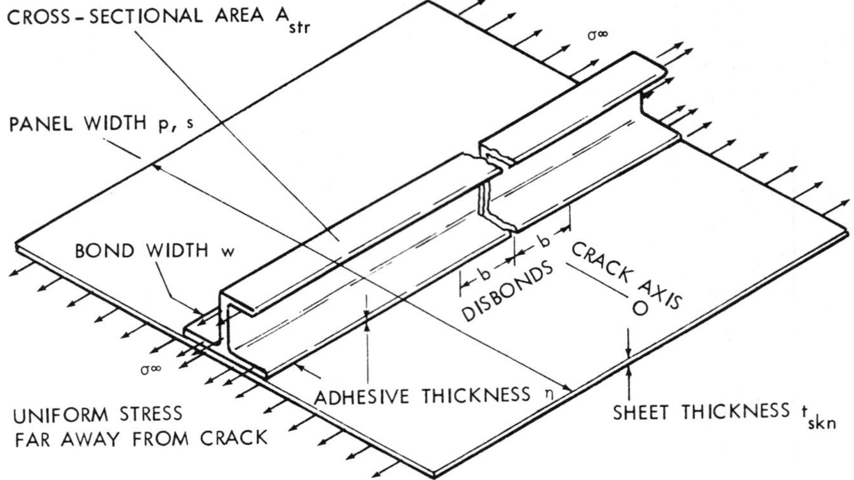

Figure 16   Sheet bonded to cracked stiffener.

The numerical solution for the problem above is σ^∞ = 31,190 psi, λ = 5.21, and T^∞ = 6238 lb in the stiffener away from the crack. The unit stiffener load is 4159 lb.

Strain compatibility away from the crack dictates that the sheet and stiffener have the same stress. The geometric influence is confined to the denominator, which should be minimized to maximize the residual strength of the structure for a given adhesive. Obviously, as long as the sheet is intact and the stiffener broken, minimizing the stiffener area maximizes the strength. However, this would lead to a deficient structure as far as restraining a sheet crack with an intact stiffener is concerned. Both cases must be considered in optimizing the geometric proportions. An increase in adhesive layer thickness will increase the value of σ^∞. Figure 17 depicts further solutions to Eq. (30) for three values of w, the bond width.

For the example above, the initial failure would be by disbonding of the stiffener if the sheet or panel stress σ^∞ is raised above 31,190 psi. The initial disbond can be propagated only by an increase in applied load. As the disbond propagates, the sheet in line with the stiffener will be able to stretch significantly over the length of the disbond, and this will have the effect of relieving the tendency of the disbond to propagate further. As the skin elongates at the end of the stiffener, the load is redistributed so that in the intact skin, the skin carries a greater percent of total load and the stringer less, hence stopping the disbond action. A more refined analysis is needed to

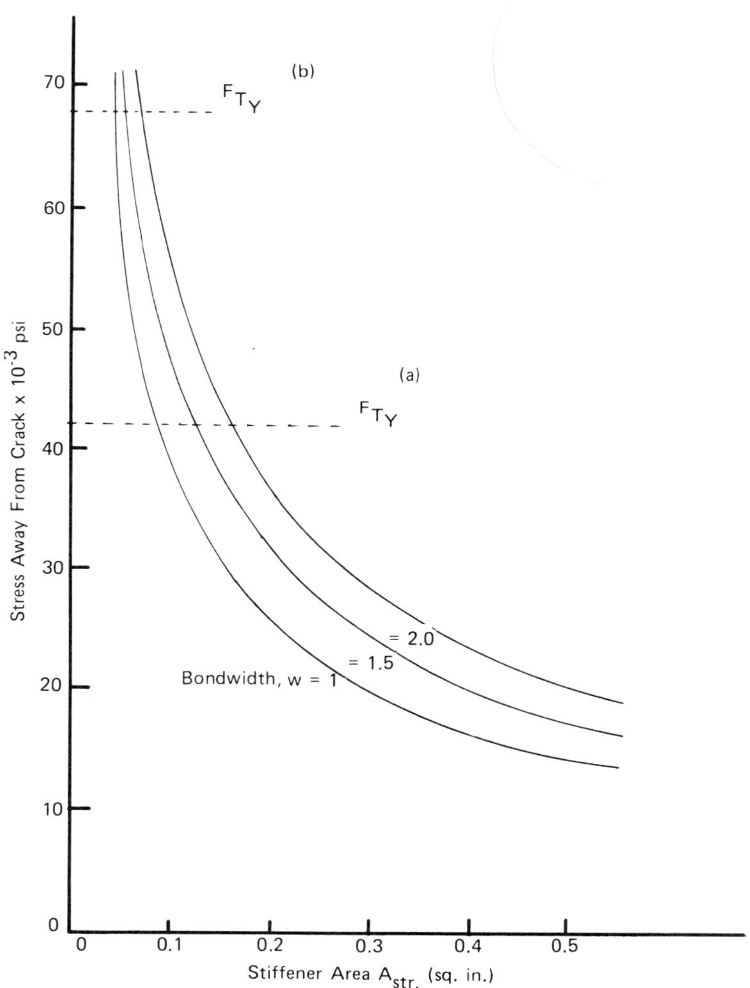

Figure 17   Solution of Eq. (30) for stiffener areas with three bonded widths and a skins thickness t_{skin} = 0.063. Stress (a) is the yield stress for 2024-T3 nonclad aluminum alloy and stress (b) is the yield stress for 7075-T6 nonclad.

identify the length of the disbond before the skin in line with the
stiffener reaches a yield stress. Once the skin yields, the disbond
propagates steadily toward the end of the stiffener as long as the
skin stress (at the tip of the disbond) is maintained at the metal
yield stress level.

From Eq. (25), solve for the plastic zone length.

$$d = \frac{1}{5.21} \left[\sqrt{1 + 2\left(\frac{0.5}{0.117}\right)} - 1 \right]$$

$$= 0.399 \text{ in.}$$

Using Eq. (20) to solve for d gives us the same answer.

$$d = \frac{4159}{7000} - \frac{1}{5.21} = 0.402 \text{ in.}$$

$$\ell_e = \frac{3}{\lambda} = 0.57 \text{ in. of the elastic zone}$$

The total length of the load-carrying bond is

$$\ell = 0.402 + 0.57 = 0.97 \text{ in.}$$

In the preceding analysis, Eq. (30) predicts the onset of failure
of the adhesive bond. However, in many instances, this will not be
immediately catastrophic. The initial disbond can be self-arresting
and will frequently require application of additional load to propagate
further. This section analyzes that aspect of the problem and illustrates how disbonds can alleviate the severity of the stress concentration in the adhesive due to the discontinuity in one of the members
bonded together.

The stress in both the sheet and stiffener far away from the
discontinuity is defined to be σ^∞. Therefore, the total sheet load
is

$$wT_2^\infty = \sigma^\infty t_{skn} p = 23{,}580 \text{ lb} \qquad T_2^\infty = 15{,}720 \text{ lb/in. bond width} \qquad (31)$$

while the stiffener load, which must be transferred through the bond,
is

$$wT_1^\infty = \sigma^\infty A_{str} = 6238 \text{ lb} \qquad T_1^\infty = 4159 \text{ lb/in. bond width} \qquad (32)$$

Now, because of the additional load being transferred in and out
of the sheet by the broken stiffener, the ends of the intact bond

separate under load a displacement given by Eq. (33), as the sheet distorts.*

$$\Delta = \frac{2(3 - \nu)(1 + \nu)q}{4\pi E} \left[\ell \ln \frac{2b}{\ell} - (2b + \ell) \ln \frac{2b}{2b + \ell} \right] \tag{33}$$

The extent ℓ of the effective load transfer zone can be approximated by considering the adhesive to be fully plastic. Then overall equilibrium requires that

$$q \ell t_{skn} = \tau_p \ell w = \sigma^\infty A_{str} = 6238 \text{ lb} \tag{34}$$

so that, in Eq. (33),

$$\ell = \frac{\sigma^\infty A_{str}}{(\tau_p w)} \quad \text{and} \quad q = \frac{\tau_p w}{t_{skn}} \quad \text{and} \quad \Delta_1 = \Delta_2 \tag{35}$$

[If the load level is not sufficiently high to cause the adhesive to yield, the bond length ℓ in Eq. (35) should be increased by replacing τ_p with the lesser value τ_{max}.] The stretching of the sheet between the stiffener break and the intact adhesive follows from Eqs. (33) and (35):

$$\Delta_1 = \frac{(3 - \nu)(1 + \nu)}{4\pi E} \frac{\tau_p w}{t_{skn}} \left[\ell \ln \frac{2b}{\ell} - (2b + \ell) \ln \frac{2b}{2b + \ell} \right] \tag{36}$$

If the load in the stiffener were to induce uniform stretching of the sheet over the same expanse, that lesser stretching would be

$$\Delta_2 = \frac{\sigma^\infty b}{E} \frac{A_{str}}{pt_{skn}} \tag{37}$$

The difference between Eqs. (36) and (37) may be looked upon as effective additional shear-strain capacity for the adhesive bond, which would be reflected in the boundary condition for γ_p in Eq. (18). It follows, in turn, from Eq. (30) that the increased stiffener load needed

---

*It is assumed here that the associated displacements at the edge of the sheet are negligible; otherwise, a more precise analysis would be needed to cover sheet segments which are not much wider than the bond width.

to cause propagation of the disbond can be approximated by adding to the adhesive bond displacement. That is,

$$\Delta_{eff} = \eta \left(\frac{\gamma_e}{2} + \gamma_p \right) + \Delta_1 - \Delta_2 \tag{38}$$

Equation (30) is then modified with (34) to read

$$\sigma^\infty = \frac{\sqrt{E} \sqrt{2\tau_p \Delta_{eff}} w}{\sqrt{A_{str}} \sqrt{1 + A_{str}/pt_{skn}}} \tag{39}$$

From the preceding set of equations and the panel geometry set forth, the following answers result, assuming the presence of an 0.5-in. disbond (b).

$$\sigma^\infty = 31,190 \text{ psi } [\text{Eq. (30)}]$$
$$\ell = 0.594 \text{ in. } [\text{Eq. (35)}]$$
$$q = 167,000 \text{ psi } [\text{Eq. (35)}]$$
$$\Delta_1 = 0.004 \text{ in. } [\text{Eq. (36)}]$$
$$\Delta_2 = 0.00039 \text{ in. } [\text{Eq. (37)}]$$
$$\Delta_{eff} = 0.0048 \text{ in. } [\text{Eq. (38)}]$$
$$\sigma^\infty = 64,671 \text{ psi } [\text{Eq. (39)}]$$

As indicated above, the results of Eq. (39) depends on (38), which used the σ^∞ from Eq. (30). Since only one value of σ^∞ can exist, an iteration process must be utilized. In the example above, the final $\sigma^\infty = 82,100$ psi.

Indeed, this stress is sufficient for gross yielding, so the initial disbond does not propagate far as the load builds up to complete sheet failure. Figure 18 shows the effect of a range of disbonds on the critical strength for this example. It is evident that the sheet will yield prior to gross disbonding, as shown by the curve in Fig. 18.

The sheet stress at the tip of the disbond is assumed to be constant over the length of the disbond and can be approximated by the maximum value, under the cracked stiffener, of

$$(\sigma_{skn})_{max} \simeq \sigma^\infty + \frac{\tau_p (d + 1/\lambda)}{t_{skn}} = \sigma^\infty \left(1 + \frac{A_{str}}{pt_{skn}} \right) = 103,800 \text{ psi} \tag{40}$$

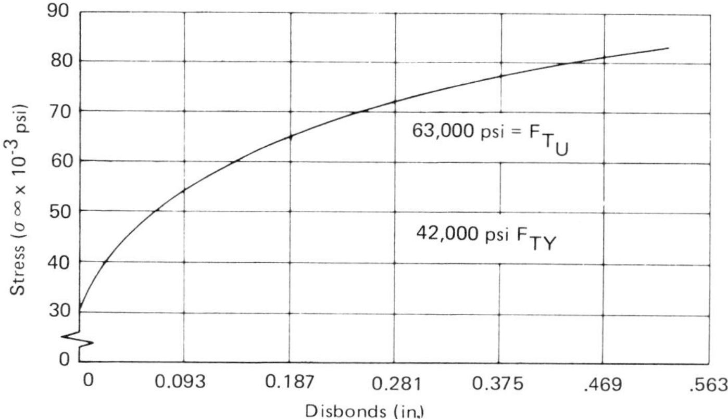

Figure 18   Effect of disbond on stress.

(This value is greater than the average value on that cross section.
Even within the bond width w, the sheet stress may exceed this value
[Eq. (40)] because the load in the stiffener web and outer cap is not
uniformly distributed across the full bond width but is concentrated
near the web.) Once the sheet yields from the stress and the load is
sustained, the disbond will not propagate further because the sheet
between the remaining intact bonds can stretch more than enough to
accommodate the stress and strain concentrations in the bond without
permitting any increase in them. The stress amplification defined by
Eq. (40) explains why a fatigue crack in the sheet would develop
rapidly in association with any discontinuity in a stiffener, such as at
the cut out (mouse holes) in frames, to permit longerons to pass
through.

The method presented in this section can be adapted to analyze
the stresses in the metal and adhesive where a finite length of stiffener
is absent, as at the mouse holes in frames (or frame/shear-tee combinations). The method is imprecise in such an application but, with sound
engineering judgement of the input and output, a good qualitative assessment of such problems can be gained. A missing length of stiffener
can be replaced mathematically by an initial disbond of the same length
because the analysis assumes a constant stress across the entire stiffener cross section. The broken stiffener is unloaded for the extent
of the disbond. Although this would be a reasonable assumption for
the outer portions of the stiffener flange bonded to the sheet, it is
far from the truth near the stiffener web. The skin fatigue cracks
induced at poorly detailed stiffener intersections or runouts start in
line with the web, where the effective stiffener thickness per unit bond

width is locally much greater than the average. Consequently, the
key to applying the present analysis to such a problem is the choice
of an effective stiffener thickness in line with the web of the stiffener.
Usually, not all of the stiffener is notched, so some load remains in the
continuous portion of the stiffener rather than passing through the
bond to the sheet. Account must be taken of this alternative load path
for some of the stress concentrations at the notch. Usually, the termination of the stiffener flange itself will not present a problem. At stiffener runouts, it is usually sufficient to cut back the web and any
flange not adjacent to the sheet progressively over several inches.
However, if the stiffener web is terminated too abruptly and has a
large cross section, the application of ultimate loads will either cause
the bond to fast fracture until enough of the sheet can stretch to relieve the stress concentrations in the bond or the sheet to yield throughout the gap in the interrupted stiffener.

In either event, the corresponding fatigue stresses will induce
fatigue cracks in the sheet in line with the cut in the stiffener web if
no alternative load path is provided. Continuous crack stoppers between frames and skin, joggling of the frame shear tees (or outer
flanges) over both longeron flanges in contact with the skin, and gus-
sets joining the shear tees to the longerons are suitable techniques for
eliminating the mouse-hole problem. The panel testing indicates that
relatively little is needed in the way of alternative load path to prevent
the initiation of such fatigue cracks at the mouse holes, but that such
cracks will initiate and grow rapidly in the absence of any alternative
load paths to reinforce the skin at the cutouts.

The next problem to be analyzed is identified in Fig. 19, which
also gives the terminology used in the analysis. The streching of the
sheet, which is concentrated near the stiffener, adds to the strain
capability of the adhesive in relieving the critical bond conditions at
the sheet crack. A uniform stiffener stress is assumed at each cross
section, and the presence of any possible disbond is initially excluded
from consideration. The key to the analysis is an approximation of
the constraints of displacement compatibility, accounting for all the
major effects.*

The sheet distortion can be approximated by Eq. (41) provided
that the plate width p is reasonably large compared with the stiffener
bond width w. The difference between displacements at the edge of
the sheet and in the middle of the bond area is

---

*Mathematically, more rigorous elastic analyses are possible, but the
results of such methods do not actually represent the critical conditions for typical structural adhesives.

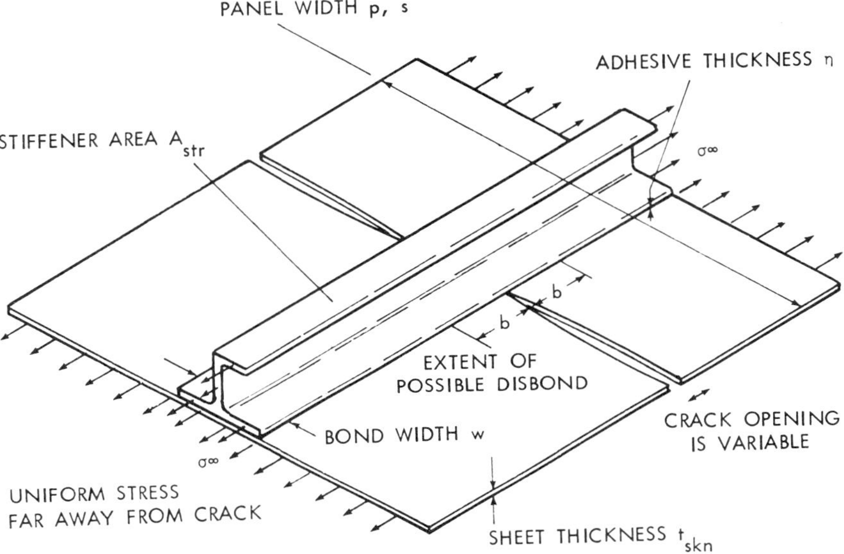

Figure 19 Stiffener bonded to fully cracked sheet.

$$\Delta = \frac{2}{\pi E} \frac{\sigma^\infty p}{w} \left(-\frac{p+w}{2} \ln \frac{2d}{p+w} + \frac{p-w}{2} \ln \frac{2d}{p-w} + w \ln \frac{d}{w} \right) \quad (41)$$

This distortion must be independent of the arbitrary constant d, which can be verified by rearrangement of Eq. (41) to the form

$$\Delta_3 = \frac{2}{\pi E} \frac{\sigma^\infty p}{w} \left(\frac{p}{2} \ln \frac{p+w}{p-w} + \frac{w}{2} \ln \frac{p+w}{2w} + \frac{w}{2} \ln \frac{p-w}{2w} \right) = 0.0782 \quad (42)$$

using $\sigma^\infty = 82{,}100$ psi. This distortion can exceed the relative movement between the stiffener and cracked sheet across the adhesive layer, and must therefore be accounted for in any analyses of this problem.

Two distinct failure mechanisms must be assessed in turn. These are failure of the bond with the stiffener behaving elastically, and yielding of the stiffener before the adhesive is loaded to its entire capacity. However, in the second case, if the load is sustained at a level that causes the stiffener to yield, the bond will progressively (or perhaps catastrophically) fracture. The effects of progressive disbonding on this problem are addressed after the analysis of the initially intact bond case. The limiting displacement implicit in Eq. (38) is now replaced by

$$\Delta_{eff} = \eta \left(\frac{\gamma_e}{2} + \gamma_p \right) + \Delta_3 = 0.001 + 0.0782 = 0.0793 \tag{43}$$

so that the stress far away from the sheet crack, which is associated with critical conditions in the bond, becomes

$$\sigma^\infty = \sqrt{\frac{2E\tau_p \Delta_{effective}}{(pt_{skn}/w)(1 + pt_{skn}/A_{str})}} = 69{,}560 \text{ psi} \tag{44}$$

The corresponding stiffener stress is

$$\sigma_{str}^0 = \sigma^\infty \left(1 + \frac{pt_{skn}}{A_{str}} \right) = 332{,}500 \text{ psi} \tag{45}$$

and, since A_{str}/pt_{skn} is typically less than 0.5, this stress amplification factor may result in the stiffener developing its yield stress prior to the adhesive bond becoming critical.

Consider again the example in Fig. 16, but with a complete sheet crack substituted for the complete stiffener crack. An iterative solution of Eqs. (44) and (43) is required, checking with Eq. (45) to ascertain whether stiffer yield is more critical than disbonding. Accounting for the sheet distortion, the initial failure is predicted to be yielding of the stiffener [from Eq. (45)] or a nominal sheet stress of $\sigma^\infty = 8800$ psi. If this distortion is neglected, however, the prediction is the disbond will not propagate until a sheet stress σ^∞ is in excess of 69,560 psi, so the bond would not be critical.

In the event that the stiffener is sufficiently stiff not to yield, a disbond will be initiated. It is appropriate now to determine whether an increase in load is needed to propagate the disbond or if the bond will fast fracture instantaneously over the length of the bonded assembly. The additional stretching of the sheet in line with the end of the bond will be diminished by any disbond. The amount of this reduction can be approximated by Eq. (33) in conjunction with the analysis above. Figure 20 explains how the analyses are used to represent this situation. Distributed loads are used to avoid the singularities associated with point loads. The distortion of Eq. (42) in Fig. 20b is reduced by an amount (c) in the figure, which is derived from Eq. (33). The latter amount is

$$\Delta_4 = \frac{(3 - \nu)(1 + \nu)}{4\pi E} \frac{\tau_p w}{t_{skn}} \left[\ell \ln \frac{2b}{\ell} - (2b + \ell) \ln \frac{2b}{2b + \ell} \right] = 0.0012 \tag{46}$$

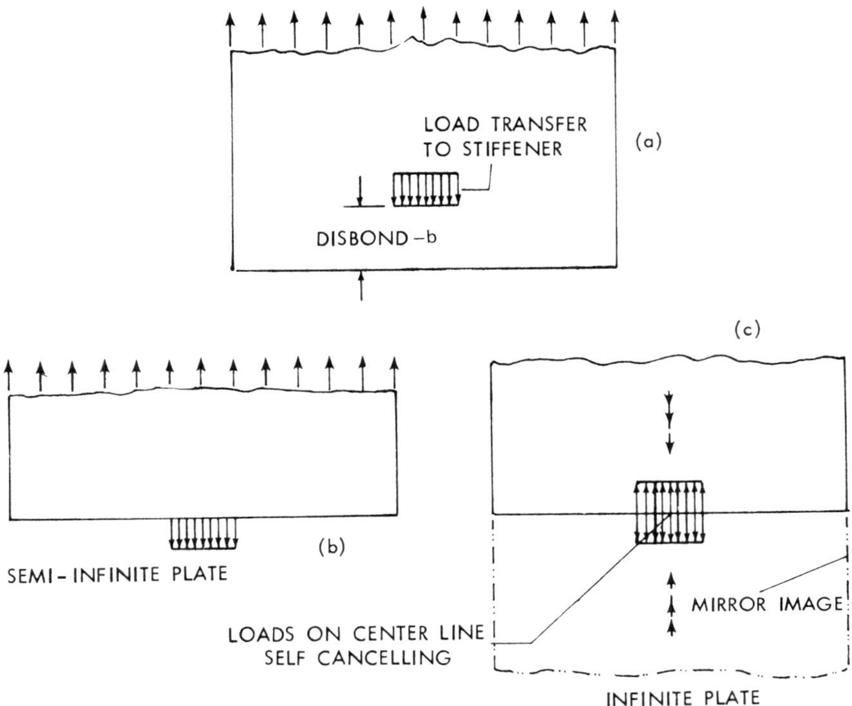

Figure 20   Forces associated with sheet distortion in the fully cracked sheet seen in Figure 18.   The sheet is uniformly loaded away from the crack.   The forces in (a) represent load transfer from the cracked sheet to the stiffener at the end of the disbond.   Approximate analysis: (a) = (b) + top half of (c).   The use of an infinite plate rather than a semi-infinite plate becomes more accurate as disbond increases.

in which the extent ℓ of the effective load zones in the adhesive is given as in Eq. (35).   Recalling that the sheet is cracked here instead of the stiffener, the value of σ^∞ equal to 8800 psi should be used.

$$\ell = \frac{\sigma^\infty A_{skn}}{\tau_p W} = \frac{\sigma^\infty p t_{skn}}{\tau_p W} = 0.634 \qquad (47)$$

(Again if the load level is not sufficient to cause the adhesive to yield, the bond length ℓ in Eq. (47) should be increased by replacing τ_p by the lesser value τ_{max}.)   The effective displacement in Eq. (43) is now replaced by

$$\Delta_{\text{eff}} = \eta \left[\frac{\gamma_e}{2} + \gamma_p \right] + \Delta_3 - \Delta_4 = 0.0781 \qquad (48)$$

It is obvious that, in this case, the influence of any disbond is to accelerate its propagation. This is in marked contrast to the situation in the preceding example with the stiffener cut rather than the sheet.

An assessment of a simultaneous solution of Eqs. (42) to (48) reveals that for practically all geometries, the stiffener is more likely to yield than to disbond. The reason is that the in-plane skin distortions [Eq. (45)] are so much greater than the bond displacements [Eq. (46)] that the latter can usually be neglected. However, once such a disbond started, it would fast fracture along its entire length instantaneously. None of the other geometries analyzed indicate such a phenomenon—their disbonds are self-arresting until some metal element is overloaded.

The explanation in the preceding paragraph explains some observations made from testing. Bonded stiffeners fell off some cracked panels. The stiffeners remained permanently bent after unloading. Everything happened too quickly during the test to establish the failure sequence. It appears that the sequence of events was: (1) propagation of the sheet crack, which may or may not have been temporarily arrested; (2) an initial disbond which was probably arrested; (3) stiffener yielding which reduced the effectiveness of the stiffeners in holding the sheet crack closed; (4) propagation of the sheet crack across the full panel width; and (5) complete failure of the bond on one or both sides of the sheet crack because of the very high stresses and strains in the stiffeners at the disbond fronts and the associated very low stresses and strains in the sheets within the extent of the disbonds. When the stiffeners were completely unbonded on one side of the sheet crack and broken where the load was introduced into the stiffener at the end of the panel, there were disbonds at least 2 in. long on the other side.

One of the test panels involved was a 36-in.-wide, 48-in.-long, 0.063-in.-thick sheet of 2024-T3 aluminum alloy with two longitudinal Z stiffeners of 0.237 in.^{2} cross section 12 in. apart, and bonded over an 0.938-in. width. Failure occurred at a gross stress of 30,600 psi. Two other panels were 20-in.-wide, 48-in.-long, 0.063-in.-thick sheets of 2024-T3 aluminum alloy with a single Z stiffener of the same cross section bonded down the middle of each panel. Failure of these panels occurred at gross stresses of 31,400 and 30,800 psi, respectively.

This same combination of a full-width crack not being arrested by bonded stiffeners which subsequently fell off after failing was also manifested in the testing of a panel having a flush splice across its middle. The panel was 24 in. wide and 48 in. long, of 0.090-in.-thick 7075-T6 aluminum alloy. A tapered splice plate 0.125 in. thick (maximum) and 6.5 in. wide bonded the two sheet segments (24 in. square)

together. Two stiffeners, of 0.237-in.^{2} cross section, were bonded
12 in. apart on the same side of the sheet as the splice plate and perpendicular to it. A small fatigue crack grew in the splice plate, in line
with the sheet junction, located about halfway between a stiffener and
the edge of the panel. This small crack, 0.7 in. long on the faying
surface side (where the crack was concealed beneath the adhesive) and
0.3 in. long on the inner surface (where it was held shut and invisible
by the bending stresses in the splice plate) fast-fractured across the
entire panel width at a sheet stress of 24 ksi. Both stiffener bonds
failed and then the stiffeners failed at their ends, where the load was
introduced far away from the splice crack.

The analyses in preceding examples indicate that the sheet distortions permitted by a full-width sheet crack alleviate most problems
associated with the bond between the stiffener and sheet until the load
level is sufficient to cause the stiffener to fail. Such large beneficial
distortions cannot occur if the crack in the sheet extends for less than
the full width of the panel. Another potential failure mode is added—
that of propagation of the sheet crack. This problem, which is analyzed
below, is described in Fig. 21.

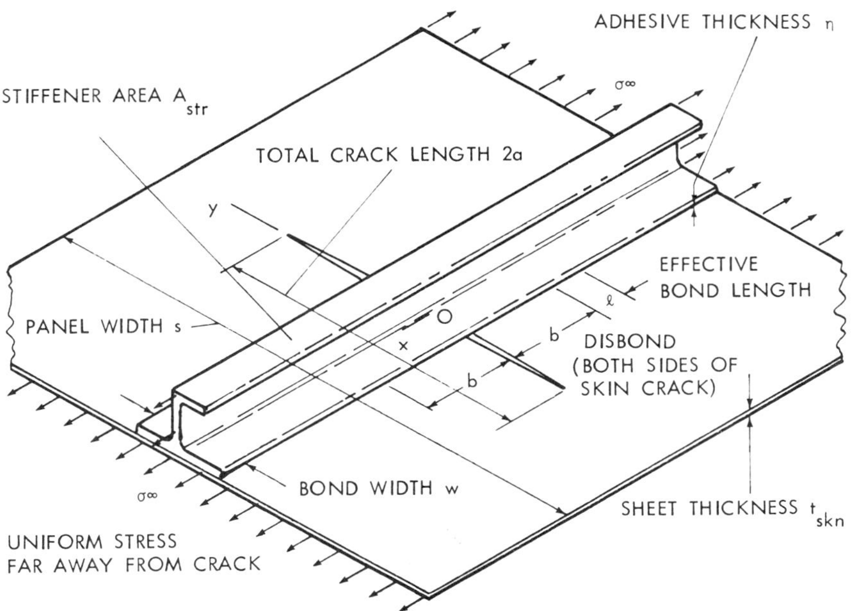

Figure 21   Two-bay sheet crack restrained by bonded central stiffener.

An analysis in Swift [12] establishes that the opening of a sheet crack in an unstiffened panel can be expressed as

$$\delta = \frac{2\sigma^{\infty}}{E} \sqrt{a^2 - y^2} \tag{49}$$

where a is half the crack width. The critical condition for propagation of such a crack in an unstiffened panel is conventionally expressed in the form

$$\sigma_{crit} = \frac{K_c}{\sqrt{\pi a}} \tag{50}$$

Finite-width effects are customarily accounted for by a modification factor of the form

$$\sigma_{crit} = \frac{K_c}{\sqrt{\pi a \, \sec(\pi a/s)}} \tag{51}$$

for a centrally located crack of extent 2a in a sheet of total width s. The present theories can be modified to account for this and other such effects in a straightforward manner.

The same provisions for sheet distortion which were developed in the form of Eqs. (42) and (46) for Δ_3 and Δ_4 are to be superimposed on the basic crack-opening expression Eq. (49) which, for a central stiffener, reduces to

$$\delta_0 = \frac{2\sigma^{\infty} a}{E} \tag{52}$$

Since there is no displacement in the x direction at the crack tip, Eq. (42) should be modified to read

$$\Delta_5 = \frac{\sigma_b^{\infty} \, 2a}{\pi E} \left(\frac{2a}{w} \ln \frac{2a + w}{2a - w} + \ln \frac{2a + w}{2w} + \ln \frac{2a - w}{2w} \right) \tag{53}$$

This distortion Δ_5 represents the amount by which the central crack opening δ_0 in Eq. (52) reduces because of the bonded stiffener. If any disbond is present, this relief is reduced by the amount given in Eq. (46):

$$\Delta_4 = \frac{(3 - \nu)(1 + \nu)}{4\pi E} \left(\frac{\tau_p^{\,w}}{t_{skn}}\right)\left(\ell \ln \frac{2b}{\ell} - (2b + \ell) \ln \frac{2b}{2b + \ell}\right) \qquad (54)$$

in which, as in Eq. (50), the plastic zone of highly loaded adhesive adjacent to the crack (or disbond, as appropriate) extends a distance

$$\ell = \frac{\sigma^\infty A_{skn}}{\tau_p^{\,w}} = \frac{\sigma^\infty 2at_{skn}}{\tau_p^{\,w}} \qquad (55)$$

The influence of any disbond is very considerable because, even though the disbond effectively stiffens the sheet (Δ_5 being reduced by Δ_4), it permits a length 2b of the stiffener to stretch far more than the sheet because of the high stiffener stress in the vicinity of the crack. One anticipates that steadily increasing load levels would be needed to propagate the disbonds in Fig. 21.

In the application of displacement compatibility of this problem, the concern is with the load $2at_{skn}\sigma^\infty$ in the sheet in line with the crack. Some of the load transfers through the adhesive bond to the stiffener and passes over the crack. The remainder remains in the sheet and is diverted around the crack tips. Accordingly, the analysis is separated into two parts. The sheet stress increments at $x = \pm\infty$ are identified by σ_b^∞ and σ_c^∞, respectively, to denote those increments passing through the bond and around the crack. The contribution of each is analyzed in turn.

For the load passing through the bond, the appropriate width of sheet to incorporate in λ, Eq. (9), is the total crack width 2a because Eq. (2) shows how the analysis is formulated in terms of strains, which extend uniformly across the entire width. The load transferred by the bond is limited by the width 2a, with the stress in the sheet beyond the crack passing straight on. Thus, if the influence of sheet distortion on σ_b^∞ is neglected, Eq. (22) becomes

$$\sigma_b^\infty = \frac{\tau_p^{\,w}}{\lambda\, 2at_{skn}} \sqrt{1 + 2\left(\frac{\gamma_p}{\gamma_e}\right)} = \sqrt{\frac{2E\tau_p^{\,}\eta[(\gamma_e/2) + \gamma_p]}{1 + 2at_{skn}/A_{str}}} \left(\frac{w}{2at_{skn}}\right) \qquad (56)$$

and the corresponding stiffener stress is

$$\sigma_{str}^{0} = \sigma_b^\infty \left(1 + \frac{2at_{skn}}{A_{str}}\right) \qquad (57)$$

The total stiffener stress over the crack is then

$$\sigma_{str}^{tot} = \sigma_{str}^{0} + \sigma_{c}^{\infty} \tag{58}$$

because the stress σ_c^{∞} must be applied to the stiffener as well as to the sheet so as not to alter the relative motion between them which gave rise to the σ_b^{∞} and σ_{str}^{0}.

The sheet stress σ_c^{∞}, which passes around the crack, tries to induce a crack opening on one side of the crack centerline given by Eq. (52) as

$$\delta_0 = \frac{2\sigma_c^{\infty} a}{E} \tag{59}$$

This will be reduced by the resultant of $\Delta_5 - \Delta_4$ given by Eqs. (53) and (54) with σ_b^{∞} replacing σ^{∞}. Similarly, the σ^{∞} in Eq. (55) refers here to σ_b^{∞}. If there be any disbonds present, the stiffener between such disbonds will stretch an amount

$$\Delta_6 = \frac{b\,\sigma_{str}^{tot}}{E} \tag{60}$$

on each side of the crack. Displacement compatibility then requires that

$$\Delta_6 + \eta(\gamma_{max}) = \delta_0 - \Delta_5 + \Delta_4 \tag{61}$$

The adhesive shear strain γ_{max} is used in place of the maximum allowable $(\gamma_e + \gamma_p)$ because the sheet may crack or the stiffener may yield prior to the adhesive failing. Equation (61) can be reexpressed as

$$\eta(\gamma_{max}) + \frac{b\,\sigma_{str}^{0}}{E} + \Delta_5 - \Delta_4 = \frac{\sigma_c^{\infty}}{E}(2a - b) \tag{62}$$

The left side of Eq. (62) can be evaluated in terms of σ_b^{∞}, while the right side is a function of σ_c^{∞}. The method of analysis is to assume trial values of γ_{max}, from which σ_b^{∞} can be determined in terms of Eq. (56). The remaining stress σ_c^{∞} follows from Eq. (62). Concurrent checks are made to ensure that σ_{str}^{tot} does not exceed the stiffener yield stress and that σ_c^{∞} is not sufficient to cause fast sheet fracture. In rare instances, a check on sheet yield is also needed. Equation (62) shows that a disbond on each side of the sheet crack which is longer than the crack itself is sufficient to nullify the crack-closing characteristics of the bonded stiffener.

At less than a critical condition, the load sharing between load being diverted around the crack tips and passing across the crack through the bonded stiffener can be assessed as follows. A value of σ_b^∞ is assumed and the associated value of γ_{max} established from Eq. (56). The other stress, σ_c^∞, then follows from Eq. (62).

Examples

Figures 22 to 27 show the results of parametric analyses using the present solution for two-bay sheet cracks held closed by a bonded central stiffener.

Figure 22 shows the influence of the length of the sheet crack on the panel stress at infinity as a function of the disbonded length. Naturally, the longer the crack, the weaker the panel. The initial failure, in every case, is a disbond which is self-arresting after an initial failure. The application of greater loads results in stable propagation of the disbond. For the shorter sheet cracks, still greater loads cause the stiffener to yield once the disbond has spread far enough to reduce the effectiveness of the stiffener in holding the crack shut. For longer sheet cracks, the increase of load leads directly to fast fracture of the sheet. The same final failure mode also prevails for shorter cracks, once the stiffener has yielded. Figure 22 can also be interpreted as a record of the growing disbond once the sheet starts to fast fracture. With a 6-in. total crack length, the disbond for sheet fracture is 3.2 in. long on each side of the crack. Once the crack grew to 12 in., the panel strength dropped from 55 ksi to 37 ksi and the disbond grew to 4.5 in. The disbond continued to grow and the crack propagated, reaching 6.5 in. once the crack grew to a total of 24 in. (12 in. on each side of the stiffener). Actually, if the load is maintained as the sheet fast fractures, the additional load must be carried in the stiffener, which will yield, causing even greater extents of disbonds for that grossly unstable condition than have been computed in Fig. 22 for the onset of instability.

Figure 23 illustrates the effect of stiffener area. The greater the stiffener area, the more effective the stiffener is in keeping the crack tips closed, so the greater the panel strength. The failure sequence, as a function of disbond length, shares the foregoing characteristics. An initial fracture of the bond at constant load is followed by stable progression of the disbond under increasing load until fast fracture of the sheet occurs. For the particular examples shown, there is no indication of stiffener yielding. However, the stiffener would yield prior to sheet fracture for small enough stiffeners.

Figure 24 shows the influence of the thickness of the adhesive between the stiffener and cracked sheet. The final sheet fracture

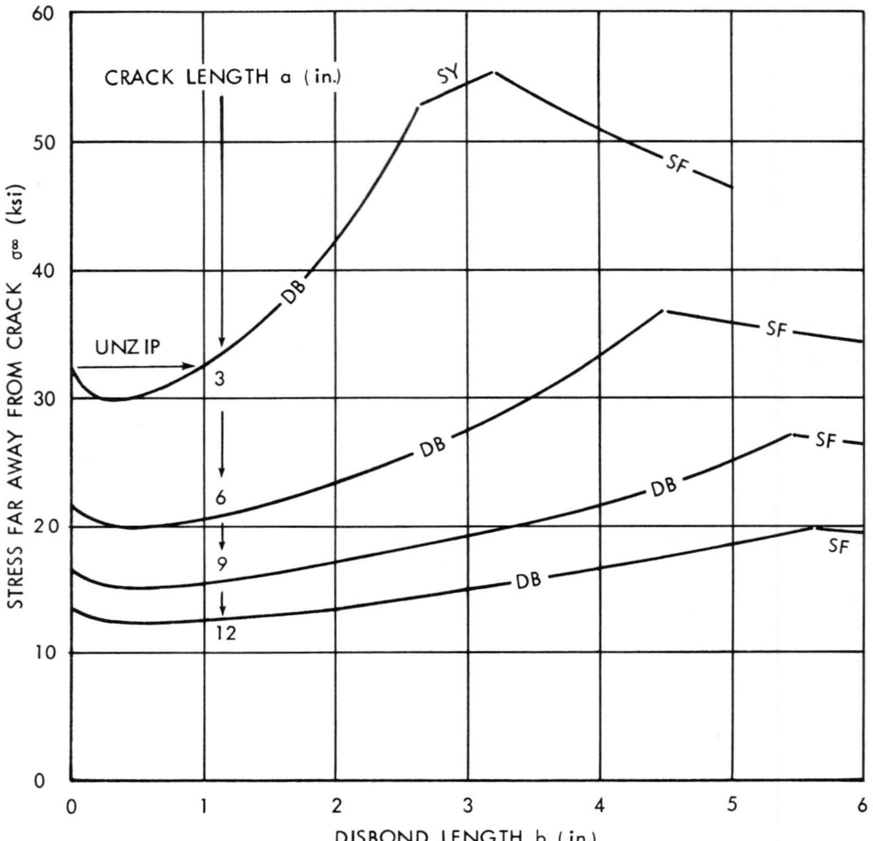

Figure 22   Parametric analysis of a two-bay crack (Fig. 21) showing the influence of crack length on panel stresses far away from crack and associated disbond length, all leading to a sheet fast fracture. Panel conditions used in the analysis: t_{skn} = 0.063., A_{str} = 0.25 in.^{2}, p = 12 in., s = 36 in., w = 1.5 in., η = 0.002 in., τ_p = 5500 psi, G = 60,000 psi, γ_p = 0.55, E = 10.5 × 10^6 psi, K_c = 134 ksi $\sqrt{in}$., $F_{y\,str}$ = 75 ksi, $F_{y\,skn}$ = 65 ksi. DB, disbond; SY, stiffener yield; SF, sheet fast fracture.

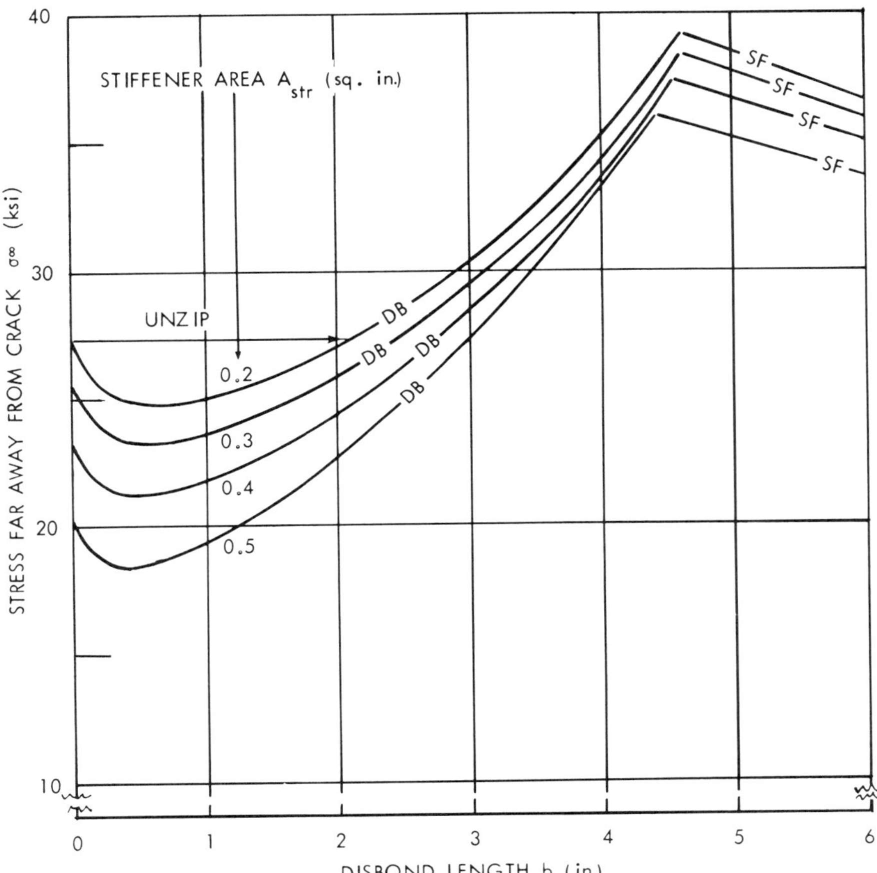

Figure 23　Parametric analysis of a two-bay crack (Fig. 21) showing the influence of stiffener cross-sectional areas on panel stresses far away from crack and associated disbond length, all leading to a sheet fast fracture. Panel conditions used in the analysis: t_{skn} = 0.063 in., 2a = 12 in., p = 12 in., s = 36 in., w = 1.5 in., η = 0.002 in., τ_p = 5500 psi, G = 60,000 psi, γ_p = 0.55, E = 10.5 × 10^6 psi, K_c = 134 ksi $\sqrt{in.}$, $F_{y\ str}$ = 75 ksi, $F_{y\ skn}$ = 65 ksi. DB, disbond; SF, sheet fast fracture.

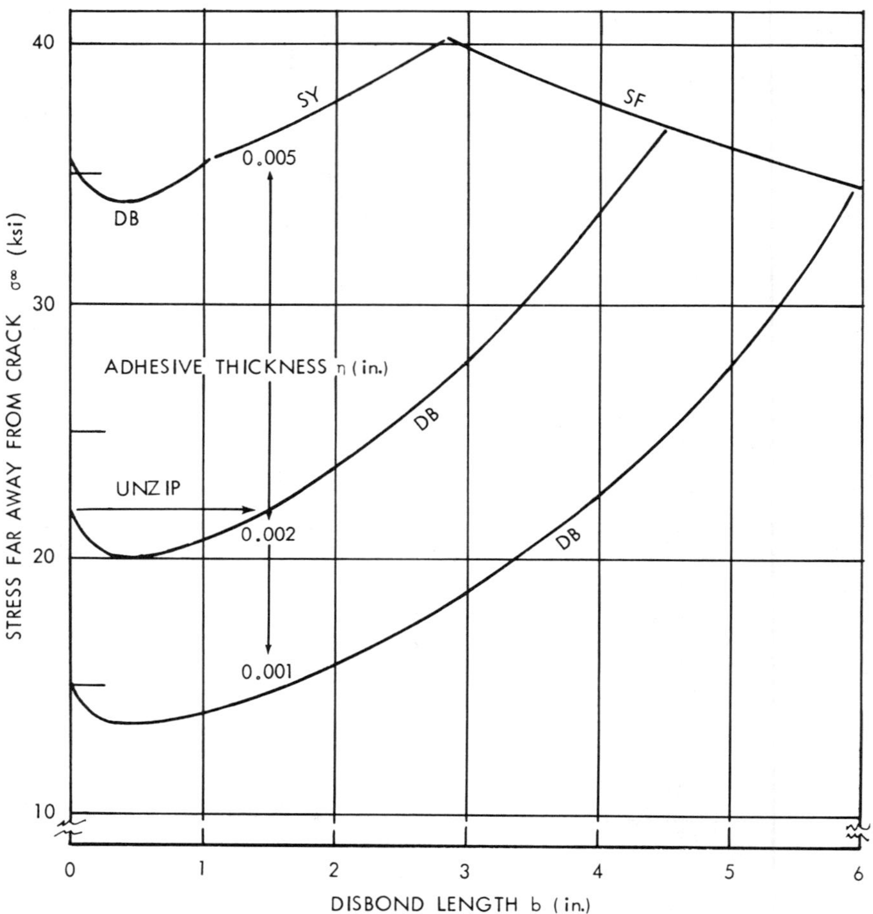

Figure 24 Parametric analysis of a two-bay crack (Fig. 21) showing the influence of adhesive thickness on panel stresses far away from crack and associated disbond length, all leading to a sheet fast fracture. Panel conditions used in the analysis: t_{skn} = 0.063 in., A_{str} = 0.25 in.^{2}, 2a = 12 in., p = 12 in., s = 36 in., w = 1.5 in., τ_p = 5500 psi, G = 60,000 psi, γ_p = 0.55, E = 10.5 × 10^6 psi, K_c = 134 ksi $\sqrt{\text{in.}}$, $F_{y\ str}$ = 75 ksi, $F_{y\ skn}$ = 65 ksi. DB, disbond; SY, stiffener yield; SF, sheet fast fracture.

characteristic is independent of the bond thickness, but the thinner
bonds permit disbonding at lower panel stress levels. The thicker
bonds permit yielding of the stiffener as a transition between disbonding and sheet fracture. The initial disbond at constant load is self-
arresting and requires additional load to be propagated. Figure 24
also shows qualitatively the effect of variations in any of the mechanical
adhesive properties. Since the influence of the adhesive is uniquely
characterized by the adhesive shear strain energy per unit bond area
$\eta\tau_p[(1/2)\gamma_e + \gamma_p]$, proportionate changes in any of the shear modulus
G, plastic shear stress τ_p, or elastic or plastic shear strains γ_e and
γ_p would have the same effect in Fig. 24 as the corresponding proportionate change in adhesive thickness η.

Figure 25 indicates the influence of the bond width for the stiffener, for a given cracked sheet and stiffener cross section. The
greater the bond width, the lower the bond strains and hence the
stronger the panel. This strength increase applies to both the sheet
fast fracture as well as to the disbonding. In some cases, the additional width can make a stiffener more prone to yielding.

Figure 26 depicts the influence of sheet thickness for a common
stiffener and crack length. The thinner sheets are stronger than the
thicker ones unless the stiffeners are reinforced for the latter. The
characteristic failure sequence for increasing disbond lengths is repeated: initial fracture, arrest, stable disbond growth, and fast
sheet fracture.

Figure 27 shows how the maximum adhesive shear strain decreases
once the disbond is great enough to make stiffener yielding or fast
fracture of the sheet crack more critical than propagation of the disbond. The figure also shows the corresponding relation between the
total sheet stress, the strength of the cracked panel devoid of stiffening, and the portion of the total sheet load which is diverted around
the crack tip.

Since the initial dip in the strength versus disbond curve is so
characteristic of these bonded stiffened panels with the sheet cracks—
and the same is found to be true in the next section for one-bay cracks
as well as the two-bay cracks considered here—it is appropriate to
further explain the phenomena. This is possible because the computer
program was developed progressively, adding each distortion contribution one at a time. With reference to the compatibility Eq. (62), it
can be stated that the contribution of stiffener stretching Δ_6 [Eq. (60)]
tends to alleviate the stress concentration at the tip of the bond and to
increase the panel strength. Similarly, the sheet distortion Δ_5 [Eq.
(53)] due to the crack closing up under the bonded stiffener tends to
increase the panel strength. Only the contribution Δ_4 [Eq. (54)],
which measures the tendency to nullify the benefits of Δ_5 because of
the gap between the crack and the intact bond, has a weakening effect.
This negative effect is eventually overpowered by the positive effects,
giving rise to the observed dips.

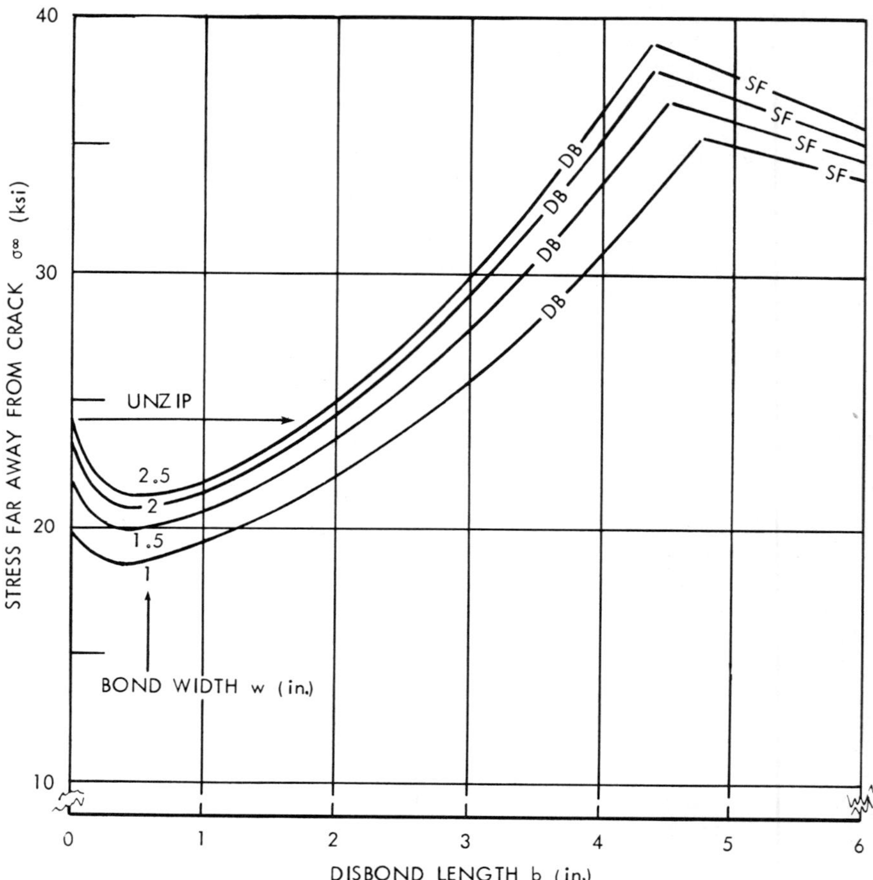

Figure 25 Parametric analysis of a two-bay crack (Fig. 21) showing the influence of bond width on panel stresses far away from crack and associated disbond length, all leading to a sheet fast fracture. Panel conditions used in the analysis: t_{skn} = 0.063 in., 2a = 12 in., A_{str} = 0.25 in.^{2}, p = 12 in., s = 36 in., η = 0.002 in., τ_p = 5500 psi, G = 60,000 psi, γ_p = 0.55, E = 10.5 × 10^6 psi, K_c = 134 ksi $\sqrt{\text{in.}}$, $F_{y\,str}$ = 75 ksi, $F_{y\,skn}$ = 65 ksi. DB, disbond; SF, sheet fast fracture.

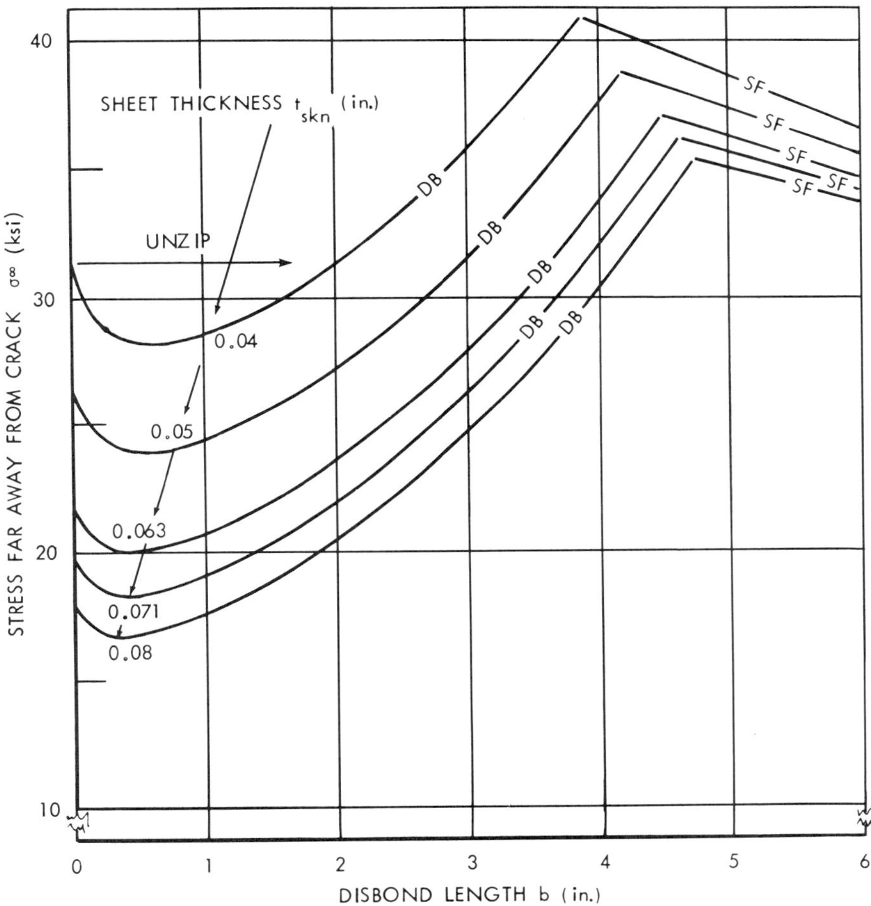

Figure 26   Parametric analysis of a two-bay crack (Fig. 21) showing the influence of sheet thickness on panel stresses far away from crack and associated disbond length, all leading to a sheet fast fracture. Panel conditions used in the analysis: $2a = 12$ in., $A_{str} = 0.25$ in.^{2}, $p = 12$ in., $s = 36$ in., $w = 1.5$ in., $\eta = 0.002$ in., $\tau_p = 5500$ psi, $G = 60{,}000$ psi, $\gamma_p = 0.55$, $E = 10.5 \times 10^6$ psi, $K_c = 134$ ksi $\sqrt{\text{in.}}$, $F_{y\,str} = 75$ ksi, $F_{y\,skn} = 65$ ksi. DB, disbond; SF, sheet fast fracture.

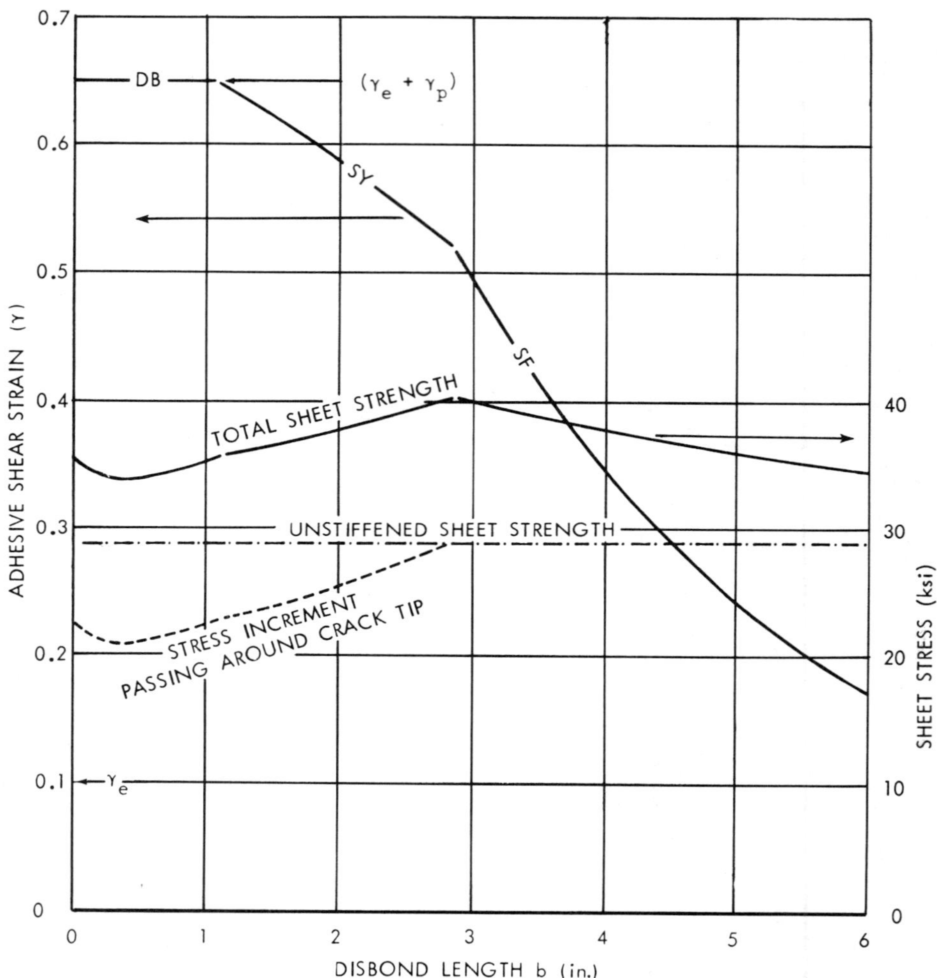

Figure 27 Parametric analysis of a two-bay crack (Fig. 20) showing reduction of adhesive shear strain as a function of disbond length. Panel conditions used in the analysis: t_{skn} = 0.063 in., A_{str} = 0.25 in.^{2}, 2a = 12 in., p = 12 in., s = 36 in., w = 1.5 in., η = 0.002 in., τ_p = 5500 psi, G = 60,000 psi, γ_p = 0.55, E = 10.5 × 10^6 psi, K_c = 134 ksi $\sqrt{\text{in.}}$, $F_{y\,str}$ = 75 ksi, $F_{y\,skn}$ = 65 ksi. DB, disbond; SY, stiffener yield; SF, sheet fast fracture.

When a sheet crack grows between two adjacent stiffeners in a panel, the crack growth is retarded as the crack approaches and grows slowly across the stiffener. The analysis methods presented here are not sensitive enough to account for the influence of a stiffener bonded beyond the crack tip. The crack tip must be trying to open up for the bonded stiffener to be able to exert any major force to keep it closed. Therefore, slow crack growth under fatigue loading is usually not terminated until after the crack has passed over all or part of the bonded crack stopper of stiffener. The analysis presented in this section therefore ignores that portion of any stiffener beyond the crack tip.

The model for the problem being solved here is defined in Fig. 28. The method of analysis follows that described in the preceding example in that the various individual displacements and distortions are balanced to ensure compatibility. Treating the cracked sheet as being unstiffened, Eq. (49) would indicate that, in line with the center of each stiffener, the crack would try to open up on each side an amount

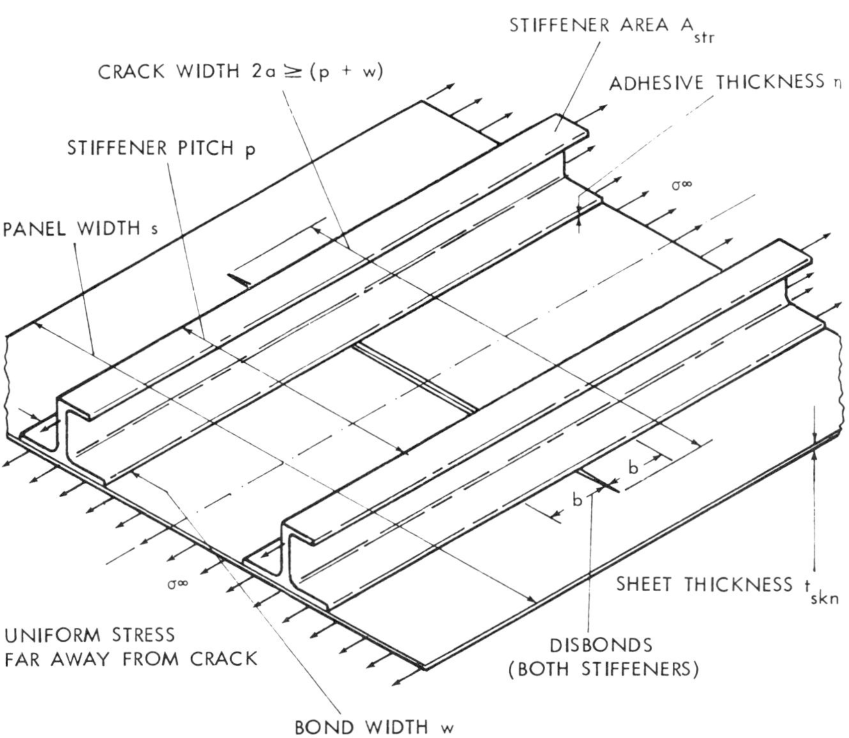

Figure 28   One-bay sheet crack restrained by bonded stiffeners.

$$\delta_0 = \frac{2\sigma_c^{\infty}}{E} \sqrt{a^2 - \left(\frac{p}{2}\right)^2} \tag{63}$$

The critical sheet stress σ_c^{∞} which must not be exceeded if the crack is not to propagate instantaneously is given in Eq. (51). This opening will be resisted by the load in the stiffeners which will cause the sheet to distort [13]. The true zero displacement reference is taken at the crack tip. The resultant closing of the crack at the middle of the stiffeners is given by

$$\Delta_7 = \frac{q}{\pi E} \left(2w \ln \frac{1}{w} + (2p + w) \ln \frac{1}{2p + w} - (2p - w) \ln \frac{1}{2p - w} \right.$$
$$- (2a - p + w) \ln \frac{1}{2a - p + w} - (2a + p + w) \ln \frac{1}{2a + p + w}$$
$$\left. + (2a - p - w) \ln \frac{1}{2a - p - w} + (2a + p - w) \ln \frac{1}{2a + p - w} \right) \tag{64}$$

The load intensity q in Eq. (64) is given as in Eq. (34) by

$$q = \sigma_b^{\infty} \frac{2a}{2w} \tag{65}$$

in which σ_b^{∞} is the sheet stress at infinity which is amplified in the ratio A_{skn}/A_{str} as it is transferred through the bond to the stiffener. If there be any disbonds present, this contraction is reduced by an amount derivable from Eq. (66) as

$$\Delta_8 = \frac{(3 - \nu)(1 + \nu)}{4\pi E} \left(\frac{\tau_p w}{t_{skn}} \right) \left(\ell \ln \frac{2b}{\ell} - (2b + \ell) \ln \frac{2b}{2b + \ell} \right) \tag{66}$$

in which the effective adhesive zone extends a distance ℓ given in Eq. (55) as

$$\ell = \frac{\sigma_b^{\infty} 2a t_{skn}}{2\tau_p w} \tag{67}$$

If the influence of sheet distortion on σ_b^{∞} is neglected, the component of the stress in the sheet away from the crack which passes through the bond is given by Eq. (22) as

$$\sigma_b^{\infty} = \frac{\tau_p 2w}{\lambda 2at_{skn}} \sqrt{1 + 2\frac{\gamma_p}{\gamma_e}} = \sqrt{\frac{2E\tau_p \eta[(\gamma_e/2) + \gamma_p)]}{1 + 2at_{skn}/A_{str}}} \left(\frac{2w}{2at_{skn}}\right) \tag{68}$$

and the corresponding stiffener stress is

$$\sigma_{str}^{0} = \sigma_b^{\infty}\left(1 + \frac{2at_{skn}}{A_{str}}\right) \tag{69}$$

Over a disbond length b on each side of the sheet crack, the stiffener will stretch an amount Δ_6 given in Eq. (60) as

$$\Delta_6 = \frac{b\sigma_{str}^{tot}}{E} = \frac{b(\sigma_{str}^{0} + \sigma_c)}{E} \tag{70}$$

Exactly the same compatibility requirement as in the preceding example must be met. Thus

$$\Delta_6 + \eta(\gamma_{max}) = \delta_0 - \Delta_7 + \Delta_8 \tag{71}$$

The adhesive shear strain γ_{max} may not exceed the maximum allowable $(\gamma_e + \gamma_p)$ but if the failure should initiate in the sheet or stiffener, γ_{max} may not attain that full value. After rearrangement, Eq. (71) is solved in the form

$$\eta(\gamma_{max}) + b\frac{\sigma_b^{\infty}}{E}\left(1 + \frac{2at_{skn}}{A_{str}}\right) + \Delta_7 - \Delta_8 = \frac{\sigma_c^{\infty}}{E}\left[2\sqrt{a^2 - \left(\frac{p}{2}\right)^2} - b\right] \tag{72}$$

Again, it is evident that a disbond twice as long as the sheet crack is sufficient to nullify the effectiveness of the bonded stiffener as a crack arrester. The method of analysis for Eq. (72) is to assume trial values of γ_{max}, deduce σ_c^{∞} from this equation, and keep increasing γ_{max} until some element of the structure becomes critically loaded. The analytical procedure for less-than-critical loads is the same as that described at the end of the preceding example.

Examples

Figures 29 to 34 illustrate some parametric effects on the strength of bonded panels with a one-bay sheet crack extending slightly past bonded stiffeners. All solutions exhibit the same characteristic form as the disbond grows; the three phases, in order, are disbond,

stiffener yield, and sheet fast fracture. The first two phases can be
absent, depending on the geometry.

Figure 29 shows the influence of how far past the stiffener the
crack has spread. As far as the sheet fracture is concerned, the
longer the crack, the weaker the panel. The same sequence holds
true for yielding of the stiffeners. However, with regard to the initial
disbond, there may be a different phenomenon when the crack extends
only immediately beyond the bond stiffeners. Figure 29 shows little
sensitivity of the disbond length to the crack size. This is because
the stiffener yielding relieves the bond stresses and strains as long
as the load is reduced while the sheet crack grows. Were the load
maintained at the 37 ksi associated with the 6.76-in. semicrack length
as the sheet crack grew, the stiffeners could stretch only as much as
the opening of the sheet crack permitted, even though they were yielding. But once the crack has extended across the entire width of the
panel (36 in. in this case), the overload stiffeners would stretch more
and more until they fractured. In association with this last phase of
failure for the stiffened panel, the disbond would try to grow along
the entire length of the stiffener if the strain rate of the loading device were not too fast. The stiffener would then fall off the panel and
break away from the sheet crack.

Figure 30 shows the influence of variations in the width of adhesive
used to bond stiffeners of common area. The general tendency of the
greater width increasing the strength is complicated by the occurrence
of yielding of the stiffener at a stress less than that needed to fast-
fracture the sheet. The reason for the change to stiffener yielding is
that the greater bond width permits the stiffener to accept more load
without breaking the bond.

Figure 31 shows how greater stiffener cross sections are effective
in increasing the panel strength.

Figure 32 shows that finite sheet widths can effect significant reductions in panel strength, indicating the need to account for this factor
in analyses.

Figure 33 shows how for a constant stiffener cross section, greater
strengths are developed by the thinner cracked sheet panels. The
failure sequence is an initial disbond at constant load which is self-
arresting, a stable propagation of the disbond under increasing load
until the stiffener yields, and eventually, the sheet fast fracture.

Figure 34 demonstrates that while the fast-fracture and disbond
characteristics can be affected significantly by change in adhesive
properties with temperature (and therefore other environments as well),
the overall effect on panel strength is slight. The reason for the relatively small changes is that the overall adhesive properties vary with
temperature and moisture in such a way that the shear strain energy
to failure changes little. As the temperature is reduced (or the adhesive dried out), the adhesives become stiffer and more brittle, while
at higher temperatures or moisture content, the adhesives become softer

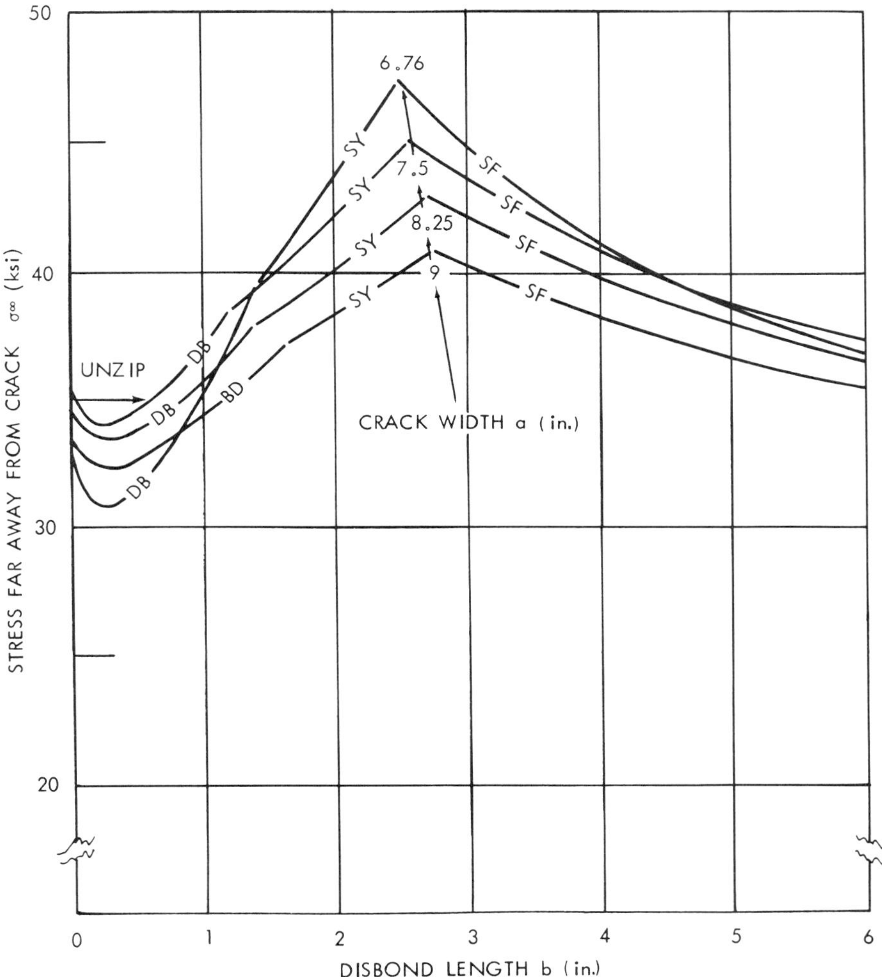

Figure 29 Parametric analysis of a one-bay sheet crack (Fig. 28) showing the influence of crack length on panel stresses far away from crack and associated disbond length, all leading to a sheet fast fracture. Panel conditions used in the analysis: t_{skn} = 0.063 in., A_{str} = 0.25 in.^{2}, p = 12 in., s = 36 in., w = 1.5 in., η = 0.002 in., τ_p = 5500 psi, G = 60,000 psi, γ_p = 0.55, E = 10.5 × 10^6 psi, K_c = 134 ksi $\sqrt{\text{in.}}$, $F_{y\ str}$ = 75 ksi, $F_{y\ skn}$ = 65 ksi. DB, disbond; SY, stiffener yield; SF, sheet fast fracture.

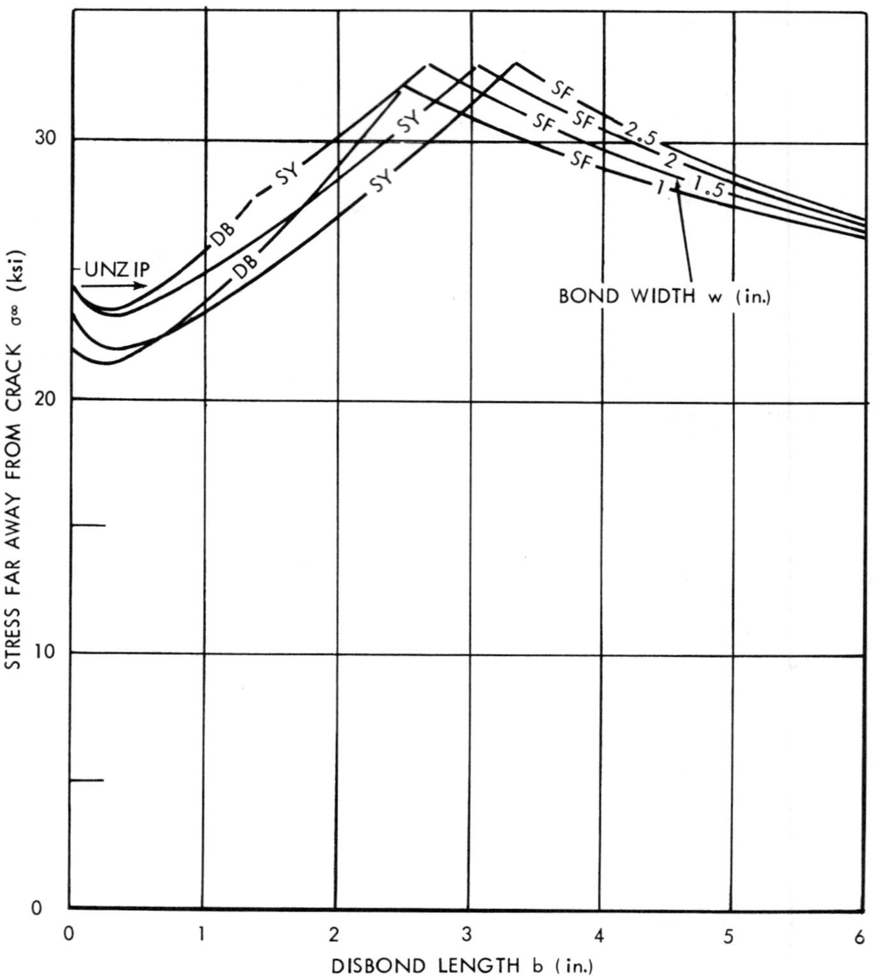

Figure 30   Parametric analysis of a one-bay sheet crack (Fig. 28) showing the influence of bond width on panel stresses far away from crack and associated disbond length, all leading to a sheet fast fracture. Panel conditions used in the analysis: t_{skn} = 0.063 in., 2a = 16.5 in., A_{str} = 0.25 in.^{2}, p = 12 in., s = 36 in., η = 0.002 in., τ_p = 5500 psi, G = 60,000 psi, γ_p = 0.55, E = 10.5 × 10^6 psi, K_c = 134 ksi $\sqrt{in.}$, $F_{y\,str}$ = 75 ksi, $F_{y\,skn}$ = 65 ksi. DB, disbond; SY, stiffener yield.

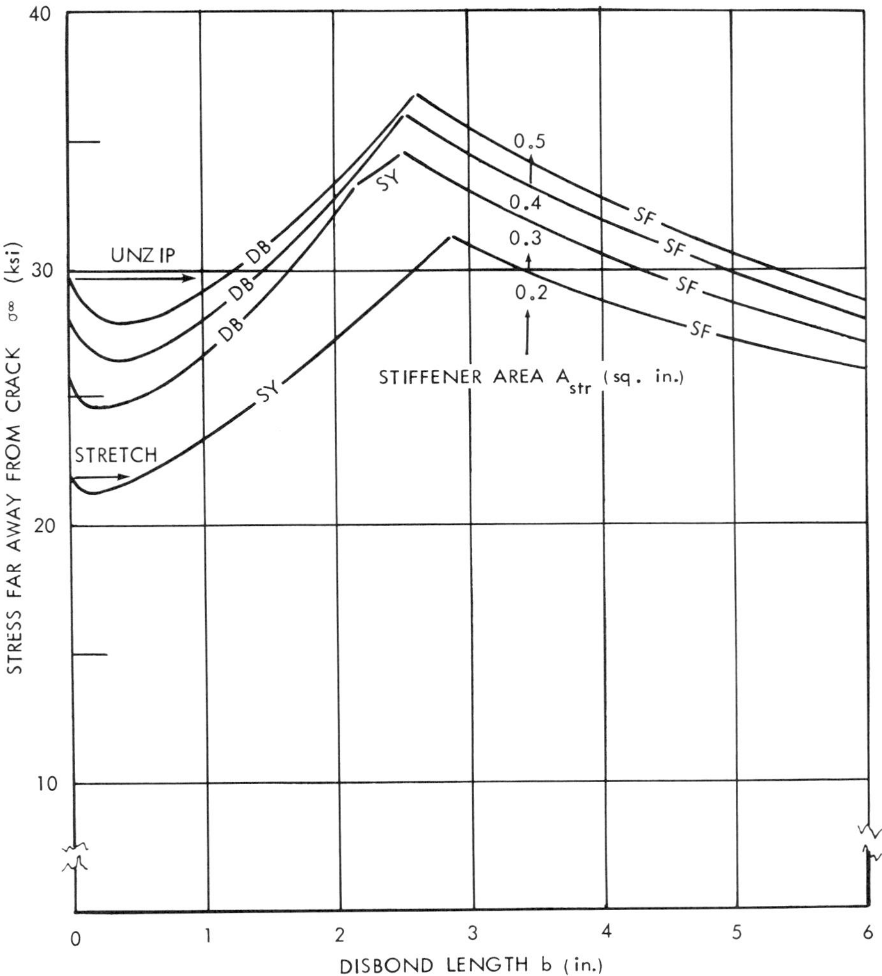

Figure 31 Parametric analysis of a one-bay sheet crack (Fig. 28) showing the influence of stiffener cross-sectional area on panel stresses far away from crack and associated disbond length, all leading to a sheet test fracture. Panel conditions used in the analysis: $t_{skn} = 0.063$, $2a = 16.5$ in., $p = 12$ in., $s = 36$ in., $w = 1.5$ in., $\eta = 0.002$ in., $\tau_p = 5500$ psi, $G = 60,000$ psi, $\gamma_p = 0.55$, $E = 10.5 \times 10^6$ psi, $K_c = 134$ ksi $\sqrt{\text{in.}}$, $F_{y\ str} = 75$ ksi, $F_{y\ skn} = 65$ ksi. DB, disbond; SY, stiffener yield; SF, sheet fast fracture.

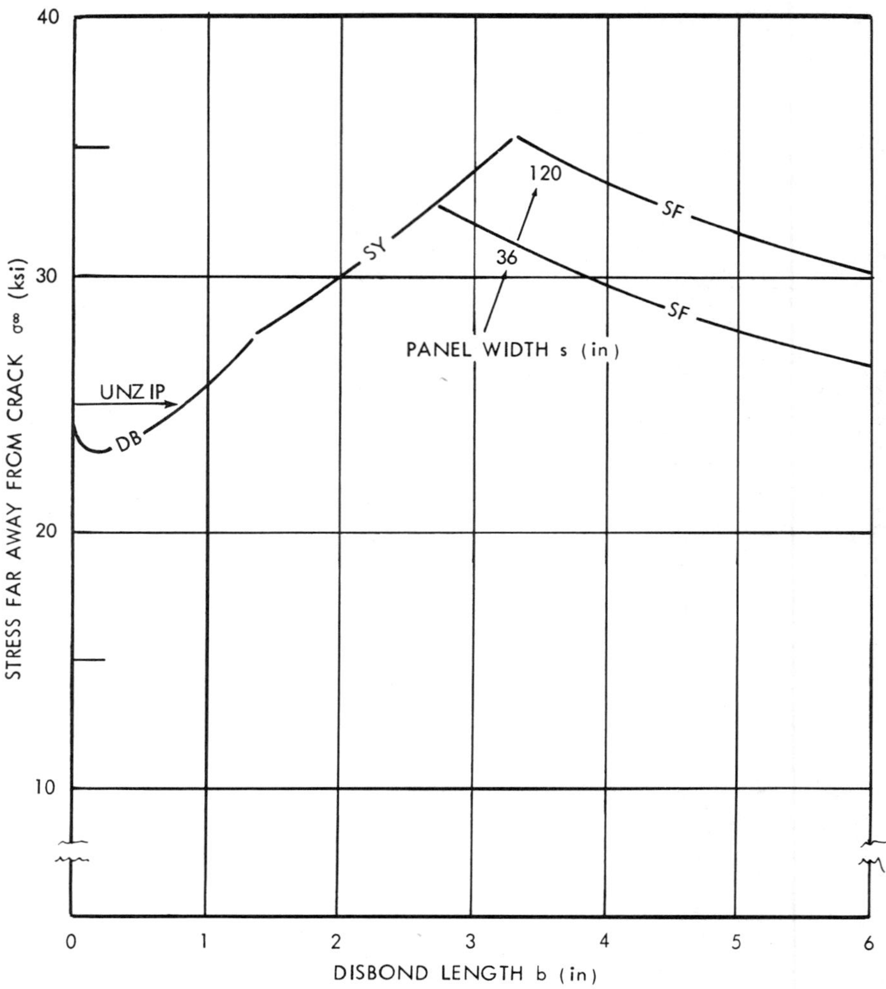

Figure 32 Parametric analysis of a one-bay sheet crack (Fig. 28) showing the influence of a panel width on panel stresses far away from crack and association disbond length, all leading to a sheet fast fracture. Peel conditions used in the analysis: t_{skn} = 0.063 in., 2a = 16.5 in., A_{str} = 0.25 in.^{2}, w = 1.5 in., η = 0.002 in., τ_p = 5000 psi, G = 60,000, γ_p = 0.55, E = 10.5 × 10^6 psi, K_c = 134 ksi $\sqrt{\text{in.}}$, $F_{y\,str}$ = 75 ksi, $F_{y\,skn}$ = 65 ksi. DB , disbond, SY, stiffener yield; SF, sheet fast fracture.

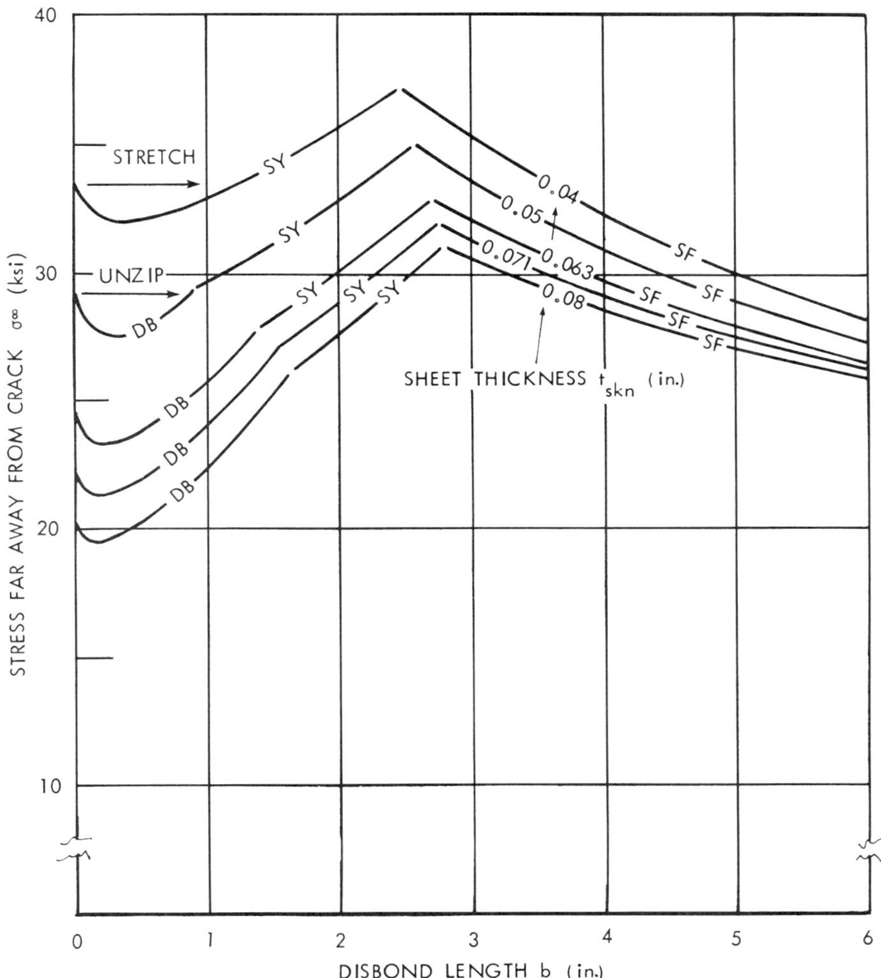

Figure 33   Parametric analysis of a one-bay sheet crack (Fig. 28) showing the influence of sheet thickness on panel stresses far away from crack and associated disbond, all leading to a sheet fast fracture. Panel conditions used in the analysis: $2a = 16.5$ in., $A_{str} = 0.25$ in.^{2}, $p = 12$ in., $s = 36$ in., $w = 1.5$ in., $\eta = 0.002$ in., $\tau_p = 5500$ psi, $G = 60,000$ psi, $\gamma_p = 0.55$, $E = 10.5 \times 10^6$ psi, $K_c = 134$ ksi $\sqrt{\text{in.}}$, $F_{y\,str} = 75$ ksi, $F_{y\,skn} = 65$ ksi. DB, disbond; SY, stiffener yield; SF, sheet fast fracture.

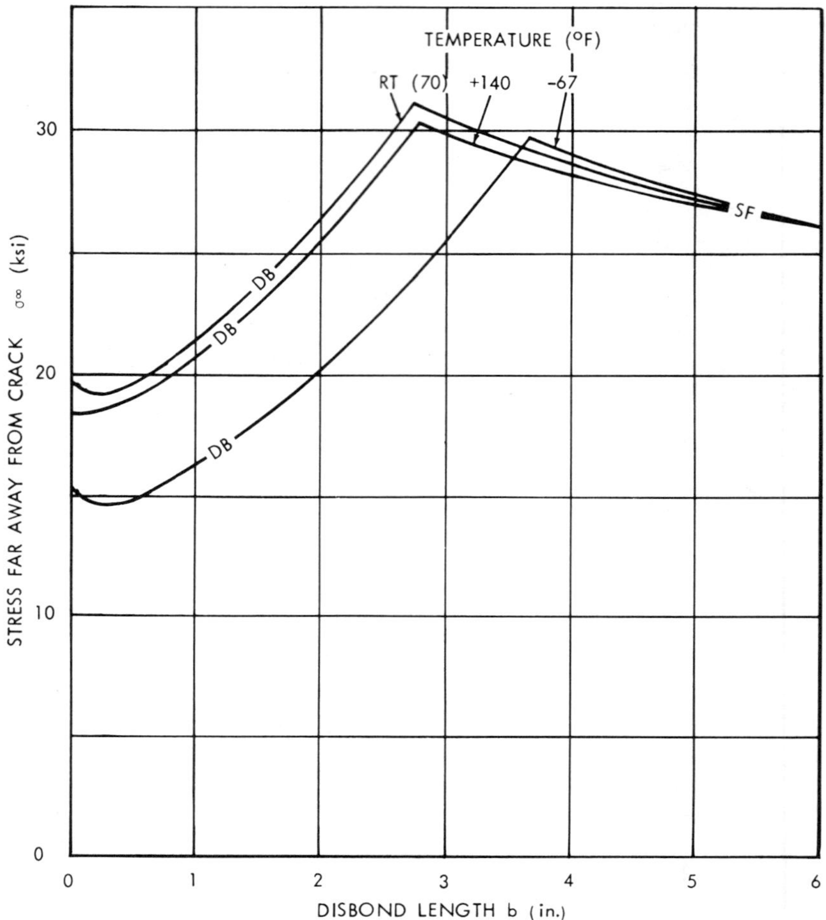

Figure 34   Parametric analysis of a one-bay sheet crack (Fig. 28) showing the influence of the adhesive properties on panel stresses far away from crack and associated disbond length, all leading to a sheet fast fracture. Panel conditions used in the analysis: t_{skn} = 0.063 in., 2a = 16.4 in., A_{str} = 0.237 in.^{2}, p = 12 in., s = 36 in., w = 0.94 in., η = 0.00175 in., E = 10.5 × 10^6 psi, K_c = 134 ksi $\sqrt{in.}$, $F_{y\ str}$ = 75 ksi, $F_{y\ skn}$ = 65 ksi. DB, disbond; SF, skin fast fracture.

Adhesive Properties			
Temp. (°F)	τ_p (psi)	G (psi)	γ_p
70	5500	60,000	0.55
140	2500	40,000	1.0
-67	7000	70,000	0.25[a]

[a] Artificially low to accent effect.

and more ductile. The curves for room temperature and +140°F have
adhesive characteristics so that the strain energy to failure of the adhesive in shear is practically identical. Consequently, those two failure characteristics in Fig. 34 lie close together. The adhesive properties at −67°F were artifically modified by representing the adhesive as
being more brittle than it really is in order to amplify the effects on
the cracked stiffened sheet failure characteristic. This artificial adhesive had only about half the shear-strain energy when it was cold
than the real adhesive had at room temperature and above. The reason
adhesive strain energy has so much more of an effect than the adhesive
failure stress τ_p or strain ($\gamma_e + \gamma_p$) is that the strain has a major effect on the load which the stiffener can carry without propagating a
disbond while the stress represents only a small perturbation on the
sheet distortion (provided that the crack tip is not too close to the
bonded stiffener).

The results of an example problem obtained from testing a stiffened
center-crack panel during a test program provide a means of assessing
the accuracy of the present analytical method. This test panel is discussed fully in Swift [12], so only a brief description is presented
here. The basic sheet was of 0.063-in.-thick 2024-T3 aluminum alloy
(nonclad), 36 in. wide by 48 in. long, with the grain in the transverse
direction. Two bonded Z stiffeners of 7075-T6 aluminum alloy extrusion, having an area of 0.237 in.^{2} and a bond contact width of 0.938 in.
were bonded longitudinally at a 12-in. pitch. A central crack was grown
to 8.2 in. on one side and 7.2 in. on the other. The adhesive thickness varied between 0.001 in. under the web of the stiffener and
0.0025 in. at the free edge of the flange. The gross stress at failure
was measured to be 30,600 psi.

Figure 35 contains the results of an analysis of this panel, using
the present methods. The analysis predicts an initial fracture of the
bond at a nominal stress of about 20 ksi. The fracture would be arrested as long as the load were not increased. With continued application of increasing load, the disbond would propagate slowly to a length
of 2.75 in., at which point the sheet crack would become unstable and
tear the panel apart at a calculated stress of 31 ksi.

Figure 36 shows the fracture surface of one of the disbonded stiffeners. Striations in adhesive identify the fracture which was arrested
at about 0.85 and 0.75 in. (not evident in the photograph), compared
with the prediction of 0.6 in., and the initiation of sheet fracture at
about 2.75 in. (clearly visable in the photograph), which agrees exactly
with the prediction. The striations are inclined because of the variation in glue thickness from 0.001 in. under the stiffener web to 0.0025
in. at the free edge of the flange. The greater disbond is correctly
associated with the thinner (and stiffer and weaker) bond.

This agreement between theory and test, which extends to the
lengths of disbonds as well as to the failure load, is sufficiently good
to lend confidence to the approximations in the theory.

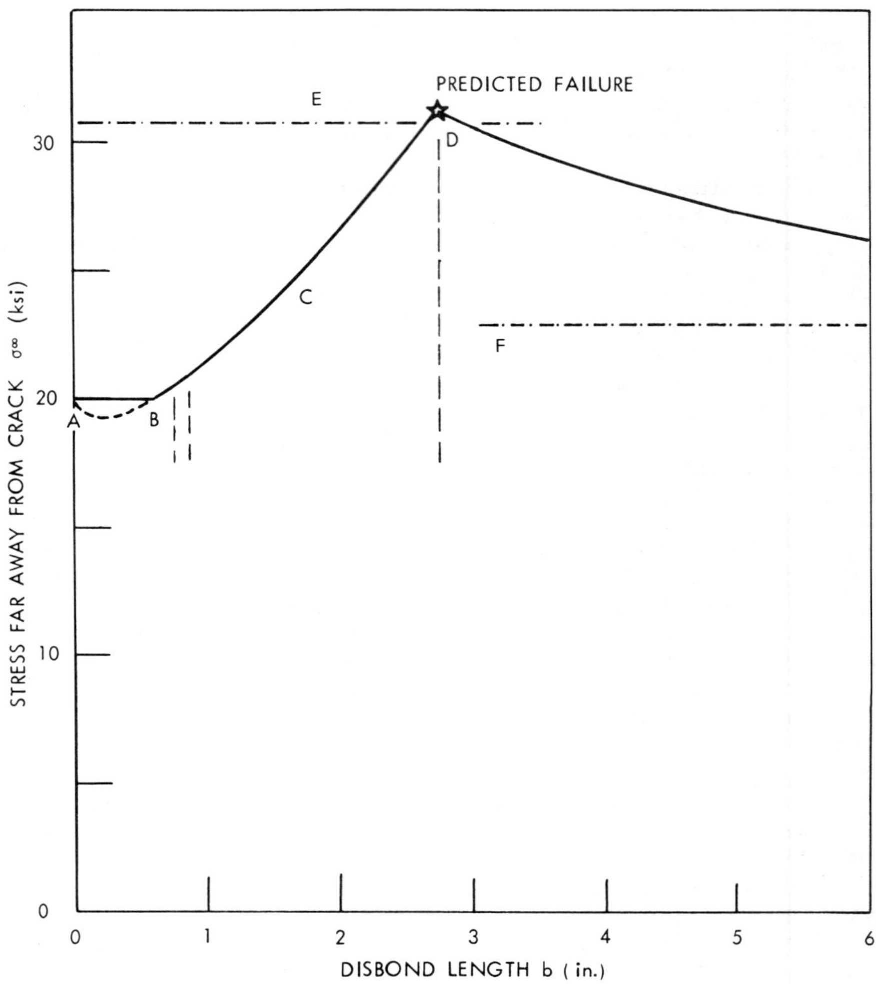

Figure 35   Test results of a one-bay sheet crack static test of panel shown in Fig. 28. Room-temperature (70°F) test. Fast disbond (unzip) at 20 ksi from A to B. As load built up from 20 ksi to 30.8 ksi a stable disbond increased from B to C to D. At the disbond length of 2.75 in. (D) there was a sheet fast fracture. E is the failure stress of 30.8 ksi. F is the theoretical strength of an unstiffened panel.

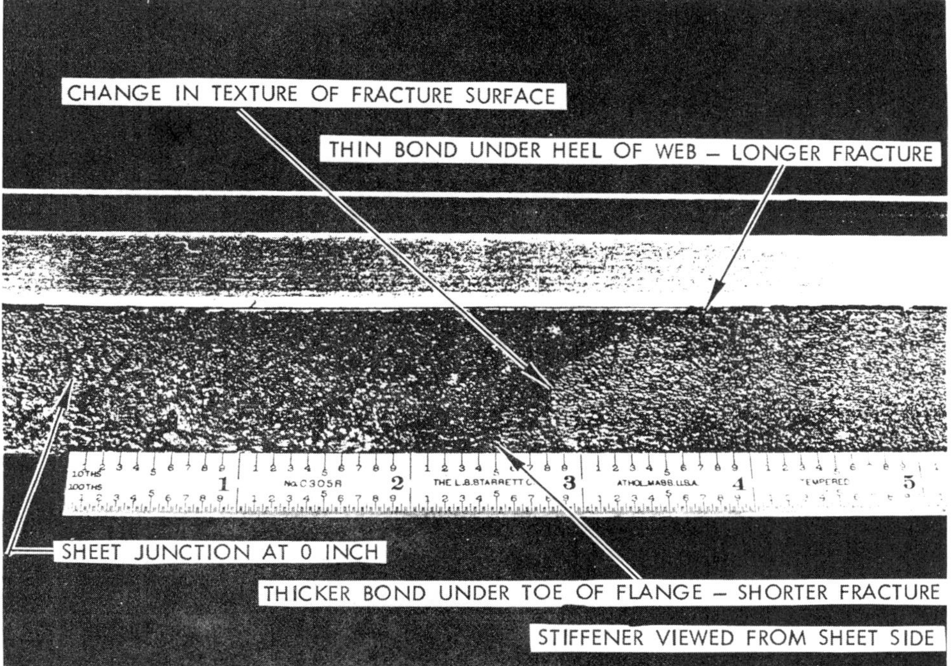

Figure 36   Fracture surface of disbond between stiffeners and sheet showing striations at changes in failure phenomena.   (Courtesy Douglas Aircraft Co.)

IX. CRACKING OF PRESSURIZED-BONDED STIFFENED FUSELAGE SHELLS

When the designer considers using a bonded structure, a prime thought should be about relatively thin formed metal details. The fuselages of all aircraft are good candidates for weight saving and therefore, the use of bonding. The fuselage has many parts (frames and longerons) attached to the relatively light skins. These are the optimum requirements for good bonding. Control surfaces, fixed and movable control surfaces and wings of light aircraft are also good candidates for bonding.

The preceding pages have presented the analysis which shows how critical the skins and frame caps are in the presence of a failure in the other member (skin with cracked frame cap or frame cap with cracked skins). In the early history of pressurized riveted fuselages, some have experienced the problem of very long skin cracks growing past

frames instead of being arrested. Further, they have not been detected on the ground during inspection, which is a dangerous situation.
The fail-safety of aircraft fuselages is strongly influenced by the
manner in which skin cracks, which exist in all structures at one time,
grow and, with proper structural design and manufacture, are retarded,
turned, or arrested by the skin-stiffening members. Some aircraft
have had to be modified in service by the addition of external belly-
bands which would arrest crack growth. The history of skin cracks
in riveted, pressurized fuselages has been random, depending on the
skin stress levels and frame cap details (cap area, rivet spacing, and
rivet size). Some cracks follow rivet lines and go right past frames,
cracks are arrested in rivet holes, cracks have gone past frames and
missed the rivet holes, and some longitudinal cracks have stopped or
turned at the frame rivet pattern to spread circumferentially.

In order to assess the durability and fail-safe characteristics of
bonded structure, both riveted and bonded flat unpressurized panels
were tested for direct comparison. The results seemed favorable to
bonded structures. From the same size initial flaw, the bonded configuration reaches a 3-in. half crack length in 230,000 cycles vs
140,000 for the riveted design (see Fig. 37). A similar pair of specimens, one bonded and the other riveted, had two stiffeners and a flaw
located centrally between the stiffeners. The initial flaw lengths were
0.25 in. and were allowed to grow to full length of 14 in. Figure 38
shows the crack growth history. Again the bonded stiffener offered
more retardation to the initial crack growth up to the time the crack
reached the stiffener and then the rate of crack growth across the
stiffener and beyond was reduced. For the riveted stiffener, the last
1.5 in. of crack growth took approximately 1000 cycles, while the bonded stiffener configuration took 11,000 cycles to go the last 1.5 in. The
residual strength failing load of the bonded panel was 13.5% greater
than for the riveted panel. Other flat unpressurized bonded panels
have had cracks grow straight across to the stiffeners and beyond.
Fail-safe loads have been applied to flawed panels and demonstrated
that fast fractures can be arrested by adjacent bonded stiffeners.

A unique loading and/or stress distribution occurs in a pressurized
fuselage structure that has thin skin and frames spaced at varying
distances. If we look at a fuselage with only frames, the sketch at the
upper part of Fig. 39 shows the "pillowing" deformation that results.
The reason for this pillowing is the higher hoop tension stress in the
skin element between frames compared to the lower stress in the frame
structure because of the greater area to carry the load. Also, the
skin attachment to frame is in tension. The lower sketch in Fig. 39
shows how the pillowing is modified by each longeron. When the skin-
to-frame joint is studied for the riveted design and the bonded one, it
can be seen that the bonded joint pinches in the skin far more severely
over each frame than does the riveted design. The bonded frame tee

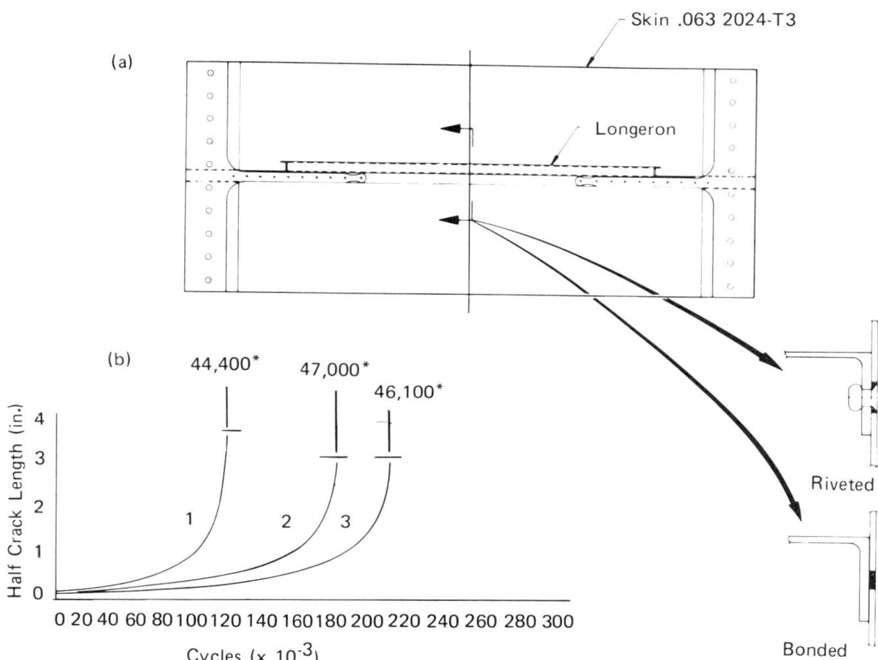

Figure 37   Comparison of initial flaw or crack growth characteristics for a riveted and a bonded stiffened panel shown in (a).   Inset at the left shows the comparative initial flaw lengths, which were 0.25 in.   Panel characteristics: t_{skn} = 0.063 in., A_{skn} = 1.26 in.^{2}, t_{str} = 0.10 in., A_{str} = 0.237 in.^{2}, Cycle σ_{max} = 14,000 psi, R = 0.05.   The graph in (b) is a plot of crack length versus test load cycles.   Curve 1 is the room-temperature test of the riveted panel. Curve 2 is the room-temperature test of the bonded panel, and curve 3 is the elevated-temperature (140°F) test of another bonded panel. The starred values indicate the residual or failing stress of each panel.

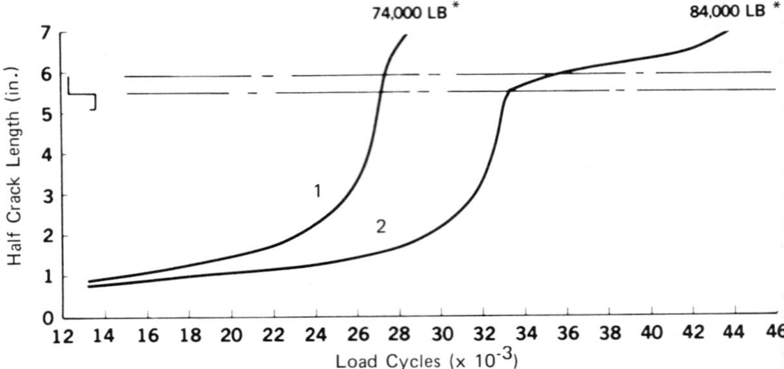

Figure 38 Comparison test results of flaw growth in a riveted and a bonded stiffened panel. Each panel had two stiffeners and an initial flaw of 0.25 in. located midway between the stiffeners. Panel dimensions and test conditions: initial gross stress τ_{max} = 14,000 psi, R = 0.05, t_{skn} = 0.063 in., A_s = 0.237 in.^{2}, t_{str} = 0.10 in. Curve 1 is the riveted panel. Curve 2 is the bonded panel. The starred values indicate the residual or failing strength of each panel.

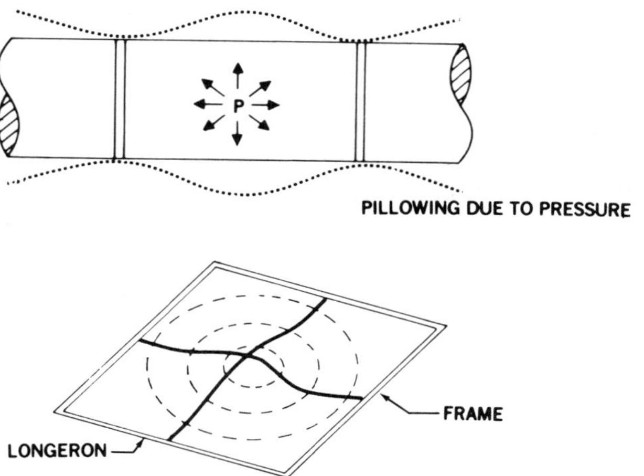

Figure 39   Stress distribution in a pressurized stiffened cylinder.

section cap has a broader base and therefore a large area of attachment to the skin, while the riveted design uses an angle, or Z, section and a single line of rivets with a very narrow base. Consequently, the skin hoop tension over the frame is much lower for a bonded structure. The more severe pinch-in increases the axial bending stresses in the skin adjacent to the frame.

Reviewing again the frame-longeron joint it should be remembered that it must have structural continuity of both the longeron and the frame cap combination where they cross. Also, this intersection represents the region of lowest skin stress in bonded designs. This is not true for riveted designs. Therefore, it seems that for structures where both longerons and frame cap members are bonded to the skin, skin cracks should be diverted away from their corners. In other words, a skin crack is more likely to go along a frame tee than to cross it. Even on the bottom of the fuselage, where the longitudinal axial membrane stress is low due to compression inertia loads subtracting from the tensile pressurization loads, the axial bonding stresses are still high enough to make the longitudinal skin cracks tend to turn and grow along the frame rather than across it.

However, comparable studies of riveted construction completely reverse this picture. When the skin is riveted to a single-flange clip, the skin is pulled or held in by tension loads on the rivets, so both the skin and flange tend to pucker there. Also, the clip flange bends between the rivet seam and web, so the skin-to-frame tie is much more flexible than it would be were it bonded (see Fig. 40) using a symetrical tee section. Consequently, the skin hoop stress over the frame is greater for riveted construction than for bonded. Furthermore, since the skin is not pinched in so severely there, the axial bending moments are reduced. The result of each of these effects is to make a skin crack more likely to cross a riveted frame clip than a bonded one. Obviously, this is a trend rather than an infallible rule, since the bonding of very light frames or the riveting of very heavy frames will induce the opposite results. Nevertheless, it must be observed that for a given design and loads, a skin crack is less likely to cross a bonded shear clip than the riveted counterpart. Far more care must be taken in the detail design and analysis of riveted structure to prevent skin cracks from growing past the frames.

Parametric studies based on the pressure pillowing analysis have also permitted an assessment of the trade-off between the lower skin hoop stresses and higher bending stresses associated with a bonded frame tee, because of the maximum pinch-in, and the higher skin hoop stresses and lower bending stresses due to the lesser pinch-in associated with riveted construction. Using the criterion of the sum of the membrane stress plus half of the associated bending stress, a range of flange flexibilities was assessed. In no case was the combination for a circumferential crack found to be more severe than for a longitudinal crack under hoop loads alone, far away from any stiffener.

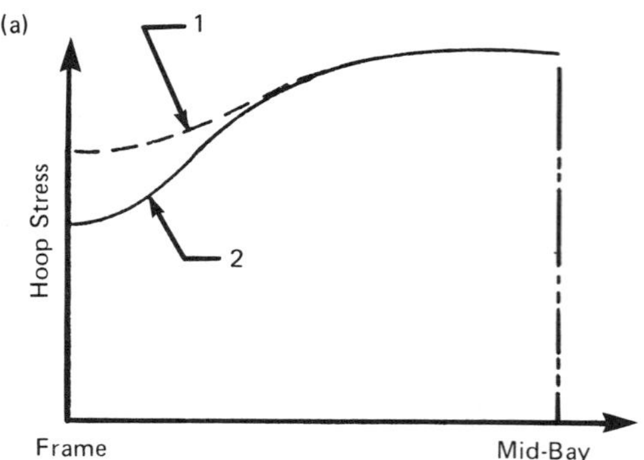

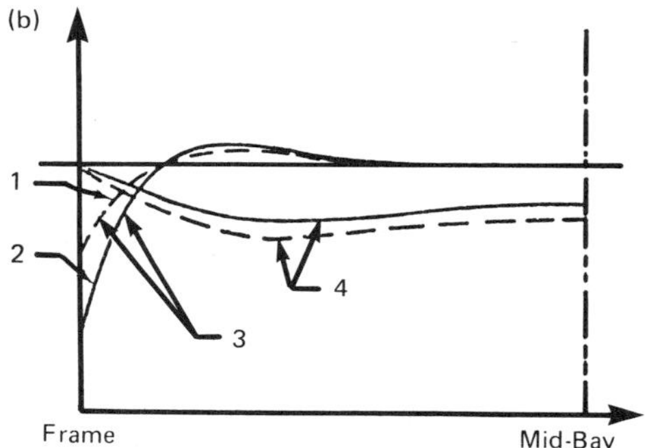

Figure 40 Comparative skin stresses resulting from aircraft cabin
pressurization load for bonded and riveted fuselage structure. Curves
1 represent the riveted structure. Curves 2 represent the bonded
structures. Curves (a) show hoop stress and curves (b) show longitudinal bending stresses. Area 3 indicates the region between longerons, and area 4 shows the region over the longerons.

The worst situations were invariably the result of a longitudinal crack
adjacent to a longeron. One is led to conclude, therefore, that the
bonding of the fuselage stiffeners is always beneficial.

So far, the discussion has been concerned with longitudinal cracks
being turned by frames. The pillowing analysis also provides insight
into the stress field away from the frames and into the bending of the
longerons. The parametric studies have also explained what happens
to circumferential cracks in terms of the relative hoop and axial stresses
and of the structural geometry. The more severe cracks are adjacent
to the frame, where there are high bending stresses, rather than in
mid-bay. A circumferential crack growing alongside a frame can do one

of three things when it approaches a longeron. If there is a structural discontinuity in the frame clip, the skin crack will probably turn toward the frame and become a longitudinal crack under the influence of the locally high skin hoop stress. Should the stiffener intersection have sufficient structural continuity not to impose a stress concentration on the skin, one of the two other results will occur, depending primarily on the relative axial and hoop stresses. If the axial stress dominates greatly over the hoop stress, the crack will grow straight across the longeron. On the other hand, if the axial stress is only just great enough to permit a crack to grow circumferentially instead of longitudinally, it is probable that the pressure pillowing will force the skin crack away from the frame in the corner before it reaches the longeron and it will then grow along the side of the longeron where there are bending stresses to augment the hoop stress. Naturally, since the role of bending stresses is so important in this issue, the amount of pinch-in influences whether or not the crack turns for a given ratio of hoop and axial membrane stresses. Consquently, the structural geometry also influences this behavior. The same situation exists with respect to a circumferential crack approaching halfway between the frames. There, the skin axial stress is aggravated by the bending of the longeron while the membrane hoop stress is supplemented by the bending hoop stress, tending to deflect the skin crack along the longeron if the axial membrane stresses are not so high as to force the crack straight across the longeron.

The flat panel tests described earlier are not sufficient to prove the crack turning or arresting feature that is predicted. Therefore, a much more sophisticated test must be developed. The test panel must be curved to match the radius of real aircraft or structure being evaluated. A pressure load must be applied to obtain the pillowing. Also, provisions must be made to apply axial loads to the panel, either tension or compression. Finally, all loadings must be applied cyclically to obtain crack growth. Another way to study crack growth is to use the complete airframe as the test vehicle and cyclically apply the ground-air-ground loadings. Granted, this is a very expensive way to do development tests, but the end or edge conditions of the test panel are correct.

Test results from a series of pressurized panel tests and a complete fuselage test of completely bonded structures give some very interesting results. A large test panel (168 in. long and 110 in. wide) was tested to observe crack growth for six cracks. Each crack, made by a thin modeling saw blade, measured 0.25 in. long. The panel had frames spaced at 12 in. and had no longerons. The panel was cycled 38,028 cycles (one aircraft lifetime) of ground-air-ground loadings. Maximum cabin pressure loads were used during "air" loads and zero pressure for the "ground" loads. Virtually no crack growth occurred during these loadings. To see if a fast fracture or crack growth could

be initiated, an 18-in. two-bay crack was sawed into the skin, leaving
the centerline frame intact. The sawcut was sharpened by pressure
cycling the panel from 0 to 4 psi until visible crack growth was observed. Next, the fail-safe load was applied as follows: Internal pressure was increased to its maximum value of 8.6 psi with the longitudinal
load held at zero. Then the longitudinal load was increased to its
maximum design value. Only during the application of the longitudinal
load was crack growth seen. Figure 41a shows the final crack growth.

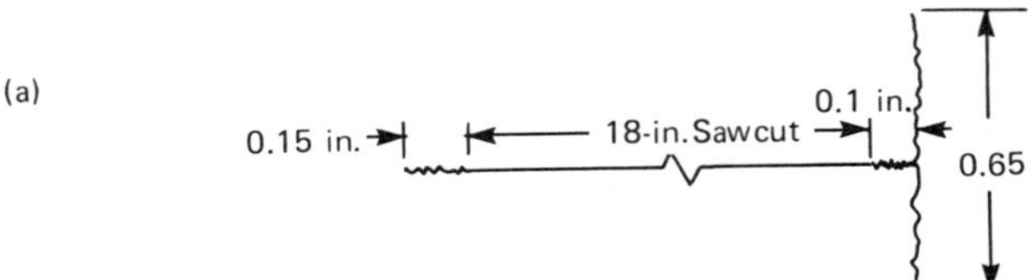

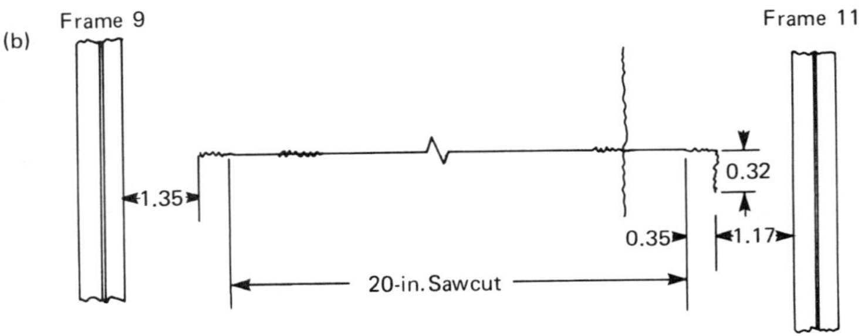

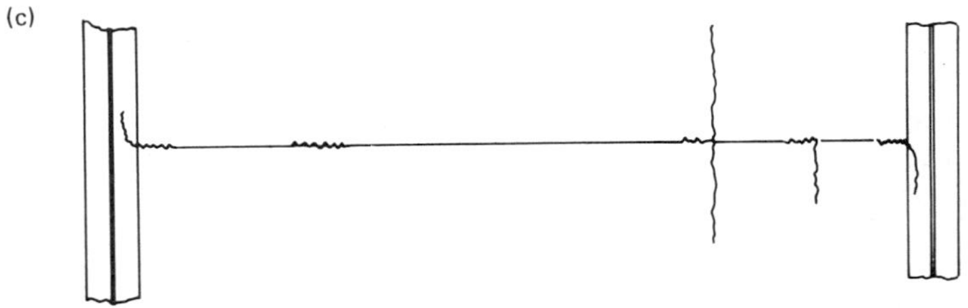

Figure 41   Presents crack growth from an initial flaw saw cut of 18-in.
to the final saw cut and crack growth. (a) Initial flaw (solid line) and
is crack growth under the static fail-safe load (wavy line). (b) Final
flaw (a) is increased to 20 in. and again a fail-safe load is applied, with
wavy line showing growth. (c) Final crack in (b) increased to within
0.25 in. of the frame bonded tee. Extreme wavy lines show fail-safe
load crack growth and final arrestment by frame tees.

A saw cut increased the crack length to 20 in., as shown in Fig. 41b. The fail-safe load was again applied in the manner described above and crack growth was seen as shown in Fig. 41b. No fast fracture developed. The ends of the cracks were cut to within 0.25 in. of the edge of each bonded tee cap. Cracks were sharpened by cycling and the fail-safe load applied in the same manner. Figure 41c shows how the crack tips, propagated to each adjacent frame shear tee, penetrated the bond line and then each crack turned circumferentially but in opposite directions. The crack growth was successfully arrested by the bonded joint of the frame shear tee and the fuselage skin.

In another panel test, where the frame spacing was 24 in. and no longerons, 0.25 in. flaws were cut into the skin. Since no crack growth was seen after one lifetime, the saw cuts were increased to 4 in. and testing continued. After an additional 53,000 cycles only slight crack growth was noted. When the crack length increased to 6 in. after another 2000 cycles, the crack growth was easy to see. As the cracks approached the bonded frame caps, each end turned, in the same direction, to follow the direction of the frame caps. Disbonding occurred in the path of the crack because of the peeling stress developed as the skin flap opened under pressure loads. Because of the final crack size and the associated opening of a skin flap, as shown in Fig. 42, it was impossible to maintain the internal pressure over the panel. Figure 43 shows a close-up of crack tips. This is a real advantage for a pressurized aircraft. With the crack arresting in a flap or window-shaped failure it will preclude pressurization of the aircraft in flight and, therefore, require a thorough inspection on the ground to locate the problem. Past experience has shown that riveted structures have had 75- to 130-in. skin cracks and no detectable loss in cabin pressure. These long cracks must be found on ground inspection and, considering the large area of skin to be inspected, the cracks are hard to find.

The test of a full-scale bonded fuselage showed crack growths very similar to the test panels above. All cracks studied had to be started by saw cuts. Some of the cracks were made near longerons and midway between frames; others were midway between the longerons and frames. Figure 44 shows the final failure of a crack that was started near and parallel to a longeron. The original cut is the straight line at the lower part of the flap. The crack grew rather slowly into the bond lines of the frame tees and then suddenly tore open, as shown. Figure 45 shows the final failure of a crack identified as DT-20. The centrally located crack went straight to the frame tee bonded joint and then the cracks grew parallel to the frame tees but in opposite directions. Continued cycle loading caused the cracks to grow in the opposite directions along the frame tee bonded joint, resulting in the failure shown in Fig. 45. Figures 46 and 47 show failures similar to the one in Fig. 45. The failure in Fig. 47 shows that the crack will turn even if the skin is not bonded to the frame tee. The heavy crosshatching in the figure denotes a disbond area that was the result of a manufacturing problem.

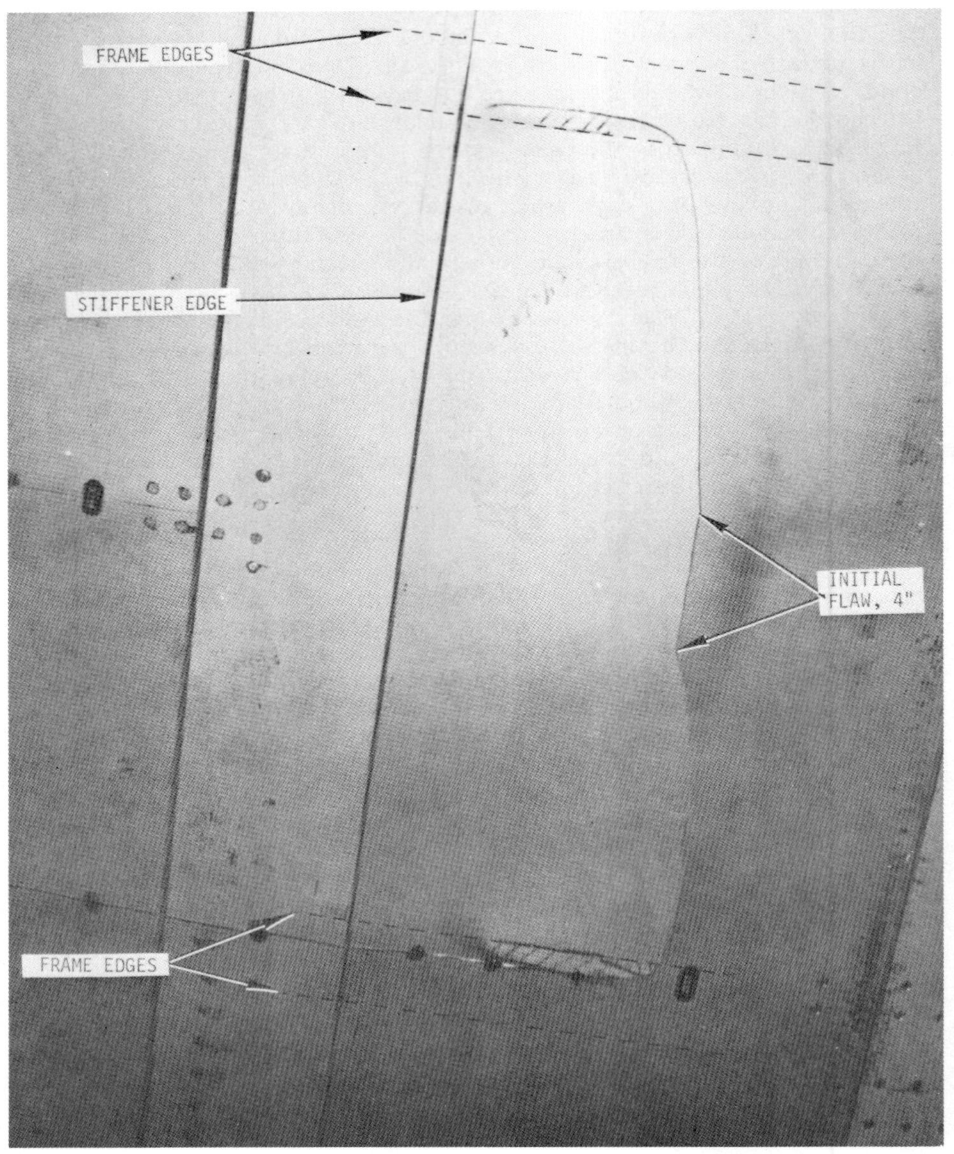

Figure 42   Bonded, curved panel crack propagation: crack turned at frame and stiffened edge. (Courtesy Douglas Aircraft Co.)

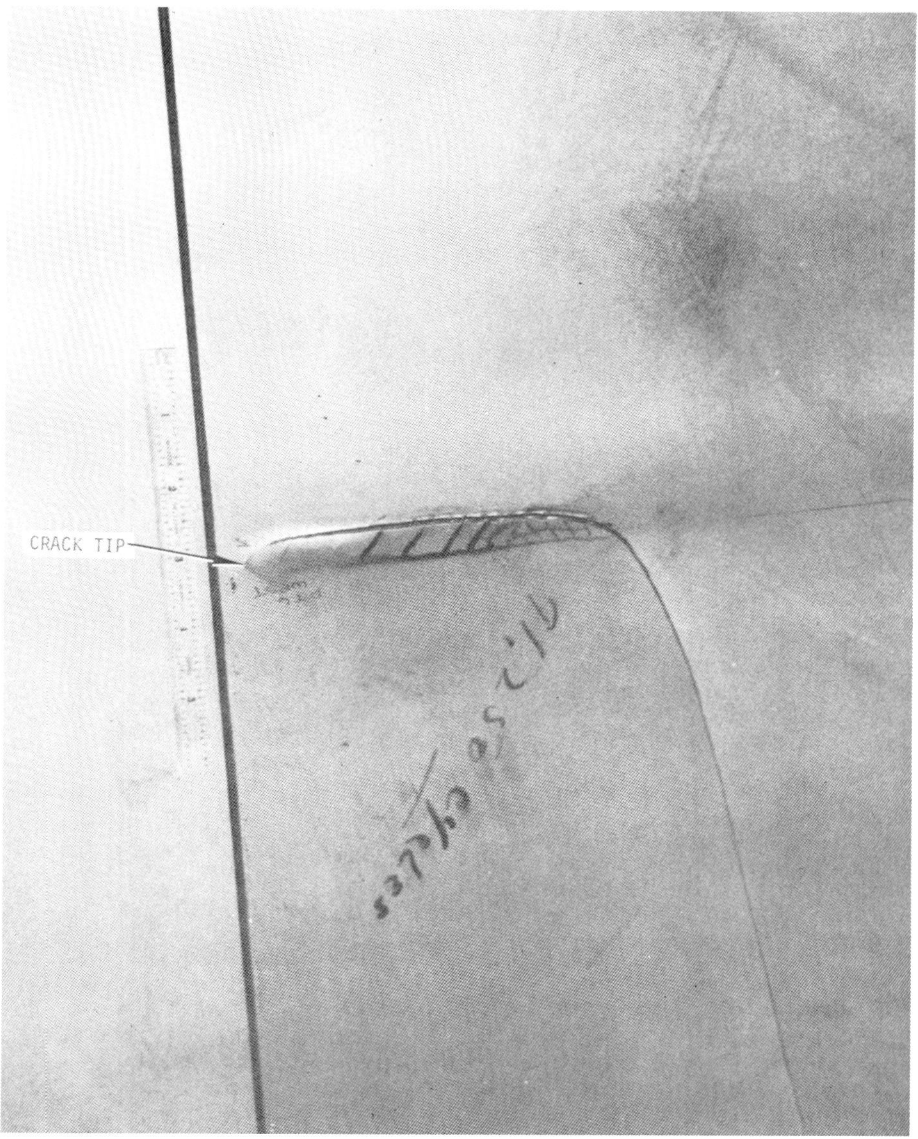

Figure 43　Bonded, curved panel crack propagation:　photo shows close-up of Fig. 42 crack tip.　(Courtesy Douglas Aircraft Co.)

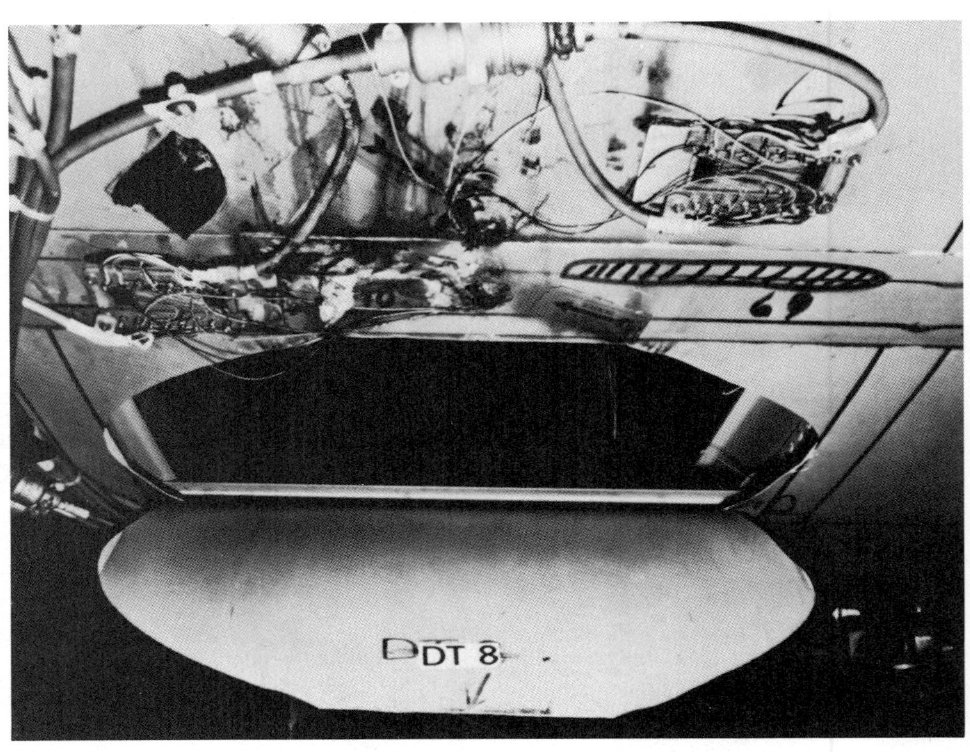

Figure 44   DT-8 failure.   (Courtesy Douglas Aircraft Co.)

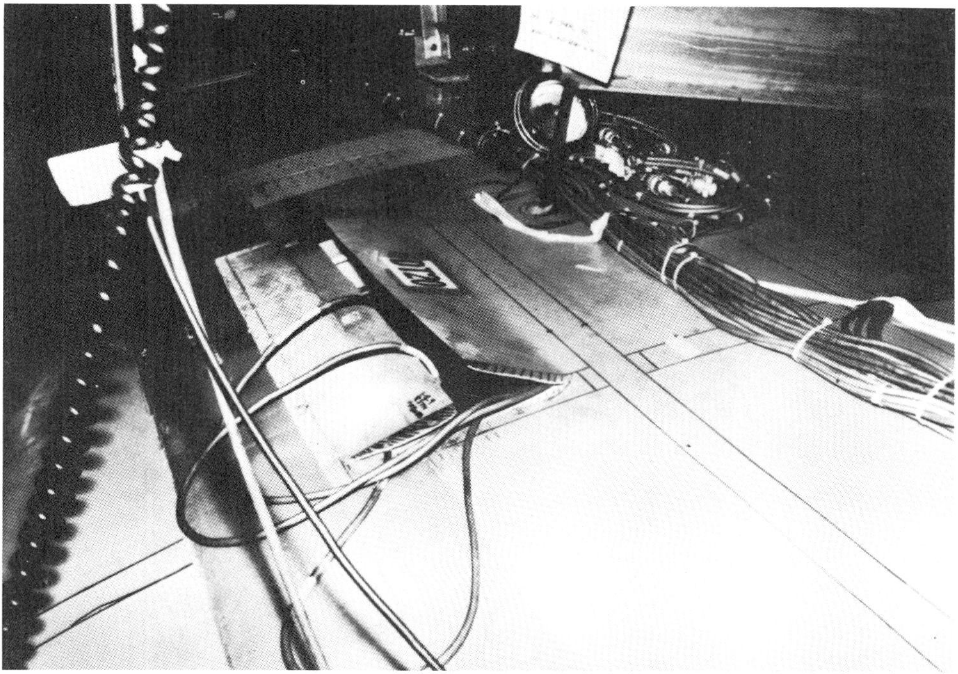

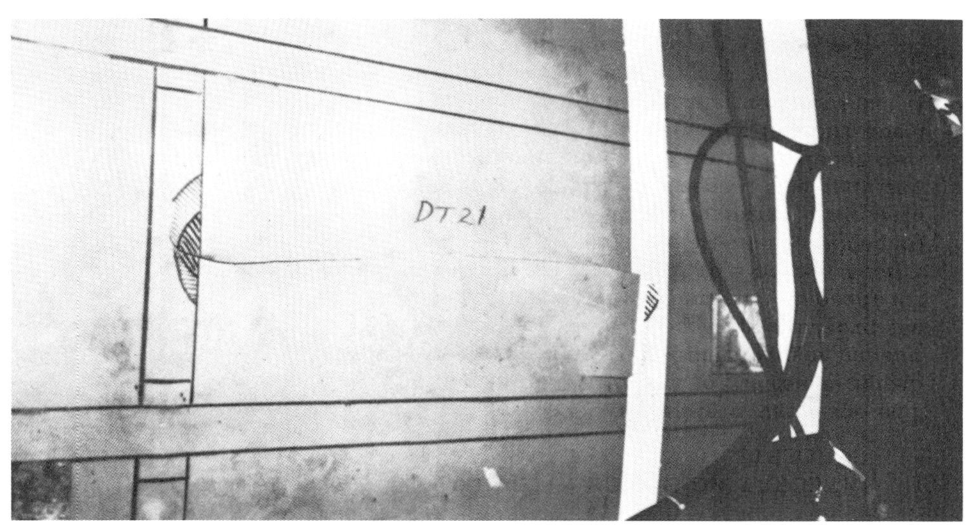

Figure 45   DT-20 failure.   (Courtesy Douglas Aircraft Co.)

Figure 46   DT-21 failure.   (Courtesy Douglas Aircraft Co.)

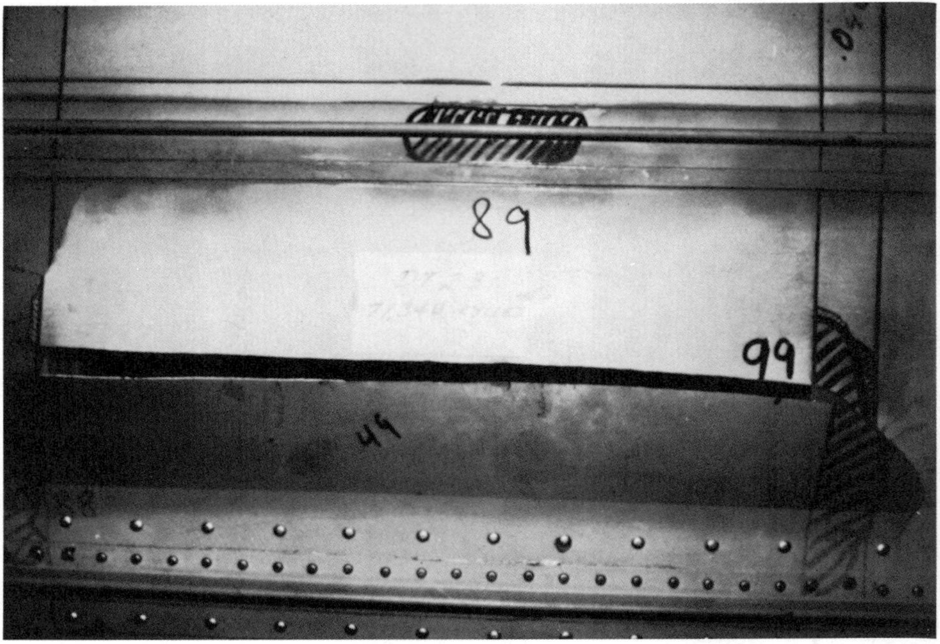

Figure 47   DT-23 failure.   (Courtesy Douglas Aircraft Co.)

The crack crossed the frame area but the remainder of the failure caused the cracked panel to open up and prevent obtaining full cabin pressure.

It can be concluded that cracks are inherently arrested in bonded structural assemblies. More developmental work needs to be done to analytically predict the way a crack turns as it approaches the bonded joint. This new analysis hopefully will predict how the crack performs near a disbond or where the adhesive properties may have been degraded due to exposure to water.

SYMBOLS

A_{str}	stiffener cross-sectional area, in.^{2}
a	half crack width, in.
b	extent of disbond on each side of discontinuity, in.
C	constant

d	extent of adhesive plastic zone, adjacent to discontinuity, in.
d	distance from discontinuity to zero-displacement reference, in.
E_1, E_2, E	Young's modulus for adherends
e	2.718281828, base of natural logarithms
F_y	yield stress, psi
G	adhesive shear modulus (elastic), psi
K_c	fracture toughness coefficient for cracked sheet
ℓ	effective extent of adhesive load zone adjacent to discontinuity, in.
$\ln$	natural logarithm
P	concentrated load, lb
p	stiffener pitch, in.
q	distributed load, lb/in.
s	panel width, in.
T_1, T_2	adherend loads per unit width, lb/width
ΔT	temperature differential ($T_{operating} - T_{cure}$), °F
t_1, t_2, t	adherend effective thicknesses per unit bond width, in.
t_{skn}	thickness of sheet, in.
w	width of bond, in.
x, y, ξ, χ	coordinates
α_1, α_2	adherend (metal) coefficients of thermal expansion
γ	adhesive shear strain
γ_e, γ_p	adhesive elastic and plastic shear strains
Δ	displacement (sheet distortion), in.
δ_1, δ_2	adherend displacements across bond line, in.
η	adhesive layer thickness, in.
λ	exponent of adhesive shear stress distribution, in.$^{-1}$
π	3.141592654
σ_1, σ_2	adherend stresses, psi

ν	Poisson's ratio
τ	adhesive shear stress, psi
τ_p	adhesive plastic (maximum) shear stress, psi

Subscripts:

b	pertaining to bond
c	pertaining to crack in sheet
skn	pertaining to sheet (skin)
str	pertaining to stiffener (stringer)

Superscripts:

0	identified with discontinuity
∞	identified with conditions far away from discontunity
tot	total

REFERENCES

1. Hart-Smith, L. J., Adhesive-Bonded Double-Lap Joints, Douglas Aircraft Company, NASA Langley Research Center Contract Rep. NASA CR-112235, Jan. 1973.
2. Hart-Smith, L. J., Adhesive-Bonded Single-Lap Joints, Douglas Aircraft Company, NASA Langley Research Center Contract Rep. NASA CR-112236, Jan. 1973.
3. Hart-Smith, L. J., Adhesive Layer Thickness and Porosity Criteria for Bonded Joints, Douglas Aircraft Company, USAF Contract Rep. AFWAL-TR-82-4172, Dec. 1982.
4. Thrall, E. W., Jr., et al., Primary Adhesively Bonded Structure Technology (PABST), Phase Ib: Preliminary Design, Douglas Aircraft Company, U.S. Air Force Flight Dynamics Lab. Tech. Rep. AFFDL-TR-76-141, Dec. 1976.
5   Thrall, E. W., Jr., Failures in Adhesively Bonded Structure, Douglas Aircraft Company, Paper 6703, in *Bonded Joints and Preparation for Bonding*, AGARD-NATO Lecture Ser. 102, Lecture No. 5, Oslo, Norway, and The Hague, The Netherlands, Apr. 1979.
6. Hughes, E. J., and J. L. Rutherford, Selection of Adhesives for Fuselage Bonding, Singer, Kearfott Division, Tech. Rep. KD-75-37, July 1975.
7. Frazier, T. B., and A. D. Lajoie, Durability of Adhesive Bonded Joints, Bell Helicopter Company, U.S. Air Force Materials Lab. Tech. Rep. AFML-TR-74-26, Mar. 1974.

8. Hart-Smith, L. J., Adhesive-Bonded Scarf and Stepped-Lap Joints, Douglas Aircraft Company, NASA Langley Research Center Contract NASA CR-112237, Jan. 1973.
9. Smith, M. K., L. J. Hart-Smith and C. G. Dietz, Interactive Composite Joint Design, Douglas Aircraft Company, U.S. Air Force Flight Dynamics Lab. Tech. Rep. AFFDL-TR-78-38, Apr. 1978.
10. Hart-Smith, L. J., Adhesive Bonded Stresses and Strains at Discontinuities and Cracks in Bonded Structures, Douglas Aircraft Company, ASME Winter Annu. Meet. Atlanta, Ga., Nov. 1977, *Trans. ASME, J. Eng. Mater. Technol.*, Vol. 100, Jan. 1978, Vol. 100, p. 16−24.
11. Hart-Smith, L. J., Further Developments in the Design and Analysis of Adhesive-Bonded Structural Joints, Douglas Aircraft Company, Paper 6922, ASTM Symp. Joining of Composite Materials, Minneapolis, Minn., Apr. 1980.
12. Swift, T., The Effect of Fastener Flexibility and Stiffener Geometry on the Stress Intensity and Stiffened Crack Sheet, in *Prospects of Fracture Mechanics*, G. C. Sih, H. C. Van Elst, and D. Broek, eds., Noordhoff, Leyden, 1974, p. 419−436.
13. Hart-Smith, L. J., Adhesive Bond Stresses at Discontuities in Cracks in Bonded Structures, Douglas Aircraft Company, Rep. MDC-J6068, Sept. 1976.

14

Basic Principles for Tooling Design and Inspection

L. J. HART-SMITH *Douglas Aircraft Company of the McDonnell Douglas Corporation, Long Beach, California*

EDWARD W. THRALL *Consultant, Aeronautical Structures, Point Arena, California*

I. INTRODUCTION

The object of this book is to promote the further use of adhesive-bonded primary structural assemblies for aircraft and other components. The preceding chapters have presented the engineering aspects of bonding aluminum alloys for use by designers and students. However, a good bonded assembly will be made economically only if the proper tool is developed. The bonding tools must be compatible with the design. For these reasons this short chapter on manufacturing and inspection is included. After properly processing the aluminum materials and selecting an adhesive compatible with the service conditions, the parts are ready to be bonded together. The bonding operation requires that the faying surfaces be held in close proximity with pressure while heat is applied at a proper rate until the curing temperature is reached. Each adhesive has a required dwell time at temperature before the assembly can be cooled to room temperature and the adhesive is fully cured.

The real crux of the tooling problem is: How can the parts be held together with the right pressure? The manner selected will determine how costly the manufacturing operation will be. Historically, the use of autoclaves (pressure chambers with heated gases) has been the most widely used pressure device. The tooling to support the parts has two requirements: to hold parts in position, and to fit inside the available autoclave. As the bonding assemblies become larger and contain a greater number of detailed parts, the placement of details on or in the tool becomes very tedious and time consuming, thus

increasing the cost. When an autoclave is used, the tool containing
the parts to be bonded must be "bagged" or placed in an envelope.
This is necessary in order that the externally applied pressure stay
external to the mating parts. If the pressure invades the "bag," no
pressure exists between the details, and, therefore, no bond. Many
successful bondments have been pressurized with clamps or hoses,
using various temperature sources.

If we stand back and review the basic requirements for bonding,
the following factors can be identified as keys to success:

1. Structural design concepts and tooling concepts must be developed simultaneously and be compatible, including production rate considerations.
2. The applied pressure should push the flexible part against the stiffer member.
3. Tooling concept—male or female?
4. Minimize contact area and mass of tool.
5. Identify structural concepts that are difficult to bond together and attempt to change the design.
6. Vacuum bag techniques.
7. Mechanical pressure techniques.
8. Inflated pressure tube techniques.
9. Autoclave versus nonautoclave bonding.
10. Inspection requirements as part of tooling concept.
11. Specific applications.

II. INTEGRATING DESIGN WITH TOOLING CONCEPTS

The one most vital key to successful adhesive bonding is to have a
design assembly of mostly flexible details. These parts will be easily
pushed together and therefore, conform to the final shape with a modest pressure application. Several tooling concepts could provide or
fill these requirements. Holding the parts in their proper relative positions can be a difficult tooling problem and may result in an impossible
inspection task. This is why the engineer-designer must consider
structural arrangement options together with inputs from the tooling
designer and inspector. Consider, for example, an aircraft pressurized fuselage: this is a cylinder that carries bending loads, shear
loads (normal and torsion), concentrated loads, and pressure loads.
If the diameter is small, it is possible to carry the bending loads by
compression of the envelope skin (a full monocoque design) without
addition of fore and aft longerons. For large-diameter fuselages the
skin gauges become too thin for monocoque design, so fore and aft stiffeners (longerons) are used to carry the compressive loads. The next
structural items in fuselage structure are frames which redistribute the

concentrated loads to the skin. The combination of frames and longerons presents the problem of intersecting structural members. The
longerons accumulate loads going from front or back into the middle
and therefore want to be continuous. The frames then become candidates to be discontinuous (i.e., to be interrupted at each longeron).
This results in a series of shear tabs between the longerons. However,
in the transfer of shear, they also become a hard spot on the skin and,
as shown in the preceding chapter, are a point of stress concentration.
Here the structural designer and tool designer must determine the best
splice that will remove the stress concentration and be easy to bond.

A design feature that is often used to eliminate stress concentration at the end of shear tabs is the use of a joggle. This would
permit a doubler to be put on the skin where the shear tab ends.
If this part is hydropress formed or a stretch die is used, the tolerance of the joggle depth may exceed the desired glue line thickness of
0.002 to 0.005 in. If the parts (frame) were pushed against the skin
in a rigid female tool, a bonding problem could arise. If the frame
were supported on a male tool and the skin was pushed against the
frame, a good frame-to-skin bond is almost guaranteed. Also, other
bonding processes would work if the more flexible skin is pushed
against the relatively stiff frame.

Using small, simple splice gusset plates will remove stress concentration at the end of the frame part. These gusset plates will deflect
and conform to the other structural details. The biggest problem might
be holding the plates in the correct position during the bonding cycle
and being able to verify by inspection that they have remained in place
throughout the layup process, bagging, and so forth.

III. TOOLING CONCEPTS

A common type of bonding tool has a solid plate or base sheet which is
used to lay the parts on. The base will take the shape of one side of
the assembly being bonded. The details are laid against the tool and
adhesive is placed in all the faying surfaces. If the assembly is to go
into an autoclave, a membrane is placed over the assembly and sealed
all around to the face of the tool. This will cause autoclave pressure
to push everything against the face of the bonding tool. As the complexity of the parts increases, for example, a compound curve skin
with matching frames, it becomes difficult to impossible to be able to
push all the parts into a female bonding tool. The buildup of tolerances
require excessive pressure to make the parts fit.

The converse from the solid-faced tool is an open tool. This tool
has supports for the stiff parts (frames) which can be fastened to the
tool and then doublers and skins can be laid on top of frames and the
complete assembly envelope bagged and pressurized in an autoclave.
This allows the relatively flexible parts to be pushed against the stiff
ones.

A. Minimizing Contact Area and Size of Bonding Tools

This introduces the next important lesson about adhesive bonding. That is that bonding tools should be as small as possible—just sufficient to apply or resist the local pressure needed to effect the adhesive bonding, which typically covers only about 10% of the total surface area. Honeycomb parts require pressure over 100% of the surface area. Such a minimum open tool has two important advantages beyond the improved fit associated with not having portions of the tool interfering with the structure in areas where it was not needed anyway. One advantage is that the heat sink capacity of the tool is considerably reduced, so that it is easier to heat up the adhesive at the correct rate. Of even greater importance is the fact that the use of the perforated tool (as opposed to a continuous surface tool) permits *visual* inspections of the fit of the details. This should be contrasted against the need for bagging and a heat and pressure cycle (in an autoclave or oven) to make a costly Verifilm check of bond thicknesses in situations where the inspection is rendered blind by an inappropriate bonding tool or design concept.

B. Integrating Design Concept with Bonding Tool Approach

The one most vital key to successful adhesive bonding is to have a design assembly of the mostly flexible details which move together and fit properly with only modest pressure application and a bonding tool which permits this to happen rather than inhibits it. It should be stressed that the design structural arrangement and tooling concept are not separate issues which can be selected independently by different departments. The choices must be made simultaneously and be compatible. It should be evident from this that one should never try to bond together two or more stiff members which are incapable of deflecting under the bonding pressure until they fit properly. These stiff members should be assembled with mechanical fasteners which have sufficient force to make adjacent details conform (albeit with induced preload). This issue introduces the notion that not all structure lends itself to efficient adhesive bonding. This should be accepted rather than resisted.

IV. PRESSURE APPLICATION TECHNIQUES

There are two basic reasons for applying pressure to assemblies while they are being bonded. The first is to hold the parts in the correct position until the adhesive has been cured (typically by application of heat), and the second is to displace any air or volatiles which would otherwise be trapped between the details and result in weakened bonds

due to porosity and voids. The highest quality and strongest bond
that can be produced from a given adhesive film is the thickest one
that can be made void- and porosity-free and without pinch-off at the
edges. Therefore, it is important not to employ design or tooling concepts that rely on excess pressure just to make the parts fit together.

A. Vacuum-Bag Techniques

Over the years, two techniques have evolved for the application of
pressure for bonding. The more versatile approach is the transmission
of pressure from the inside of an autoclave through a nylon film or
rubber sheet which is first suched down onto the part by vacuum and
then vented to atmosphere outside the autoclave as the cure progres-
ses. A modification of this, which is often employed for light-bonded
honeycomb panels, is the use of vacuum bag pressure alone in an oven.
However, since the exposed edges of the adhesive are then exposed
to vacuum pressure, any minute amount of entrapped air or volatiles
tends to produce disproportionately large porosity in the adhesive
bond layer. Very few structural adhesives can be processed satisfactorily by such a vacuum technique. The other basic approach to pressure application actually preceded the use of bagging, which is associated with the low-volatile epoxy adhesives, because the first structural adhesives for metal bonding were phenolics, which generate
considerable steam during cure. For that reason, it is desirable to
use heated dies or platens in a press in such cases. Quite apart from
the consideration of the volatiles produced during cure, the shape and
size of some bonded assemblies tend to favor one approach over the
other. The basic difference between these two approaches is that the
autoclave can be looked upon as a universal pressure and heat application tool with proportionally low nonrecurring costs and high labor
costs associated with the bagging of each and every panel, whereas
the out-of-autoclave (or press) work has relatively low recurring costs
but needs more capital to fabricate dedicated tools which can do only
one job each. Each approach can still be considered appropriate for
some applications today.

Of the lessons learned in regard to pressure application through
vacuum bags, the most important is that the simultaneous inclusion of
two orthogonal sets of intersecting stiffeners (longerons and frame
shear tees) under a single bag makes the bagging operation extremely
tedious and troublesome, at best, and probably inadvisable in virtually
all cases. The excess bag material needed to drape over intersecting
stiffeners results in bridging across the corners and sometimes punc-
turing of the bag as the vacuum is being drawn early in the cure cycle.
It also leads to local areas in which no pressure is applied. It is so
labor intensive as to be uneconomical for production. Also, it impedes
the use of visual fit checks prior to bonding, thereby placing more

reliance on Verifilm checking, which, in turn, necessitates even more autoclave and bagging operations.

A bagging problem exposed by the use of the female tool arose because the intersecting longerons and frame shear tees formed a kind of egg crate condition that was difficult to bag over. A solution was to cover the skin and fill the pockets with small hollow aluminum beads (about 1/4 in. in diameter) up to a depth slightly above the shear tee webs. The nylon bag was then draped over the much more even surface of the beads rather than the deeply pocketed surface of the structure itself. The belief had been that the beads would transmit the pressure through to the skin and stiffeners and push them out to the inside of the mold, to which the perimeter of the bag was sealed. What actually happened was that the beads locked up solid under the vacuum packing and transmitted pressure only to the skins, actually reducing the load on the stiffeners. Far from causing the skins and stiffeners to move together under pressure, the use of the beads resulted in the skins moving outward, away from the stiffeners. Consequently, the fit was very poor, with gaps of 0.040 in. revealed by the Verifilm (in comparison with a typical bond line thickness of 0.005 in.) and an excess of voids and porosity. In an attempt to minimize this problem pressure plates were installed to lay over the outer flanges of the stiffener. Thus the beads pushing on the plate would transmit more radial load to the stiffeners than could be exerted directly on the flanges alone. The bonds so obtained still had many voids but were structurally adequate.

B. Problems with Bridging Using a Male Tool

A problem encountered with the use of a male tool merits special mention, since it caused bad mark-off on a panel. Because the skin and stiffeners were envelope-bagged, it was assumed that the load on each side of the skin would inevitably be the same. That was simply not so because of the use of excess bleeder and packing material to protect the vacuum bag against tearing on some sharp corner. This resulted in bridging across the length of the frame. A review and analysis made it clear that the portions of the skin between the frames and longerons would be subjected to a net inward force if there were any bridging of the vacuum bag in the corners between the skin and stiffeners. On one panel, made on a male tool, there was pronounced mark-off for that reason, and even the longerons were pushed in with the skin because they had not been tied to the frames or the contour boards across the tool. Once this problem was understood, the subsequent panels were made with only imperceptible mark-off by using greater care in lay-up and bagging and much less bleeder cloth.

C. Mechanical Pressure Application

There is a difference between vacuum bag and platen press application of pressure. With a vacuum bag, air bubbles and volatiles can be trapped within an adhesive bond, particularly in wide-area doublers. Once the air bubble has been compressed to the external pressure (in autoclave) there is little, if any, differential pressure to displace that bubble to anywhere else. Only in the immediate vicinity of the edges of the overlap is there direct access to the suction of the vacuum source or atmospheric vent. That is why most bonded joints tend to be inherently free from voids and porosity right where there is the greatest need for high-quality bonds—in the areas where the load is actually transferred. Unfortunately, the situation is not all good, since there tends to be excessive pinch-off of the glue at those same edges, right where the structural considerations would demand that the glue be locally thickened. Platen press work presents quite a different situation than prevails with the vacuum bag. The adhesive is virtually not pressurized at all, until the pressure plates have deflected sufficiently to bottom out somewhere, at least. Consequently, there are differential forces to displace the air bubbles and the edges are not pinched off. However, in complex stiffened bonded assemblies, the cost of matched dies and details which do not hang up locally somewhere and prevent pressure application everywhere else is a potentially severe problem.

D. Use of Inflated Rubber Tubes to Apply Local Bonding Pressure

The resolution of the vacuum bag problems would be to apply pressure only over the areas actually to be bonded, ensuring that there is no pressure whatever on either side of the skin elsewhere. One vital characteristic of the so-called fire-hose technique is that it provides a means of absorbing quite large tolerances in the bonding fixture and still applying a controlled uniform pressure over all areas to be bonded. The other, equally important, merit of inflated tubes is that they permit separate control over the pressure and temperature of the bond. The electric heaters in the support fixture are found to be more reliable than the rubber strip heaters outside the skin (with molded-in resistance heater wires), but each technique is effective. A defective electric cal-rod heater can even be replaced while the pressure is being applied and all the other heaters are bonding the rest of the structure. The separate control over heat and pressure means that the glue will not be squeezed out by excessive premature pressure. Neither is there any chance of the glue setting up before the pressure is applied. A further benefit, discussed below, is that it permits a complete visual fit check of the conditions that actually prevail during the cure cycle.

V. AUTOCLAVE VERSUS NONAUTOCLAVE BONDING

The discussion above of pressure application techniques should make it clear that there is no universal answer to the question of whether to bond inside or outside an autoclave. Each case must be judged on its

merit. The actual choice must be influenced by the availability of time
on a large enough autoclave, of course, but structural considerations
do have a substantial influence on which approach should be preferred.
At one extreme is the bonding of skin-doubler combinations without any
stiffeners. It is obvious that the most efficient way of doing this is to
envelope bag the skin and doubler(s) without any tooling or, if necessary, including a rolled caul plate to define the shape. Then several
such panels can be nested side by side in a single autoclave cure cycle.
There is no way of performing that particular operation more efficiently
outside an autoclave. At the other extreme is the simultaneous bonding of intersecting frame tees and longerons. It seems clear that the
most economical approach to that problem lies outside the autoclave,
and techniques for doing so are described in the remainder of this
chapter. It is quite conceivable that the best solution will sometimes
be a combination of autoclave bonding for some of the bonding on a
panel followed by out-of-autoclave bonding for those details which
would have made the first operation impractical. As a general rule,
large-area bonded doublers and honeycomb assemblies are more compatible with autoclave work, while stiffeners that require pressure and
heat over quite a small area are usually more effectively bonded outside, without having to move or heat relatively large assembly tooling.

Of every bit as much importance as the cost of actually bonding
the structure together is the associated cost of inspection for proper
fit and for the absence of voids. Autoclave bonding tends to compli-
cate the prefit inspection of all but the simplest of assemblies (because
of the inability to perform the inspection visually while the parts are
under pressure), while out-of-autoclave bonding at least offers the
potential of so simplifying the prefit inspection that the postbonding
inspection can be minimized. Thus the cost of inspection should be
included when deciding on the manufacturing technique (these costs
are not independent).

VI. INSPECTION REQUIREMENTS

A. Quality Assurance

Quality assurance can be considered as the single most important
function of a manufacturer of tangible products. The inspection operation is to determine how well the manufactured details and assemblies agree with the specific dimensions and material requirements
called out on the drawings. Actually, the inspection operation has
been expanded today to the point of reviewing the engineering drawings in development to make sure that the bonded assemblies can be
inspected, and then the tooling design is reviewed for the same reasons. The engineer must come up with a detailed design that will
not only accomplish the desired end task, but it must be inspectable
during the various steps of process and assembly. It is disastrous

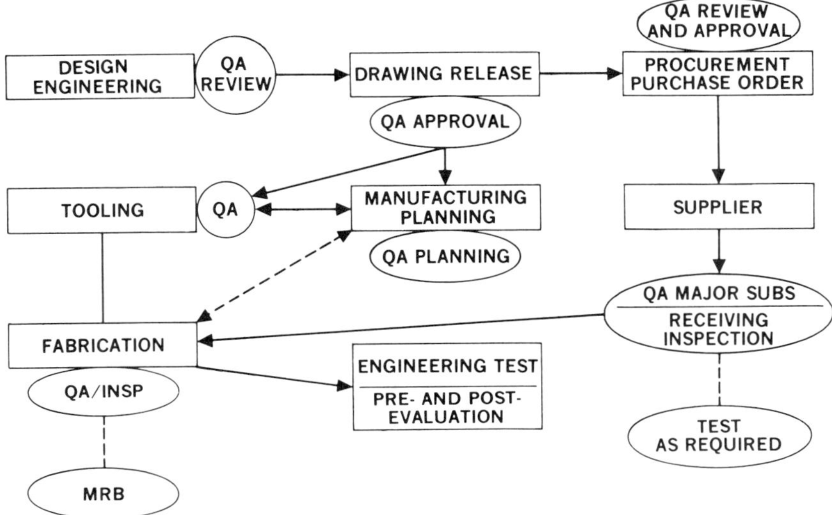

Figure 1 Flowchart of quality assurance design and manufacturing coverage.

to have the completed end product, for example a large bonded assembly of 100 or more parts, rejected because of an unrepairable bond or a mismatch of parts.

Rather than try to discuss specifically each inspection area and how the work should be done, the outlines in Figs. 1 and 2 cover the subject fairly well. Also, the preceding chapters identify specifically the inspection requirements. The issue of concern here is that the quality of the bonded panels and, consequently, the cost of inspection depend just as much on the compatibility of the structural design and the bonding tool as on the workmanship of the manufacturers.

B. Problems with Verifilm Fit Checks

It should be evident from the description of how the panels were buried between external tools on one side and over 2 in. of beads on the other side that total visual prebonding fit checks are often impossible. Consequently, it was typically necessary to subject each assembly of detail parts to three cure cycles. That alone should have been sufficient to render that approach unfit for production. Consider the following situation. After all the details had been formed and checked for contour, they were assembled in the 30-ft-long by 10-ft-wide tool for the first time. However, instead of the layer of adhesive film, strips of Verifilm were inserted between the parts to be bonded. This is a

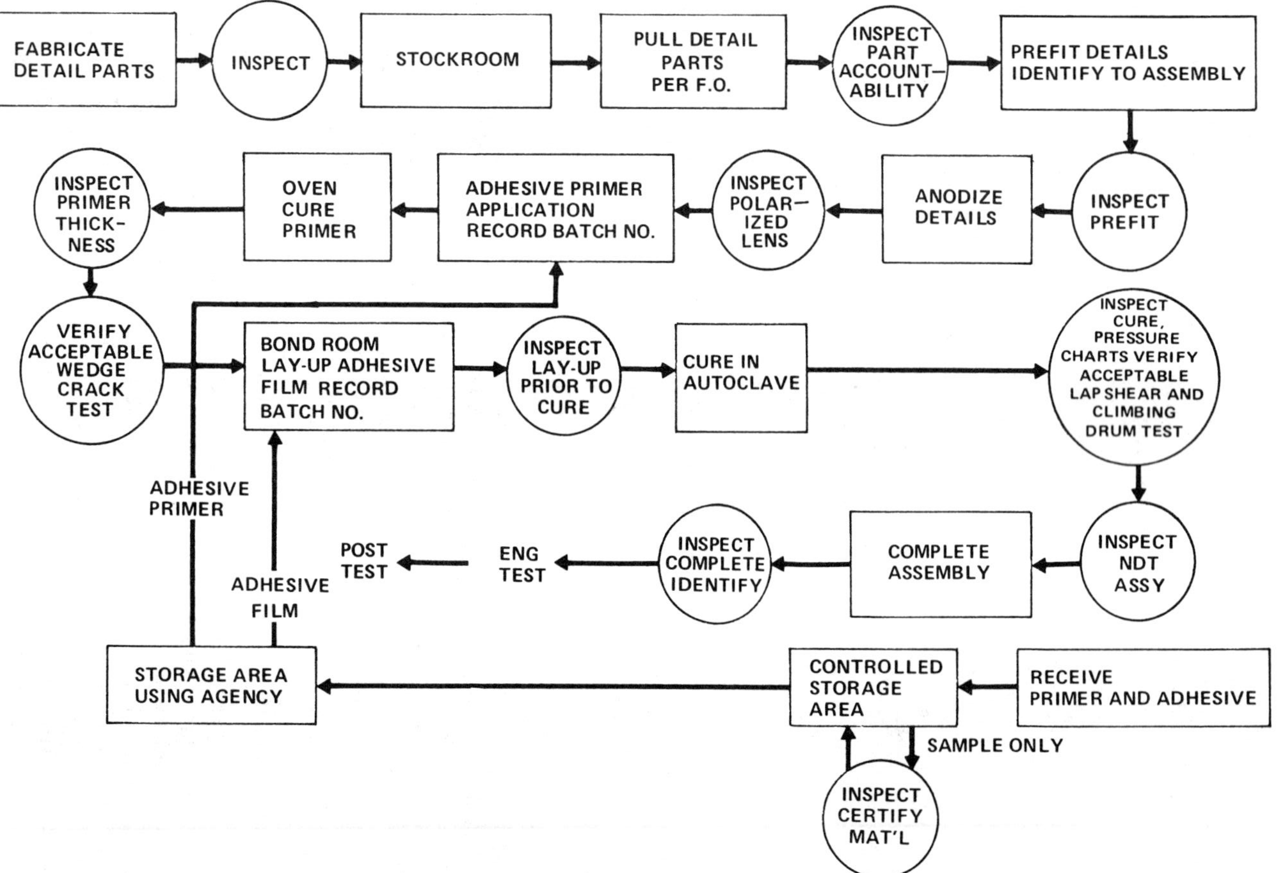

Figure 2   Inspection requirements for the bonding operation.

plastic film which will not stick to the surfaces but softens, changes thickness, and then sets under the cure cycle of heat and pressure to form a permanent record of what the adhesive thickness would have been had it been there. The Verifilm is identified as to location and then measured for a check of the expected thickness of the final "glue" layer. Based on these measurements, the detail parts would be re-worked to improve the fit by decreasing the gap between the parts or areas which need two layers of adhesive to fill up the gaps. Depending on the degree or amount of misfit, the assemblies might need to be re-Verifilmed after rework. Only then were the parts anodized and primed and the assemblies laid up with adhesive and bonded.

C. Use of Adhesive Film as Verifilm

Two major problems have been encountered with the use of the Veri-film. The results are not reproducible when consecutive runs (new layups) are made on precisely the same assembly of details. Where hollow beads are used it is probable that different degrees of locking up of the beads under vacuum contribute to nonreproducibility. When laying up a large number of details in a large tool, the positioning of parts is difficult to reproduce exactly. A thick vacuum layer on the inside of the tool must have been less than ideal for the bonding cycle as well as for the Verifilm cycle. Part of the lack of reproducibility was attributed to the vastly different flow characteristics of the Veri-film and the adhesive, which had a much lower minimum viscosity and squeezed out more. Therefore, Verifilm checks can be made with a layer of uncured adhesive film sandwiched between sheets of Mylar plas-tic to eliminate the latter problem.

Even with the male tool, prefit inspection is still a problem because the envelope bag and the considerable bleeder material obstruct the view right where inspection is needed to ensure a proper fit.

D. Bonding Tools that Permit Visual Fit Checks

The prebonding inspections described above were rendered impractical by all the obstructions to a direct visual or feeler gauge check. From the inspection point of view, the best possible bonding tool would be an out-of-autoclave tool which could be pressurized independently of the heat application without obstructing the view, permitting a simple overhead inspection or one in the vertical plane in front of the tech-nicians. By using a design that permits the supporting of the stiff stretch-formed frame shear tees and pushing the flexible skins and doublers in to make them fit, it is probable that visual inspections would uncover accidentally kinked or dinged parts as well as gross mis-fits. With only a little finesse, the contour boards that support the

shear tees, or any other stiffeners, could serve double duty as a check
gauge on the parts themselves even before pressure is applied. With
an adequate supply of parts for quick replacement of the odd defective
detail, such a further saving on inspection is not at all unreasonable.
Thus the benefits from doing things logically cascade and grow beyond
the original objectives.

Visual checks can still be performed on some simple envelope-
bagged assemblies, provided that there are no intersecting stiffeners
to wrinkle the bag and that there is no breather or bleeder material or
flash-breaker tape inside the bag to obstruct the view. Indeed, such
inspection has been practiced by Fokker in the Netherlands for decades.
Glue squeeze-out is not a problem unless there is a wrinkle in the bag,
running over the edge of a stiffener. This condition is easy to observe
and correct.

If the parts fitted together properly before bonding, there should
be no need to inspect them after bonding. Conversely, if they did not
fit together properly before bonding, there is no point in bothering to
inspect them after bonding. It is a bit late to try to build quality into
the assemblies after they have been manufactured.

E. Importance of Visual Inspection of
Adhesive Bond Fillet

The most underrated of the bonded structure acceptance tests is a visual inspection of the glue fillet at the edges of all bonded overlaps. It
is the most effective of all bond inspection techniques. The fillet should
show evidence of adhesive flow and of wetting the adherend surfaces.
Since most of the load in bonded joints is transferred in the immediate
viscinity of the edges of the various overlaps, the glue in the fillet is
indicative of the glue in the most important areas. If the heat-up rate
were too slow, the adhesive would not form a nice, smooth fillet but
would tend to have stayed on any excess adhesive carrier. If the heat-
up rate had been excessive, the acceptable fillet would be replaced by
a porous spew. (Care must be taken in this respect with phenolic adhesives, which generate steam during cure.) Surface contamination is
revealed by the adhesive being repelled by the adherend surfaces, just
as in the water-break test. Further, a complete lack of filleting or a
visible gap reveals a lack of pressure due to possible poor fitting details or bonding tools, or a leak in a vacuum bag, or the inclusion of
a foreign particle. In short, visual inspection is the best of all non-
destructive inspection techniques for bonded joints. It does not even
contaminate the surface and leave it unrepairable as sometimes happens
with ultrasonic inspection couplant fluids.

Perhaps the next most underrated nondestructive inspection technique for bonded assemblies is the coin-tap test. Its great virtue is
that it cannot find flaws that are so small that they should not be repaired anyway.

In concluding this section on inspection of bonded assemblies, it is appropriate to reiterate that the best value received for effort expended is gained by concentrating on prebonding inspections, to minimize the need for postbonding repairs.

15

Nondestructive Inspection

DONALD J. HAGEMAIER *Douglas Aircraft Company of the McDonnell Douglas Corporation, Long Beach, California*

I. INTRODUCTION

The preceding chapters have explained carefully how to provide adherends with a proper anodize, how to select the adhesive to be used, and manufacturing techniques required to achieve good bonded assemblies. Chapter 13 presented an analysis of the bonded joints and discusses what happens if voids (lack of bond), inclusions, or variations in glue line thickness are present and the effect on the strength. This chapter addresses the problem of how to inspect bonded assemblies so that all discrepancies are identified. Once the flaws are located, an engineering judgment and analysis can determine if adequate strength exists or the parts need to be reworked or scrapped. When very large and costly assemblies have been bonded, the importance of good nondestructive inspections is fully appreciated.

This chapter describes several techniques that can be used and presents drawbacks and limitations where they exist. No single method or procedure can find all the flaws, so it is important to know what is available and what the limitations are. Inspections of bonded assemblies are made immediately after the adhesive cure cycle, after the parts are cleaned up and removed from the manufacturing jigs or fixtures. The next important time for inspection is after the assemblies have been in use for a while and the condition of the parts is known. Have any new delaminations occurred, or any metal cracks developed? Inspections in the field require that the instruments can be easily transported and can be used in tight places.

The acronyms commonly used describe what is to be done. For example: NDT is nondestructive testing and comprises the testing principles methodology; NDI is nondestructive inspection, inspection to meet an established specification or procedure; NDE is nondestructive evaluation, the actual examination of materials, components, and assemblies to define and classify anomalies or discontinuities in terms of size, shape, type, and location. This chapter specifically addresses the three acryonyms as they apply to adhesive-bonded joints within structures.

Inspection reference standards must be developed for each instrument, and the standards made available to and used by inspectors to standardize their equipment and permit them to detect anomalies that will indicate a flaw. Inspectors must also be familiar with the internal details of the assembly being inspected so that they can distinguish between flaws and legitimate structural details. For example, how many layers of metal were bonded at any one spot?

II. DESCRIPTION OF DEFECTS

There are a wide variety of flaws or discontinuities which can occur in adhesive-bonded structures. A list of possible generic flaw types and their producing mechanisms were prepared by Clark [1] and is shown in Table 1. Metal-to-metal voids or unbonds are the most frequently occurring rejectable flaws, as shown in Table 2.

Interface defects are the result of errors made during the pretreatment cycle of the adherends prior to the actual bonding process. In practice, pretreatment flaws are reduced by careful process control and adherence to specification requirements and inspection before proceeding with the bond cycle. Controls generally include the water-break test and measurement of the anodic layer and primer thickness. Interface defects can be caused by improper or inadequate degreasing, deoxidizing, anodizing, drying, damage to the anodizing layer, or excessive primer thickness.

Interface defects are generally not detectable by state-of-the-art NDT methods. Therefore, test specimens are processed along with production parts and sent to the laboratory for evaluation. Applicable wedge crack specimens, lap shear specimens, or honeycomb flatwise tension specimens are fabricated and tested to determine if the process meets specification requirements before the bonding cycle starts.

Considerable effort at Fokker in the Netherlands led to the important discovery that the ideal oxide configuration for adhesion on aluminum alloys can be detected by inspection with an electron microscope at suitable magnification [3–5]. In order to inspect with the electron microscope, a piece of the structure must be removed. As a consequence, the electron microscope became a useful tool for adhesion quality control. Another physical parameter that was used as a basis for NDT of surfaces for ability to bond was contact potential, which is measured by a proprietary method developed by Fokker and known as a contamination tester [2,4] (see Fig. 1). This recently developed instrument is based on Kelvin's dynamic-condenser method but avoids

Table 1 Generic Flaw Types and Flaw Producing Mechanisms

Flaw producing mechanism	Generic flaw type				
	Metal-to-metal	Metal-to-core	Core	Surface	Adhesive
1. Disbonds, internal	X				
2. Disbonds, part edge	X	X			
3. Disbonds, high core	X				
4. Porosity	X	X	X		X
5. Unremoved protective release film from adhesive	X	X			
6. Foam adhesive in film adhesive bond line	X	X			
7. Cut adhesive	X	X			X
8. Adhesive gaps	X	X			
9. Missing adhesive	X	X	X		
10. Weak bonds	X	X	X		
11. Extra layers of film adhesive	X	X			
12. Foreign objects	X	X	X		X
13. Double drilled or irregular holes	X			X	
14. Disbonds, low core		X			
15. Void or gap, chemical milled land		X			

Table 1   (continued)

Flaw producing mechanism	Generic flaw type				
	Metal-to-metal	Metal-to-core	Core	Surface	Adhesive
16. Void or gap, doublers		X			
17. Missing fillets		X			
18. Voids, closure-to-core		X			
19. High density inclusions, (chips, etc.)	X	X			X
20. Voids, foam joint		X	X		
21. Disbond, shear ties	X	X			
22. Lack of sealant at fasteners		X			
23. Thick foam adhesive		X			
24. Broken fasteners	X	X			
25. Crushed core			X		
26. Wrinkled core			X		
27. Condensed core			X		
28. Distorted core			X		
29. Blown core			X		
30. Node bond spearation			X		

31. Missing core (short core)	X	X	
32. Cut core	X	X	
33. Water in core		X	
34. Cracks			X
35. Scratches			X
36. Blisters			X
37. Protrusions			X
38. Indentations (dents/dings)			X
39. Wrinkles			X
40. Pits			X

Table 2 Frequency of Rejectable Flaws in Adhesively Bonded Assemblies

Defect	Number of defects	Percent of total
Metal-to-metal voids and disbonds	378	74
Skin-to-core voids and disbonds	19	3
Gap in core-to-closure bond	9	2
Lack of foaming adhesive or voids in foaming adhesive	22	4
Difference in core density	6	2
Lack of fillets	1	1
Crushed or missing core	32	6
Short core	40	8
	507	100

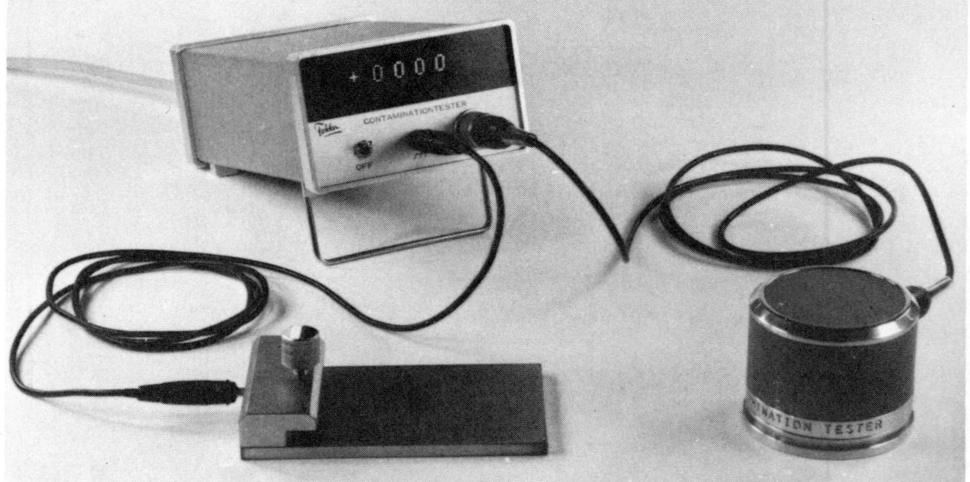

Figure 1 Fokker contamination tester. (Courtesy Fokker—VFW).

the disturbances usually associated with it. There is now sufficient evidence that the contamination tester is able to detect nondestructively the absence of the optimum oxide configuration arising from incomplete anodizing and/or subsequent contamination [5].

More recently, Couchman et al. [6] at General Dynamics in Ft. Worth, Texas, developed an adhesive bond strength classifier algorithm that can be used to build an adhesive-bond-strength tester. Lap shear specimens were fabricated using Reliabond 398 adhesive. The test specimens include (1) a control set with optimum bond strength, (2) an undercured set, (3) a weak bond produced by an unetched surface, and (4) a thin-bond adhesive that was cured without a carrier. The weakest bond was observed to fail at 105 psi, while the strongest held to 2270 psi. Tabulated results showed the following:

Undercured set: 105 to 930 psi
Unetched surface set: 840 to 1100 psi
Control set: 1940 to 2150 psi
Thin adhesive set: 2050 to 2270 psi

The accept/reject value was set at 1900 psi and all specimens were classified correctly. The important factor is that an interface defect (unetched surface), which results in poor adhesion, was detected.

Defects within the cured adhesive layer can be one or more of the following:

1. Undercured or overcured adhesive
2. Thick adhesive resulting in porosity or voids due to improper bonding pressure or part fit-up
3. Frothy fillets and porous adhesive caused by too fast a heat-up rate
4. Loss of long-term durability due to excessive moisture in the adhesive prior to curing

In normal cases the curing time is very easily controlled. The curing temperature and temperature rate are controlled by proper positioning of the thermocouples on the panel and regulating heat-up rate.

Thick glue lines occur in a bonded assembly due to inadequate mating of the facing sheets or blocked fixing rivets, and result mostly in porosity and voids. However, a thick glue line made with added layers of adhesive is usually free of porosity. Porosity has a significant effect on strength of the adhesive, with higher porosity related to a greater loss in strength and a void condition resulting in no strength. The frequency of these defects is quite large, as shown in Table 2. Porosity can also be caused by the inability of volatiles to escape from the joint, especially in large-area bond lines. Excessive

moisture in the adhesive prior to curing can be prevented by controlling the humidity of the lay-up room. The entrapped moisture, after
curing, cannot be detected by NDT methods unless it results in
porosity.
Other defects that occur during fabrication can include:

1. No adhesive film
2. Protective film left on adhesive
3. Foreign objects (inclusions)

In practice, these conditions have to be prevented by process
control and training of the personnel engaged in the bonding operation.
The first two conditions occur infrequently. Shavings, chips, wires,
and so on, can result in porosity or voids. Honeycomb core assemblies
have been found with all types of foreign material.

A. Metal-To-Metal Defects

Voids

A void is any area that should contain, but does not contain, adhesive. Voids come in a variety of shapes and sizes and are usually
at random locations within the bond line. Voids are generally surrounded by porosity if caused by a thick bond line and may be surrounded
by solid adhesive if caused by entrapped gas from volatiles.

Unbonds or Disbonds

Areas where the adhesive attaches to only one adherend are termed
"unbonds." These may be caused by inadequate surface preparation,
contamination, or improperly applied pressure. Because both adherends
are not bonded, the condition is similar to a void and has no strength.
Unbonds or disbonds are generally detectable by ultrasonic or sonic
methods.

Porosity

Many adhesive bond lines have some degree of porosity, which may
be either dispersed or localized. The frequency and/or severity of
porosity is random from one assembly to the next. Porosity is defined
as a group of small voids clustered together or in lines. Figure 2 shows
reproductions made from x-ray negatives showing scattered linear and
dendritic porosity which is usually found in adhesives supported with
a matte carrier. Linear porosity generally occurs near the outer edge
of a bonded assembly and in many cases forms a porous frame around
a bonded laminate. Porosity is usually caused by trapped volatiles
and is also associated with thick (single-layer adhesive) bond lines
which did not have sufficient pressure applied during the cure cycle.

(a)

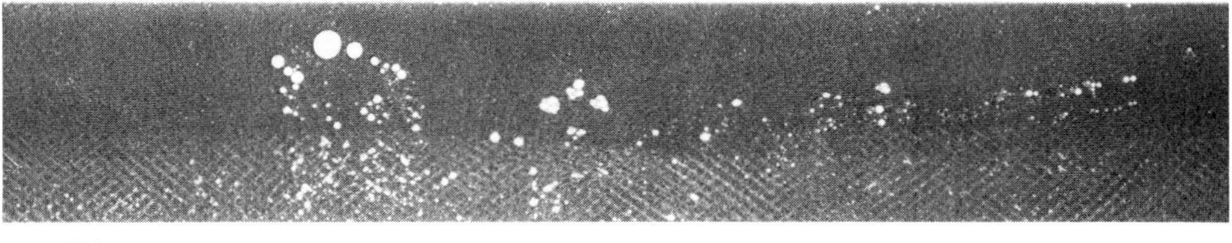

(b)

(c)

Figure 2   Positive radiographic examples of porosity: (a) irregular scattered linear porosity; (b) rounded scattered linear porosity; (c) scattered dendritic porosity. (Courtesy Douglas Aircraft Co.)

The reduced bond strength in these porous areas is directly related
to its density frequency and/or severity.

Porous or Frothy Fillets

This condition results from too high a heat-up rate during curing.
The volatiles are driven out of the adhesive too rapidly, causing bub-
bling and a porous bond line, which is distinguished by the frothy
fillets. This defect is visually detectable and should also be seen in
the test specimens processed within the production parts.

Lack of Fillets

Visual inspection of a bonded laminate may reveal areas where the
adhesive did not form a fillet along the edge of the bonded adherends
or sheets. In long, narrow joints a lack of fillet on both sides generally
indicates a complete void. This defect is considered serious because
the high stresses near the edges of a bond joint can cause a cracked
adhesive layer due to shear or peel forces. A feeler gauge can be used
to determine the depth of the defect into the joint. If the gap is too
tight for a feeler gauge, ultrasonic or radiographic techniques may be
used to determine the depth of the edge void.

Fractured or Gouged Fillets

These defects are visually detected. Cracked fillets are usually
caused by dropping or flexing the bond assembly. Gouges are usually made with tools such as drills or by impact with a sharp object.
Fractured and gouged bond lines are considered serious for the reasons
stated earlier for lack of fillets.

Adhesive Flash

Unless precautions are taken, adhesive will flow out of the joint
and form fillets plus additional adhesive flow on mating surfaces. Although the condition is not classified as a defect, it is considered unacceptable if it interferes with ultrasonic inspection at the edges of the
bonded joint where stresses are highest.

Burned Adhesive

The adhesive may be burned during drilling operations or when
bonded assemblies are cut with a band saw. The burned adhesive is
essentially overcured, causing it to become brittle and to separate from
the adherend. Also, the cohesive strength of the burned adhesive is
drastically reduced. Figure 3 shows burned adhesive around hole 9
caused by improper drilling and bond delamination along the edge of
the panel adjacent to holes 16 through 20 caused by band sawing.

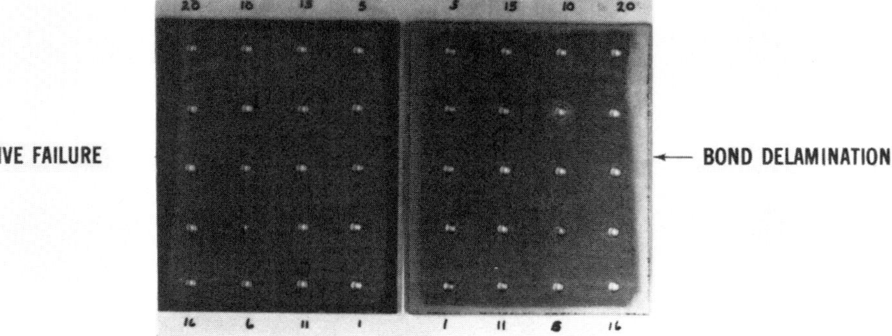

Figure 3   Examples of burned adhesive.   (Courtesy Douglas Aircraft Co.)

Improper drill speed or feed coupled with improper cooling can cause these types of defects. Burned adhesive is detectable by ultrasonic C-scan recording methods.

B. Adherend Defects

These defects are visually detectable and do not include the processing procedures.

Fractures (Cracks)

Cracks in the adherend, whatever the cause, are not acceptable.

Double-Drilled or Irregular Holes

Some bonded assemblies may contain fastener holes. When holes are drilled more than once, have irregular shapes, or are formed with improper tools, they are considered defects. The load-carrying capacity of the fasteners is unevenly distributed to the adherends, resulting in high local stresses which may fracture during service.

Dents, Dings, and Wrinkles

These defects are serious only when extensive in nature, as defined by applicable acceptance criteria or specifications. They are most detrimental close to, or at, a bond joint. Dents are usually caused by impact with blunt tools or other objects and are usually rounded depressions. Dings result from impact with sharp objects or when an assembly is bumped at the edge. Dents or dings may cause bond line or adherend fractures. Wrinkles are bands of distorted adherends and are usually not important.

Scratches and Gouges

A scratch is a long, narrow mark in the adherend caused by a sharp object. Deep scratches are usually unacceptable because they can create a stress riser which may generate a metal crack during service. On the other hand, gouges are blunt linear indentations in the adherend surface. Deep gouges, like scratches, are generally not acceptable.

C. Honeycomb Sandwich Defects

The most prominent defects found or generated in honeycomb sandwich assemblies are summarized in Table 1. Adhesion and/or cohesion defects may also occur in bonded honeycomb sandwich assemblies. The metal-to-metal closure areas for honeycomb panels may exhibit the types of defects discussed in the preceding section. In addition, sandwich assemblies can have defects in the honeycomb core, between the core and skins, between core and closure, at chemically milled steps, and in core splices. These bond areas are shown in Fig. 4 for a typical honeycomb assembly.

Water in Core Cells

Upon completion of the bonded assembly some manufacturers perform a hot-water leak test to determine if the assembly is leakproof. If the assembly emits bubbles during the leak test, the area is marked

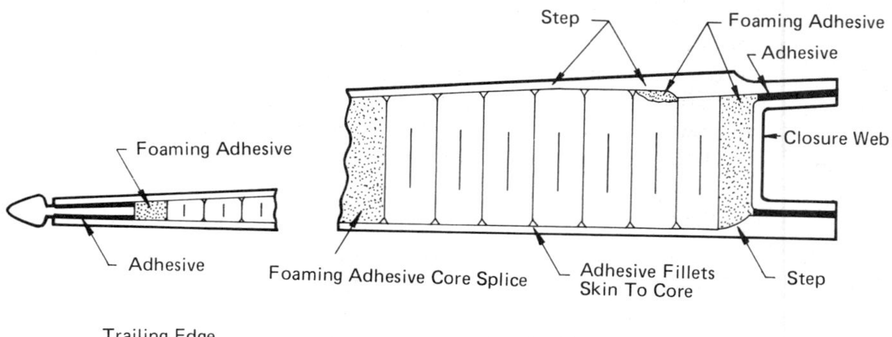

Figure 4 Typical configuration of bonded honeycomb assembly. (Courtesy Douglas Aircraft Co.)

and subsequently repaired. To assure that all bonded areas are inspected and that no water remains trapped in the assembly, it is then radiographed. This is important since water can turn to ice during operational service and rupture the cells, or it may initiate corrosion on the skin or core. Water in the core can be detected radiographically when the cells are filled to at least 10% of the core height. Also, x-ray detection sensitivity is dependent on the sandwich skin thickness and radiographic technique. An additional problem is the ability to determine whether the suspect area has excessive adhesive, filler, or water. Water images usually have the same film density from cell to cell or for a group of cells, whereas adhesive or filler images may vary in film density within the cells or show indications of porosity. A radiographic positive print of moisture in honeycomb is shown in Fig. 5.

Crushed Core

A crushed honeycomb core may be associated with a dent in the skin or caused by excessive bonding pressure on thick core sections.

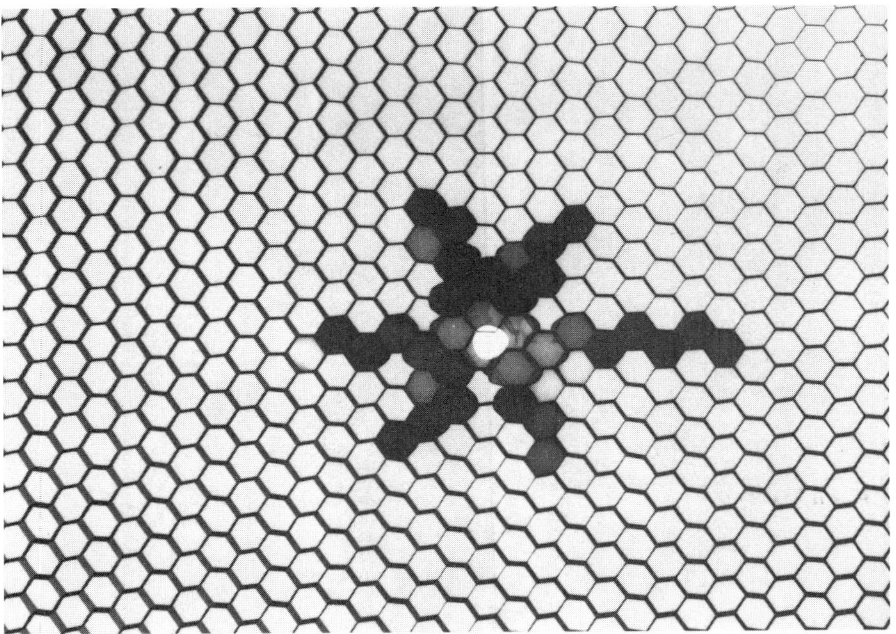

Figure 5   Positive print from x-ray negative showing water intrusion into honeycomb cells. (Courtesy Douglas Aircraft Co.)

Crushing of the core greatly diminishes the core's ability to support
the facing sheets. Figure 6 shows an x-ray positive print of crushed
core. Generally, crushed core is most easily detected using angled
x-ray exposures. Crushed core is defined as localized buckling of the
cell walls at either face sheet, when associated with the halo effect on
a radiograph. On the other hand, for wrinkled core the cell walls are
slightly buckled or corrugated. Radiographically, the condition appears
as parallel lines in the cell walls. A wrinkled core is generally acceptable.

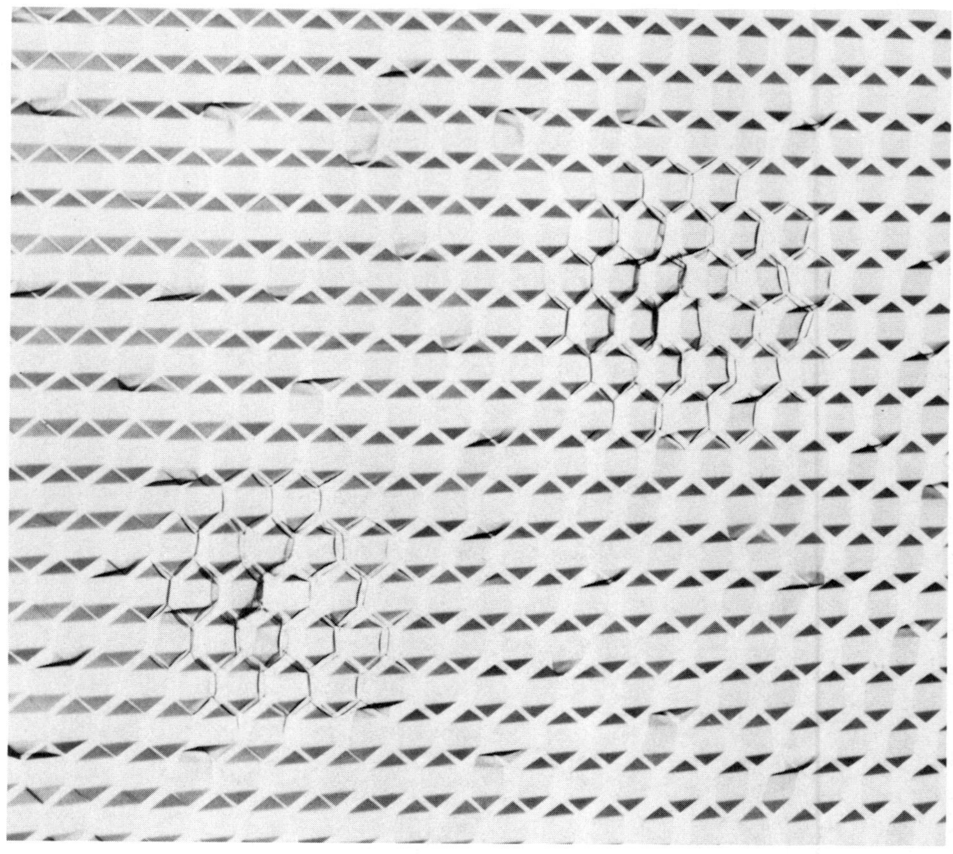

Figure 6   Positive print from x-ray negative showing crushed honeycomb core. (Courtesy Douglas Aircraft Co.)

Condensed Core

This condition occurs when the edge of the core is compressed laterally. This may result from bumping the edge of the core during handling or lay-up, or slippage of detailed parts during bonding. The condition most often occurs near honeycomb edge closures. Figure 7 shows a positive radiograph of various degrees of condensed core.

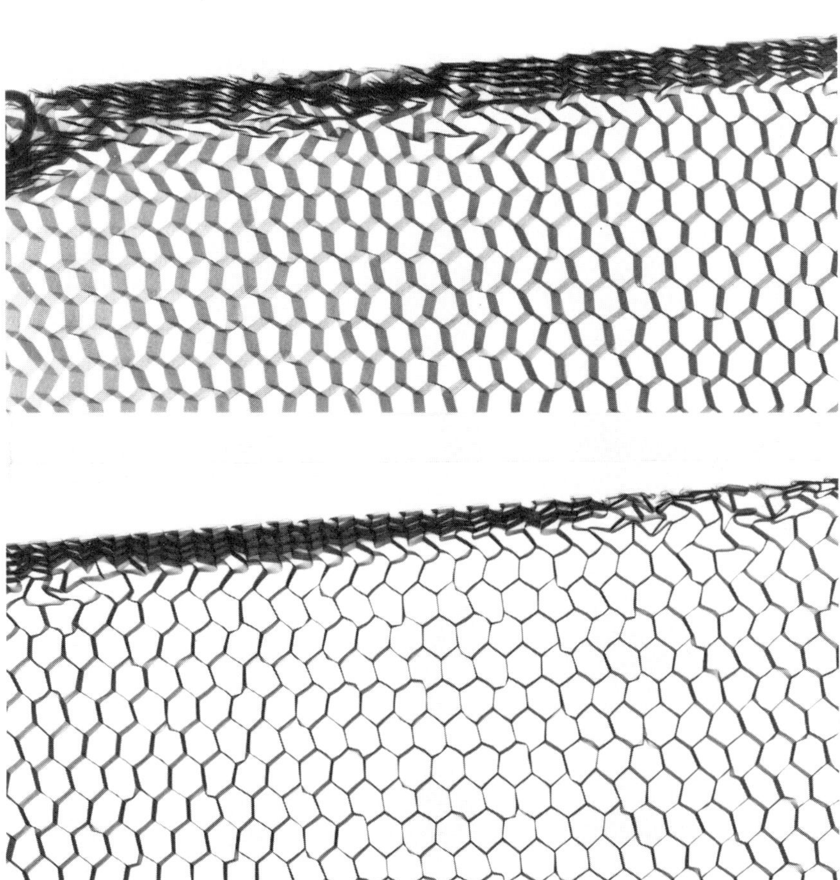

Figure 7   X-ray positive radiograph showing various conditions of condensed core.   (Courtesy Douglas Aircraft Co.)

Node Separation

This condition results when the foil ribbons are separated at their connecting points or nodes, as shown in Fig. 8. This usually occurs during core fabrication. It also may result from pressure buildup in cells as a result of vacuum bag leaks or failure, which allows the pressurizing gas to enter into the assembly and core cells.

Blown Core

This occurs as a result of a vacuum bag leak or due to a sudden change in pressure during the bonding cycle. The pressure change produces a side loading on the cell walls that can either distort the cell walls or break the node bonds. Radiographically this is indicated as (1) single-cell damage, usually appearing as round or elliptical cell walls with partial node separation; or (2) multicell damage, usually appearing as a curved wavefront of core ribbons that are compressed together.

The blown core condition is most likely to occur at the edge of the assembly in an area close to the external surface where the greatest effect of sudden change in pressure occurs. This condition is most prevalent whenever there are leak paths, such as gaps in the closure ribs to accommodate fasteners, or chemically milled steps in the skin where the core may not fit properly. When associated with skin-to-core unbonds, the condition is detectable by pulse-echo and through-transmission ultrasonic techniques. The condition is very readily detectable by radiography when the x-ray beam centerline is parallel to the core cell walls, as shown in Fig. 9.

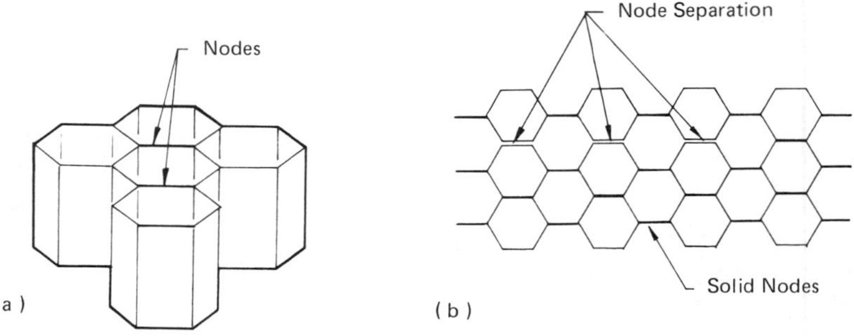

Figure 8   Examples of honeycomb core separation: (a) joined and solid notes; (b) node separation.

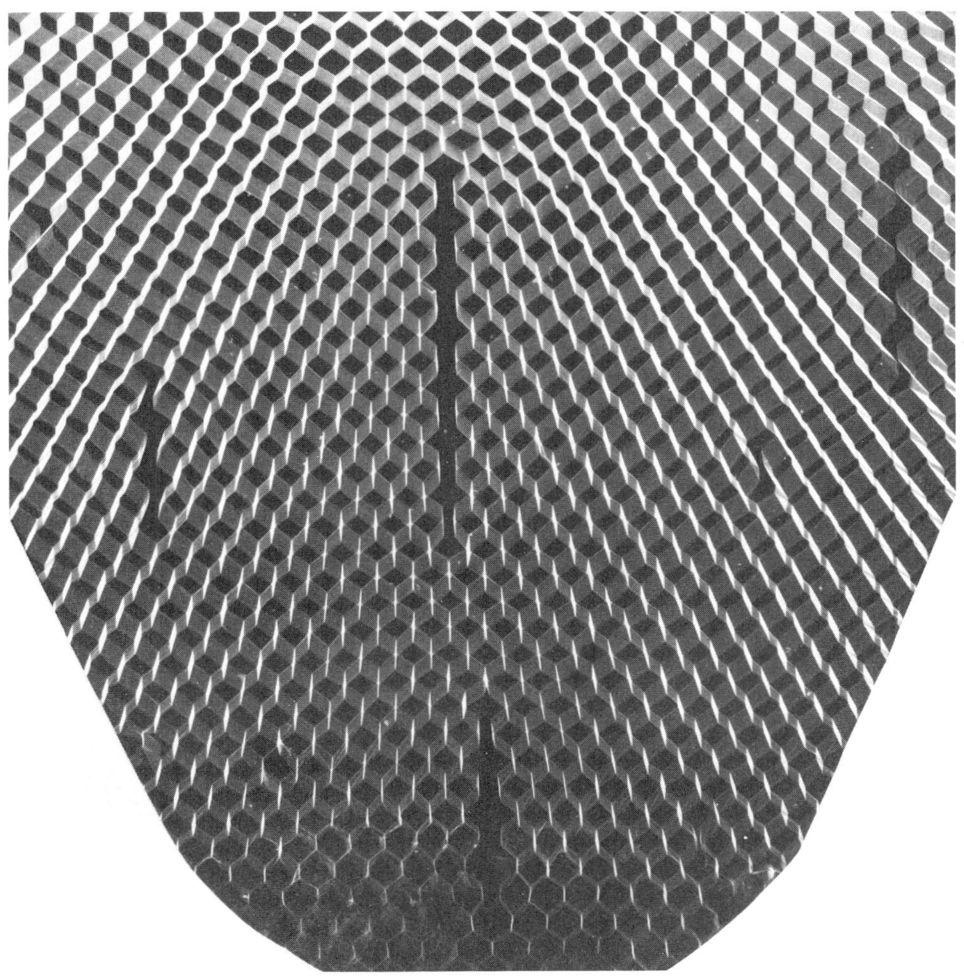

Figure 9   Radiographic illustration of blown core. (Courtesy Douglas Aircraft Co.)

Voids in Foam Adhesive Joints

Defects in core-to-closure, core-to-core splice, core-to-trailing edge fitting, and chemically milled steps (see Fig. 4) in foam adhesive joints can result from several conditions:

1. The foam adhesive can slump or fall, leaving a void between the core and the face skins. This particular condition is most readily detected by ultrasonics.
2. The core edge dimension may be cut undersize and the foam does not expand uniformly to fill the gap between the closure and core. This condition is clearly depicted by the x-ray positive radiographs in Fig. 10.
3. The foaming adhesive can fail to expand and surround the core tangs, as illustrated in Fig. 11.

Protective Film Left on Adhesive

Protective films are usually given a bright color so that they can be seen and removed from the adhesive before bonding. If they remain on the adhesive, the adhesive is prevented from contacting one of the adherends. This condition is very difficult to detect by ultrasonic techniques and x-ray radiography, where it would appear as a mottled condition. It generally produces porosity, especially at the perimeter, which aids in its detection. Figure 12 is a positive neutron radiograph showing rectangular pieces of protective film left on the adhesive prior to bonding the honeycomb assembly.

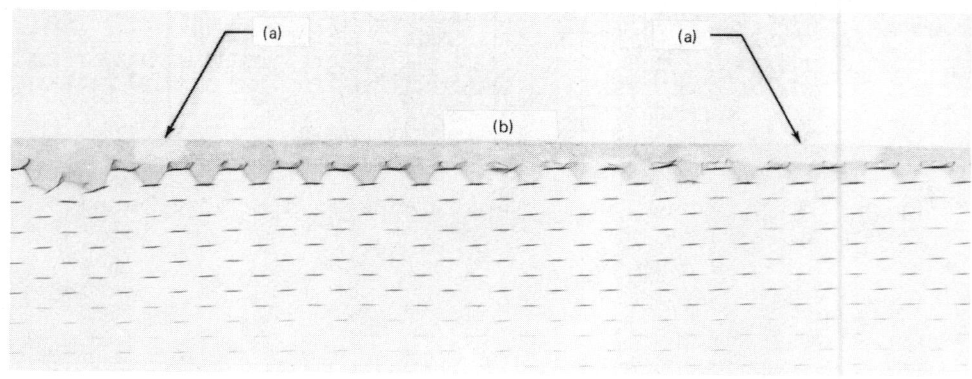

Figure 10   Positive print from x-ray negative showing voids in foaming adhesive at closure. (a) Void. (b) Good closure. (Courtesy Douglas Aircraft Co.)

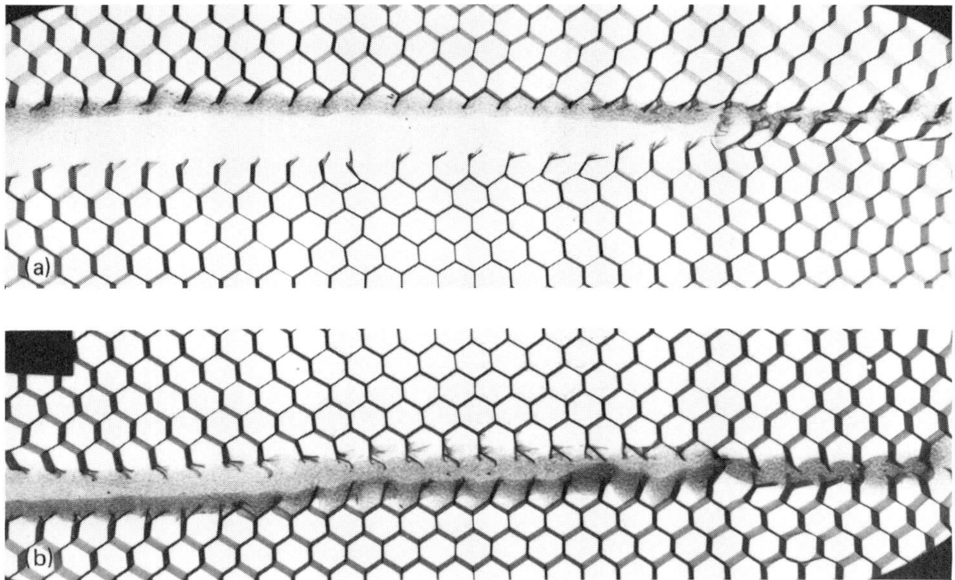

Figure 11 Positive print from x-ray radiograph of foaming adhesive at core splice: (a) lack of or unacceptable foaming at core splice; (b) acceptable foaming at core splice. (Courtesy Douglas Aircraft Co.)

Metal-to-Metal Defects

The most prominent defects are voids and porosity. Disbonds can occur when the core is slightly higher than a closure member, lack of applied pressure from tooling, by entrapped volatiles in the bond joint prior to cure, by excessive moisture in the adhesive prior to cure, and/or contaminated honeycomb core.

Voids and porosity are detectable by low-kilovoltage (15 to 50 kV) x-ray techniques using a beryllium-widow x-ray tube, by thermal neutron radiography, and by ultrasonic C-scan techniques employing small-diameter or focused search units operating at 5 to 10 MHz. If the flaw is the result of insufficient pressure, the adhesive will be porous, as shown in Fig. 2. The lower the kilovoltage and/or the thicker or denser the adhesive, the higher the resolution of the flaw image. In general, the flaw size detectable by radiography is smaller than that detected by ultrasonics. Also, some adhesives, such as AF-55 and FM-400, are x-ray opaque, which yields a much higher contrast between voids, porosity, and solid adhesive. Disbonds (separation between adhesive and adherend) are best detected by ultrasonic techniques.

Figure 12   Positive neutron radiograph showing protective film left
on adhesive. (Courtesy Douglas Aircraft Co.)

     One of the most important types of voids in honeycomb assemblies
is the "leaker"-type void, which is oriented normal to the metal-to-
metal bond line and penetrates to the core. Moisture can penetrate
such a void during operational service and cause corrosion or ice damage to the core. This type of void is illustrated in Fig. 13.
     Another type of defect that can cause problems for radiographic in-
spectors is foam intrusion into a metal-to-metal joint near a closure.
This type of flaw is also illustrated in Fig. 13. Ultrasonically, it will

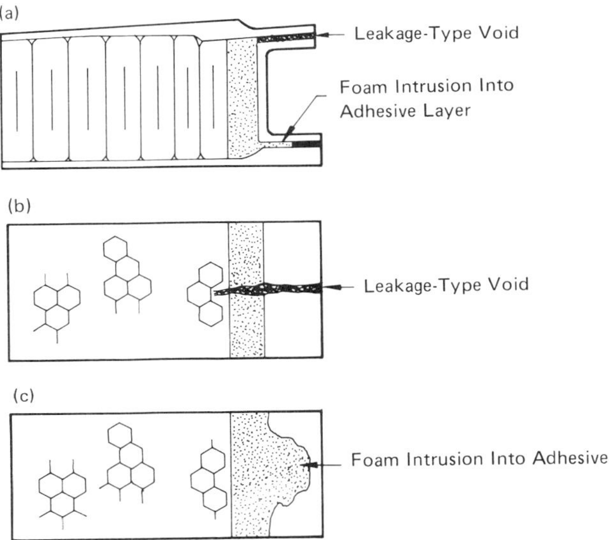

Figure 13 Representations of leakage type void and foam intrusion in metal-to-metal joints: (a) foam intrusion into adhesive layer caused by excessive gap between extrusion and skins; (b) leakage type void; (c) foam intrusion into adhesive.

appear bonded; hence ultrasonics cannot confirm the radiographic findings. Since this may or may not be a defect, it must be determined by engineering analysis.

Skin-to-Core Voids at Edges of Chemically Milled Steps or Doublers

This condition occurs when the adhesive fails to bridge the gap at the edges of chemically milled or laminated steps or doublers (see Fig. 4). This is detected radiographically as a dark line or an elliptically shaped dark image. Ultrasonically, it will appear as a linear void along the chemically milled step position.

Missing Fillets

As pressure is applied during the bonding cycle, adhesive fillets are formed at the edges of each honeycomb cell. Fillets will not be formed if pressure is not maintained. If the adhesive is x-ray opaque, this condition is readily detected by directing the radiation at an angle of approximately 30° with respect to the centerline of the core or

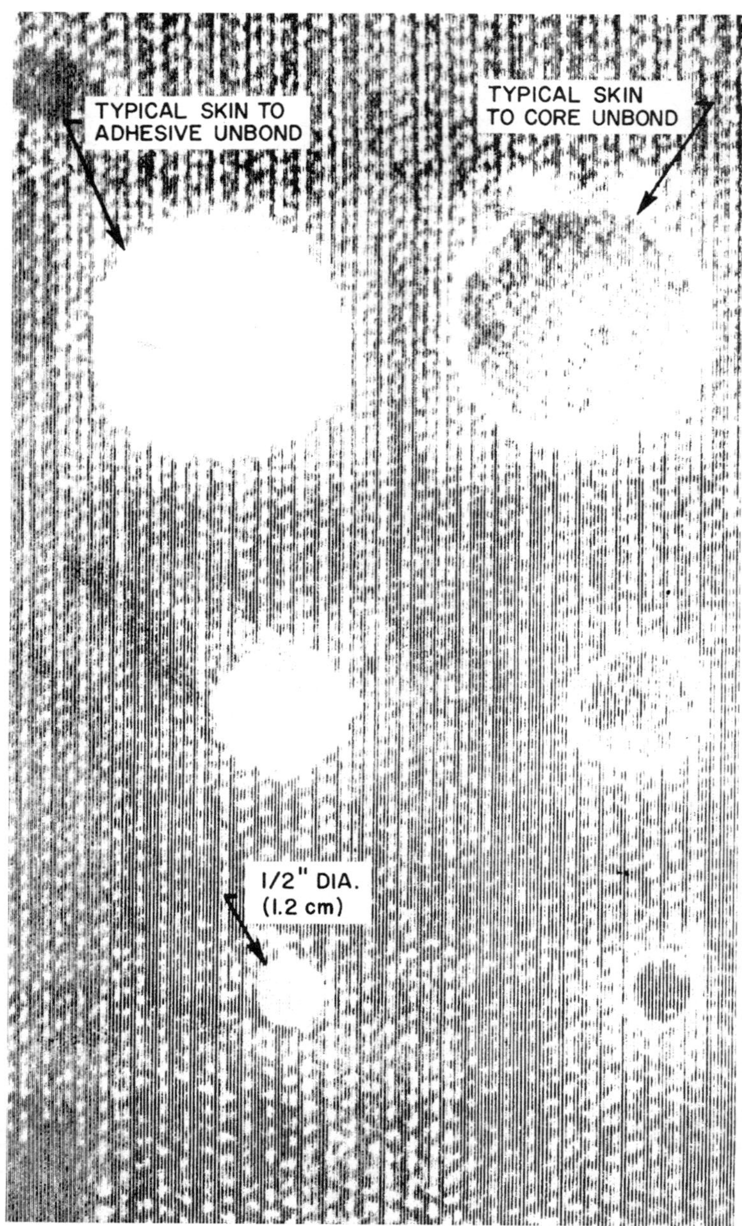

Figure 14   Ultrasonic C-scan recording of adhesive-bonded honeycomb structure.   (Courtesy Douglas Aircraft Co.)

closure web. If the adhesive is not x-ray opaque, then ultrasonic,
eddy sonic, and tap tests can be used to locate the area having unbonded cells. Figure 14 is an ultrasonic C-scan recording of skin-to-
adhesive and adhesive-to core voids. Note that the adhesive-to-core
voids are more difficult to detect.

Short Core

This condition can exist if the core edges are cut shorter than
the assembly into which it will be placed and bonded. It is detectable
by x-ray radiography and is evident as core edges which do not tie
into the closure via the foaming adhesive.

Foreign Objects

Honeycomb assemblies may contain foreign objects as a result of
poor fabrication practices. Nuts, rivets, small fasteners, metal chips,
and similar items can be left in the honeycomb cells before bonding or
as a result of drilling operations for fastener installations near hinge
points. These objects are easily detected by x-ray radiography. They
are not usually detrimental if they can be potted in place using a room-
temperature curing adhesive.

D. Repair Defects

All the flaws or defects defined previously for metal-to-metal or honeycomb sandwich assemblies can occur during salvage or repair of these
components. Repairs must be of good quality in order to maintain the
reliability of the bonded structure during continued service.

E. In-Service Defects

Most defects or flaws caused during service originate from the following
basic causes:

Impact Damage

Many bonded assemblies are made from thin materials and are susceptible to damage by impact. Damage can be caused by stones or other
debris being thrown up by the aircraft wheels, small arms projectiles,
work stands, dropped tools, personnel walking on "no-step" assemblies,
and similar damage. Impact imposes strain on the adhesive, causing it
to crack or separate from the adherends. Impact can cause crushed
honeycomb core, resulting in a loss in strength. The crushed core
can resonate during service and slowly degrade the adhesive by fatigue
until it debonds from the adherend or cracks occur adjacent to the
skin-to-core fillets. Fortunately, impact damage will generally leave a

mark in the surface of the part. These surface marks pinpoint or
indicate possible subsurface damage which can be evaluated by NDT
inspections.

Corrosion

The record of adhesive bonding has ranged from excellent to
totally worthless. The corrosion can be found in all structural concepts. A good example of bonding excellence can be found in the
honeycomb acoustic panels for the DC-10 and 747 engine inlets. These
applications are quite demanding due to the combination of sonic and
ambient environment being introduced into the perforated sandwich
structure. A stable oxide surface preparation is an essential part of
the bond foundation. Improper surface preparation can result in an
unstable oxide layer which may allow ingress of moisture, delamination,
and/or crevice corrosion, illustrated in Fig. 15.

Figure 16 is a typical example of moisture ingress to the core of a
honeycomb assembly with resultant corrosion of the core. This is possible if moisture moves along the bond line to individual cells. Moisture
moves more rapidly in an assembly if the core is perforated. Fortunately, this moisture problem is detectable by NDT inspection methods.
Water is detectable by x-ray radiography (Fig. 16) or acoustic emission

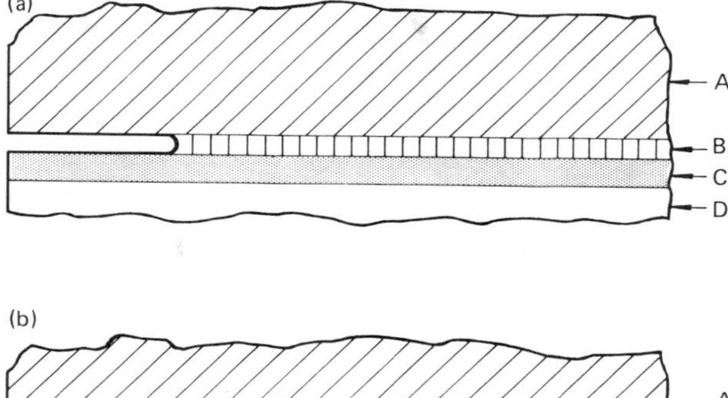

Figure 15   Causes of adhesive delamination for metal adherend: (a)
results of moisture ingress in the unstable oxide: (b) corrosion of
cladding and base aluminum.   (A) is the adhesive primer system,
(B) oxide, (C) alcladding, (D) base aluminum.   (Courtesy U.S. Air
Force.)

Figure 16 Radiographic (x-ray) of corroded honeycomb panel. (Courtesy Fokker VFW.)

testing using a hot-air gun or heat lamp to cause boiling or cavitation of the water [7—9]. Core corrosion and subsequent core-to-skin delamination is detectable by a variety of NDT methods: x-ray, contact ultrasonic ringing, sonic bond testers, eddy sonic, tap test, and acoustic emission with a heat source.

Figure 17 shows a section from a commercial aircraft wing trailing-edge panel which had moisture ingress at the leading and trailing edge, causing bond line corrosion. The corrosion in this particular panel was detected by x-ray radiography (Fig. 17), Fokker, Harmonic, and NDT-210 bond testers, tap test, neutron radiography (which looked much like Fig. 17), and acoustic emission using a heat source.

Poor Fabrication (Flaws or Weak Bonds)

In-service failures can occur from weak bonds (adhesion failures) caused by poor surface preparation, unstable oxide failure, and corrosion of alcladding and base aluminum alloy (Fig. 15). Many tests show that the alclad on 7000 series aluminum sheet is very susceptible to corrosion attacks in the bond line and must be thoroughly tested before being considered for bonding operations. The 2000 series alcald aluminum alloys are not as susceptible to this condition. Both series of

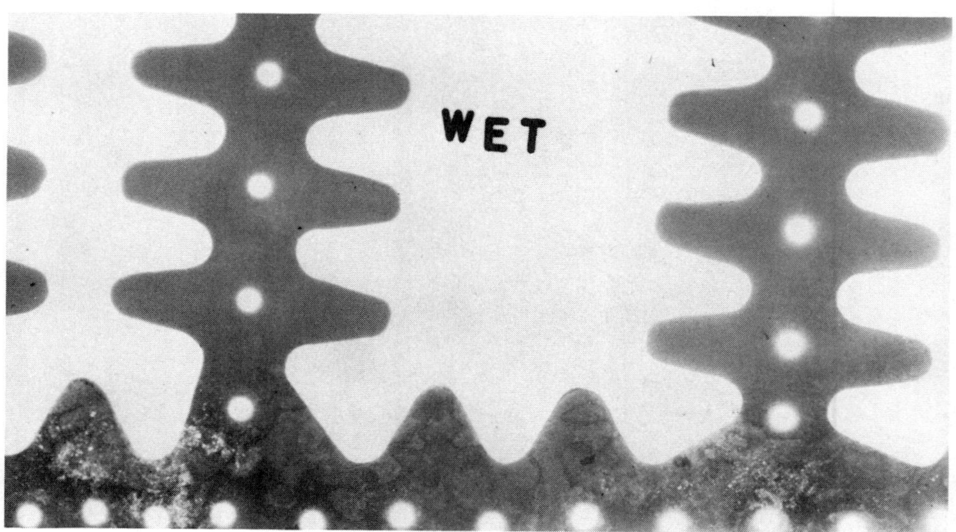

Figure 17   Positive print from x-ray radiograph showing interface corrosion of an adhesive bonded aluminum laminate. (Courtesy Douglas Aircraft Co.)

alloys with no alclad surface but good surface preparations, adhesive, and corrosion-inhibited primer will not have corrosion problems.

Another manufacturing condition that leads to adhesion failures is caused by improper cleaning of the honeycomb core prior to bonding. This condition results if glycol (which is often used to support the core when it is machined) is not completely removed prior to bonding. This will cause a weak bond to exist and in-service stresses may cause a skin-to-core delamination. This condition is presently being controlled by the addition of a fluorescent tracer to the glycol. After the core is machined and cleaned, it is inspected, using an ultraviolet (black) light, for any residual glycol prior to bonding.

III. APPLICATIONS AND LIMITATIONS OF NDT TO BONDED JOINTS

There are a variety of NDT methods available for inspection. Only the methods applicable to the inspection of bonded structures will be presented. All the methods or techniques presented can be used in fabrication inspections, while only a limited number are applicable to on-aircraft or in-service inspections. The following descriptions are for methods which have proved to be most successful in detecting flaws in bonded laminates and honeycomb assemblies.

A. Visual Inspection

All metal details must be inspected to assure conformity with design. Before large assemblies are ready for bonding, the details are assembled in the bonding jig as though bonding were to occur. In place of the adhesive, a sheet of Verifilm is used. This material has nearly the same flow characteristics as the adhesive, but it is prevented from bonding by release film so that it will not stick to the details. The whole assembly is placed in an autoclave or press just as it would were it being bonded. After the heat-pressure cycle, the parts are disassembled and the Verifilm is inspected visually to see that the pressure marks are uniform throughout. A uniform marking gives good assurance of proper pressure at the bond lines. All areas showing no pressure must be inspected and parts modified to obtain proper fit and pressure during the cure cycle. A poor showing on the Verifilm is cause to rerun the check.

When the details go through an anodize cycle a visual check can be made for phosphoric acid anodize but not other types of anodize. For visual verification of phosphoric anodize the inspector looks through a polarizing filter at an angle of approximately 5 to 10° from parallel to the surface of the anodized part. The surface of the panel is well lighted by a fluorescent tube and the inspector rotates the polarizing filter while looking at the anodized surface. If the panel has been properly anodized, the inspector will see a change in hues equivalent to the colors of the rainbow (see Fig. 18.)

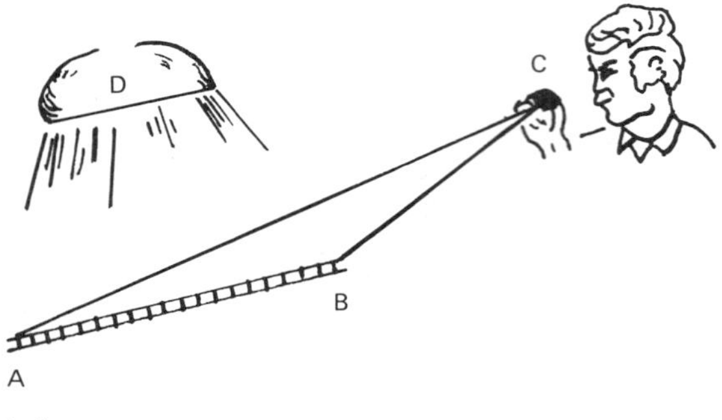

(b)

Figure 18   Using visual means to inspect for the presence of phosphoric acid anodize on surface. (a) Graphical representation of relative position of details. AB is surface being inspected, C is a polarizing filter, and D is a light source. Angle CAB should be between 500 and 100. (b) Actual inspection being made on a rack full of anodized parts. (Courtesy Douglas Aircraft Co.)

From the opened wedge crack specimen, the inspector looks at the failed surface to determine the type of failure. Areas of poor adhesion are where the adhesive separated from the substrate. This condition is manifested by variations in color and texture. A cohesive failure occurs through the adhesive and is uniform in color and texture.

After an assembly is bonded, it is given a visual inspection for scratches, gouges, dings, dents, or buckles. The adhesive flash and fillets can be inspected for cracks, voids, or unbonds at the edge. Feeler gauges can then be used to determine the depth of the unbond at the edge.

In-service visual inspection of bonded joints can reveal cracked metal or adhesive fillets; delamination or debonding due to water intrusion or corrosion; impact or foreign object damage; and blisters, dents, or other mechanical damage.

B. Ultrasonic Inspection

A number of different types of ultrasonic inspections can be applied to bonded structures. Following is a brief description of the various ultrasonic techniques being used to inspect bonded structures.

Contact Pulse Echo

In this technique, the ultrasonic beam is transmitted and received by a single search unit placed on one surface of the part (Fig. 19A). The sound is transmitted through the part and reflections are obtained from voids at the bond line. If the bond joint is of good quality, the sound will pass through the joint and be reflected from the opposite face, or back side. Bonded aircraft structure is usually composed of thin skins, which result in multiple back reflections appearing on the cathode ray tube (CRT) screen of the pulser/receiver. The appearance of multiple reflections appearing on the CRT has prompted some operators, or inspectors, to call this the "ringing" technique. When a void is present, the reflection pattern changes on the CRT and no ringing is seen. When inspecting bonded parts having skins of different thickness, inspection should be conducted from the thin skin side.

Contact Through Transmission

This technique is also illustrated in Fig. 19B. It is useful for inspecting flat honeycomb panels and metal-to-metal joints. Special search unit holding devices have been fabricated so that the test can be performed by one inspector. Figure 20 shows such a device used for inspecting the metal-to-metal closures of a bonded honeycomb panel. Longer and wider-spaced holders have been fabricated from tubing. The holding tool should be custom designed for the assembly being inspected. To perform the test, liquid couplant must be applied to both sides of the assembly.

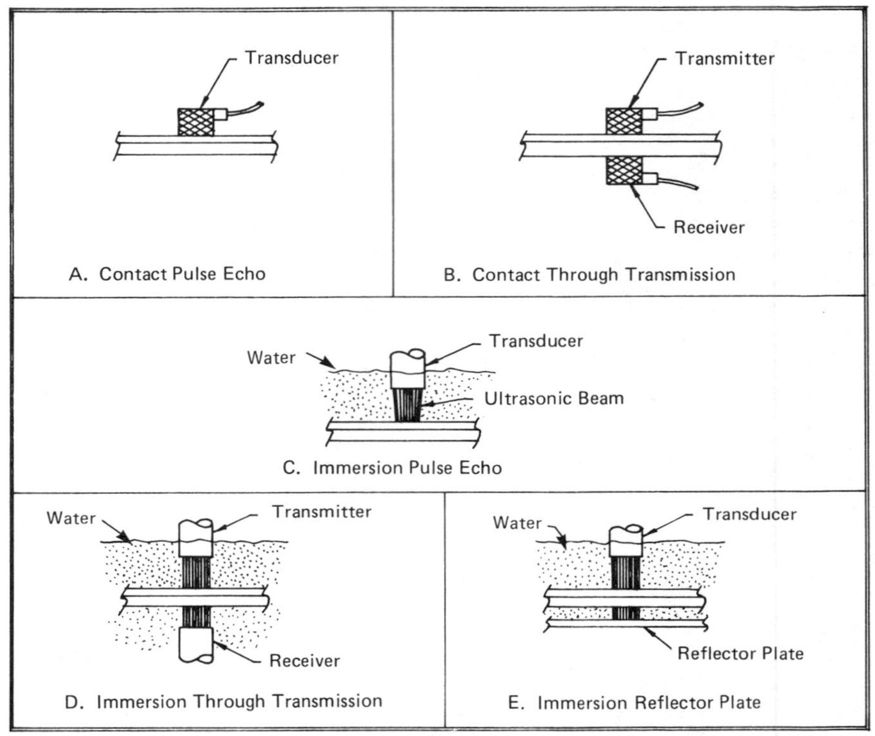

Figure 19   Ultrasonic inspection techniques. (A) Contact pulse echo
with a search unit combining a transmitter and receiver. (B) Contact
through transmission. Transmitting search unit on top and receiving
search unit on bottom. (C) Immersion pulse echo with search unit
(transmitter/receiver) and part being inspected under water. (D)
Immersion through transmission with both search units (transmitter
and receiver) and part being under water. (E) Immersion reflector
plate. Same as (C) but search unit requires a reflector plate below
part being inspected. (Courtesy Douglas Aircraft Co.)

Immersion Method

This method requires that the assembly be immersed in a tank of
water or that water squirters be used to act as a couplant for the ultrasonic beam. The method may be subdivided into three fundamental
techniques: pulse echo (Fig. 19C), reflector plate (Fig. 19E), and
through transmission (Fig. 19D). These techniques are applicable to
bonded laminates and honeycomb structures. The choice of technique

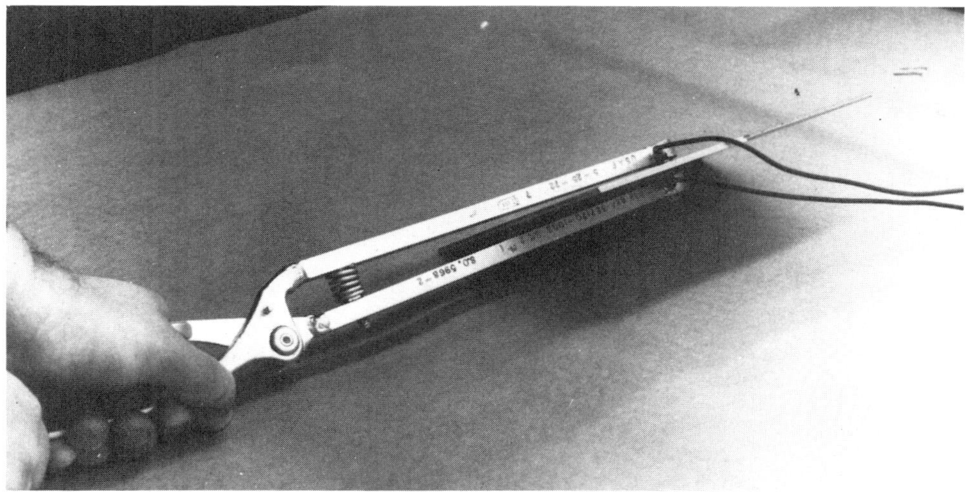

Figure 20 Ultrasonic pliers for contact through transmission testing of bonded joints.

is partially based on the thickness and configuration of the bonded assembly as well as physical size.

Bonded structures are generally inspected by the immersion method using a C-scan recorder. A C-scan recorder is an electrical device which accepts signals from the pulser/receiver and prints out a plan-view record of the part on a wet or dry facsimile paper recording. To obtain the recording, the bonded panel is placed under water in a tank and the ultrasonic search unit is automatically scanned over the part. The ultrasonic signals for bond-unbond conditions are determined from built-in defect reference standards. The signals are displayed on the pulser-receiver CRT screen and the signal amplitude is used to operate the recording alarm after setting the electronic gate around the signal of interest (see Fig. 21). Generally, high-amplitude signals will record, whereas low-amplitude signals will not record. The recording level is adjustable to select the desired signal amplitude. Figure 21 illustrates a typical ultrasonic C-scan recording system.

Ultrasonic immersion techniques, employing a C-scan recorder, are used extensively by aircraft manufacturers to perform inspection of adhesive-bonded assemblies. The C-scan recordings provide a permanent record with information on the size, orientation, and location of defects in bonded assemblies (Fig. 14). The C-scan systems are designed to inspect particular parts and hence vary considerably in size and configuration. Computer-operated controls have been incorporated into

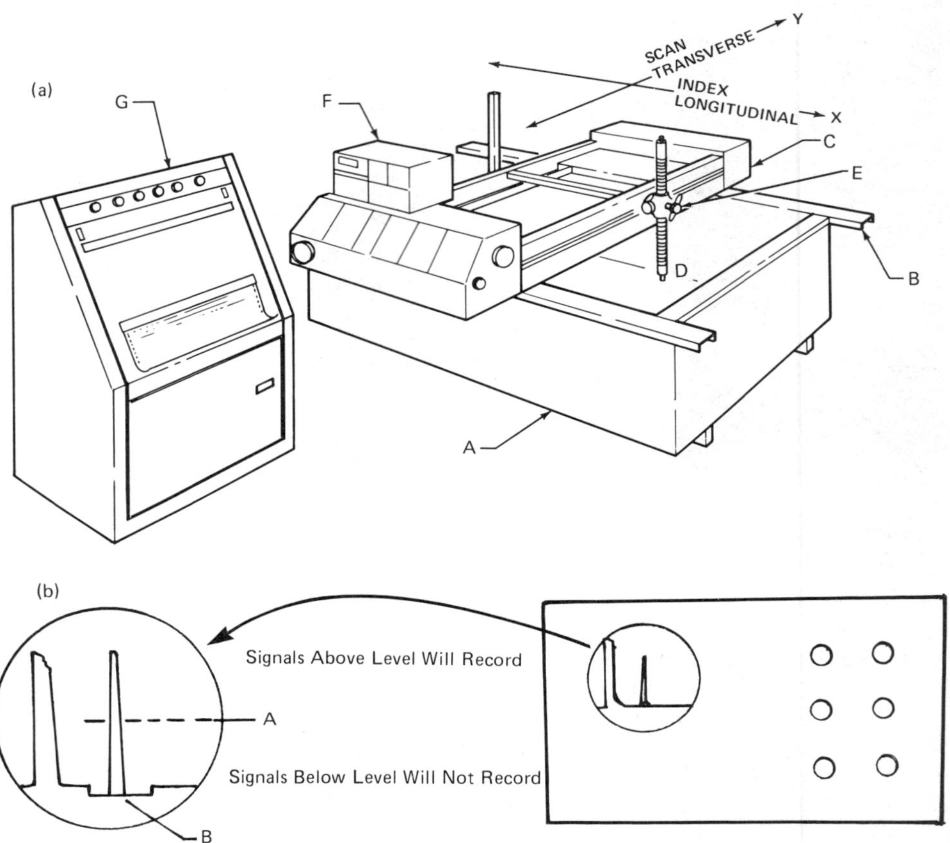

Figure 21   Ultrasonic C-scan immersion system. (a) Overall arrangement for the system. Consists of large water tank (A) with side rails
(B) which support the movable bridge (C). The search unit (D) is
held at proper distance above part by scanner arm (E) and moves
across the part and then indexed along longitudinal axis. The visual
pulsar/receiver unit (F) is mounted on the bridge. A permanent record
is made by the C-scan recorder (G). (b) Close-up pulser/receiver on
the right and a close-up of the CRT presentation on the left. A recording level (A) is set on the instrument and all signals above will record
on the C-scan machine and signals below will not show. Item (B) is an
automatic recording gate associated with the mechanics of record making.
(Courtesy Douglas Aircraft Co.)

some systems to control the scanning motions of the search units and also to change instrument gain at changes of thickness in the assembly.

C. Ultrasonic Bond Testers

A wide variety of ultrasonic bond testers have been developed over the past 20 years. As a consequence, this mode of bond inspection probably requires the most study because of the number of instruments available and the claims of the manufacturers. Various independent studies have been conducted in an attempt to clarify the situation [10, 3, 4, 11–20]. The results of these studies revealed that all the instruments are capable, in varying degrees, of determining the quality of the bond. None of the instruments are capable of establishing the adhesive quality of a bond that is defective due to (1) poor surface preparation of the substrate, (2) insufficient cure temperature, or (3) contamination of the adhesive or substrate prior to bonding. The conclusions indicate that ultrasonic bond test instruments are most reliable and sensitive for detecting voids, porosity, thick adhesive, and corrosion at the bond line, and can be used to inspect metal-to-metal and honeycomb structures.

Instruments such as the Coinda Scope, Stub Meter [13,14], Sonic Resonator [11], and Arvin Acoustic-Impact Tester [15] have not found general acceptance and are no longer on the market. Following is a short description of the ultrasonic bond testers which are currently being used to inspect adhesive-bonded assemblies.

The Fokker Bond Tester (Fig. 22a)

It is based on analysis of the dynamic characteristics of the mass-spring and dashpot system formed by the combination of the bonded assembly with a piezoelectric transducer having known mass and resonance characteristics. Changes in viscoelastic properties of the adhesive layer are detected as variations in the resonance frequency and impedance of the system. The calibrated body acts as a transducer that can be driven at different frequencies. The dimensions of the transducer are chosen in relation to the total thickness of the metal sheet to be tested and to the required mode of resonance. The response of the total system, as shown by the impedance curve over the swept frequency band, is displayed on a CRT display (A scale) (Fig. 23) and the peak-to-peak amplitude of the curve is shown by a microammeter (B scale).

For inspecting metal-to-metal bonded joints, the probe containing the calibrated body is placed on top of a piece of sheet material with the same thickness as the upper sheet material of the bonded laminate. The central frequency of the oscillator is selected such that the lowest point of the impedance curve is in the center of the A scale. Simultaneously, the B scale is adjusted to 100 (see Fig. 23). Calibration of the instrument on a nonbonded sheet ensures that,

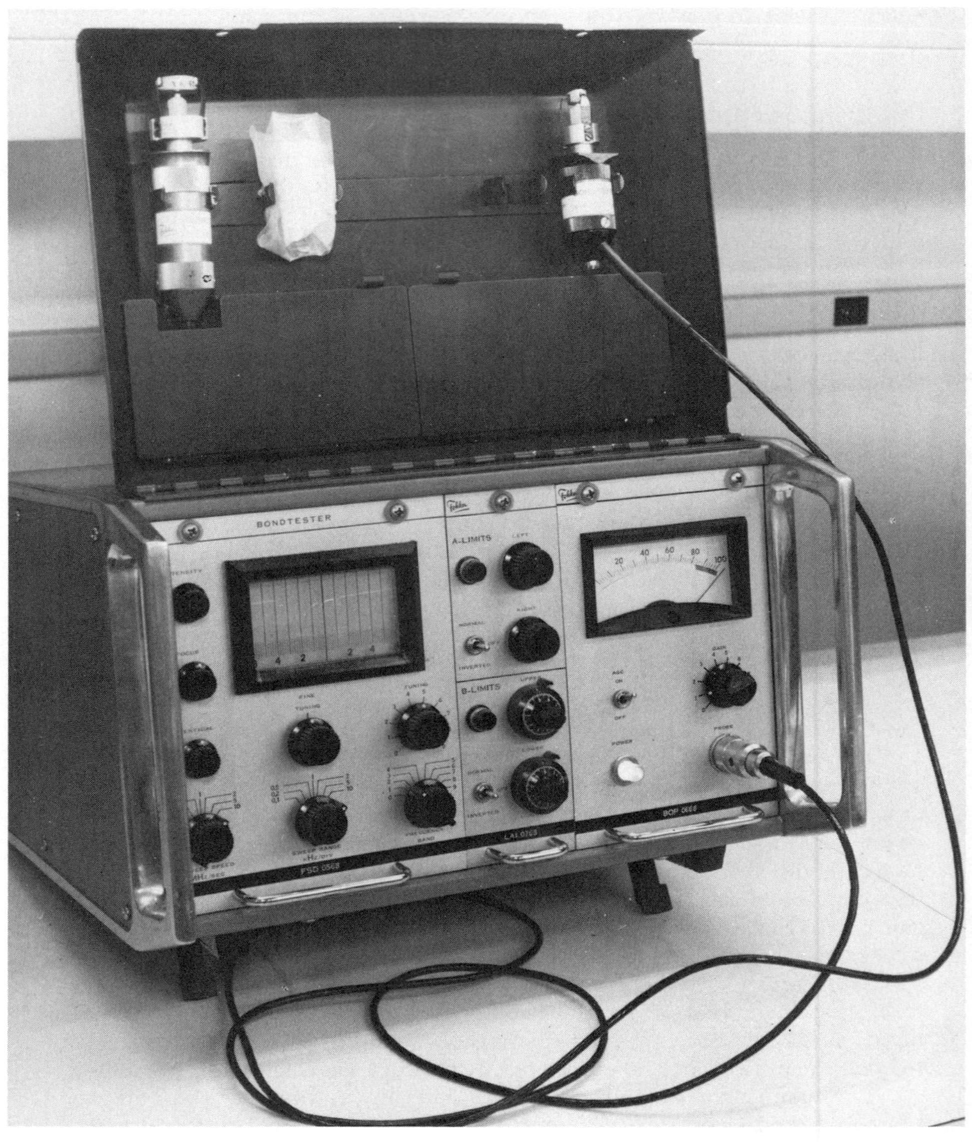

(a)

Figure 22   Three bond tester instruments: (a) Fokker bond tester, (b) 210 bond tester, and (c) Surtonics Harmonic bond tester.

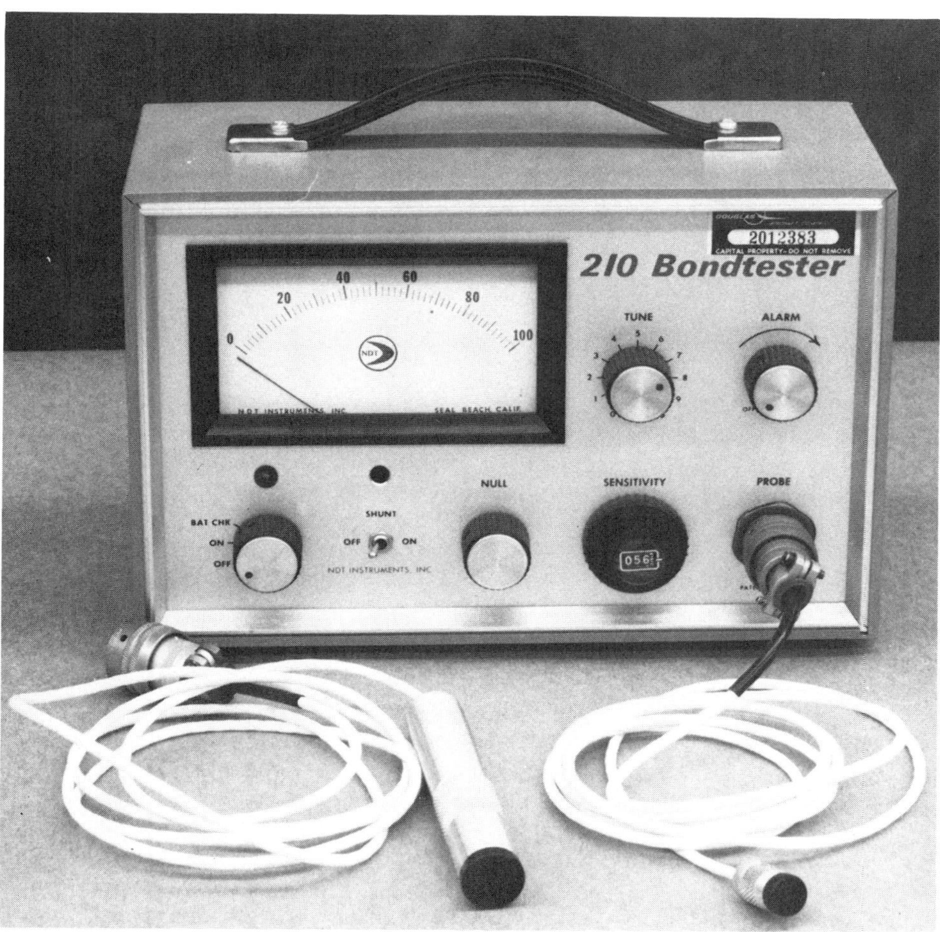

(b)

Figure 22 (continued)

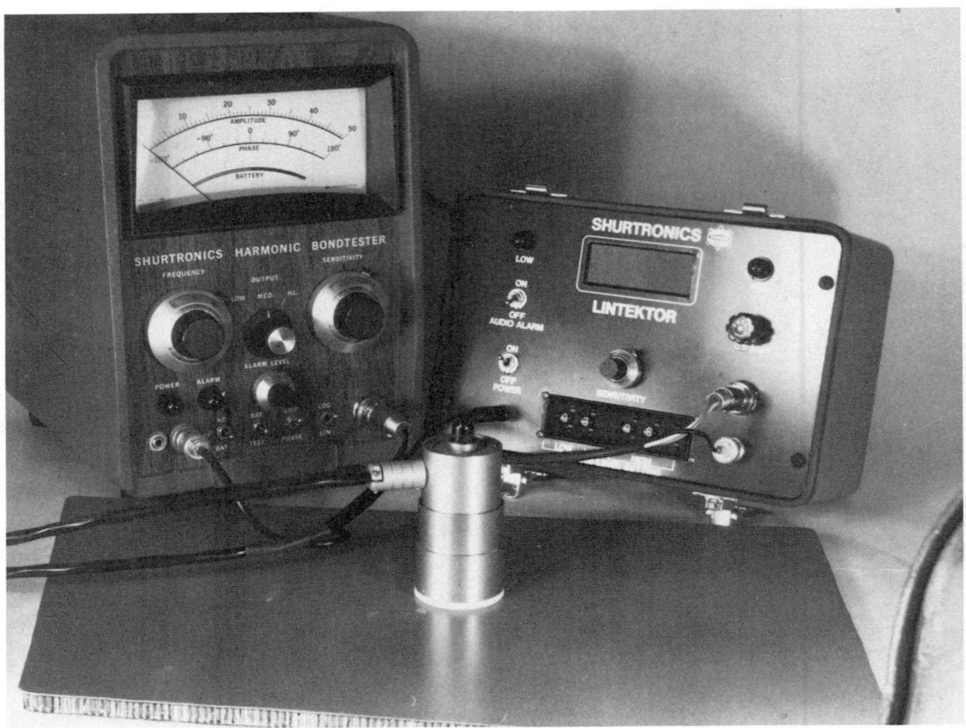

(c)

Figure 22   (continued)

in all cases of a complete void, the peak position will return to the center
of the A scale and the B scale reading will be 100 (Fig. 24). The next
calibration places the probe on a piece of metal sheet equivalent in thickness to all the metal sheets in the subject bonded laminate. The peak
obtained in this test is the resonance frequency to be expected of an
ideally bonded laminate (see Figs. 23 and 24).

The quality of cohesion from any test may be accurately determined
by comparing the instrument reading with established correlation curves.
In practice, the acceptance limits are based on the load or stress requirements of the adhesive for each joint. The accuracy of the prediction of
quality depends mostly on knowing the manufacturing variables and the
accuracy of the nondestructive and destructive tests conducted in accordance with MIL-STD-860 [23]. The destructive test for metal-to-
metal joints uses the lap shear specimen and, for honeycomb panels,

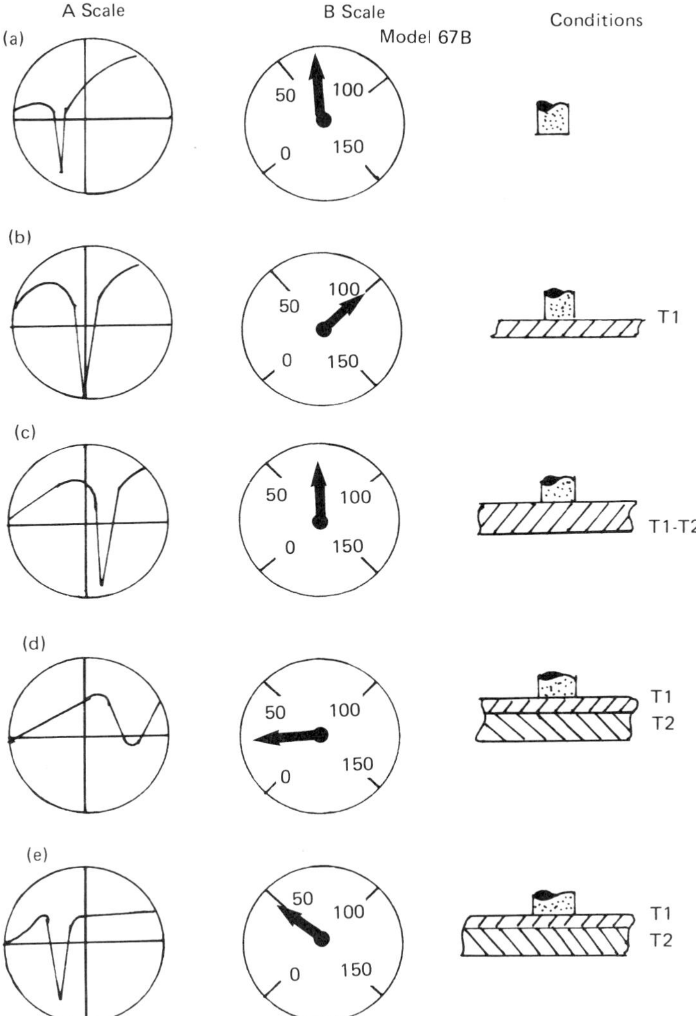

Figure 23   Several typical Fokker bond tester readings for specific bond conditions. Quality (a) results from probe held in air; (b) results when probe is on top of the upper adherend (no bonds) and is a calibration for unbond condition; (c) places probe on a single piece of metal as thick as combined parts being bonded and is calibrated for ultimate quality; (d) places probe on bonded joint with good-quality bond (no voids) and shows high-strength bond; (e) places probe on bonded joint with porosity and records low-strength bond. (Courtesy Douglas Aircraft Co.)

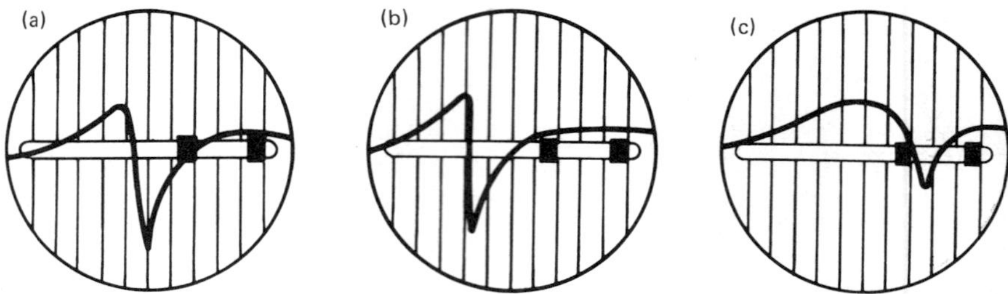

Figure 24   Typical displays of Fokker bond tester A scale for various
qualities of adhesive-bonded joints: (a) central frequency, no strength;
(b) higher frequency, low strength; (c) lower frequency, high strength.
(Courtesy Douglas Aircraft Co.)

uses the tension specimen for bond strength correlations. Comparative
tests by several investigators [10,2—4,16,18,21,22] indicate Fokker
bond tester to be more reliable in quantitatively measuring bond strength
as related to voids and porosity in the joint. For inspection of honeycomb, calibration is accomplished in the same manner except that the
micrometer, or B scale, on the instrument is used. The degree of
quality is reflected on the B scale. Low-quality bonds will give a high
B reading, and good-quality bonds will give a low B reading.

The Fokker bond tester has been successfully applied to a wide
variety of bonded sandwich assemblies and overlap-type joints of various
adherends, adhesive materials, and configurations. To date, the method
has proved suitable for joints having reasonably rigid adherends, including metallic and nonmetallic materials. Highly elastomeric or porous
adherends attenuate ultrasonic response. The method is most sensitive
to properties of the adhesive, and particularly to voids, porosity, and
incomplete wetting (unbond). Fokker tests readily detect voids in either
adhesive or nonmetallic adherend materials, cracks and delaminations in
adherends, and unbonded, flawed, ruptured, unspliced, or crushed
honeycomb core. Data indicate the test to be capable of measuring bond
degradation caused by factors such as moisture, salt spray, corrosion,
heat aging, weathering, and fatigue.

NDT-210 Bond Tester (Fig. 22b).

Sound waves from a resonant transducer are transmitted into, and
received from, the bonded structure. Unbonds or voids in the structure
alter the sound beam characteristics, which, in turn, affect the electro-
acoustic behavior (impedance) of the transducer. A frequency range
extending from less than 100 Hz to more than 6 MHz is provided for
maximum performance. The 210 automatically adjusts to the probe's

resonant frequency. A precision AGC (automatic gain control) oscillator maintains a constant voltage at all frequencies, eliminating the need for manual adjustment. Test response is displayed on a 4 1/2-in. meter. The operator can set an adjustable audible alarm to trigger when the response exceeds a preset level.

Metallic, nonmetallic, or a combination of joints can be inspected for voids and unbonds. The joints can be adhesively bonded, brazed, or diffusion bonded.

Shurtronics Mark I Harmonic Bond Tester (Fig. 22c)

The eddy-sonic bond tester is a portable, low-frequency, instrument using a single transducer for inspection of thin metal laminations and metal-to-honeycomb bonded structures. The instrument involves the use of a coil which electromagnetically vibrates the metal face sheet. It induces a pulsed eddy current flow in a conductive material by an oscillating electromagnetic field in the probe. The alternating eddy current flow produces an accompanying alternating magnetic field in the part. The attraction of the magnetic fields causes acoustomechanical vibrations in the part. When a structural variation is encountered, the change in acoustic response is detected by a wide-band receiving microphone located inside the eddy current coil.

An ultrasonic oscillator generates a 14- to 15-kHz electric signal in the probe coil, creating an oscillating electromagnetic field. The resulting acoustomechanical vibration in the test part is detected by a microphone with a bandwidth of 28 to 30 kHz. The microphone signal is filtered, amplified, and displayed on a microammeter. The instrument is calibrated to read just above zero for a condition of good bond, and full scale for unbonds over 1/2 in. in diameter or larger. This method does not require a liquid couplant on the part surface. It is useful for quick detection of unbonds in large-area bonded metal laminates, metal-to-nonmetal laminates or honeycomb structures. It is highly effective for detecting skin-to-core unbonds or voids in acoustic-honeycomb panels utilizing perforated facing sheets. The instrument's sensitivity decreases rapidly for skin thickness over 0.080 in. and near the edge of the part.

The new Mark IIB Harmonic Bond Tester has the following improvements:

1. *1% alarm accuracy*: Vernier control assures test repeatability with zero hysteresis.
2. *Phase comparator*: Detects defects in nonmetallic composite materials and provides additional information on metallic structures.
3. *Portable operation*: Self-contained batteries and built-in recharger.

4. *Modular construction*: Plug-in circuit boards for fast field servicing and maintenance.
5. *Lintektor*: The Lintektor provides a linear scaling technique with the same dynamic range as the HBT. It provides a digital readout with a numerical value directly proportional to panel signal response. By setting the Lintektor to a midrange value on a good bond area, moisture will be indicated as a low number and an unbond as a high number on the digital display. Thus the linear technique is essential to differentiate simultaneously between moisture, bond, and unbond areas. The Mark IIB and Lintektor are shown in Fig. 22c.

Sondicator

The Sondicator (Fig. 25) is a pulsed transmit/receive ultrasonic portable instrument that is capable of operating in a very low (25 to 50 kHz) acoustic frequency range. The instrument operates within this range at a selected single frequency obtained by manual turning for best instrument performance. The Model S-2 contains two meters and associated electronic circuitry, and a test probe containing two Teflon-tipped transducers mounted approximately 3/4 in. apart. One transducer imparts vibration at 25 kHz into the surface. Vibrations travel laterally through the material to the other transducer. The second transducer detects the amplitude and phase relationship of the vibrating surface, directing the associated signals to the two meters of the test equipment. If the probe encounters a delaminated or unbounded area, it will, in effect be introducing vibration into an area that is thinner than the bonded area. The amplitude of vibration will increase and the phase of vibration will shift as the thinner section vibrates more vigorously. The needles on the two meters will move toward each other.

The instrument is used primarily to detect delaminations of the bond or laminar-type voids in metal and nonmetal structures. The inspection may be performed from one face of the structure or by through transmission. It requires no liquid couplant. Multiple bond lines and part edges will reduce the sensitivity of the instrument. Because of the directionality of the sound from the transmitter to the receiver, the part must be scanned in a uniform manner. The Sondicator is capable of detecting defects 1 in. in diameter and larger in most materials. Metallic and nonmetallic materials can be inspected without changing probes. Audible and visible alarms are activated by received signals. The Sondicator can detect internal delaminations and voids in bonded wood, metal, plastic, hard rubber, honeycomb structures, Styrofoam, and composites. It can measure skin or face sheet thickness in composite structures, as a change in amplitude or phase, as illustrated in Fig. 26.

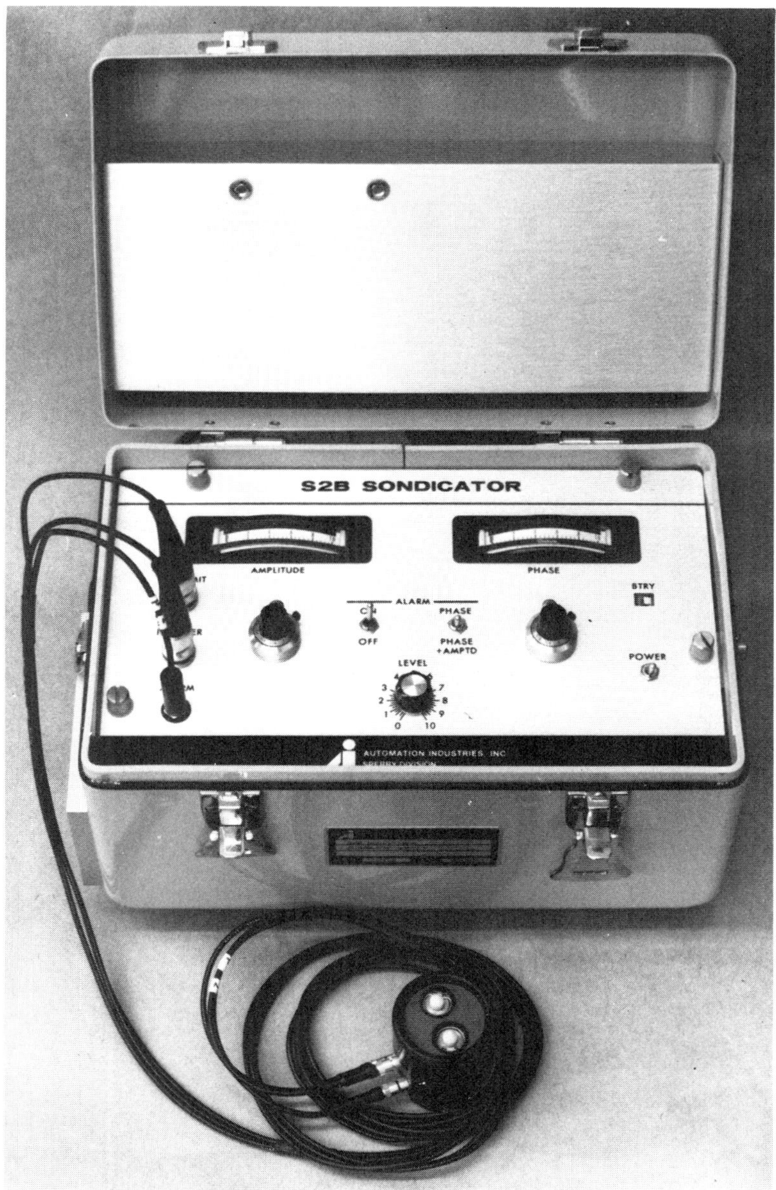

Figure 25   Sondicator Model S-2 meter readout.

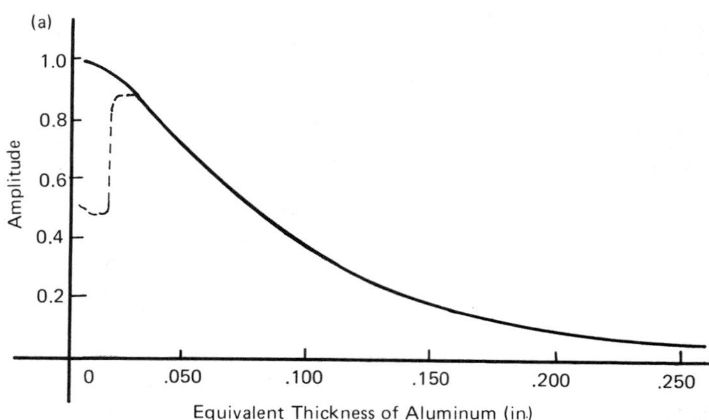

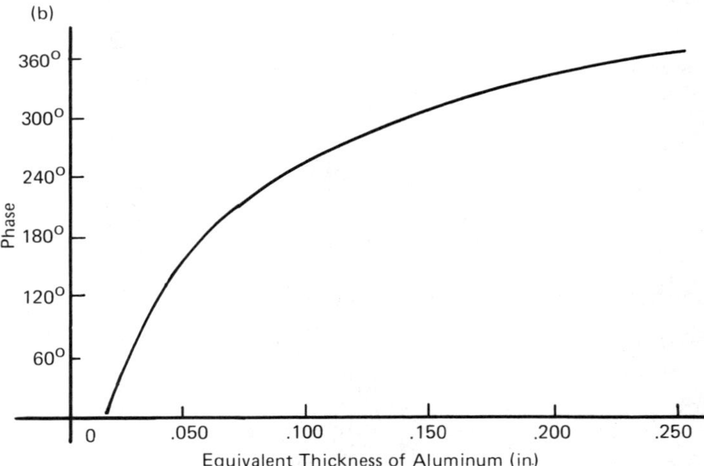

Figure 26   Sondicator—variation of (a) amplitude and (b) phase. (Courtesy Douglas Aircraft Co.)

Bondascope 2000

The bondascope is an advanced microprocessor-based device which operates on the principle of ultrasonic impedance analysis. This technique allows the total ultrasonic impedance vector for the material to be monitored as a flying dot on a scope display. Using the center of the scope screen as the origin (reference), phase changes as impedance are displayed circumferentially, while amplitude changes are displayed radially. Thus the position of the dot on the scope immediately reveals the phase and amplitude of the material's impedance. It then follows that defects at different bond lines each possess a characteristic dot position on the scope display.

Figure 27 is a sketch showing the typical Bondascope response to unbonds at different bond lines (depths) in a multilayered adhesive-bonded laminate. The instrument was calibrated so that the dot was in the center of the screen (origin) when the probe was placed on a well-bonded section of the laminate. The numerically labeled dots represent the signals obtained when the probe was placed over regions which had unbonds located at different respective bond line depths in the laminate. The dot labeled "air" is the signature obtained when the probe was off the sample. Thus improper ultrasonic coupling into the material is quickly recognized.

A meter on the Bondascope, by means of selectable pushbottoms, can display the signal amplitude, phase, or the vertical resolved component of the amplitude. By means of a phase-rotator control, the signal response can be positioned so that irrelevant signals are suppressed from the meter/alarm readings.

A keyboard pushbutton matrix allows the operator to digitally program calibration-sample reference dots on the scope display to aid in

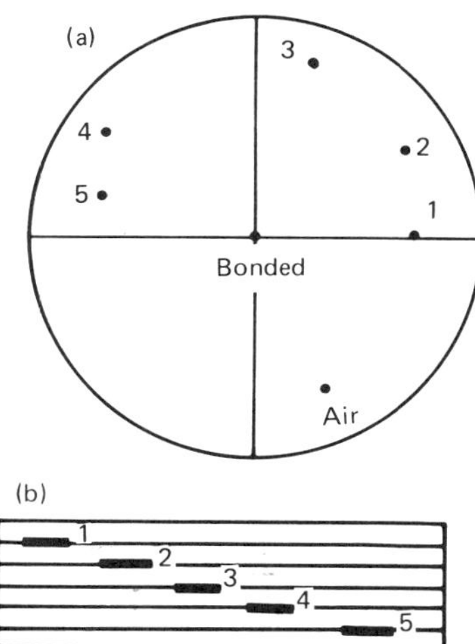

Figure 27 Bondascope ultrasonic impedance plane presentation for multilayer laminar: (a) orientation of Bondascope display of the unbonds shown in (b); (b) multilayer bonded laminar with unbonds located at 1, 2, and so on. (Courtesy NDT Instruments, Inc.)

interpreting signals obtained during actual testing. This type of
storage display not only aids in operation and interpretation, but
eliminates the confusing permanent streaks which occur if a conventional storage scope were used to display the flying dot.

The Bondascope is suitable for inspecting both metallic or nonmetallic bonded structure (multilayered laminates or honeycomb). Graphite-
epoxy and other types of fiber-matrix composites are also suitable for
testing with this device.

Ultrasonic and Bond Tester Method Sensitivity

Defect detention sensitivity of ultrasonic techniques with respect
to defect size and location is highly dependent on changing conditions
of part complexity, number of bond lines, operator experience, and so
on. Most instruments detect major defects most of the time. However,
in certain instances special techniques and skills are needed to conduct
a reliable inspection. Whenever possible, bonded structure should be
inspected from opposite surfaces to detect small defects. In the ultrasonic bond test methods, different-size search units or probes are
available to the inspector. Experience indicates that the smallest flaw
which can readily be detected is of the order of one-half the diameter
of the search unit or probe element.

The second item to consider is the test frequency. The higher the
frequency, the smaller the size flaw that can be detected. Ultrasonic
testing at 5 or 2.25 MHz will detect smaller falws than can be detected
with the Sondicator (25 to 50 kHz) and Harmonic bond tester (15 kHz).
Third, with multiple bond lines in a structure, smaller flaws become
progressively more difficult to detect from the surface through suc-
ceeding bond lines. This statement does not apply to ultrasonic through
transmission or reflector plate testing where the small-diameter sound
beam passes completely through the part. Figure 28 shows the relative
defect sensitivity of ultrasonic and bond tester techniques or methods.

D. X-Ray Radiography

Radiography is a very effective method of inspection which allows a view
into the interior of bonded honeycomb structures. The radiographic
technique provides the advantage of a permanent film record. On the
other hand, it is relatively expensive and special precautions must be
taken because of the potential radiation hazard. With the radiographic
method, inspection must be conducted by trained personnel. This
method utilizes a source of x-rays to detect discontinuities or defects
through differential densities or x-ray absorption in the material.
Variations in density over the part are recorded as various degrees of
exposure on the film. Since the method records changes in total density
through the thickness, it is not a preferred method for detecting defects (such as delaminations) that are in a plane normal to the x-ray
beam.

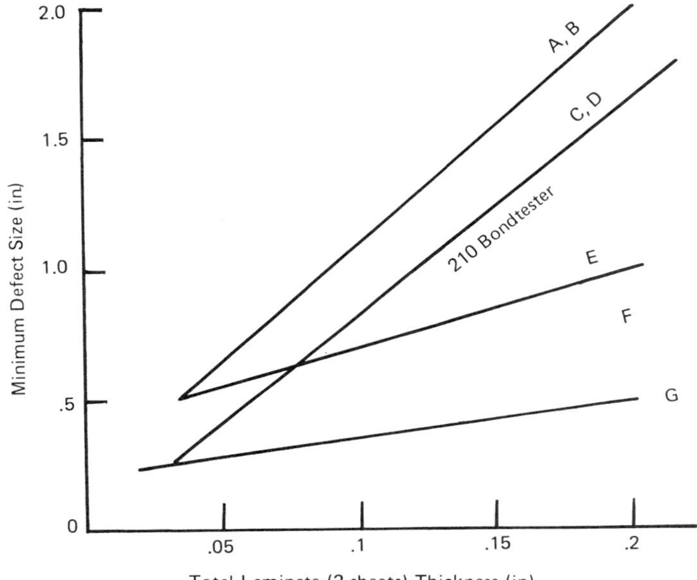

Figure 28   Relative sensitivity of instruments inspecting bonded laminates of increasing total thicknesses: (A) Sondicator, contact method; (B) Harmonic bond tester; (C) Fokker bond tester; (D) 210 bond tester; (E) Sondicator, through-transmission method; (F) ultrasonic pulse echo; (G) ultrasonic through-transmission. (Courtesy Douglas Aircraft Co.)

Some adhesives (such as AF-55 and FM-400) are x-ray opaque, enabling voids and porosity to be detected in metal-to-metal bond joints (see Fig. 2). This is extremely advantageous, especially for complex geometry joints which are difficult to inspect during fabrication. The x-ray inspection should be performed at low kilovoltage (25 to 75 kV) for maximum contrast. A beryllium window x-ray tube should always be used when performing radiography of adhesive-bonded structure. To speed up exposures, medium-speed, fine-grained film should be used. Selection of the film cassette should be given special consideration since some cassette materials produce an image on the film at low kilovoltages.

E. Neutron Radiography

Neutron radiography is very similar to x-ray or gamma-ray radiography in that both depend on variation in attenuation to achieve object contrast

on the resultant radiograph. There are significant differences in the
effectiveness of the two methods, especially when certain combinations
of elements are examined. The mass absorption coefficients of the different elements for x rays assume a near-linear increase with atomic
number, whereas the coefficients for thermal neutrons show no such
proportionality. Attenuation of x rays is determined largely by the
density of the material being examined. Thus thicker and/or denser
materials appear more opaque. The absorption for thermal neutrons
is a function of both the scattering and capture probabilities for each
element. The density or thickness of a material or component is less
important in determining whether it will be relatively transparent or
opaque to the passage of neutrons. For example, x rays will not pass
through lead easily, yet they will readily penetrate hydrocarbons. In
contrast, neutrons will penetrate lead and are readily absorbed by the
hydrogen atoms in the adhesive or hydrocarbon material.

Neutrons are produced primarily from accelerator, radioisotope, or
reactor sources. Neutrons, like x rays, can be produced over an enormous energy range with large differences in attenuation at the various
energy levels. The major efforts in neutron radiography have been
performed using thermal neutrons because it is in this energy regime
that the best detectors exist. Neutron sources generally produce gam-
ma rays of moderate intensity, so that a neutron detector sensitive to
gamma-ray radiation has a gamma-ray image superimposed on the neutron
image.

The most widely used imaging method is conventional x-ray film
with converter screens. The rate of radioactive emission of the converter screen divides the photographic imaging into prompt emission or
delayed emission. The prompt emission converters (gadolinium or
rhodium) require that the film be present during neutron exposure.
This process is referred to as the "direct method." The delayed emission converters (indium or dysprosium) allow for activation of the converter and transfer of the induced image to the film after neutron exposure. This technique is referred to as the "transfer method." In
the original film, the neutron opaque areas appear lighter than the surrounding material. Image contrast can be increased by making a contact positive print from the film negative. When this is done, the neutron
opaque area will be darker than the surrounding material (see Fig. 12).

Hydrogenous (adhesive) materials inspected by N radiography can
be delinated from other elements in many cases where x-ray radiography
is inadequate. However, the N-radiography inspection method does not
appear to be cost-effective for routine inspection of adhesive-bonded
structures. It is extremely useful for evaluating the quality of built-in
defect reference standards or for failure analysis. If the adhesive is
not x-ray opaque, neutron radiography may be used to detect voids
and porosity. The hydrogen atoms in the adhesive absorb thermal neutrons rendering it opaque.

F. Tap Test

Probably the simplest inspection method to assure that a bond exists between the honeycomb and facing sheet is that of coin tapping. An unbond is readily apparent by a change in the tone or frequency of sound when tapped with a coin or rod as compared to the sound for a bonded area. Coins such as silver dollars or half-dollars are used for this test. For standardized production testing, a rod (1/2-in.-diameter solid nylon or aluminum, 4 in. in length, with the testing end smoothly rounded to a 1/4-in. radius) is presently used. Another version of a tap tester is the aluminum hammer shown in Fig. 29.

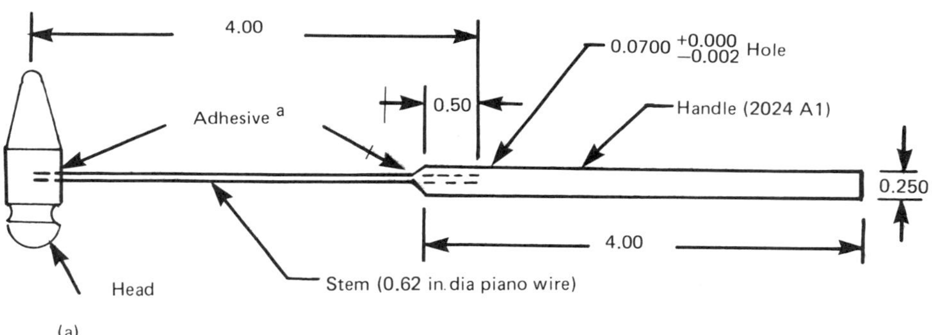

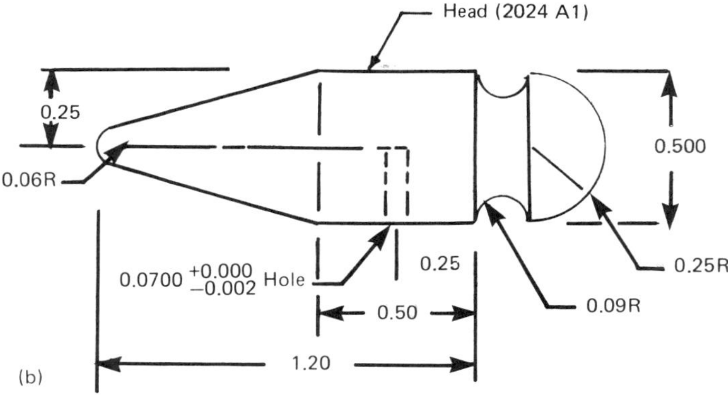

Figure 29 Construction details for an inspection-tap hammer: (a) complete assembly; (b) dimensional information is given in inches. [a]liquid/paste adhesive may be used if desired. (Hole in handle/head may be reduced to provide an interference fit and preclude the need for the adhesive.) (Courtesy The Boeing Airplane Co.)

Tap testing with a coin or small aluminum rod or hammer is useful for locating large voids or disbonds of the order of 1 1/2 in. diameter or larger in metal-to-metal or thin facing-sheet honeycomb assemblies. It is limited to detection of upper facing sheet to adhesive disbonds or voids. It will not detect voids or disbonds at second, third, or deeper adherends. The test method is subjective and may yield variation in test results. On thin face sheets, the coin tap will produce undesirable small indentations. In this case, an instrument like the Shurtronics Harmonic bond tester can be used effectively.

G. Acoustic Emission

In certain instances, acoustic emission (AE) techniques have been more effective than x-ray or conventional ultrasonic methods in detecting internal metal corrosion and moisture-degraded adhesive in bonded panels. The principle is based on the detection of sound or stress-wave signals created by material undergoing some physical or mechanical transformation. Regarding the detection of bond line corrosion, the acoustic signals apparently arise from cavitation or boiling of moisture within the joint. If the corroded joint is dried out before performing the AE test, no acoustic response is obtained.

The equipment consists basically of an amplifier and a piezoelectric sensor with a resonant frequency around 200 kHz. The emission level is recorded on a chart or counter. The inspection process is illustrated in Fig. 30. The equipment consists of a signal processor, a search unit, a 50-dB preamplifier, an X-Y chart recorder or counter, and a hot-air gun. Simple heating methods employing a hot-air gun or heat lamp are used to increase emissions from active corrosion sights or boiling of moisture.

The panel to be tested is heated to about 150°F by holding the hot-air gun within 2 to 3 in. of the surface of the panel for about 15 to 30 sec. Immediately after heating, the transducer or search unit is placed a short distance from the heated spot. The transducer is held in position for 15 to 30 sec to obtain a complete record of any emission in the heated area. The inspection is conducted on a 6-in. grid. An important consideration during the test is the manner in which the transducer is held against the part surface. Since movement of the transducer can produce appreciable noise, care must be exercised in its placement and holding.

Corrosion has been detected in a number of adhesive-bonded honeycomb structures by Lt. Rodgers and Lt. Moore, at McClellan AFB, in Sacramento, California [7–9]. They claim a direct inspection cost savings of over 75% over comparative x-ray inspection.

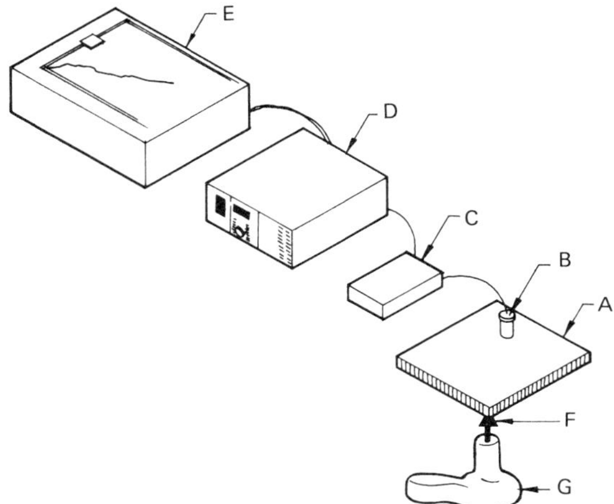

Figure 30   Acoustic emission detection of active corrosion in adhesive bonded structures: (A) specimen being inspected; (B) search unit; (C) preamplifier; (D) acoustic emission monitor; (E) X-Y recorder; (F) heat; (G) hot-air gun (heating may be done from search unit side). (Courtesy Douglas Aircraft Co.)

H. Special NDT Methods

Holographic Interferometry

Defects such as core-to-skin and core-splice voids, delaminations, crushed core, and bond line corrosion can usually be detected by holographic nondestructive test (HNDT) techniques [4,24]. A hologram of the test specimen is first recorded by means of laser light reflected from its surface and superimposed on a mutually coherent reference beam in the plane of a high-resolution photographic emulsion. The hologram provides a complete record of all visual information about the entire illuminated surface of the specimen, including phase as well as amplitude of the reflected wavefront.

The specimen is then stressed in one of several possible ways, including heat, pressure change, evacuation, and acoustic vibration. A surface displacement differential of only a few microinches between a defective and a good-quality bond is adequate to produce a holographic image. Differential surface displacements, caused by subsurface anomalies, are observed or recorded in one or more of the following three HNDT techniques: (1) real time, in which the stress illuminated

specimen is viewed through its developed hologram (made in a different
state of stress); (2) time lapse, in which two holographic exposures
are made on the same plate, with the specimen in two states of stress,
and then reconstructed; and (3) time average, in which the hologram
is recorded during many cycles of sinusoidal vibration of the specimen.

A fluid-gate photographic plateholder [24] provides for in-place
development of holograms and enhances the speed of operation. Because of the acute sensitivity of holographic interferometry to small
disturbances, it is necessary to reduce spurious and unwanted acoustic
noise and temperature gradients within the environment of the testing
system. Floor vibrations arising from heavy traffic, loading/unloading
vehicles, and heavy-duty machinery generate noise that must be minimized.

Systems for testing sandwich structures generally use He-Ne gas
lasers which normally deliver between 60 and 80 mW of power at 6328 Å.
This power is sufficient to examine at least 2 ft^2, and the light's deep
red color is close enough to the sensitivity peak of the eye for good
fringe-to-background contrast.

A typical real-time hologram showing a skin-to-core void revealed
by thermal stressing is shown in Fig. 31. In large measure, the basic
limitation of HNDT is hinged to the stressing techniques utilized. Holography cannot be used without a surface manifestation of a defect during stressing. If the material thickness precludes detection of a resolv-
able fraction of the fringe spacing, then HNDT is ineffective. It is not
useful for inspecting complex laminates because of the inability to stress
the void areas. Thermal stressing of aluminum is unsatisfactory because of its high thermal conductivity (heat is transferred laterally
rather than vertically through the joint). It is satisfactory for inspecting honeycomb, but few voids exist at the skin-to-core interface. It
is limited in locating voids at honeycomb closures or multilaminate metal-
to-metal areas. Numerous flaws in closure areas of bonded honeycomb
assemblies (voids and porosity) detected by ultrasonics and x-ray
methods were not detected by holographic interferometry.

At Fokker-VFW, a holographic installation for testing bonds was
developed and put to use for production inspection [4]. The installation uses a single hologram for components up to 6 m in length. It has
adjustable optics for optimum sensitivity and component scanning. The
defect area can be magnified and the results presented on a highly sensitive video monitor which can record the information. The bonded components can be deformed by either vibrations or thermal effects.

Schliekelmann reports [3,4,21] that interference holography shows
much more than just bond defects and requires training to learn to
distinguish between areas which are defective and those which are
merely not ideal. He also reported that core-splice location and quality
can be inspected and core machining anomalies are visible. Also, the
thermal method is the most practical deformation method for sandwich

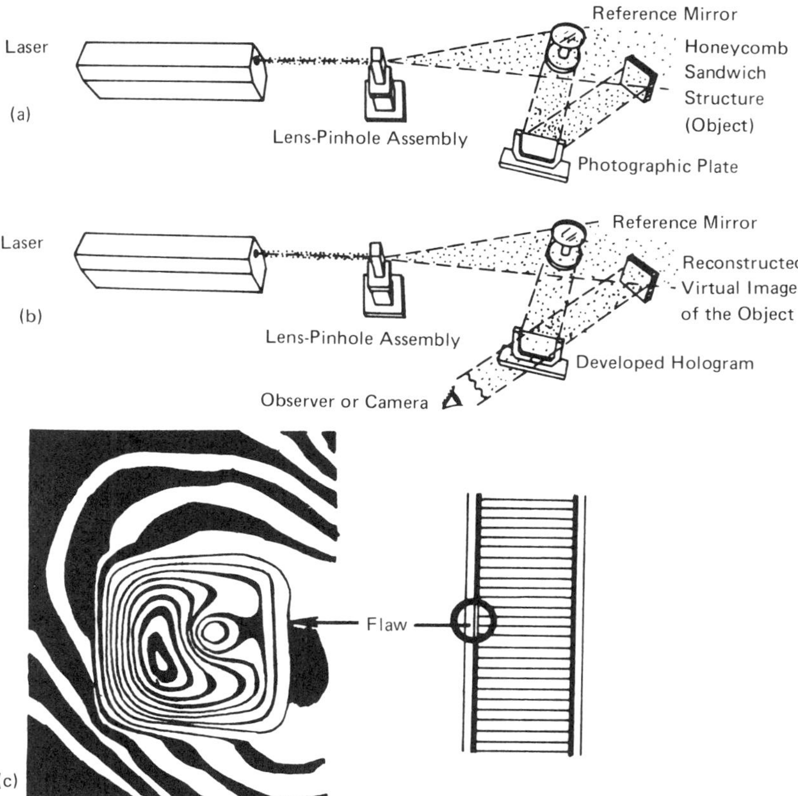

Figure 31 Holographic recording and viewing: (a) setup for recording hologram; (b) setup for real-time viewing of hologra; (c) example
of real-time hologram of a flaw (void in adhesive) in a honeycomb core
bonded assembly. (Part was heated to give thermal stressing.)
(Courtesy Douglas Aircraft Co.)

structures, especially for graphite-epoxy skins bonded to aluminum
core. Extensive studies in the Netherlands showed that thermal deformation was difficult to use for metal-to-metal structures [21]. These
difficulties led workers at Fokker to concentrate on vibration testing
because of the relation between resonance properties and the quality of
cohesion. It will take considerable time to develop a universal system
of quality testing based on interference holography. Such a system
has been developed by Grant, at Industrial Holographic, Inc., for the
inspection of truck and aircraft tires [25]. A wide variety of tire

defects have been detected and catogorized for both new and retread
tires and related to durability.

Thermal or Infrared

Thermal NDT of adhesive-bonded structures has been performed
by a variety of techniques, including the following:

1. Infrared radiometer [22,26]
2. Thermochromic or thermoluminescent coatings [22,27]
3. Liquid crystals (cholesteric) [28,29]

For radiometer testing, the detector employs a moving heat source
and records variations in heat absorption or emission while scanning
the part surface. The scanners, or detectors, used in infrared testing
are called radiometers. The radiometer generates an electrical signal
exactly proportional to the incident radiant flux. Since scanning and
temperature sensing are performed without contact, the observed surface is not disturbed or modified in any way. The thermal pattern
from the part under test may be observed on a cathode ray tube (CRT),
storage (memory) tube, or X-Y recorder.

Tests are performed by heating the sample with a visible-light heat
source (quartz lamp or hot-air gun), then observing the surface heating
effects with the radiometer. Heat is applied to both bonded and unbonded areas. The bonded areas will conduct more heat than the unbonded areas because of good thermal conduction to the honeycomb
structure. The test panel is first coated with flat black paint to assure
uniform surface emissivity. The sample is then scanned as shown in
Fig. 32. Graphs are obtained by connecting the horizontal movement
of the sample to one axis of an X-Y recorder and the radiometer output
to the other axis. Typical line scan and area scan test results are
shown in Fig. 32. Similar results are obtained using the AGA Thermo-
vision (in the "pitch or catch" or "through-transmission" modes) [22].
Good results are generally obtained when testing graphite-fiber or
boron-fiber composite face sheets bonded to aluminum core. Honeycomb
structures fabricated exclusively from aluminum (skin and core), or
aluminum skin bonded to plastic core, are difficult to inspect by infrared
methods because of the lateral heat flow in the aluminum face sheets.

The use of an ultraviolet-sensitive coating containing a thermoluminescent phosphor that emits light under excitation by ultraviolet
radiation (black light) permits the direct visual detection of disbonds
as dark regions in an otherwise bright (fluorescent) surface [22,27].
The coating is sprayed on and dried to a 3- to 5-mil-thick plastic film.
Defects show as darkened areas when the panel is heated to 140°F and
viewed under ultraviolet light. The defective regions can then be
marked on the plastic film with a felt-tip pen. The rate of fluorescent

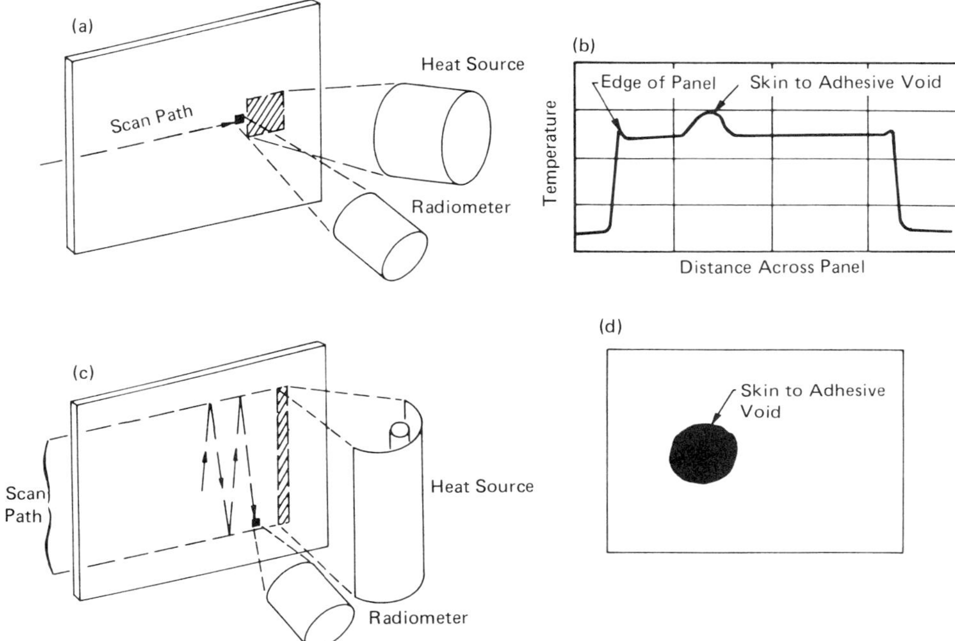

Figure 32 Schematic drawings of infrared radiometer tests: (a) arrangement for line scan test; (b) example of readout and how a void is detected; (c) arrangement for area scan test; (d) example of readout and how a void is detected. (Courtesy Douglas Aircraft Co.)

reversal and retention depends on the thermal conductivity of the underlying structure.

Thermochromic paints are composed of a mixture of temperature-indicating materials which have the ability to change color when certain temperatures are reached. These paints were developed in Germany and were sold under the trade name Detecto Temp. Thirty-six materials covering the temperature range 104 to 2962°F were developed, with the low-temperature materials being used for bond inspection. The low-temperature paint changed from light green to vivid blue on reaching a temperature of 104°F. The color change will extend (in the defect area) for about 15 min based on the relative humidity. The dark areas can be made to revert back to the original color by applying moisture in the form of steam. The tests may be repeated a number of times without destroying the properties of the paint. The paints are easy to apply and remove.

Liquid crystals are a mixture of cholestric compounds that change color when their temperature changes as little as 1.5°F and always attain the same color at a given temperature for a specific crystal composition. After suitable surface preparation or cleaning, a thin coating of liquid crystals is applied with spray or brush to the test object surface. When the object is heated correctly (relatively low temperatures) using a heat lamp or hot-air gun, the defects are shown by differences in color [28,29]. Unfortunately, the color keeps changing through a specific color band as it is heated and cooled. Therefore, the defects must be marked on the surface of the part as they appear and disappear. Photographs may be taken as a record of the test results after specific defects are located. The test may be repeated a number of times without destroying the liquid crystals. In some cases, black paint is required under the liquid crystal coating to obtain uniform emissivity and color contrast of the defect areas. The theory and results obtained using liquid crystals are virtually the same as for thermoluminiscent and thermochromic coatings (Fig. 33).

Leak Test

A hot-water leak test is generally performed on all bonded honeycomb assemblies immediately after fabrication or repair. The test is performed by immersing the part in a shallow tank of water heated to about 150°F. The heat causes entrapped air to expand and if there are any leakage paths, bubbles will be generated at the leakage site. The panel is visually monitored for escaping bubbles and the leakage site is marked on the panel for subsequent sealing. A part that leaks after fabrication will generally develop problems in service. After the leak test, the assembly is radiographed for water that may have become entrapped during the leak test.

Acoustical Holography

This method provides a way to visually observe the interior properties of composite laminates or adhesive-bonded joints. This acoustical technique employs ultrasound to obtain three-dimensional information on the internal structure of the test specimen. The acoustical hologram is the converter, or recorder, which allows acoustic information to be visualized much as film is the converter or recorder for light. Holosonics Incorporated manufactures acoustical holography systems for a variety of applications. The acoustic imagers employ different holographic recording technqiues depending on the specific application of the instrument [30]. One acoustical holography technique uses a liquid surface. The liquid surface acts as a dynamic film for momentary storage of the hologram while it is converted to a visual image through use of coherent laser light. This technique (Fig. 34) allows real-time acoustical imaging, providing the operator with an instantaneous view of all

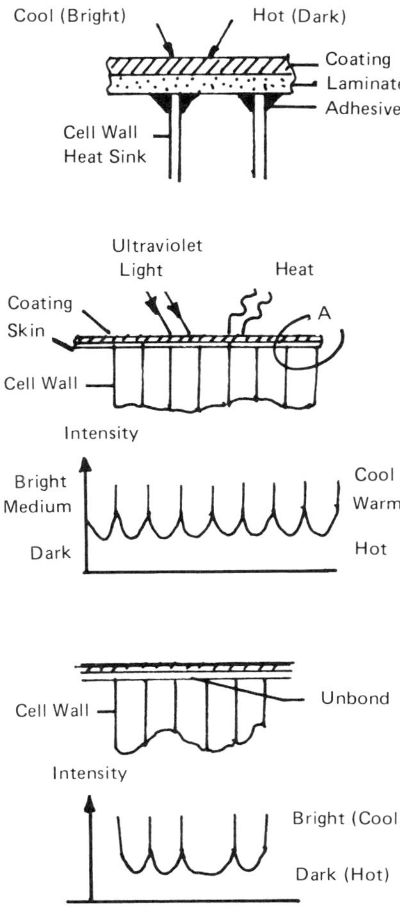

Figure 33   Thermoluminescent coating technique on boron composite aluminum honeycomb flap assembly.   (Courtesy Douglas Aircraft Co.)

internal structures. Because the part must be moved through the fixed acoustic beam, part size is a limiting factor using this technique.

A second method employs a scanning technique to construct the hologram. The hologram is then recorded on either transparency film or a storage oscilloscope. This technique provides a permanent record of the holographic information and allows the operator to observe the reconstructed image at any later time. By illuminating either the liquid surface hologram or transparency film with laser light, the

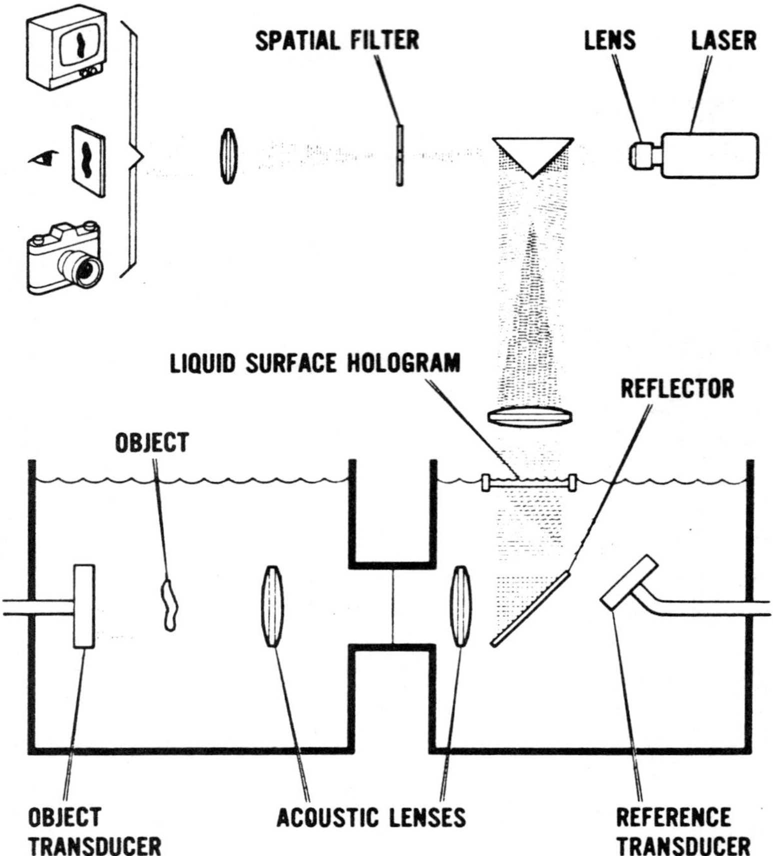

Figure 34   Schematic diagram of liquid surface imaging systems.
(Courtesy Acoustic Imager Series 100, Holosonic, Inc.)

operator immediately observes all interior properties of the test sample.
Figure 35 schematically illustrates the scanning holographic system
series 200, manufactured by Holosonics. The capability of acoustical
holography to reveal flaws in bonded structure is adequately illustrated
in Ref. 31. One drawback to the use of scanned acoustical holography
is the cost of the equipment (i.e., $50,000 to $100,000). It also requires considerable time to reconstruct the various holograms of the
entire part.

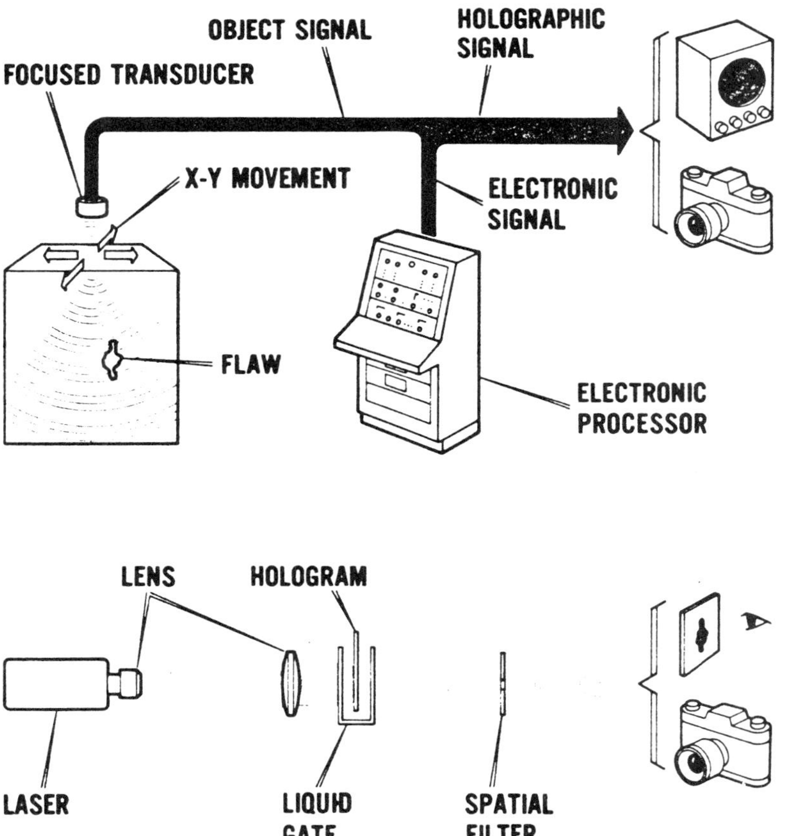

Figure 35 Schematic diagram of pulse echo acoustical holography systems and holographic reconstruction unit. (Courtesy Acoustic Imager Series 200, Holosonic Inc.)

I. Method Summary

A universally accepted single test method for evaluating bonded structure has not been developed. Thus a combination of test methods are required for complete and reliable inspection. Generally, the selection of the test method is based on (1) part configuration and materials of construction, (2) types and sizes of flaws to be detected, and (3) accessibility to the inspection area and availability of equipment/personnel.

The application of NDT methods in the production cycle for honeycomb sandwich panels generally includes the following:

1. Material property tests
2. Surface preparation checks (wedge crack, T-peel, etc.)
3. Verifilm (prebond) tooling check
4. Visual inspection
5. Hot-water leak check
6. Radiographic check for water and other internal discontinuities
7. Ultrasonic C-scan for high-resolution inspection of skin-to-core bond
8. Manual inspection of doublers, close-out members, and so on, using acoustic methods such as pulse-echo ultrasonic, Shurtronic Harmonic bond tester, or Fokker bond tester.

Built-in defects were produced in typical laminate and honeycomb specimens. These specimens were evaluated by state-of-the art NDT methods to determine which methods reliably detected the different defects. The results of this investigation are shown in Table 3 for adhesive-bonded laminates and in Table 4 for adhesive-bonded honeycomb

Table 3 Correlation of NDT Results for Built-in Defects in Laminate Panels

In the header below, the three IMMERSION C-SCAN columns are grouped under the heading ULTRASONIC (5).

LAMINATE DEFECTS	NDE METHODS	LOW KV X-RAY	FOKKER BONDTESTER	SONDICATOR	HARMONIC BONDTESTER	210 SONIC BONDTESTER	CONTACT PULSE-ECHO	CONTACT THROUGH-TRANSMISSION	IMMERSION C-SCAN PULSE-ECHO	IMMERSION C-SCAN THROUGH-TRANSMISSION	IMMERSION C-SCAN REFLECTOR PLATE	COIN TAP TEST	NEUTRON RADIOGRAPHY	REMARKS
1. VOID	(1)	D	D	D	D	D	D	D[2]	D	D	D	PD[4]	D	1, 2, 4, 6
2. VOID (C-14 REPAIR)	(1)	D	D	D	D	D	D	D[2]	D	D	D	PD[4]	D	2, 4
3. VOID (9309 REPAIR)	(1)	D	D	D	D	D	D	D[2]	D	D	D	PD[4]	D	2, 4
4. LACK OF BOND (SKIN TO AHDESIVE)	(1)	D	D	D	D	D	D	D[2]	D	D	D	PD[4]	ND	2, 4
5. MFGR'S SEPARATOR SHEET (FM123-41)	(1)	D	D	ND	PD	ND	PD	D[2]	D[3]	D[3]	D[3]	ND	PD[3]	2, 3
6. THICK ADHESIVE (1, 2, 3 PLY)	(1)	D	ND	ND	D	ND	ND	ND[4]	PD[2]	PD[2]	D[2]	ND	ND	2, 4
7. POROUS ADHESIVE	(1)	D	ND	ND	D	D	D	D	D	D	D	ND	D	1
8. BURNED ADHESIVE	(1)	D	ND	ND	D	D	D	PD	D	D	D	ND	ND	7
9. CORRODED JOINT	D	D	PD	PD	D	D	D	PD	D	D	D	PD	D	8

ND – NOT DETECTED   PD – PARTIAL DETECTION   D – DETECTED

1. PANELS WERE MADE USING FM-73 WHICH IS NOT X-RAY OPAQUE. WITH X-RAY OPAQUE ADHESIVE, DEFECTS 1, 2, 3, 5, 6, 7, ARE DETECTED
2. METHOD SUFFERS FROM ULTRASONIC WAVE INTERFERENCE EFFECTS CAUSED BY TAPERED METAL DOUBLES OR VARIATIONS IN ADHESIVE THICKNESS.
3. MANUFACTURER'S SEPARATOR SHEET NOT DETECTABLE BUT DEVELOPED POROSITY AND AN EDGE UNBOND DURING CURE CYCLE WHICH WAS DETECTABLE.
4. MIL-C-88286 (WHITE) EXTERNAL TOPCOAT AND PR1432G (GREEN) PLUS MIL-C-83019 (CLEAR) BILGE TOPCOAT DAMPENED THE PULSE-ECHO RESPONSE.
5. CONTACT SURFACE WAVE WAS TRIED BUT DID NOT DETECT ANY BUILT-IN DEFECTS.
6. MINIMUM DETECTABLE SIZE APPROXIMATELY EQUAL TO SIZE OF PROBE BEING USED.
7. CAUSED BY DRILLING HOLES OR BAND SAWING BONDED JOINTS
8. MOISTURE IN BOND JOINT (ARMCO 252 ADHESIVE, FPL ETCH)

Table 4   Correlation of NDT Results for Built-in Defects in Honeycomb

HONEYCOMB DEFECTS	NDE METHODS LOW Ku X-RAY	FOKKER BONDTESTER	SONDICATOR	HARMONIC BONDTESTER	210 BONDTESTER	ULTRASONIC CONTACT PULSE-ECHO	CONTACT THROUGH-TRANSMISSION	CONTACT SHEAR WAVE	IMMERSION C-SCAN PULSE ECHO	IMMERSION C-SCAN THROUGH-TRANSMISSION	COIN TAP TEST	NEUTRON RADIOGRAPHY	REMARKS
11. VOID (ADHESIVE TO SKIN)	ND[1]	D	D	D	D	D	D	ND	D	D	D	D	REPLACEMENT 1 STANDARD
12. VOID (ADHESIVE TO SKIN) REPAIR WITH C-14													NO VOID IMPROPERLY MADE
13. VOID (ADHESIVE TO SKIN) REPAIR WITH 9309													NO VOID IMPROPERLY MADE
14.* VOID (ADHESIVE TO CORE)	ND	D	—	D	D	ND	D	ND	D	D	D	D	REPLACEMENT STANDARD
15. WATER INTRUSION	D	ND	ND	ND	ND	ND	D	ND	ND	D	D	D	
16. CRUSHED CORE (AFTER BONDING)	D[2]	PD[4]	ND	PD[4]	ND	ND	PD[4]	ND	PD[4]	PD[4]	D	PD	2, 4
17. MFGRS SEPARATOR SHEET (SKIN TO ADHESIVE)	ND[3]	ND	ND	PD	ND	PD	D	ND	D	D	PD	D	3
18. MFGRS SEPARATOR SHEET (ADHESIVE TO CORE)	ND[3]	ND	ND	ND	ND	D	ND	ND	D	PD	D	D	3
19. VOID (FOAM TO CLOSURE)	D	D	D	D	D	D	D	ND	D	D	D	D	
20. INADEQUATE TIE-IN OF FOAM TO CORE	D	D	D	D	D	ND	D	ND	D	ND	D	D	
21. INADEQUATE DEPTH OF FOAM AT CLOSURE	D	D	D	ND	PD[5]	ND	ND	ND	ND	ND	PD	D	
22. CHEM-MILL STEP VOID	ND[3]	ND	ND	ND	ND	ND	ND	ND	ND	ND	ND	D	3

ND — NOT DETECTED   PD — PARTIAL DETECTION   D — DETECTED

1.   PANELS WERE MADE USING FM-73 WHICH IS NOT X-RAY OPAQUE.
2.   THE 0.005-AND 0.010-INCH CRUSHED CORE DETECTED BY STRAIGHT AND BETTER BY ANGLE SHOT.
3.   HAS BEEN DETECTED BY X-RAY WHEN ADHESIVE WAS X-RAY OPAQUE (FM-400).
4.   DETECTS 0.010 CRUSH CORE.
5.   DISCLOSES DEFECT AT A VERY HIGH SENSITIVITY.

structures [6]. The NDT methods described in this section are being
used to test sandwich structures having aluminum or titanium skins
and aluminum core. Most of the methods may be applied effectively to
structures with metal facing sheets and nonmetallic core and structures
with graphite fiber, boron fiber, and fiberglass face sheets bonded to
aluminum core. Only a limited number of the methods are applicable
for structures having both core and facing sheets made from nonmetallic materials.

IV. NDT IN THE PRODUCT CYCLE

A. Quality Zoning and Quality Grades

NDT serves as a basis for process control and to establish quality control standards. The acceptance or rejection of a bonded assembly is
related directly to the quality level desired by the designer. The
quality level, in turn, is based on the importance of the part or component in terms of performance and safety. The frequency/severity
flaw criteria and sensitivity of the test should be based on the desired
quality level. To avoid unwarranted inspection costs, it is advisable
to divide the bonded joint into zones based on stress levels or criticality of the part function. Bonded joints are generally designed to withstand shear loads. Stress analysis of bonded joints reveals that the
outer edges of a bonded joint will be subjected to higher stress levels
than the center of the joint. Therefore, higher quality may be desired
at the edges, and parts should be zoned accordingly. The zoned area
should be dimensioned so that the inspector can define the two different
zones prior to inspection. Figure 36 shows an example of quality zoning

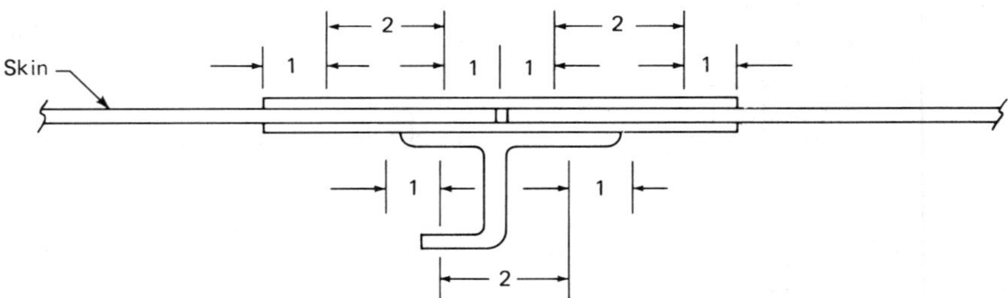

Figure 36   Example of how a bonded joint can be divided into quality
zones which are based on design shear stresses. The example shown
is a longitudinal skin splice with an internal longeron. Quality grade 1
is within 1/2 in. of the edge of any bonded member and grade 2 is the
bond between any two grade 1 bonds. (Courtesy Douglas Aircraft Co.)

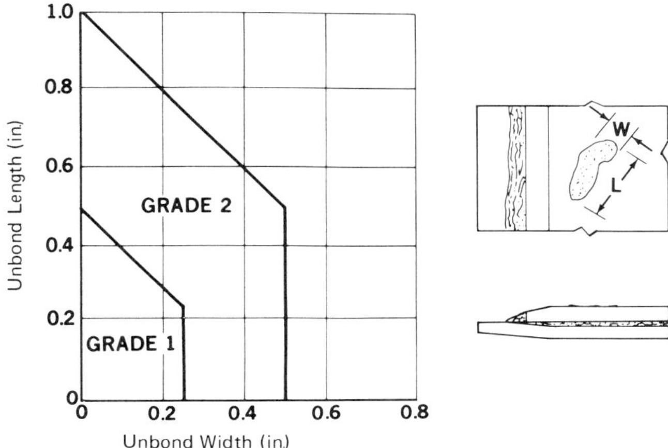

Figure 37   Definition of acceptance grades for voids or unbonds. (Courtesy Douglas Aircraft Co.)

for bonded laminate skin splice. A definition of inspection grade numbers versus allowable void sizes is illustrated in Fig. 37.

For multiple bond line joints, it may not be possible to inspect to a high quality at the edge of all adherends because of the loss of test sensitivity at each successive bond line below the surface. If the edges of a bond joint are stepped or staggered, high-quality inspection is possible at the edges of all bond lines. The width of the step should be large enough to accommodate the test instrument probe. Joint edges that are not stepped should be inspected from opposite sides.

B.   Acceptance/Rejection Criteria

Before inspection can be performed on a bonded structure, accept/reject criteria must be established by engineering. These criteria are usually in the form of frequency (number of flaws per unit area) and/or severity (maximum allowable size flaw data). To avoid unnecessary rejections, engineering must not specify unrealistic or overly conservative criteria. The frequency and severity criteria should be based on sound engineering judgement as related to calculated levels. It should be noted that flaws do not grow under cycle loads if good, durable adhesives are used and generous overlaps exist.

For military contracts, the guidelines or requirements for preparing such criteria are specified in MIL-A-83377 [32]. Specific acceptance criteria for adhesive-bonded aluminum honeycomb structure is specified in MIL-A-83376 [33]. MIL-A-83377 states that the contractor shall prescribe nondestructive tests and a complete NDT process specification when required by MIL-I-6780 [34] to evaluate the quality of adhesive-bonded structures. The NDT process specification must be submitted for approval by the government prior to bonded assembly production. The contractor is required to specify the type and size of defects that are acceptable. The size of the defect must be consistent with the capability of the NDT method to be used, realizing that NDT methods and instrumentation may have different capabilities. The application of the various NDT methods is limited by material types and design configuration. For aluminum honeycomb sandwich, the allowables of MIL-A-83376 may be used, if determined by the contractor to be acceptable minimums for the particular application.

MIL-A-83377 also requires that NDT of adhesive-bonded structures be performed only by qualified NDT personnel. Personnel qualification requirements for NDT are specified in MIL-STD-410 [35].

C. Records and Specifications

It is recommended that the four basic inspection records listed be prepared and maintained:

Specifications

Test method specifications should be obtained for each production inspection instrument to be used (i.e., Fokker bond tester [5], 210 bond tester, Harmonic bond tester, Sondicator, etc.). Each specification should specify controls for the particular method and define reference standard configurations. In addition, an acceptance criteria specification should be prepared.

Inspection Log

The production inspection personnel should maintain a daily inspection log. This log should contain the part number, part dash number, part serial number, fabrication outline serial number, inspector name or stamp, date and method of inspection, and a remarks column to indicate pass or reject. The rejection tag number can be entered as well as a discription of the flaw, noting its exact location.

Detailed Written Test Procedure

NDT inspections should follow a detailed written procedure for each component tested. The written procedure must comply with the test method specification requirements and, as a minumum, include:

1. A sketch of the part or configuration showing in detail part thickness, alloy and temper, and adhesive system used.
2. Manufacturer and model number of instrumentation to be used.

3. Model or designation number of probe to be used.
4. Any recording, alarm, or monitoring equipment and test fixtures or inspection aids that are to be used.
5. Applicable reference standard(s) serial number and design description.
6. Type of fluid couplant that is to be used.
7. Instrument calibration procedure, if required.
8. Testing plan: the part surface from which the test will be performed, the testing technique(s) for each joint, and the maximum probe index or shift per scan should be described.
9. Acceptance limits for each joint of the assembly.
10. Reference point or identifying mark, which is placed on each part prior to inspection. This mark is used for orientation when reviewing inspection results.
11. Outline of the areas where multiple layers of adhesive have been used.
12. Identification of section area changes not visible to the inspector.
13. Verification of the test procedure on a production part prior to approval.

Rejection or Nonconformance Record

All nonconforming flaws should be described on a rejection tag. The inspection agency should outline on the part surface the location and shape of unacceptable bond line flaws. If it is necessary to mark the part permanently, the inspector must remove the couplant and wipe with solvent the area to be marked, and then reapply the marks. Then a very light coat of clear epoxy enamel can be used to overcoat the markings. This procedure is very helpful in looking for problem areas after a part has been in service.

D. Reference Standards

Reference test standards to calibrate and standardize the bond inspection instruments are very essential. Even with adequate reference standards, inspection often produces conflicting information or a deviation of results. In these instances, more than one inspection method should be used to improve the accuracy of the data and to define the deviating condition. In ideal conditions, the test standard with built-in defects of selected sizes should closely duplicate the structure adherends to be inspected. The laminate skin should be of the same thickness and the honeycomb core should be of the same size and density. Other variations to be encountered in the test specimen, such as tapered core, chemically milled skins and doublers, and so on, should be incorporated in the test standard. Except for built-in defects, the test standard for adhesive-bonded structures should be

fabricated in the same manner as the production assembly. The
void or defect should be introduced in the same bond line as that to
be inspected in the structure. The standard may be a series of
simple test specimens, composed of details identical to the several
areas of the assembly to be inspected, having a bond of good quality
but with controlled or known defect locations and size.

Attempts to produce voids or controlled understrength bonds by
local application of grease or other foreign materials have been found
to be ineffective. All adhesives and primers vary and may not respond to the following suggested methods of standard preparation.
Therefore, the finished standards should be evaluated by various
NDT methods prior to validating their use as an inspection standard.
Neutron radiography is an excellent method for verifying if voids
and/or porosity have been generated in the standard. The standard
must then be inspected by the test instrument chosen for the particular part to confirm its ability to locate the flaw.

Metal-To-Metal Reference Standards

Various methods have been developed for producing simulated
voids, porosity, or unbonds in metal-to-metal bonded laminates.
Hagemaier and Fassbender [10] have described numerous ways to
produce reference standards for metal-to-metal bonded joints or assemblies. Because of the variety of such standards, it is not possible to describe all of them in this chapter, and it is recommended
that the reader consult the references for additional information.
Fig. 38 shows two methods for producing standards for simulated
skin-to-adhesive unbonds. The method 1 or 2 standards are useful
for all NDT application methods provided the Teflon or Mylar inserts
do not bond to the adherends. Voids or unbonds of specified sizes
are difficult to produce because free-flowing adhesive may fill the
anticipated void or unbond area. Clark [1] was successful in producing natural flaws in the adhesive layer of metal-to-metal bond
joints. Figure 39 is a positive reproduction from an x-ray radiograph of AF-55 adhesive showing a void produced by removing a
section of adhesive, a void produced by inserting a 0.025-in.-diameter
wire between adherends, and porosity caused by inserting a 0.015-in.-
diameter wire between adherends prior to bonding. By trail-and-
error methods, these defect conditions can be produced at specific
locations in a bonded joint and also with desired sizes.

The methods 3, 4, and 5 standards (Fig. 40) are useful for manual bond test methods, such as Fokker, Harmonic, and 210 bond test
applications, and pulse-echo ultrasonics. The holes at the edge and
center of the method 3 standard are required if pinch-off of adhesive
occurs near the edge with thicker adhesive near the center of the
standard. For maximum sensitivity, a bond joint should be inspected

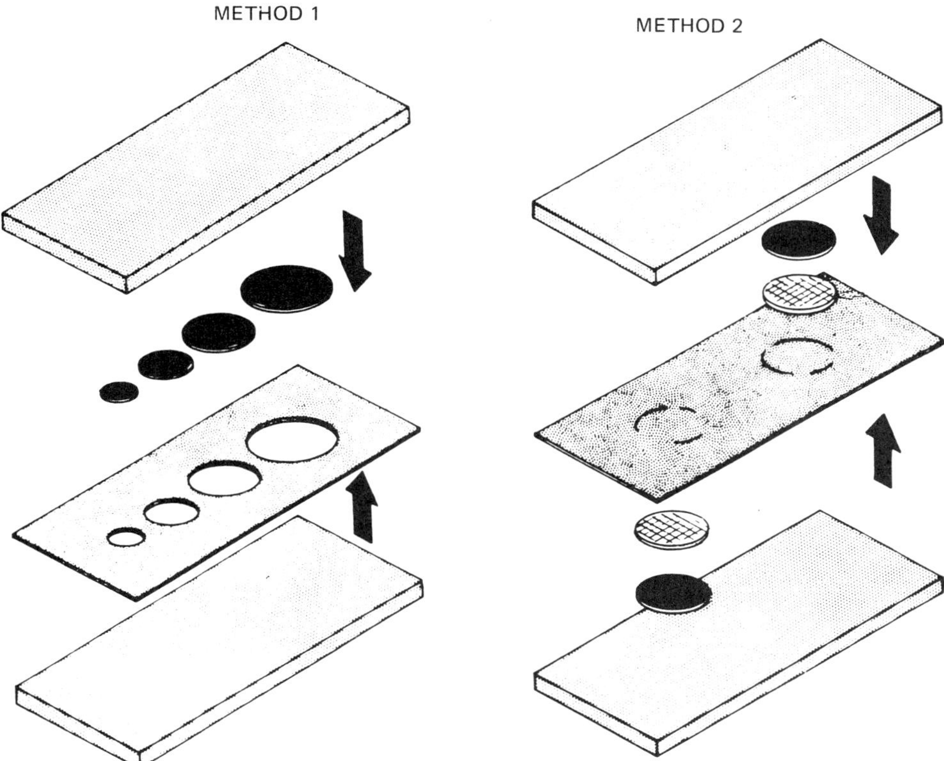

Figure 38   Two methods of fabricating unbond reference standards. Method 1 uses two pieces of aluminum sheet that have been anodized and primed and then bonded together with a piece of adhesive which has various-size cutouts filled with 0.005 in. thick Teflon disks filling the holes. Method 2 uses metal as in method 1, but the voids are produced by placing a disk sandwich of Mylar and scrim cloth of desired size on uncured adhesive. The assembly is then bonded. (Courtesy Douglas Aircraft Co.)

from both sides and method 3 and 4 standards are used. If the joint can be inspected from only one side, the method 5 standard is used. Holes are drilled after part is bonded.

Standards containing porosity can be made by the wire insertion method shown in Fig. 39 or by bonding a large-area panel (2 ft × 2 ft minimum) at low pressure (40 psi maximum). When bonding with a mat (nonwoven) carrier the adhesive pinches off at the outer edge,

producing a picture frame of gross porosity close to the edge. Scattered porosity and some small voids are produced throughout the remainder of the panel. After ultrasonic C-scan or radiographic inspection, representative specimens or small standards may be removed
from the panel. The methods for producing voids and porosity
(shown in Fig. 39) are useful for creating defects at specific locations
in slow-cycle fatigue test specimens. This type of specimen is used
to study the effects of defects under load, cycle rate, or environment.

Method 6 (Fig. 41) is used to fabricate a standard representing
burned adhesive caused by improper drilling or sawing of bonded
panels. If improper drills, high drill speed and feed, or no coolant
is used, burning of the adhesive adjacent to the hole may occur.

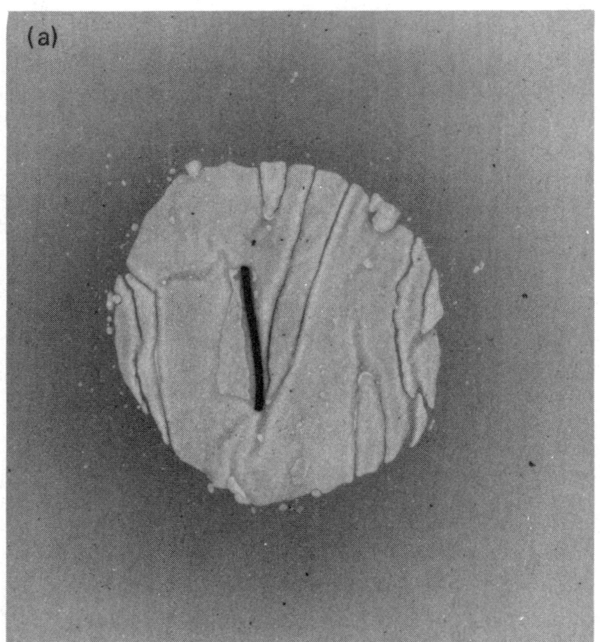

Figure 39   Additional ways of producing natural flaws in bonded
laminates: (a) a void produced by inserting 0.025-in.-diameter
wire between adherends before bonding; (b) example of porosity produced by inserting a 0.015-in.-diameter wire; (c) void produced by
removal of adhesive prior to bonding. (Courtesy Douglas Aircraft Co.)

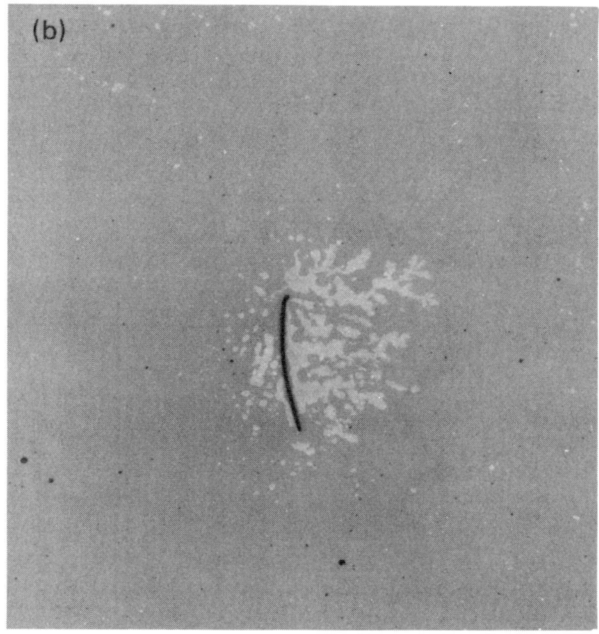

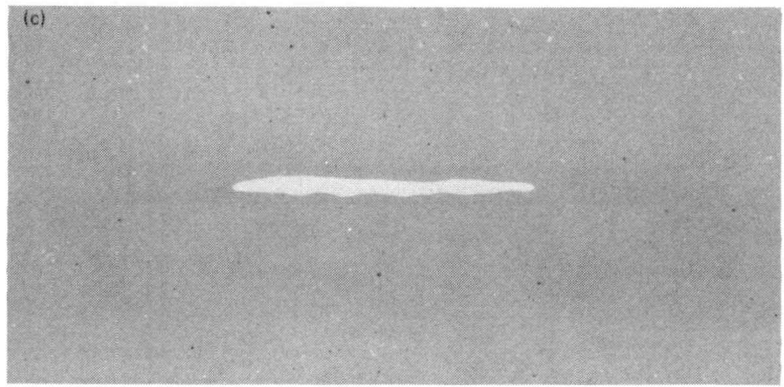

Figure 39 (continued)

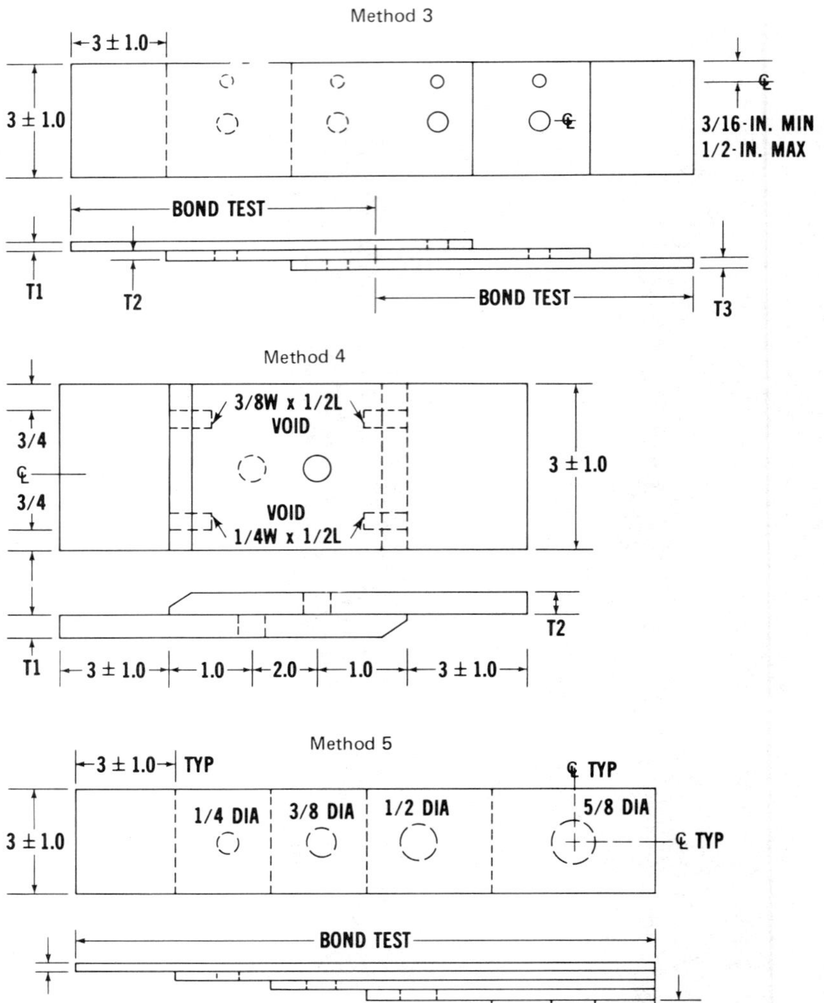

Figure 40   Additional methods for fabricating unbond reference stand-ards.   Method 3 uses three flat unchamfered aluminum sheets of differ-ent thicknesses and can be inspected from either side.   Hole diameters can vary from 1/4 to 1/2 in. and are free of adhesive.   Method 4 uses two chamfered aluminum sheets of different thicknesses with drilled holes 3/8 or 1/2 in. in diameter (free of adhesive) and edge voids made by inserting Teflon shins of the size shown.   Shins must be re-moved after bonding.   Inspect from either side.   Method 5 is a flat step standard using a buildup of various sheet thicknesses which match structure to be inspected.   No adhesive should be at the base of the holes.   Inspect from the top side only.   All dimensions are in inches.   (Courtesy Douglas Aircraft Co.)

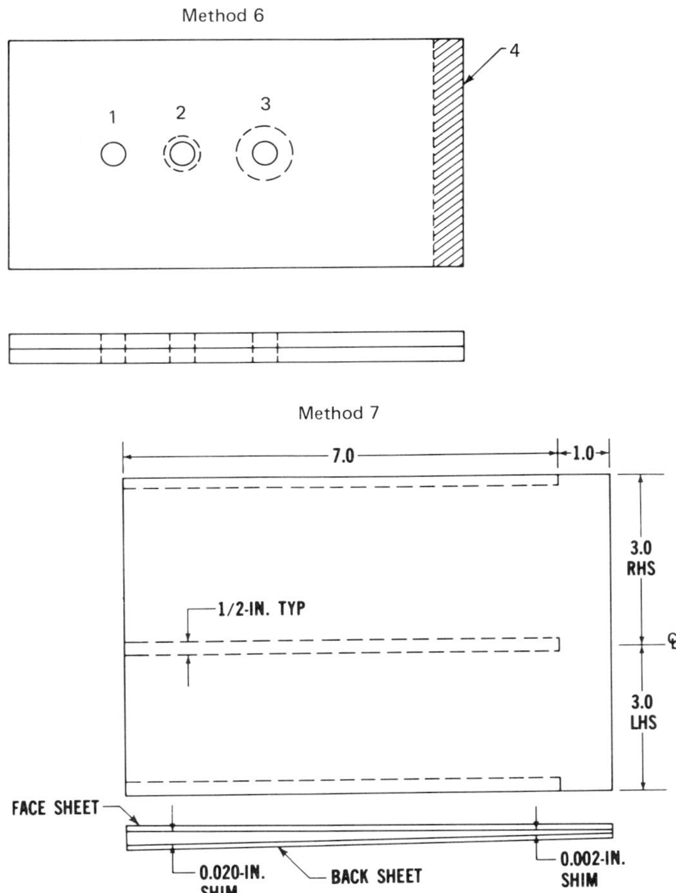

Figure 41   Additional methods for fabrication unbond reference standards.   Method 6 presents a burned-adhesive standard where (1) is a good-quality hole, (2) slightly burned adhesive, (3) grossly burned adhesive, and (4) a grossly burned edge caused by abusive band sawing.   Method 7 gets a variable quality standard by inserting tapered shins between face and back sheet before bonding.   Sketch shows three 1/2-in.-wide shins in place.   Tapered shins cause increasing porosity going from right to left.   One-half of the face sheet can be removed after inspection to reveal condition of adhesive. All dimensions are in inches.   (Courtesy Douglas Aircraft Co.)

Method 7 (Fig. 41) is a way to fabricate a variable-quality standard.
By using tapered shims, good quality is obtained at the thin end and
poor quality at the thick end. Figure 42 shows a positive reproduction made from an x-ray negative of AF-55 adhesive in a tapered
standard. For visual correlation, only the right-hand upper skin
need be removed from the panel, thereby revealing the variable adhesive quality. The left-hand side is kept intact to determine variations in NDT instrument response to different adhesive quality conditions.

When the size of the unbond or void is not of primary concern,
or to calibrate instruments to a bond/no-bond response, the following standards prove adequate. These standards are especially useful
for resonant-type instruments such as the Fokker, Harmonic, Sondicator, and 210 bond testers. They relate well to the bonded part when
a 250 to 350°F curing epoxy or epoxy phenolic resin system, which
does not attenuate the ultrasonic energy, is used. Basically, the

Figure 42   Positive print from x-ray negative of tapered shin variable
quality standard shown as method 7 in Fig. 41. (Courtesy Douglas
Aircraft Co.)

standards are made by cutting 1.5-in.-diameter holes in a 1/2-in.-thick plywood panel and then bonding 3 in. × 3 in. aluminum sheets, using room-temperature curing adhesives, so that the center of the sheet is centered over the holes in the plywood. The thickness of the aluminum sheets is chosen to match the thickness of the aluminum adherends of the test part. The number of sheets is governed by the number of adherends in the test part. When the aluminum sheet is bonded to a wooden base, it does not resonate freely and reacts similarly to unbonds in the part being inspected.

Honeycomb Reference Standards

To create a void or unbond reference standard for honeycomb, use method 1 or method 2 as shown in Fig. 38. For honeycomb standards, the unbonds are created between the adhesive-to-skin and adhesive-to-core interfaces. Skin-to-adhesive voids or unbonds are usually easier to detect than are adhesive-to-core unbonds. Another way to prepare a void or unbond standard for a honeycomb panel is illustrated in Fig. 43. This standard is useful when attempting to detect unbonds on both sides of the panel when inspection can be performed from only one side.

Honeycomb crushed core standards are made by inserting 1-in.-diameter, 0.005- and 0.010-in.-thick metal disks between the skin and core and applying pressure. The disks are removed and the panel is bonded. The 0.005-in. crushing is very subtle and difficult to detect radiographically. The 0.10-in. crushing is easily detectable. Water intrusion standards are produced by hot-bonding one skin to the core, adding various quantities of water into the cells, and then cold-bonding the second facing sheet in place. Core splice standards are made by cutting the edges of the core with a taper and then applying foaming adhesive to one side of the splice before bonding. Voids in the foaming adhesive are detectable by low-kilovoltage radiography. Core-to-closure voids are made by eliminating foaming adhesive at small areas along the joint. Separator sheet standards are produced by bonding-in a desired size of the material at the skin-to-adhesive and adhesive-to-core interfaces. A blown core standard may be made by mechanically damaging the core prior to bonding. Corroded core or core-to-adhesive corrosion standards may be made by bonding only one skin to the core (open-face honeycomb) and the alternately spraying the honeycomb side in a salt spray system. The degree of corrosion may be monitored visually. If it is desired to close up the panel, cold-bond the second facing sheet to the other side of the core.

Substitute Standards

When the ideal test standards are not available, other standards of construction similar to that of the test part may be used. However,

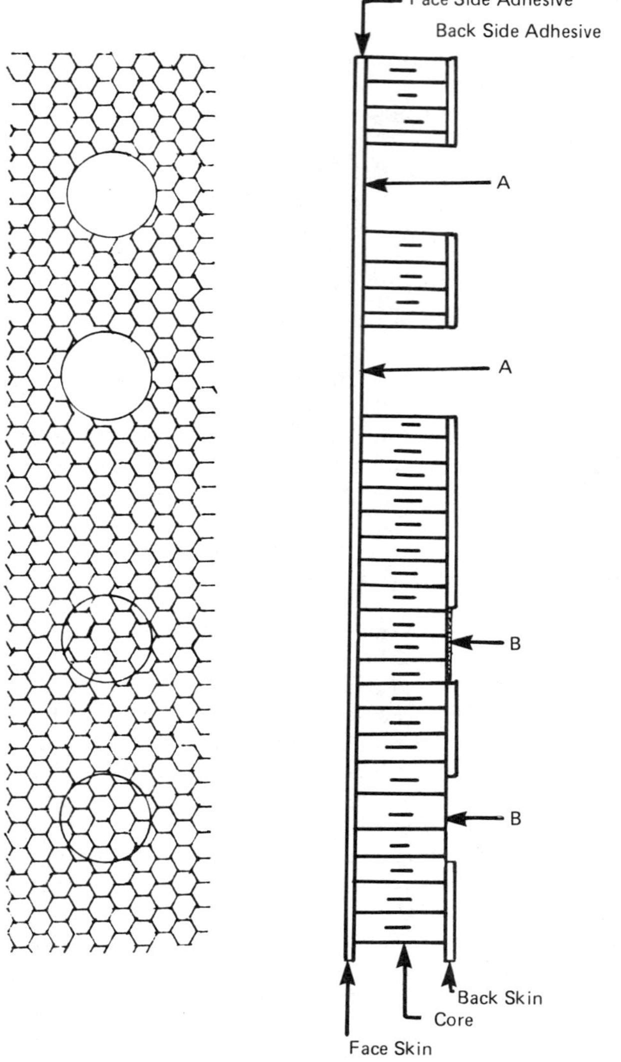

Figure 43   Construction details for making honeycomb standard for voids and unbonds. (a) Back skin and core cut out to desired diameter. Cutting through face sheet adhesive shows skin-to-adhesive unbond, cutting to the adhesive shows adhesive-to-core unbond. (b) Cutting through back skin first to the adhesive to show skin to adhesive unbond and then through the adhesive to show adhesive-to-core unbond. Note that Teflon disks may be used to help separate adhesive from skins at cutouts. All tests are to be made from the face skin side. (Courtesy Douglas Aircraft Co.)

honeycomb standards should not be used for metal-to-metal laminate structure, and vice versa.

Inspection Without Standards

When standards are not available, a known undamaged area can be used as the standard. To inspect repaired areas, compare them to an unrepaired area. The instrument reading may change due to structural changes resulting from the repair. Repeated inspection scans must be conducted under various instrument settings and the inspection should be verified by other methods and/or instruments. A knowledge of the part configuration (i.e., number of bond lines, adhesive thickness, skin thickness, etc.) is essential for interpretation of the test results.

V. EVALUATION AND CORRELATION OF INSPECTION RESULTS

Experience, confidence, and effectiveness of NDI technicians and engineering personnel is gained by correlating NDI results with destructive testing. Comparisons may be made between the NDI results and the actual size, shape, location, and type of defect. Mistakes will occur in the early stages of the bonding program, but if accurate records, photographs, and sketches are kept, they can serve as effective training devices. Sections of defective parts should be kept for training purposes. All of these items add to the confidence of the inspector in evaluating and accurately reporting defects. They also add to the engineer's confidence in the inspector and in the data reported by the inspector.

A. Procedure

Before new adhesive-bonded assemblies are fabricated on a mass production basis, the first assembly is evaluated by numerous methods. Inspections are performed and NDT results are evaluated to detect any variations in the part which are indicative of discontinuities. The part is then cut up and special precautions are taken to section through the discontinuities or to separate the joint at the discontinuity to identify and verify the NDI results. In order to determine which NDT method(s) detected the discontinuity, the NDI technicians use different color markings (i.e., blue for x-ray, green for ultrasonic, red for model X bond tester, etc.). These markings may be affixed to the surface of the part or on a transparent overlay of the part. When the part is sectioned, the results will identify which NDT method(s) correctly identified the location, size, shape, and type of discontinuity. Although it is expensive to cut up adhesive-bonded assemblies, sometimes it is the only way

to determine the nature of the discontinuity or to verify its existence.
All inspection results should be carefully documented for future reference.

Following the establishment of a highly confident inspection technique, remaining defective assemblies can be used to develop repair
procedures. The original NDI techniques can then be used to determine if the repairs are effective. In some instances defective parts
are consistently produced and the materials, production process, or
design must be changed to correct the condition. This condition is
usually evident when the same type of defect occurs in the same location in all manufactured parts.

Reliable feedback from manufacturing personnel is beneficial to
the NDI technician. This information can include such items as: the
oven temperature was too high or too low, the heat-up rate was incorrect; there were leaks in the vacuum bag; additional layers of adhesive
were added at a particular spot in the assembly, and so on. With this
information available, the NDI technician will be better prepared to inspect a particular group of parts.

B. Destructive Correlations of NDI Results

Following is a typical example of correlating NDI results by destructive
testing. In this example, a metal-to-metal joint containing a bonded
doubler and longerons and frames was rejected for numerous voids and
porosity. The initial inspection was performed using the Fokker bond
tester and the defective areas were marked on the surface of the part
as shown in Fig. 44. The inspection results were verified using the
210 bond tester. The part was then sent out for thermal neutron radiography. There was excellent correlation between the Fokker bond test
results and the neutrograph (Fig. 44). The panel was then ultrasonically C-scanned using the through-transmission technique. Again, the
correlation was excellent, as shown in Fig. 45. A plastic overlay was
made to record the Fokker bond tester results. The part was then subjected to chemical milling to disolve away the aluminum adherends, leaving the adhesive layer intact. The adhesive thickness was measured
and recorded. The destructive to nondestructive correlation was again
excellent, as shown in Fig. 46. This figure reveals that the adhesive
thickness in and around the defects is greater (0.013 to 0.018 in.) than
it is in the nondefective areas (0.007 to 0.011 in.). In addition to the
correlations noted above, lap-shear specimens could have been made
through the different defect conditions and variations in bond strength
could have been determined.

Because of the variety of defects that can occur in honeycomb assemblies, defects detected by one test method should be verified by an
additional, but similar test method. Time and facilities permitting, the
assembly should also be x-rayed in the defective area.

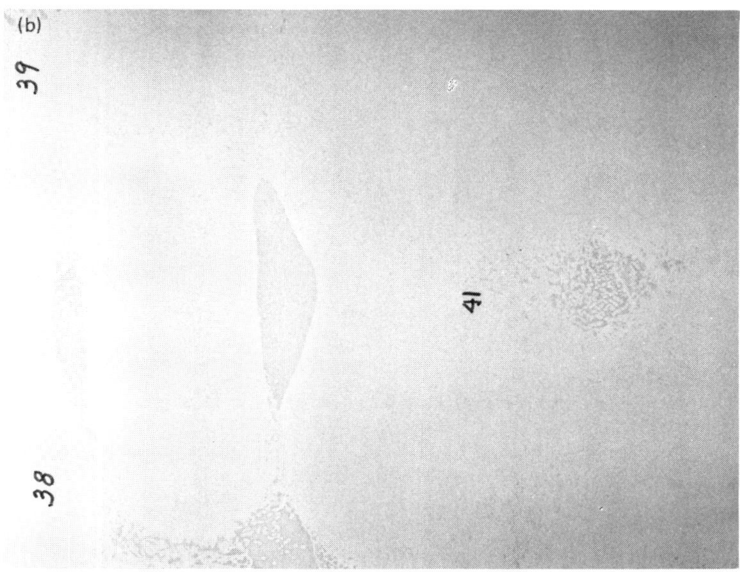

Figure 44   Correlation of NDI results.   The same specimen was inspected by (a) Fokker bond tester and (b) neutron radiograph.

Figure 45   Ultrasonic C-scan recording. (Courtesy Douglas Aircraft Co.)

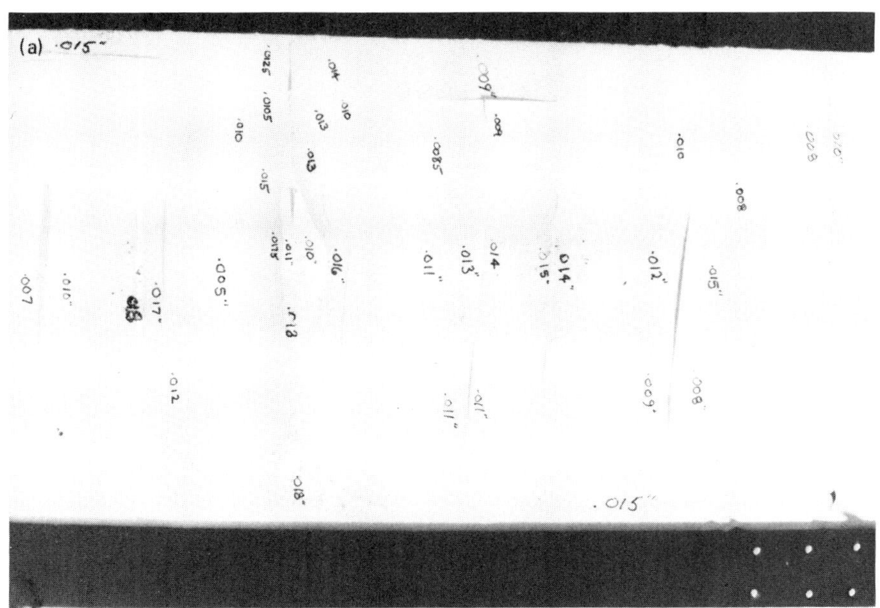

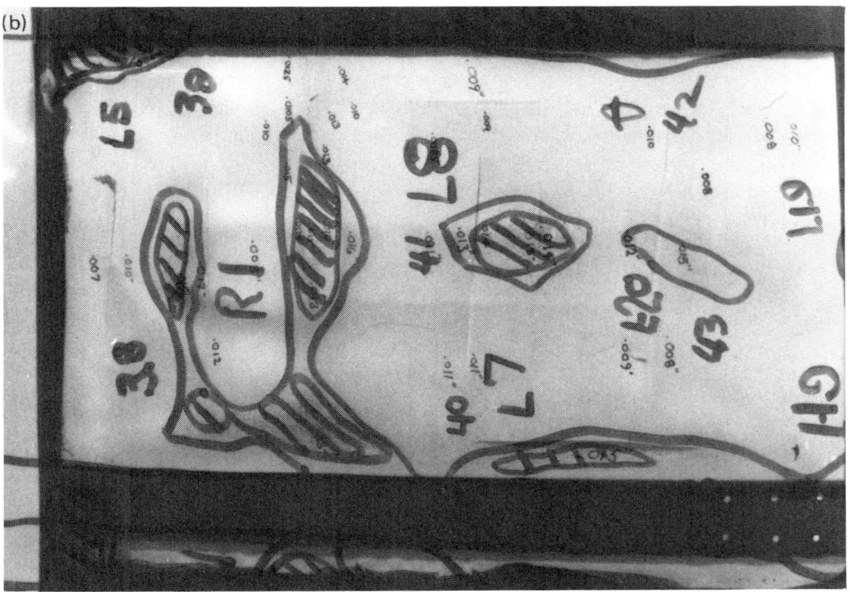

Figure 46   Correlation of NDT to destructive test results: (a) measured thickness of adhesive after the adherends were removed by chemical milling; (b) how plastic overlay of Fokker bond tester results fit measured thicknesses.

C. Ultrasonic Wave Interference Effects

When performing ultrasonic inspection of adhesive-bonded joints, the
following two conditions may cause ultrasonic wave interference effects:
(1) when inspecting bond lines under tapered metal doublers; and (2)
when inspecting close to the edge of a bonded laminate and adhesive is
thin due to pinch-off [10]. Ultrasonic theory predicts that constructive
interference (increased amplitude) occurs when the transmitted and
reflective waves are in phase with one another [36,37]. Conversely,
destructive interference (decreased amplitude) occurs when the transmitted ad reflected waves are 180° out of phase with one another. Destructive wave interference causes a drop in signal amplitude and the
C-scan recorder will not record in the area of interference. These
areas can be interpreted as voids or unbonds to an inexperienced operator. However, interference effects (caused by tapered adherends) are
uniform across the part, resulting in bands of light and dark on the
C-scan recording [10]. When destructive interference effects are
shown in C-scan recordings of bonded splices having tapered adherends,
the bond joints are given another inspection by a different method, such
as the Fokker bond tester.

Destructive wave interference effects may also be seen in C-scan
recordings of adhesive-bonded laminates made with an adhesive having
a "mate" carrier. It is less evident in laminates bonded with adhesive
having a "woven" carrier. The destructive interference effects are most
evident near the outer edges of the assembly where the adhesive is
pinched off to thicknesses of 0.003 in. or less. The interference effects
are frequency dependent and hence will change position if the test frequency is altered [36,37]. Less wave interference is experienced using
through-transmission testing than using pulse-echo testing at similar
frequencies. Bonded laminates exhibiting destructive wave interference
effects require inspection by another test method in the area in question
to verify bond quality.

D. Correlation of Bond Strength and
Fokker Bond Tester Readings

The ability of the Fokker bond tester to measure bond strength is completely dependent on the accuracy of the correlation curves, which plot
Fokker readings versus either lap-shear strength (for metal-to-metal
joints) or flatwise tensile strength (for honeycomb joints). For metal-
to-metal joints, the correlation is made between the Fokker A-scale
readings and tensile lap-shear strength. The correlation is made between Fokker B-scale readings and flatwise tensile strength for honeycomb structure.

Flatwise tensile specimens are preapred in accordance with MIL-STD-
401 [38], except that the adherends and core should stimulate those of
the bonded test assembly being tested. Fabrication procedures should

be identical with those used in preparing the assembly to be tested. Specimens are prepared with various degrees of bond quality by the introduction of porosity in the cured adhesive. The degree of porosity increases as the thickness of the adhesive layer is increased. Figure 47 illustrates the procedure for honeycomb-to-skin joints.

Nondestructive testing of the panels is performed in accordance with procedures outlined in MIL-STD-860 [23], or as specified in the Fokker bond tester training manual. Instrument readings should be taken within predetermined areas of the test specimen. Each reading should be recorded and correlated with the probe location on the specimen. The specimens are then destructively tested and tensile strength values recorded. Flatwise tensile strength should be plotted versus average ultrasonic readings for each specimen and a representative curve drawn through the plotted points. Two correlation curves should be drawn for specimens with dissimilar thickness adherends. A typical correlation curve for Fokker B-scale readings versus flatwise tensile strength is shown in Fig. 48.

A relationship can be established between Fokker bond tester A-scale frequency shift and lap-shear strength as related to porosity. Tests have shown that the percent porosity increases with increased adhesive thickness (single layer) [2,10]. Porosity first appears at an adhesive thickness of 0.010 in. and becomes almost completely void at 0.020 in. thickness.

The first step is to produce high-quality lap joints (1.0 in. wide by 1/2 in. overlap) containing 100%, 80%, and 50% width bond lines, as shown in Fig. 49. The test specimens are approximately 7 to 8 in.

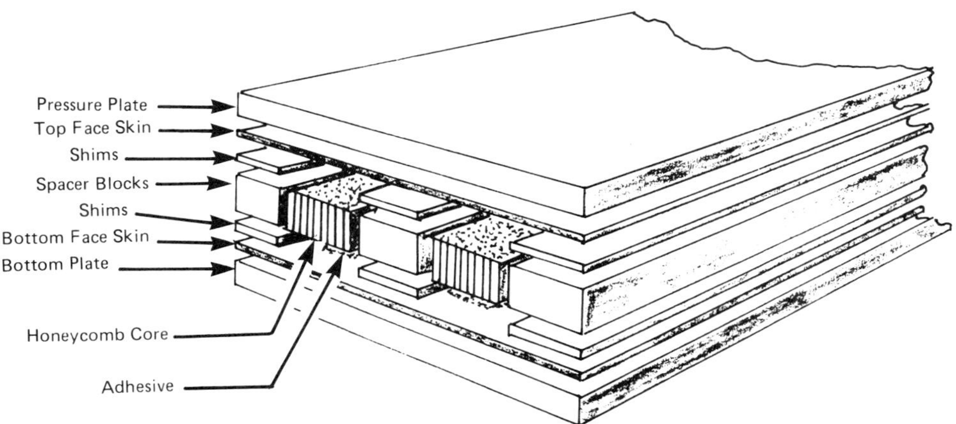

Figure 47   Honeycomb sandwich test panel. (Courtesy Fokker-VFW.)

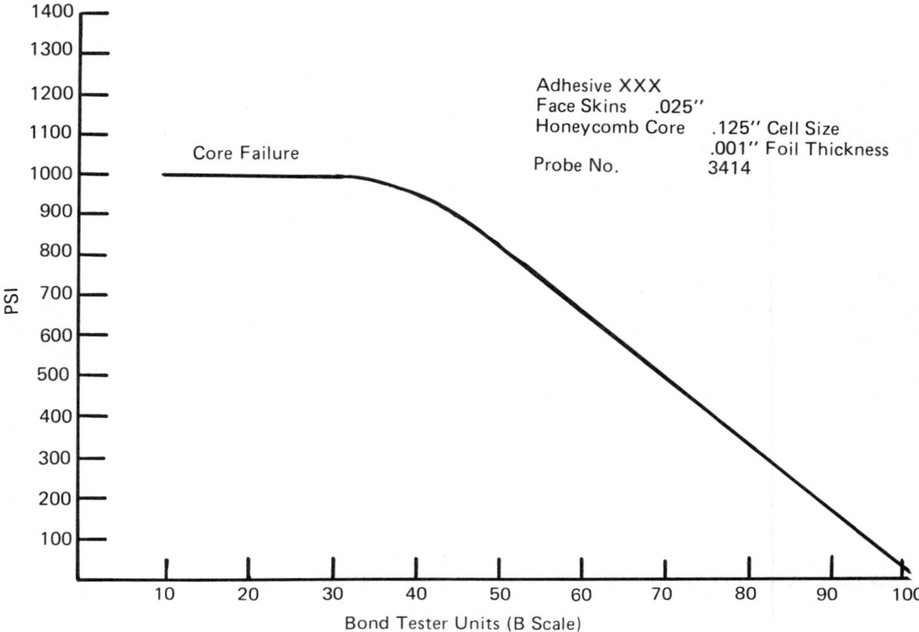

Figure 48   Typical correlation curve—honeycomb flatwise tensile. (Fokker–VFW.) (Courtesy Fokker–VFW.)

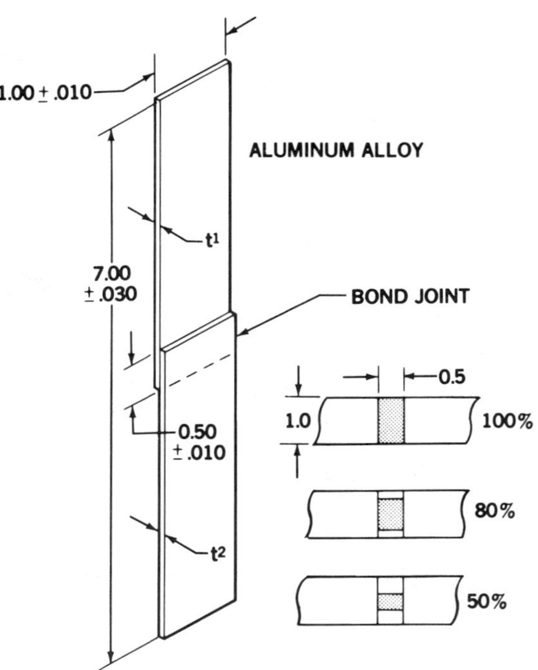

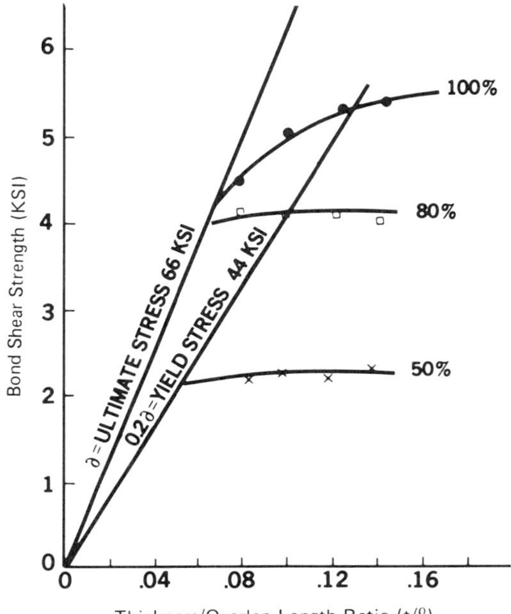

Figure 50   Variation in lap shear strength with t/ℓ ratio and bond length using alloy 2024-T3 and adhesive FM-73. t_1/t_2 (in.): 0.040/ 0.040; 0.050/0.050; 0.063/0.063; 0.071/0.071.

long, and a minimum of six specimens should be made for each percentage bond and thickness (i.e., 0.040/0.040, etc.). Typical average lap shear values for 2024-T3 and FM-73 adhesive are shown in Fig. 50.

The second step is to make specimens with increasing thickness of the single adhesive layer. This will produce specimens of variable quality resulting from porosity. To obtain the quality variation, spacer shims are used, as shown in Fig. 51. The thickness of the spacers is increased in steps (0.004 in. per step) from 0.004 in. up to 0.020 in., resulting in a highly porous bond joint. The lap shear specimens are numbered and neutron radiographed. The specimens are then Fokker bond tested and the A-scale readings are recorded for each specimen. The lap shear coupons are cut from each panel and the adhesive thickness (for each specimen) is accurately measured. The actual thickness

---

Figure 49   Lap shear test coupon design. Dimensions in inches. (Courtesy Douglas Aircraft Co.)

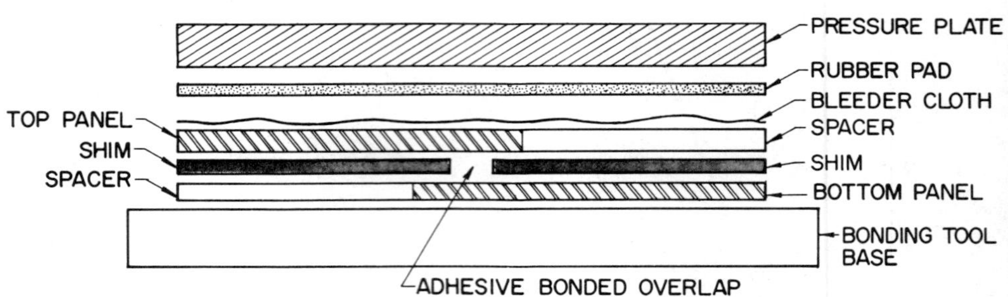

Figure 51   Metal-to-metal lap shear test panel. (Coutesy Fokker—VFW.)

of the adhesive will generally be found to be slightly thicker than that of the associated shim. The lap shear test is performed in accordance with Federal Test Method Standard No. 175, method 1033, or Federal Specification MMM-A-132. The neutron radiographs indicate areas of nonuniformity of porosity and results from these nonuniform areas should be discarded.

The third step is to prepare shimmed specimens using additional layers of adhesive. This will produce good-quality specimens of thick adhesive. The panels are neutron radiographed, Fokker bond tested, cut into specimens, and lap shear tested, with the data subsequently evaluated. Plots of lap shear strength versus adhesive thickness for the single porous layer and multiple nonporous layers of FM-73 are shown in Fig. 52.

The fourth step is to plot lap shear strength versus Fokker A-scale readings for each adherend thickness (Fig. 53). If the thickness of the adherends is not equal, the back (unprobed) thickness is plotted. Unequal thickness adherend specimens are tested from both sides and two quality curves are established. By referring to the previous figures, the bond strength can be determined for 100%, 80%, and 50% specimens having various thickness or t/ℓ ratios. By evaluating all the data, the following classes for single-layer (porous) and multiple-layer (nonporous) joints can be specified:

Class A: greater than 80%
Class A/B: greater than 65%
Class B: greater than 50%
Class C: greater than 25%

The large scatter in results for the 25% quality level specimens can be attributed to the extensive porosity in these specimens. This does

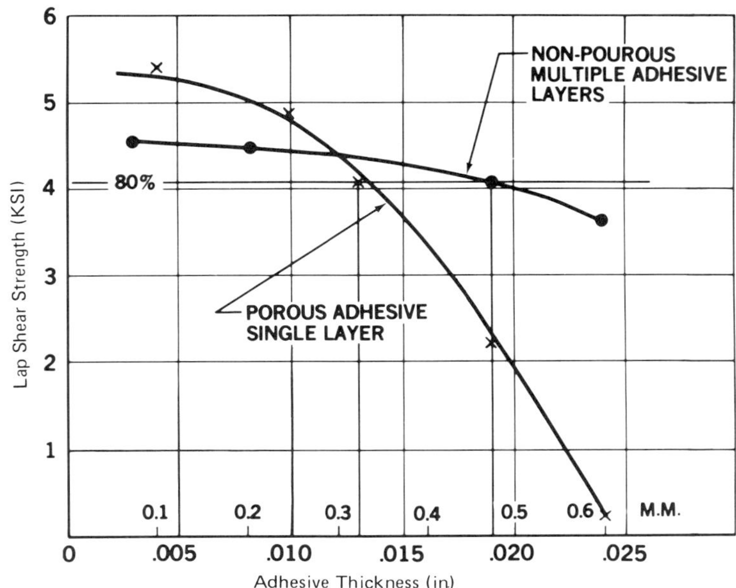

Figure 52   Lap shear strength versus adhesive thickness for porous and nonporous adhesive.   Adherends, 0.063/0.063; adhesive FM-73 with mat carrier; primer, BR127.   (Courtesy Douglas Aircraft Co.)

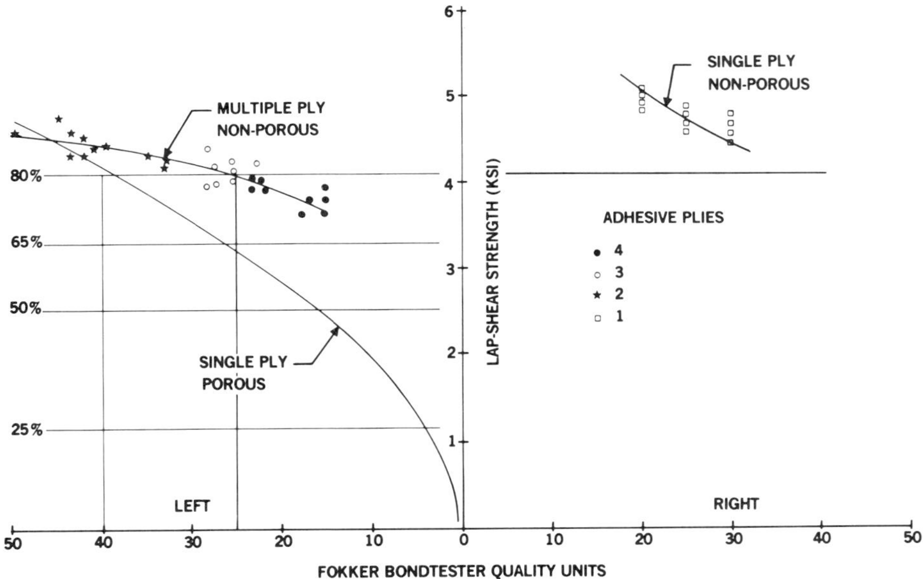

Figure 53   Lap shear strength versus Fokker bond tester quality units. Bond tester Model 70; probe, 3814; adherends, 0.063/0.063; adhesive FM-73; primer, BR127.   (Courtesy Douglas Aircraft Co.)

Table 5   Fokker Cohesive Bond Strength Acceptance Limits—Single Bond Line (FM-73)

Lower sheet thickness (in.)	Model 70 probe	Upper limit	Lower limits[a]			
			Class A 80%	Class A/B 65%	Class B 50%	Class C 25%
0.040	3814	>R8	>L54	>L40	>L25	>L10
0.050	3814	>R10	>L48	>L36	>L20	>L9
0.063	3814	>R12	>L42	>L30	>L18	>L8
0.071	3814	>R16	>L40	>L28	>L18	>L7
0.081	3814	>R18	>L38	>L25	>L15	>L6

[a]Skin/stringer and skin/frame bonds shall be Class minimum; doubler/skin bonds shall be Class A/B minimum.   Additional adhesive layers shall not yield readings less than the lower limits by the following factors: 2 layers (−15); 3 layers (−25); 4 layers (−30).

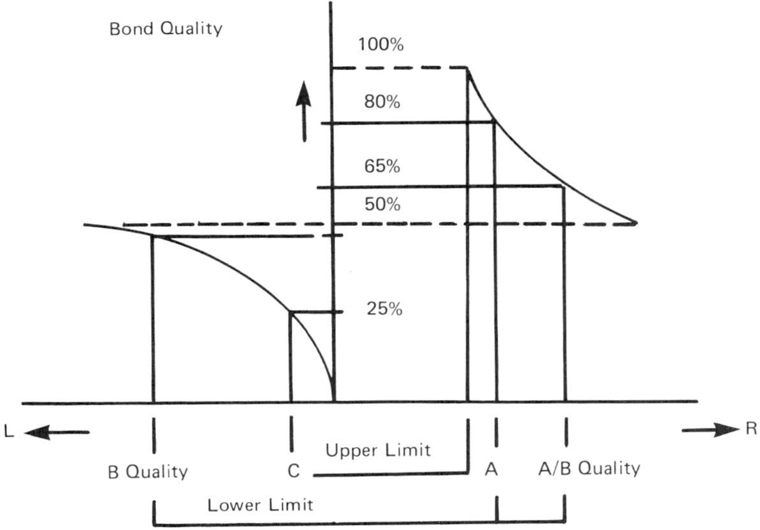

Figure 54   Establishment of acceptance limits (sheet to sheet). (Courtesy Douglas Aircraft Co.)

not pose a problem to the inspector since quality levels below 50% are not usually acceptable. The actual relationship between the A-scale frequency shift (as seen by the inspector) and the bond quality (class) is illustrated in Fig. 54.

Finally, the acceptance limits are incorporated into a table for use by the inspector. Table 5 lists the acceptance limits for single bond line joints with FM-73 adhesive and specific thickness adherends. Establishing Fokker quality diagrams for multiple bond line joints is possible but complicated. The reader should contact Fokker-VFW in the Netherlands to obtain the recommended procedures for developing such quality diagrams.

REFERENCES

1. Clark, M. T., Definition and Non-destructive Detection of Critical Adhesive Bond-Line Flaws, U.S. Air Force Materials Lab. Tech. Rep. AFML-TR-78-108, 1978.
2. Rienks, K. J., The Resonance/Impedance and the Volta Potential Methods for the Non-destructive Testing of Bonded Joints, 8th World Conf. Nondestructive Test., Cannes, France, 1976.

3. Schliekelmann, R. J., Non-destructive Testing of Adhesive
    Bonded Metal-to-Metal Joints, *Non-Destructive Test.*, Apr. 1972.
4. Schliekelmann, R. J., Non-destructive Testing of Bonded
    Joints—Recent Developments in Testing Systems, *Non-destructive
    Test.*, Apr. 1975.
5. Bijlmer, P. and R. J. Schliekelmann, The Relation of Surface
    Condition After Pretreatment to Bondability of Aluminum Alloys,
    SAMPE Q., Oct. 1973.
6. Couchman, J. C., B. G. W. Yee, and F. M. Chang, Adhesive
    Bond Strength Classifier, *Mater. Eval.* Vol. 37, No. 5, Apr.
    1979.
7. Rodgers, J. and S. Moore, *Applications of Acoustic Emission to
    Sandwich Structures*, Acoustic Emission Technology Corp.,
    Sacramento, Calif., 1980.
8. Rodgers, J. and S. Moore, The Use of Acoustic Emission for
    Detection of Active Corrosion and Degraded Adhesive Bonding
    in Aircraft Structures, Sacramento Air Logistics Center (SM/
    ALC/MMET), McClellan AFB, Calif., 1980.
9. The Sign of a Good Panel Is Silence, *Aviation Eng. Maint.*, Vol.
    3, No. 4, Apr. 1979.
10. Hagemaier, D. and R. Fassbender, Nondestructive Testing of
    Adhesive Bonded Structures, *SAMPE Q.*, Vol. 9, No. 4, 1978.
11. Botsco, R., The Eddy-Sonic Test Method, *Mater. Eval.*, Vol.
    26, No. 2, 1968.
12. Kraska, J. R. and H. W. Kamm, Evaluation of Sonic Methods
    for Inspecting Adhesive Bonded Honeycomb Structures, U.S.
    Air Force Materials Lab. Tech. Rep. AFML-TR-69-283, 1970.
13. Miller, N. B. and V. H. Boruff, Evaluation of Ultrasonic Test
    Devices for Inspection of Adhesive Bonds, Final Rep. ER-1911-
    12, Martin Marietta Corp., Baltimore, Md., 1962.
14. Arnold, J., Development of Non-destructive Tests for Structural
    Adhesive Bonds, Tech. Rep. WADC-TR-54-231, 1957.
15. Schroeer, R., et al., The Acoustic Impact Technique—A Versatile
    Tool for Non-destructive Evaluation of Aerospace Structures and
    Components, *Mater. Eval.*, Vol. 28, No. 2, 1970.
16. Gonzales, H. M. and C. V. Cagle, Non-destructive Testing of
    Adhesive Bonded Joints, ASTM Spec. Tech. Publ. 360, 1964.
17. Smith, D. F. and C. V. Cagle, Ultrasonic Testing of Adhesive
    Bonds Using the Fokker Bond Tester, *Mater. Eval.*, Vol. 24,
    July 1966.
18. Gonzales, H. M. and R. P. Merschell, Ultrasonic Inspection of
    Saturn S-II Tank Insulation Bonds, *Mater. Eval.*, Vol. 24, July
    1966.
19. Botsco, R. J., Nondestructive Testing of Composite Structures
    with the Sonic Resonator, *Mater. Eval.*, Vol. 24, Nov. 1966.

20. Newschafer, R., Assuring Saturn Quality Through Non-destructive Testing. *Mater. Eval.*, Vol. 27, No. 7, 1969.
21. Schliekelmann, R. J., Non-destructive Testing of Adhesive Bonded Joints, in *Bonded Joints and Preparation for Bonding*, AGARD-NATO Lecture Ser. 102, Oslo, Norway, and The Hague, Netherlands, Apr. 2–6, 1979.
22. Hagemaier, D. J., Bonded Joints and Non-destructive Testing: Bonded Honeycomb Structures, 2, *Non-destructive Test.*, Feb. 1972.
23. MIL-STD-860 (USAF), Fokker Ultrasonic Adhesive Bond Test, 1978.
24. Wells, D., NDT of Sandwich Structures by Holographic Interferometry, *Mater. Eval.*, Vol. 27, No. 11, 1969.
25. Grant, R. M., Conventional and Bead to Bead Holographic Non-destructive Testing of Aircraft Tires, 1979 ATA NDT Forum, Seattle, Wash., Sep. 11–13, 1979.
26. Vettito, P. R., A Thermal I.R. Inspection Technique for Bond Flaw Inspection, *J. Appl. Polym. Sci. Appl. Polym. Symp. No. 3*, 1966.
27. Searles, C., Thermal Image Inspection of Adhesive Bonded Structures, *Proc. Symp. NDT of Welds and Materials Joining*, 1968.
28. Woodmansee, W. and H. Southworth, Detection of Material Discontinuities with Liquid Crystals, *Mater. Eval.*, Vol. 26, No. 8, 1968.
29. Brown, S., Cholestric Crystals for Non-destructive Testing, *Mat. Eval.*, Vo. 26, No. 8, 1968.
30. Collins, D. H., Acoustical Holographic Scanning Techniques for Imaging Flaws in Reactor Pressure Vessels, *Proc. 9th Symp. NDE*, San Antonio, Apr. 25–27, 1973.
31. A Sample of Acoustical Holographic Imaging Tests, Holosonics Inc., Richland, Wash.
32. MIL-A-83377, Adhesive Bonding (Structural) for Aerospace Systems, Requirements for.
33. MIL-A-83376, Adhesive Bonded Aluminum Honeycomb Sandwich Structure, Acceptance Criteria.
34. MIL-I-6870 (ASG), Inspection Requirements, Nondestructuve: For Aircraft Materials and Parts.
35. MIL-STD-410, Nondestructive Testing Personnel Qualification and Certification.
36. Martin, B. G., et al., Interference Effects in Using the Ultrasonic Pulse-Echo Technique on Adhesive Bonded Metal Panels, *Mater. Eval.*, Vol. 37, No. 5, 1979.
37. Flynn, P. L., Cohesive Bond Strength Prediction for Adhesive Joints, *J. Test. Eval.*, Vol. 7, No. 3, 1979.
38. MIL-STD-401, Sandwich Construction and Core Materials; General Test Methods.

16

Adhesive-Bonded Aluminum Structure Repair

MURRAY H. KUPERMAN   *United Airlines Maintenance Operations Center,*
San Francisco, California

RAY E. HORTON   *The Boeing Commercial Airplane Company, Seattle,*
Washington

I. INTRODUCTION

Repairs to adhesive-bonded aluminum aircraft structure are required
as a result of mechanical damage on the manufacturing or flight line
and also as a result of in-service problems. Most manufacturers elect
to substitute a new part from their production process for the damaged
unit and either recycle or scrap the old part. Airline operator preference usually dictates the purchase of "new" structure, so except for
repair of cosmetic damage, the offending article is removed. However,
mechanical damage is a small fraction of the overall problem necessitat-
ing bonded structure repair. This chapter is devoted mainly to fixing
in-service problems. The techniques, surface preparations, materials,
and processes that repair in-service bonded structure are also applicable to manufacturer-based units. The end result of the repair process
will be to restore the structural integrity of the bonded article and in
some cases to increase the service period over and above the initial
design life.

II. REPAIR CONSIDERATIONS

A. Alternatives

The in-service repair of adhesive-bonded aluminum aircraft structure
has, as its primary concern, the safety and dependability of the aircraft. Within this framework, there is an optimum schedule to perform
repair work, consistent with aircraft scheduling requirements. The

aircraft operator is given a structural repair manual by the aircraft
manufacturer, which usually conservatively estimates the damage tolerance of the adhesive-bonded structures. This tolerance is a function
of the design, weight, performance criteria, and service environment
of the aircraft. As a rule, balanced flight control surfaces are more
critical than other units and require more stringent controls. Keeping
the design tolerance in mind, an operator might elect (time permitting)
to repair the damage on the aircraft in a permanent manner. Also, time
permitting and given that the surface is interchangeable, the operator
might elect to replace the structure with a spare unit. The damaged
surface would then be repaired in the aircraft operator's shop or other
repair station. A third option would be to perform a temporary repair,
again within the constraint envelope of damage tolerance. This temporary repair would be life limited, inspected at suitable frequencies for
deterioration, and replaced by a permanent repair at a scheduled aircraft maintenance opportunity. Hence, with the extent of damage the
primary consideration and the aircraft's schedule in mind, the determination is made to perform a temporary or a permanent repair.

B. Damage and Assessment

The types of damage that aluminum aircraft adhesive-bonded structures
most often see in service are as follows:

Dents

A dent is usually defined as a skin depression with the adhesive
bond left intact (undelaminated).

Scratches and Gouges

A scratch is surface damage not resulting in material removal. A
gouge is defined as damage resulting in material removal and is considered to be the same as a crack.

Crack

A crack is any break in the surface of a honeycomb panel, usually
somewhat linear in nature.

Puncture

A puncture is defined as damage resulting from physical penetration of the skin surface.

Delamination

Delamination is an adhesive-type failure either between (if a honeycomb panel) the core and adhesive or between the skin and adhesive.

If skins are primed before bonding, the separation may be between the adhesive and primer or between the primer and skin. In any case, most delaminations result in visible skin deformations or they may be detected by tapping with a blunt metallic object. A dull thud indicates a delaminated area; a crisp ring indicates a sound bond. In some cases, particularly in multiple-layer bonded structure, delamination cannot be detected by simple tapping techniques. More sophisticated methods, such as ultrasonic, Fokker bond tester, or similar instruments are required. Service experience with the part in question dictates the type of inspection requirement.

Water Contamination (or other aircraft fluid)

If the honeycomb panel skin is damaged by punctures or cracks, with sealant deterioration or adhesive delamination, water may be present. Water contamination cannot be tolerated in honeycomb panels, for many reasons. Some of the more important ones are: water freezes in the panel, with the resulting expansion causing core damage and delamination; the weight of the panel increases, causing an imbalance in flight control surfaces; the presence of moisture in the metal honeycomb core causes accelerated corrosion. Water can also enter a honeycomb panel by absorption when a fillet seal becomes damaged, or moisture may be absorbed by or diffused through the adhesive. X-ray inspection can be used to detect the physical presence of water and is a common nondestructive inspection technique. Weight-history methods are also used on parts that can easily be removed and replaced.

Environmental Deterioration

Usually defined as failure of the adhesive bond due to interaction between the adhesive or primer, metal surface, honeycomb, and the aircraft environment. Aluminum adhesive-bonded structure on aircraft in commercial service rarely fails in a cohesive (within the adhesive) manner. Except for mechanical damage, most failures start from an interface with the aircraft environment. This interface may be a tooling hole in which the sealant has failed, allowing moisture to penetrate the honeycomb. It could be the edge of a part (leading/trailing) a rivet or fastener hole through the structure, cracked honeycomb potting compound, or edge close-out material. On U.S. commercial transport aircraft fabricated prior to 1970, interface failures resulted in deteriorated or missing honeycomb and debonded skins. The debond mechanism in the skins was, and still is, alclad dissolution. Honeycomb, once perforated to allow ease of bonded structure fabrication, was changed to nonperforated material. This reduced the rate of moisture propagation through the core. This honeycomb was later given a corrosion-inhibiting treatment and the primary failure mode became skin-to-core disbond. The Boeing Company postulates the environmental attack of the bonded surface as follows:

1. Moisture penetrates through the adhesive by diffusion and reaches the aluminum oxide bonded surface.
2. A weak hydrated oxide forms and then fails under stress. This action forms a crevice.
3. Corrosion continues within the crevice, which further accelerates the adhesive disbond and destroys the surface.

New developments within the United States to ensure the environmental stability of the bonding surface consist of anodizing in a solution of phosphoric acid. Chromic acid anodizing has been used overseas for years as a method of obtaining an environmentally durable bonding surface. These surface treatments are discussed in Chapters 2 to 4. The discussion on surface preparation that follows looks at this aspect from a repair standpoint.

III. SURFACE PREPARATION

A. Relationship to Service Life

The importance of properly preparing the aluminum surface prior to bonding can hardly be overemphasized. This is especially true if the bonds are to be made with a room-temperature or 250°F curing adhesive system. It is, however, also important in general.

Satisfactory service life as a function of bond quality specifically related to surface preparation has been demonstrated conclusively by tests on several high-life aircraft [1]. Evaluations were made on bonded structure where no problems were evident and comparisons made of locations adjacent to areas where delaminations had occurred. Although the quality of the areas was indistinguishable when lap shear or portashear testing procedures were employed, the difference was readily apparent when specimens were loaded in cleavage in a hot (140°F), high-humidity environment by use of the wedge test (Fig. 1). The measured rate of crack growth for the specimens was high where test coupons were taken adjacent to the failed areas and typically low for a satisfactorily performing structure. Additionally, the former specimens typically failed adhesively, leaving one of the metal surfaces shiny and bare while the specimens from the satisfactory areas failed cohesively through the center of the bond line.

The results were important for two reasons. First, they indicated that the weak link was typically the bond of the adhesive to the metal, not the internal strength of the adhesive itself. Second, the evaluations established the wedge test as the much-sought-after short-time test that could meaningfully indicate whether a particular surface preparation/adhesive combination could be expected to have long-time service durability.

Those working at a repair facility might use the wedge test to rate their current metal cleaning and bonding procedures. If a typical

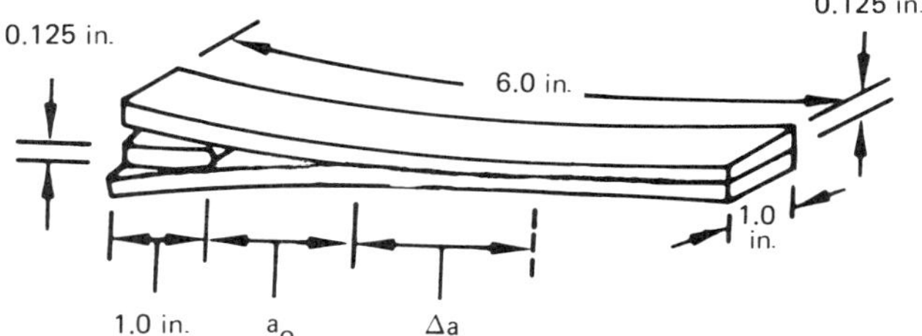

Figure 1   Wedge test specimen configuration.

250°F curing adhesive [e.g., AF-126 (3M Company)] is used, the initial crack opening, should typically be 1.2 to 1.5 in. Growth of the crack after 1 hr of exposure to 120°F and 100% relative humidity should not exceed 0.2 in. A 350°F curing adhesive will usually have a longer initial crack opening, however, its subsequent crack growth rate should essentially be zero. More important, when the specimens are split apart, the failure should be cohesive and show no evidence of separation from the base metal.

A less quantitative but more dramatic demonstration of surface preparation quality can be attained by use of the peel specimen detailed in Fig. 2. Exert a constant but moderate force to pull the pieces apart while a drop of water (or coffee) is placed in the crack at the start of

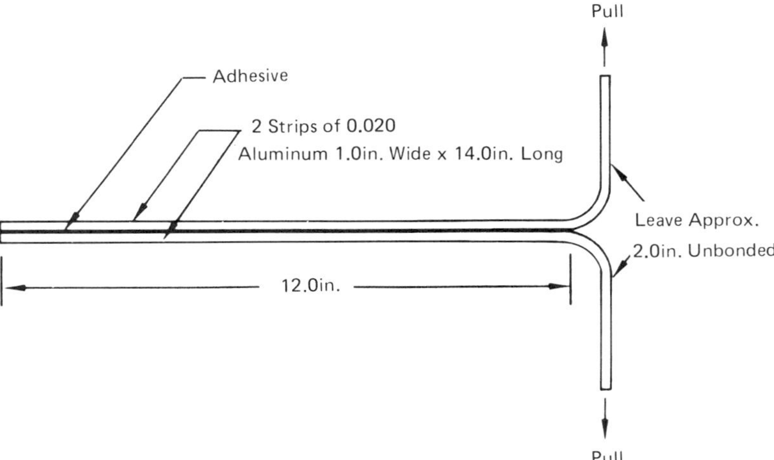

Figure 2   Metal-to-metal peel test specimen.

the bond. Typically, a bond with poor or marginal surface preparation
will unzip with minimum effort, leaving a shiny metal surface. The
bond to a well-prepared surface will remain tough and failure will be
through the center of the adhesive.

Another method of illustrating surface preparation quality is to
press a piece of 3M 250 masking tape on the cleaned surface. On a
well-prepared surface, such as one prepared by phosphoric acid hand
anodizing, the adhesive will strip from the tape as it is removed.

B. Qualitative Comparison of Several Surface Preparation Procedures

A rather comprehensive program to evaluate various quality levels of
surface preparation was conducted by Boeing under the Air Force contract cited as Ref. 2. As part of the program, AF-126, a 250°F curing
epoxy adhesive, was used with EC-2320 (3M Company) primer to bond
wedge specimens. These were cleaned by a variety of commonly used
methods. These included solvents [i.e., Freon (duPont), methyl ethyl
ketone (MEK), and trichloroethylene (TCE)]; hand-applied acids [i.e.,
sulfuric-dichromate (FPL etch), 2% hydrofluoric acid, and PasaJell 105
(Semco Sales and Service Company)]; and acid tank processes, including the conventional and an optimized sodium dichromate/sulfuric acid
etch and phosphoric acid anodize [the latter is a Boeing proprietary
process (designated BAC-5555)]. Additionally, the solvent procedures
were evaluated with and without a supplemental conversion coating
treatment. In this case, Alodine 1200 (Amchem Products) was used.
The results are illustrated graphically in bar chart form in Fig. 3.
Comparative results can also be seen quite well by noting the differences in the surfaces of failed specimens shown in Fig. 4.

The results indicated that, at least for the AF-126/EC-2320 system,
hand cleaning with solvents was an inadequate surface preparation
method. Solvent cleaning with conversion coating prior to bonding
gave considerable improvement. Similar results were obtained in Boeing in-house tests; these data are shown in Table 1. Note that methods
6 and 7, where Alodine 1200 was not used, gave much more rapid crack
growth. Returning to Figs. 3 and 4, progressively more improvement
can be seen with the use of the hand acid procedures. Evidence of
adhesive failure almost totally disappeared when tank cleaning was
used.

It should be mentioned that the results are also quite dependent on
the particular adhesive system. The AF-126/EC-2320 system has a comparatively wide tolerance to surface preparation quality. Most room-
temperature curing adhesives, by contrast, have a narrow tolerance.
This is illustrated in Fig. 5, where wedge specimen failure surfaces
are shown for four typical room-temperature curing systems. Surfaces were tank cleaned using the optimized FPL etch procedure.

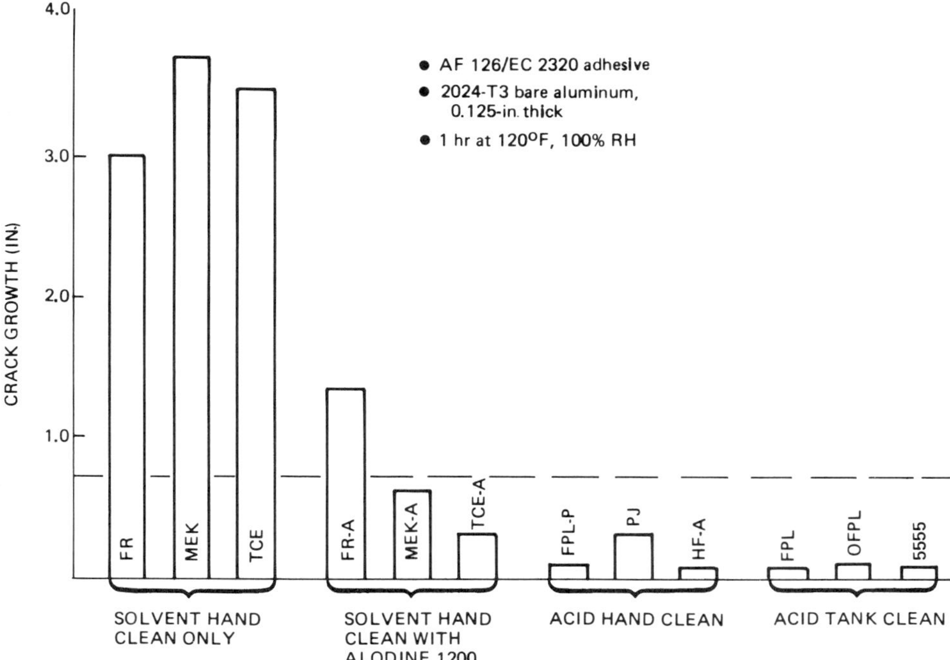

Figure 3   Wedge test data showing comparative quality of various surface preparation methods.   FR, freon, MEK, methyl ethyl ketone; TCE-trichloroethane; A, chromate conversion coating (Alodine 1200); FPL, Forest Products Laboratory etch; PJ, PasaJell 105; HF, 2% hydrofluoric acid; OFPL, optimized FPL etch; 5555, phosphoric acid anodized; FPL-P, FPL paste.

The results can be compared with equivalent conditions for AF-126 shown in Fig. 4D.

Wherever practical, tank-type surface preparation facilities should be used. The advantages of tank equipment in having precise control over solution chemistry, temperature, immersion time, and access to rapid and complete rinsing cannot be equaled by hand procedures.

Of the tank procedures, phosphoric acid anodize has proven superior for use with 250°F curing epoxy systems [2–4]. Its use, where possible, is also recommended for room-temperature curing adhesives. The 350°F curing adhesives, at least those evaluated in Ref. 6, are seemingly less discriminating and perform equally well with the optimized FPL etch, chromic acid, or phosphoric acid anodizing.

At locations where adequate surface preparation facilities are not available, consideration should be given to stocking prepared skins or

(a)

Figure 4   (a) Adhesive failure of solvent hand-cleaned wedge specimens after 1 hr of exposure to 120°F/100% relative humidity, AF-126
adhesive. (b) Failure surfaces of solvent plus Alodine 1200 hand-
cleaned wedge specimens after 1 hr of exposure to 120°F/100% relative
humidity, AF-126 adhesive. (c) Failure surfaces of acid hand-cleaned
wedge specimens after 1 hr of exposure to 120°F/100% relative humidity,
AF-126 adhesive. (d) Cohesive failure of tank-cleaned wedge-specimens
after 1 hr exposure to 120°F/100% relative humidity, AF-126 adhesive.
(Courtesy The Boeing Airplane Co.)

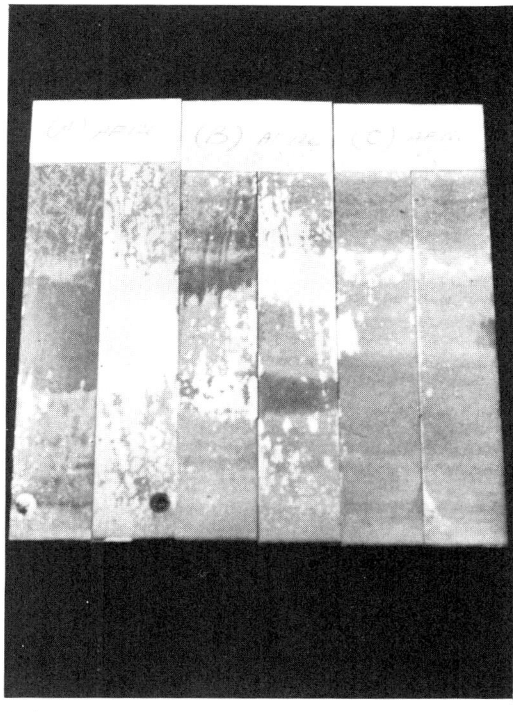

(b)

Figure 4 (continued)

sheet stock. This material can be cleaned using good-quality tank procedures. The skin is primed and a nylon peel ply bonded to the surface with an approved adhesive film. The prepared skin is then stored
until needed. At that time the suitable patch or splice plate configuration can be cut, the nylon peel ply stripped from the bond surface,
and the skin piece bonded in place. The bond surface on the part to
be repaired must still be cleaned, but the total amount of cleaning has
been minimized. Cleaning of the remaining surface can be accomplished
using one of the nontank surface preparations discussed in the following sections.

C. Evaluation of the Phosphoric Acid
Nontank Anodizing Procedure

In many instances, tank cleaning is not practical or possible. For
example, although repair details such as patch plates can be immersed

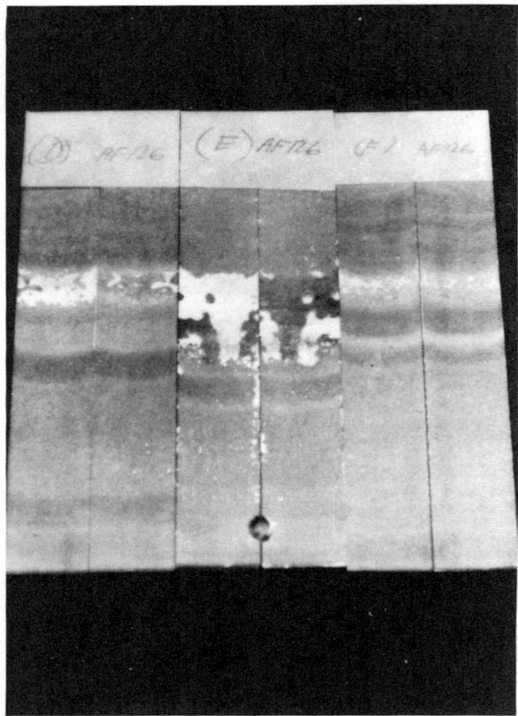

(c)

Figure 4   (continued)

in an acid tank, typically the part to be repaired cannot. Also, cleaning or surface preparation tanks may not be available at a particular
repair location. In such cases, a nontank or hand cleaning procedure
must be used.

The wide relative quality range of several nontank methods has
been discussed. Understandably, their quality is almost directly
inverse to their ease of application. It should be noted, however,
that even the best and hence most time consuming methods involve
only a minor part of the repair task. As such, their use is strongly
recommended.

Relative to the above, a phosphoric acid hand anodize procedure
has been developed which is decidedly superior for typically sensitive
room-temperature and 250°F curing adhesives. The method is not
overly difficult. The acid in paste form is applied to the surface by
saturating it in gauze. This is overlaid with a stainless steel screen.

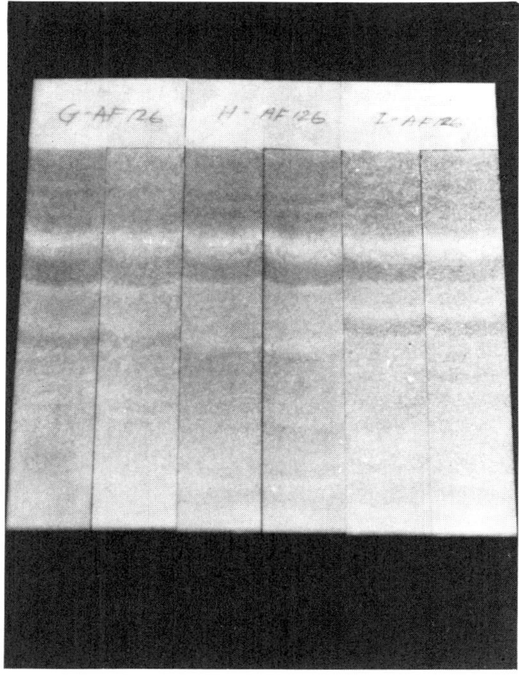

(d)

Figure 4   (continued)

Anodizing is accomplished quickly with low voltage (10 min at 6 V dc).
The area, after rinsing and drying, is ready for bonding. Handling
of freshly anodized surfaces will result in poor bond strength, so care
must be taken prior to adhesive application.

The phosphoric acid nontank anodize method has been quite thoroughly evaluated and optimized in a Boeing/Air Force contract [5].
The prime advantage is that the durability of the bonds, as indicated
by the wedge test, is closely comparable to those using the best tank
cleaning procedure. This can readily be seen in Table 2, where anodizing is compared with three other nontank and two tank methods.

Another convincing comparison of the nontank anodize with a tank
and a commonly used nontank procedure is shown in Fig. 6. Note that
all three methods gave closely comparable results until water was introduced; then the peel strength of the bond prepared with PasaJell 105
dropped to nearly zero while the strength of the phosphoric acid nontank anodized specimen remained high.

Table 1   Effect of Conversion Coating (Alodine 1200) on Bond
Stability (Boeing Data)

Surface preparation method[a]	Salt spray wedge test; days to failure	120°F/100% relative humidity wedge test; days to failure	140°F water wedge test; crack growth (in.)
1. 2% HF, Alodine 1200, 1-hr dry	85	85	0.25
2. 2% HF, Alodine 1200, 4-hr dry	79	85	0.25
3. Alodine 1200 only, 1-hr dry	72	77	0.38
4. 2% HF, Alodine 1200, baked 60 min at 160°F	114	92	0.12
5. 2% HF, Alodine 1200, 1-hr dry plus DeSoto 513-707 primer	No failure	No failure	0.38
6. Same as 1. except no Alodine 1200	7	10	0.75
7. Same as 5. except no Alodine 1200	–	–	Failure in 5 sec

[a] 2024-T3 nonclad aluminum, AF 126 adhesive, EC 2320 primer except
as noted.

Morphology studies also indicate that the nontank phosphoric acid
anodize should perform quite comparably to the tank process. The surface oxide growths are closely similar, as seen by the two comparative
scanning electron microscope (SEM) photos in Fig. 7.
The practicality of using the nontank anodize procedure for repair
of real structure has been quite thoroughly demonstrated. Numerous
aircraft parts were received from the Air Force and Navy and satisfactorily repaired using this process in the Air Force/Boeing program [5].
The anodize procedure used for repair of damage to an AWACS wing is
shown in Fig. 8. The process was also demonstrated in making large
bonded repairs to fuselage structure in a program supportive of the
McDonnell Douglas/Air Force Primary Adhesively Bonded Structures
Technology (PABST) program [7]. Surface anodizing is shown being

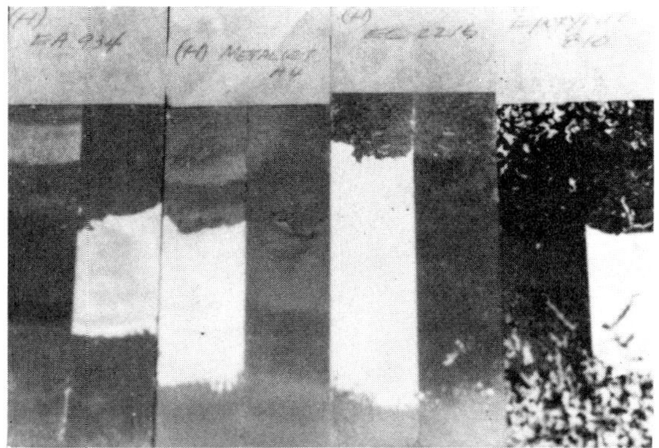

Figure 5   Typical wedge specimen adhesive failure surfaces for four room-temperature cure adhesives.   Optimized FPL etch.   (Courtesy The Boeing Airplane Co.)

accomplished for this program in Fig. 9 as a result of simulated damage to the skin and frame.   After repair, specimens were removed for lap shear and wedge testing.   The results were satisfactory.   It was learned, however, that when large complex areas are to be anodized, special care must be taken to assure proper contact of the anodizing components.

D. Procedures for Accomplishing Several Nontank Surface Preparation Methods

Since tank surface preparation procedures such as the optimized sodium dichromate/sulfuric acid etch, phosphoric acid anodize, and chromic acid anodize are covered in Chapters 2 to 4, their descriptions are not repeated here.   Their use is recommended where facilities are available.

Before describing the individual cleaning procedures, it is cautioned that many of the materials used for this purpose are hazardous in that they may be toxic, flammable, or corrosive.   Exercise care in their handling.   Adhere to regulatory safety rules as appropriate, including the following:

Keep repair area well ventilated.
Minimize breathing of cleaning solvent fumes.
Wear rubber gloves, goggles, and protective clothing when
    handling acid and caustic cleaning solutions.   If skin or eyes
    come in contact with these solutions, immediately flush the area
    with water and seek medical attention.

Table 2 Comparison of Various Tank and Nontank Surface Preparation Methods Using the Wedge Test[a]

Alloy	Surface preparation	Adhesive	Primer	Initial crack length (in.)	Crack growth (in.)		
					1 hr	4 hr	30 days
2024-T3B	TCE-A	EA 9601	EA 9261	1.58	Failure		
2024-T3B	PJ	EA 9601	EA 9261	1.61	2.39	Failure	
2024-T3B	HF-A	EA 9601	EA 9261	1.58	1.69	Failure	
2024-T3B	OFPL	EA 9601	EA 9261	1.52	0.13	0.13	0.43
2024-T3B	5555	EA 9601	EA 9261	1.56	0.08	0.13	0.18
7075-T6C	NTA	EA 9601	EA 9261	1.52	0.13	0.13	0.22

[a]All tests at 140°F/100% relative humidity. B = nonclad, C = alclad. TCE-A, Trichloroethylene plus alodine; PJ, PasaJell 105; HF-A, 2% hydrofluoric acid plus alodine; OFPL, Optimized FPL etch (tank); 5555, Phosphoric acid anodize (tank); NTA, Phosphoric acid anodize (nontank).

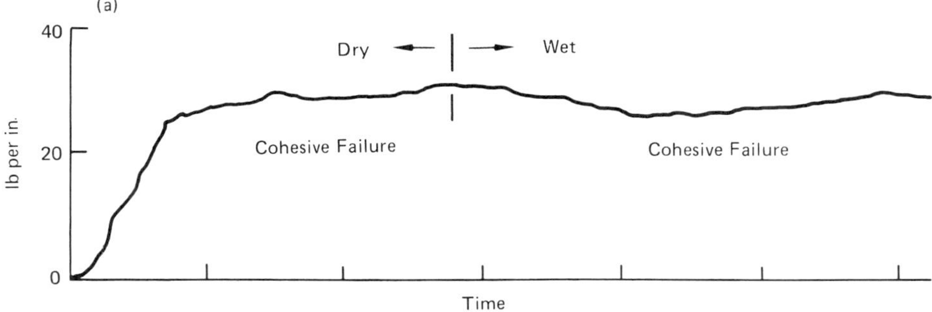

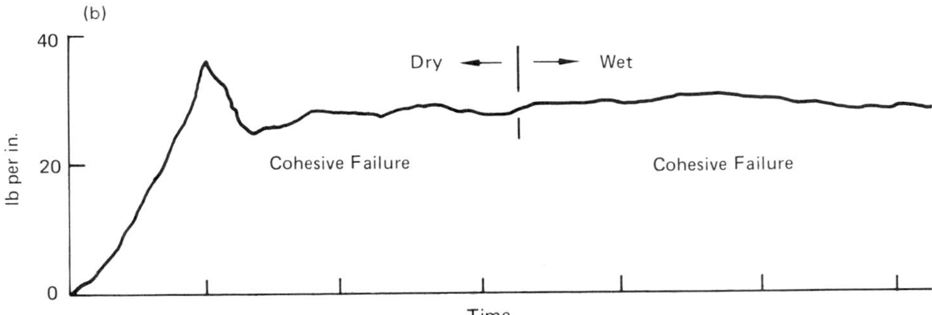

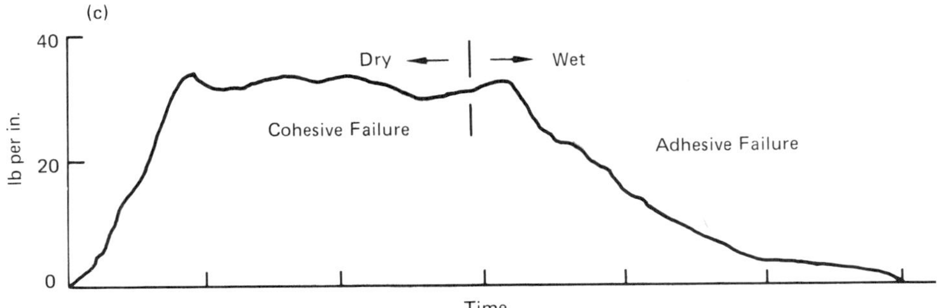

Figure 6 Comparative dry/wet peel test results using three surface preparation procedures, FM-73, 7075-T6 nonclad. (a) Optimized FPL etch; (b) phosphoric acid, nontank anodized; (c) PasaJell 105. (Courtesy The Boeing Airplane Co.)

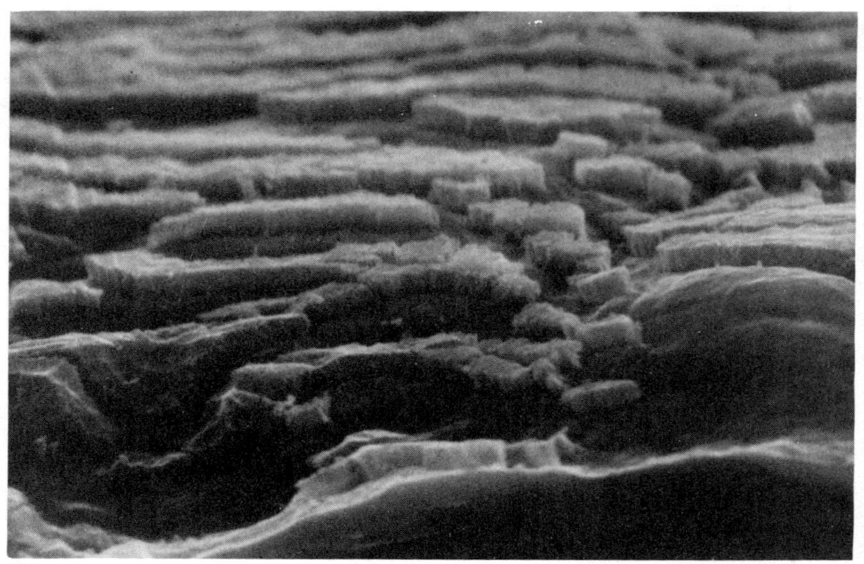

(A)

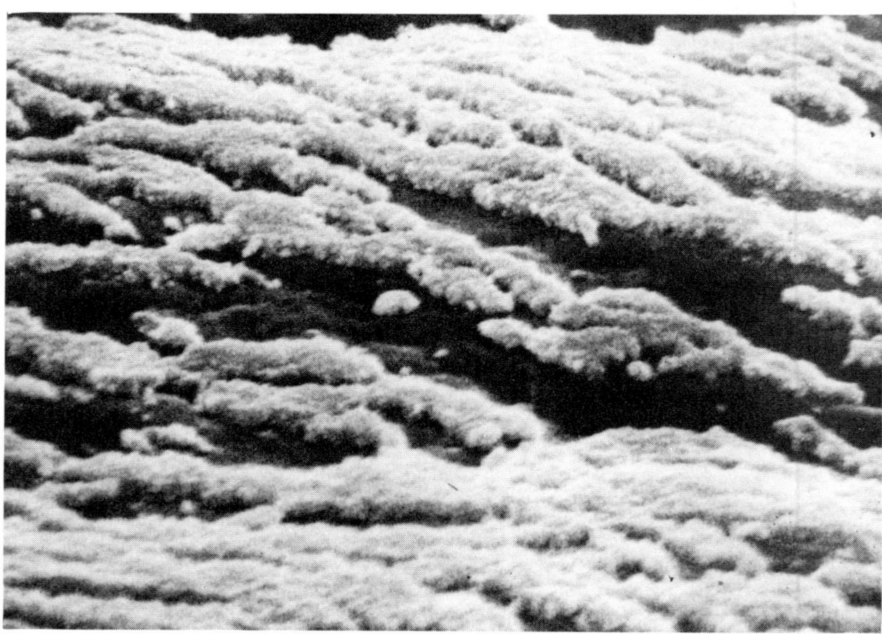

(B)

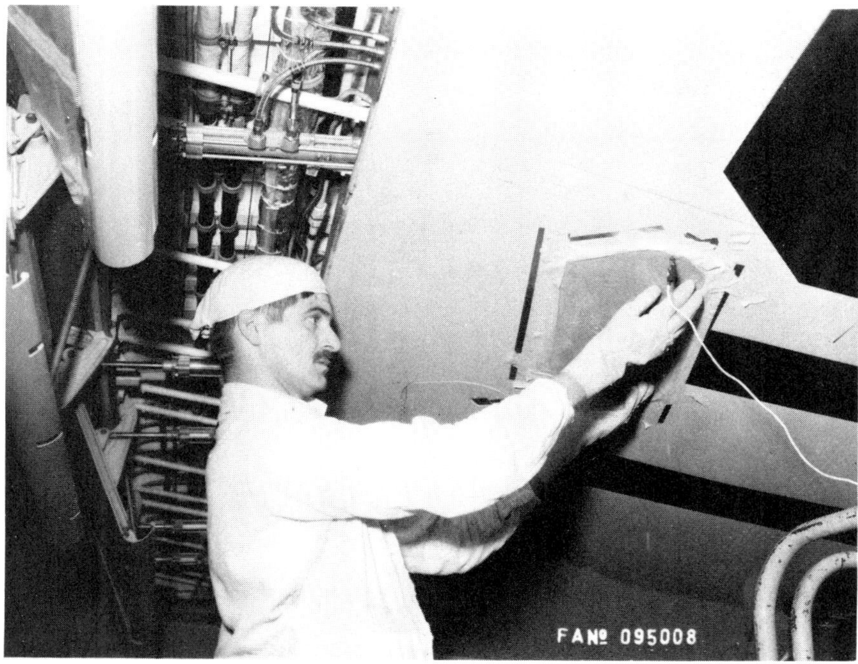

Figure 8   Anodizing to install a patch on an AWACS aircraft lower wing surface. (Courtesy The Boeing Airplane Co.)

Protect surrounding areas by masking and do not allow acid or
   caustic solutions to enter cracks or crevices.
Do not allow acids to come in contact with the surface of high-
   strength steels.
Remove all cleaning solutions by thorough rinsing.
Keep solvents and cleaners in safety containers.
Use explosion-proof electrical equipment.

Nontank cleaning procedures are given in order of preference relative to bonding with room-temperature or 250°F curing adhesives.

Figure 7   (A) SEM of oxide layer formed by phosphoric acid anodizing , tank process. (B) SEM of oxide layer formed by phosphoric acid anodizing, nontank process. (Courtesy The Boeing Airplane Co.)

Figure 9   Anodizing the surface to be bonded to repair a large
fuselage panel.   (Courtesy The Boeing Airplane Co.)

PasaJell 105 appears equal to anodizing for the 350°F systems and is
preferred because of its ease of application. Precautions must be
taken, however, because of its extreme corrosivity. For convenience,
an abbreviation of the processing steps is given in Table 3.

Precleaning

The following procedure should be used for precleaning prior to
starting one of the subsequently described surface preparation methods.
Note: Prior to the start of processing, be sure to protect undamaged
areas, crevices, and fasteners from acid contamination by masking off
these areas with suitable tape and plastic film.

1.   Wipe the surfaces surrounding the repair area with solvent to
     remove oil, dirt, grease, and so on.
2.   Remove the organic finish with solvents and mechanical means
     or with approved strippers. The former is preferred. *Note*:
     Prevent stripper from entering existing bond lines, as stripper
     may have a deleterious effect on the bond.

Table 3   Summary of Nontank Surface Preparation Procedures for Aluminum

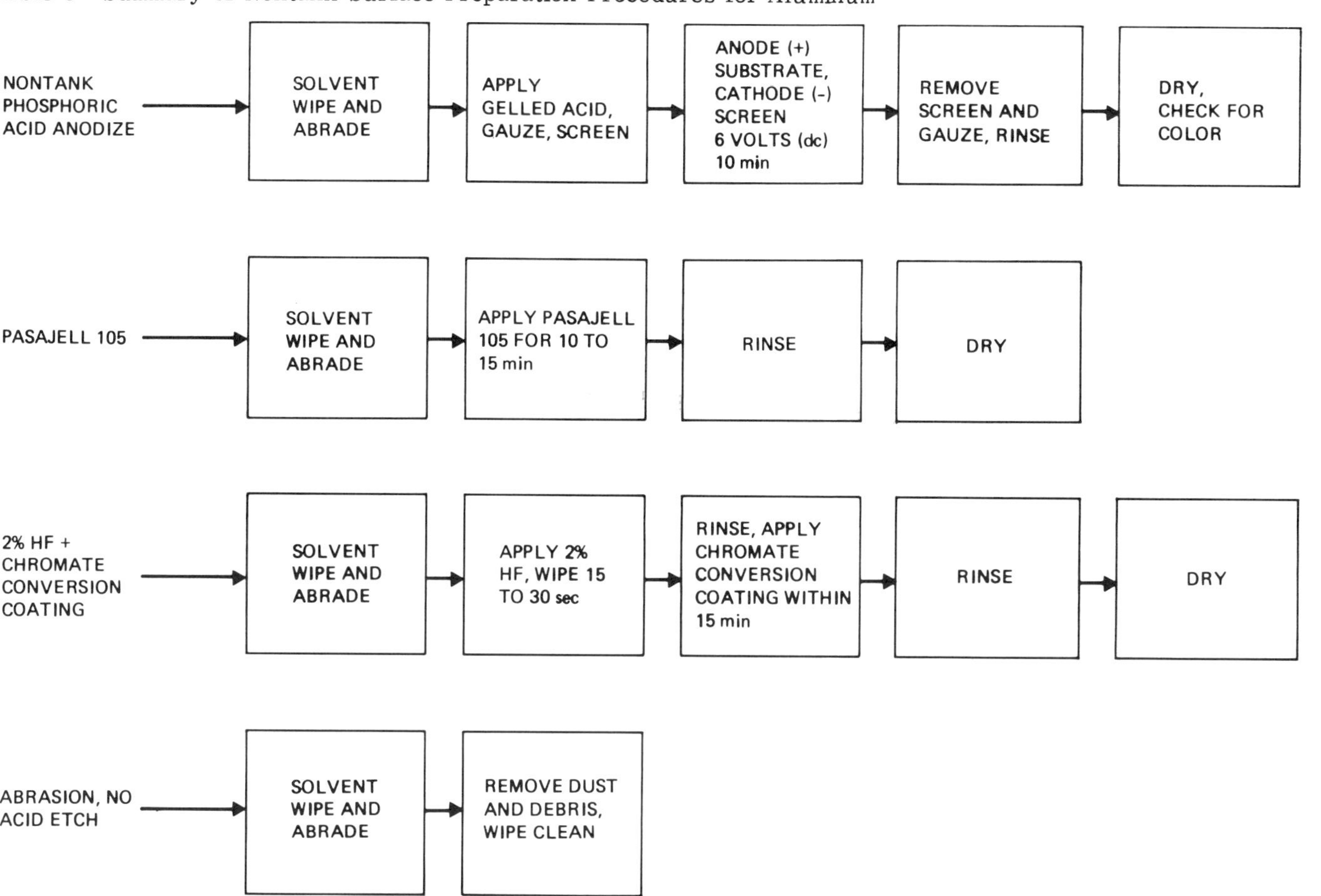

Phosphoric Acid Nontank Anodize

Caution: *This procedure requires the use of acid. See the safety notes listed above.*

1. Solvent-wipe with MEK, trichlorethane, or equivalent.
2. Roughen the surface with a nonwoven abrasive such as nylon abrasive pads or equivalent.
3. Dry-wipe with clean gauze to remove dust and debris.
4. Apply a uniform coat of gelled 12% phosphoric acid or PR 50* to aluminum surface.
5. Place three to four layers of gauze over the coating; apply another coat of gelled phosphoric acid to completely saturate the gauze.
6. Embed a piece of stainless steel screen over the coated gauze. Apply another coating of the gelled phosphoric acid. *Note:* Be sure that the stainless screen is in continuous contact with the gauze but *does not contact* any part of aluminum surface being anodized.
7. Connect the screen as the cathode (−) and the aluminum as the anode (+). *Note:* Check this setup before proceeding: screen as the cathode (−) and aluminum as the anode (+).
8. Apply a dc potential of 6 V for 10 min. *Note:* A rectifier may be used to supply the voltage and current during anodizing. Current density should be in the range 1 to 7 A/ft^2. In an emergency, a fresh or fully charged dry or wet cell battery may be used to anodize small areas.
9. At the end of the anodizing time, open the circuit and remove the screen and gauze.
10. Moisten clean gauze with water. Lightly wipe off the gelled acid with the moistened gauze without delay. The rinse delay time is limited to less than 5 min. *Do not rub the anodized surface.* Immersion or spray rinsing should be used if possible.
11. Air-dry a minimum of 30 min at room temperature or force-air oven dry at 140 to 160°F.
12. Check the quality of the prepared surface. A properly anodized surface will show an interference color when viewed through a polarizing filter rotated 90° at a low angle of incidence to fluorescent light or daylight.

---

*Gelled phosphoric acid compound can be obtained from Products Research Corporation or made by adding a fumed silica powder to 12% phosphoric acid until thickened.

13. If no color change is observed, repeat steps 4 through 12. *Note:* Machined or abraded surfaces are sometimes difficult to inspect for color. Rotation of the polarizing filter is required because some pale shades of yellow or green are so close to white that without a color-change inspection, they might be considered "no color." This would falsely indicate the absence of anodic coating.

Caution: *Do not touch the dried anodized surface. Do not apply tape to the surface.*

PasaJell 105 Method

Caution: *This procedure requires the use of acid. See the safety notes listed above.*

1. Solvent-clean with MEK, trichloroethane, or other approved solvent.
2. Abrade with a nylon abrasive pad or 400-grit aluminum oxide abrasive paper.
3. Dry-wipe with clean gauze pads.
4. Apply PasaJell 105 to the surface with a spatula, brush, or gauze.
5. Leave PasaJell 105 on the surface for 10 to 15 min.
6. Lightly wipe off the PasaJell. Do not wipe PasaJell onto adjacent areas or crevices, or allow it to enter honeycomb core.
7. Dampen a clean gauze pad with clean water and lightly wipe the treated area. Repeat as necessary and check with litmus paper to be sure that all traces of acid have been removed. Do not rub the surface.
8. Allow the area to air-dry before applying primer or bonding. Do not handle freshly etched surface.

Two Percent Hydrofluoric (HF) Acid Method

Caution: *This procedure requires the use of acid. See the safety notes listed above.*

1. Solvent-clean with MEK, trichloroethane, or other approved solvent.
2. Abrade the surface thoroughly with nylon abrasive pads (MIL-A-9962 Type A) or 400-grit aluminum oxide abrasive paper to remove contaminants.
3. Dry-wipe with clean gauze to remove abrasive residue.
4. Moisten (do not saturate) clean gauze with 2% hydrofluoric acid solution. Wipe the abraded area briskly with the pad. Treat this procedure as though the HF were a solvent.

5. After the acid has worked for 10 to 30 sec, dampen a clean
    gauze pad with clean water and wipe the acid-treated areas.
    Do not drag gauze through nontreated areas onto treated
    areas.
6. Within 15 min, apply chromate conversion coating (MIL-C-5541).
    Instructions for application of the conversion coating are given
    below.

Solvent-Wipe and Abrade

Note: This method should be used only on areas internal to the
honeycomb panel or in areas where contamination of adjacent structure
cannot be prevented if acid were used.

1. Solvent-wipe with MEK, trichloroethane, or other approved
    solvent.
2. Abrade with a nylon abrasive pad (MIL-A-9962 Type A) or
    400-grit aluminum oxide abrasive paper.
3. Gauze-wipe to remove dust or vacuum off dust. Solvent-wipe
    again.
4. Following the conventional solvent-wipe and abrade procedure
    (i.e., steps 1 through 3), it is recommended that a chromate
    conversion coating be applied to the surface as described be-
    low.

Application of a Chromate Conversion Coating, MIL-C-5541 Class 1A

It has been demonstrated that the application of a chromate conversion coating as a final step to some surface preparation procedures
significantly improves bond durability. For this reason, inclusion
of the conversion coating is recommended if these surface preparation methods are to be used.

The conversion coating also has a second use. In cleaning for the
repair bond, it may be necessary to bare some of the peripheral surface
area. Treatment of these surfaces with the chromate coating provides
an excellent paint base and acts as a corrosion inhibiter. Prior to application of the coating to these surfaces, they should be cleaned as
previously described.

The application procedure for a chromate conversion is described
below. *Note*: The procedure pertains specifically to the application of
a common conversion coating, Alodine 1200. This is not intended to
indicate its preference. If another coating is used, the manufacturer's
recommended procedures should be followed.

1. Apply Alodine 1200 evenly and liberally with a fiber or nylon
    brush, clean cheesecloth, or a cellulose sponge.

2. Allow the solution to remain for 3 to 4 min to form a bronze or golden brown coating. Keep the treated area from drying by gently blotting with cheesecloth moistened with Alodine 1200. Do not allow the alodine surface to become powdery.
3. Rinse with clean water by gently contacting the treated surface with wet (not saturated) clean cheesecloth. Contact (blot) for 1 to 2 min and repeat.

Caution: *Exercise care when rinsing and contacting treated surface to avoid scratching or removing the freshly formed coating.*

4. Gently contact the surface with clean dry cheesecloth to absorb excess liquid. Repeat as necessary.
5. Check crevices with blue litmus paper for possible acid contamination. If litmus paper turns red, acid is still present and steps 3 to 5 should be repeated.
6. Air dry at room temperature or use hot air (140°F maximum).

Caution: *Cloths or other materials used to apply the conversion coating should be rinsed out thoroughly in water after use and deposited in an approved safety container. Dispose of chromate conversion and rinse solutions following local safety and health regulations.*

IV. REPAIR MATERIALS

A. General Considerations

Selection of proper repair materials involves many alternatives and details. Since it is usually the intention to restore the structure to its prior-to-damaged condition, consideration should be given to using the original materials. However, if examination proves that the original material has been deficient or if its use is not compatible with the repair situation, a more suitable replacement must be substituted. The former is especially the case where a part is being rebuilt. A light-density core in a damage-prone area, for example, can usually be replaced with one of a heavier density if the balance of the part is not critical.

If the aircraft has been in the fleet for some time, the manufacturer is a good source for an improved-material recommendation. The manufacturer is constantly evaluating new products and it is impossible to keep the repair manual minutely up to date. An inquiry will also alert the manufacturer of a problem area to which they should give their attention.

When selecting repair materials, consideration must also be given
to the structural criticality of the damaged area, the time that is available to make the repair, and the availability of tools and facilities at
the repair location. In many cases, temporary repairs can be quickly
made to secondary structure using fiberglass-resin patches and a heat
lamp to accelerate the cure. Where practical, however, if the damage
is significant, the part should be replaced. The old part can be sent
to the shop for repairs of a more permanent quality.

B. Metal Sheet Materials

Attention to permanent good-quality repairs requires that metal repair
parts should usually be replaced. The manufacturer's repair manual
is used as a source of material identification. The engineering staff
should be consulted where it is necessary to make substitutions. In
general, for small repairs, substitutions are allowable where a stronger
material is the replacement. For example, 7075-T6 sheet may be substituted for 2024-T3, or 5056 aluminum honeycomb core for 5052. The
substitution may be reversible if the consideration is one of stiffness
or bond line corrosion resistance and not strength. In the latter case,
a strength substitution becomes permissible by increasing the material's
gauge or density. Such substitutions allow maximum utility of material
inventories. Where larger repairs or rebuilding are to be accomplished,
however, original materials should generally be used. Selection of the
initial material involves many properties in addition to strength and
stiffness (e.g., toughness, corrosion resistance, etc.). General substitutions, therefore are not recommended without detailed analysis of
the problem.

Where design information is required on aluminum facing materials,
reference should be made to MIL-HDBK-5B or specification QQ-A-250.
Where possible, bonds should be made to nonclad faces. The use of
fiberglass/resin composites to accomplish large aluminum repairs is
usually not recommended, primarily because of the material's low modulus. However, the relative ease and speed with which these repairs
can be made often make them useful where stiffness is not a critical
criterion and stresses are low. Fiberglass/resin repair procedures are
especially adaptable to areas having compound curvatures or for honeycomb edge closeouts. Either wet layup or preimpregnated glass plies
(prepregs) can be used. Typical material data are given in MIL-C-
9084. Mechanical strength and design information is given in MIL-
HDBK-17.

C. Honeycomb Core

Aluminum core is available in a wide range of standard cell sizes and
densities. The replacement core should match the core cell size and

density of the original. The core ribbon should be oriented in the same direction. If the original core size or density is not available, use of the next-higher core density or next-smaller cell size is permissible. Corrosion-resistant core should be used as a replacement in all repairs regardless of the original aluminum core type that was specified. This honeycomb should be nonperforated. Aluminum 5056 alloy core may be substituted for 5052 core. Aluminum core information is given in MIL-C-7483.

Nonmetallic cores such as glass-fabric-reinforced or fibrous-nylon-base cores can also be used for repairs. Where they are substituted for aluminum core, they should equal the aluminum in strength and stiffness. Information on fiberglass core is given in MIL-C-8073. Fibrous nylon base (i.e., Nomex) core data are given in AMS specification 3711.

D. Adhesives and Sealants

Mechanical property data for adhesive systems are available from several sources. Examples of these information sources are the adhesive supplier and the airframe manufacturer that uses the particular material. General information can also be obtained from military specifications such as MIL-A-25463 and MMM-A-132. Care must be taken in using these data to allow for subsequent effects of environmental degradation or strength loss of the material at elevated temperature. It should also be noted that these data typically apply to 0.5-in.-lap-length splices. The strengths, particularly the 350°F curing adhesives, may show drastic reductions for longer laps.

It is very important that the processing procedures used to determine the design strengths duplicate those that will be used for the repair. Many of the commonly used adhesives were tested with cleaning and bonding methods representative of repair procedures in Air Force contracts F33615-73-C-5171 and F33615-76-C-3137. These data are reported in Ref. 1 and 5. Typical strength plots for various bonded joint overlap lengths are given for several 250°F and 350°F curing adhesives. Note that these are typical values. It is recommended that they be reduced to a minimum of two-thirds for design utility.

Data for several room-temperature curing adhesives tested with 0.50-in. lap splices are given in Ref. 1. Two other excellent sources of test data are Refs. 2 and 3. This information is valuable for adhesive comparison purposes. Again, the data are typical and values must be reduced substantially for design. The values are for lap lengths of 0.50 in. Values for other lap lengths, if required, will have to be determined.

The use of primers is recommended to improve durability and to minimize contamination prior to adhesive cure. Both noncorrosion-inhibiting and corrosion-inhibiting primers are available and may be

applied by spray or wipe-on techniques. The use of corrosion-inhibiting primers is especially recommended for the 250°F curing adhesives
where proper equipment and necessary personnel skills are available
for application. Bonds incorporating corrosion-inhibiting adhesive
primers (CIAPs) have considerably increased durability over non-CIAP
bonds when the primer is applied correctly to a well-prepared surface.
A disadvantage is that the CIAP must be baked prior to cure of the
primary bond. Offsetting this disadvantage, however, is that the
aluminum pieces with the CIAP coating can be stored for long periods
of time if properly protected from sunlight and dirt. This gives the
option of stocking precleaned and primed sheet material at a repair
facility. The material can then be cut to size as required and merely
solvent-wiped prior to bonding.

The application of corrosion-inhibiting primers is best suited to a
large repair station. The surface must be well prepared by tank cleaning or hand anodizing and the application and thickness must be carefully controlled. The use of non-corrosion-inhibiting primers is
recommended where conditions are less optimum and where the additional
time required to bake cure CIAP primer is not permissible. The use of
primer with 350°F curing adhesive is recommended. However, it is not
considered as important as with 250°F systems. In general, the 350°F
systems are in themselves more environmentally stable and more tolerant
of surface preparation.

The adhesives to be used for permanent repairs should be of the
same class as those used in the original structure. Typically, a 250°F
curing epoxy is used for subsonic aircraft where temperatures may reach
a maximum of 180°F (e.g., as encountered on the ground in desert
areas). Where higher temperatures are encountered, such as around
the auxiliary power unit (APU), adhesives with higher-temperature
capability, usually curing at 350°F, may be required. The 350°F curing systems are also typically required for supersonic aircraft.

Where operating temperatures permit, 250°F systems are preferred.
They are typically stronger, tougher, and easier to process. By contrast, 350°F system strength is lower and use is more applicable to
honeycomb core/face sheet bonds. The strength versus lap length
curve is typically quite flat (Fig. 10), and little strength gain can be
obtained by lengthening a splice. In many cases, consideration must
be given to supplementing the bond strength using mechanical fasteners.

Room-temperature curing adhesives have application for temporary
repairs to lightly loaded structure. However, their use is not recommended for permanent repairs or temporary repairs in critical areas.
Even the "best candidate" adhesives carefully screened for evaluation
in Ref. 1 gave poor wedge test results and had limited strength at the
moderate upper surface temperature ranges (Fig. 5). Results of
stressed load tests performed on two 250°F curing adhesives and a
candidate room-temperature curing system are shown in Fig. 11. The

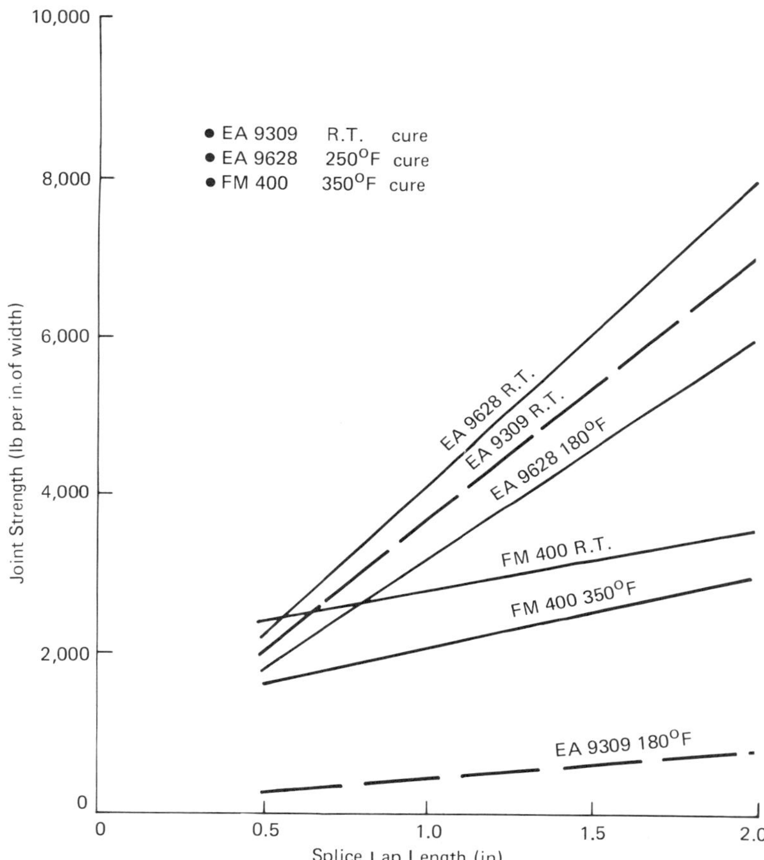

Figure 10 Comparative effect of lap strength and temperature on the strength of three classes of adhesives. Cyclic conditions: (1) 140°F/ 100% relative humidity; (2) 15 min at maximum load, 5 min at zero load, 3 cycles/hr.

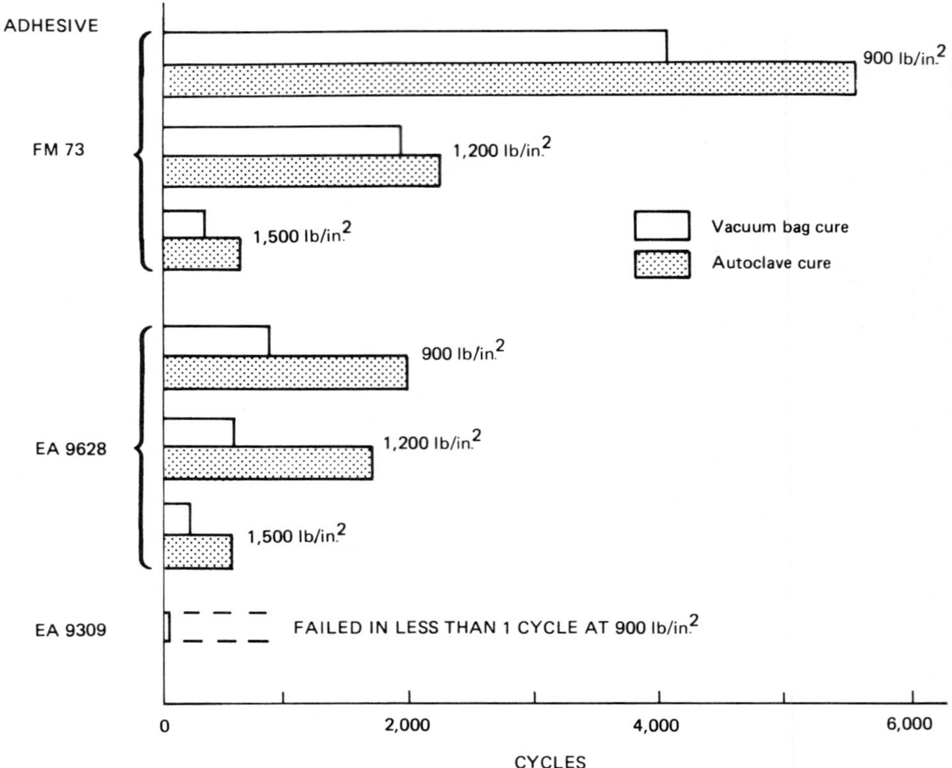

Figure 11 Comparison of cyclic load-environmental exposure data for three adhesive systems.

exposure temperatures at 140°F was not high, yet the room temperature curing system failed quickly under sustained load. Use of an autoclave is compared to vacuum bagging for 250°F systems in Fig. 11. Other comparative data of interest are shown in Table 4.

After cure of the adhesive, sealing of the assembly is important. The adhesive flash fastener holes through bond lines and all bare edges exposed by trimming, drilling, machining, or scuffed through the surface finish should be protected. Two suitable sealants are MIL-S-8802 polysulfides and corrosion-inhibiting MIL-S-81733 polysulfides. Silicone sealants are usually used for sealing applications in excess of 300°F. Painting the entire structure with a filaform corrosion-resistant polyurethane primer and enamel system will greatly increase the environmental durability of the 250°F cured bond.

Table 4 Lap-Shear and Peel Test Results

		0.5-in. lap shear (psi)				M/M peel (lb-in./in.)		
Adhesive	Cure	Room temperature	180°F	14-day salt spray	Stressed[a] 60 days at 600 psi	Room temperature	−67°F	14 days, 250°F/ 100% relative humidity
EA-9628	Autoclave[b]	4598 S[e] = 92	2896 S = 38	4606 S = 154	4610 S = 66	61.1 S = 5.6	11.6 S = 1.0	62.7 S = 3.9
	Vacuum[c]	4012 S = 93	2272 S = 72	4056 S = 38	4285 S = 64	None	None	None
FM-73	Autoclave[b]	3986 S = 87	2616 S = 88	4012 S = 92	3990 S = 125	56.4 S = 5.8	15.8 S = 5.5	56.7 S = 2.5
	Vacuum[c]	3778 S = 110	2388 S - 153	3920 S = 114	3820 S = 348	None	None	None
EA-9309	Vacuum[d]	3486 S = 442	576 S = 34	3462 S = 593	Failure in 5, 11, 14 and 18 days	55.8 S = 5.9	4.5 S = 1.2	43.9 S = 3.9

[a]Exposure at 120°F/100% relative humidity.
[b]35 psi, 250°F, 90 min.
[c]25-28 in. Hg, 200°F, 2 hr.
[d]25-28 in. Hg, room temperature.
[e]Standard deviation.

V. TYPE OF REPAIR

A. Concepts

From a facilities and techniques standpoint, repairs can be grouped within five categories:

1. Small-area line maintenance work, where fiberglass overlay patches, aluminum skin/aluminum honeycomb plug patches, or riveted aluminum sealed with polysulfide sealant will suffice.
2. Life-limited repairs using rivets and room temperature curing adhesives, necessary for aircraft dispatch but requiring short-time inspection intervals and replacement at the first available maintenance opportunity.
3. Partial repairs using a heating blanket and vacuum bag technique for localized areas or when the chance of damaging the original bond with more extensive heat application is a factor. Perforated metal repair skins are used in some instances.
4. Partial repairs using the autoclave to replace honeycomb core and up to one side of the damaged structure. If surface preparation techniques are less than optimum, the parts are monitored for deterioration and rebuilt as time permits.
5. Fully repaired or rebuilt bonded structure where the old spar, end ribs, and so on, are used as the basis for building a reconditioned part. All surfaces are given the optimum preparation prior to bonding, and corrosion-inhibiting honeycomb, corrosion-inhibiting adhesive primer, film adhesive, and new aluminum skins are utilized.

B. Line Maintenance Repairs

Small-area line maintenance repairs will be classified as follows:

Surface dent repair
Puncture and gouge repairs up to 1 in. in diameter using room-temperature curing adhesive and serrated rivets
Skin damage or delamination, no honeycomb damage—typical wet lay-up repair
Skin and honeycomb core damage—wet lay-up potted core repair
Skin and honeycomb core damage—wet lay-up and core replacement repair technique
Skin and honeycomb core damage on two surfaces—aerodynamically flush patch—repair access from both surfaces
Skin and honeycomb core damage on two surfaces—nonaerodynamic patch—repair access from both surfaces
Trailing-edge damage repair

Core and skin repair using prefabricated repair plug—non-aerodynamic

Core and skin repair using prefabricated repair plug—aerodynamic flush patch

Surface Dent Repair

As a rule of thumb, dent filling is never allowed where bond separation could allow the filler material to be ingested into an engine or accessory intake opening. Some typical requirements for the extent of allowable dent filling are as follows (excluding balanced panels, where weight additions are minimized without rebalancing):

1. The skin must not be cracked.
2. The depth of any gouges or scratches must not exceed 20% of the skin gage.
3. The maximum dent depth must not exceed 1/8 in.
4. The size of a single dented area must not exceed 25 in.^{2} and all dented areas shall not exceed 20% of the total skin area on one surface of the part.
5. No delaminations are allowed in the dented area and around the periphery. Nondestructive testing must be utilized to determine if delamination exists.
6. Dent filling is required only on aircraft external surfaces.
7. Minor dents less than 1 in. in diameter or less than 0.035 in. in depth do not require repair provided that they are smooth and free of punctures, cracks, or scratches.

The usual surface preparation for a dent repair is to sand or strip (using paint stripper) off the paint and primer, then hand-wipe with a grease-free solvent. The area is then lightly and evenly abraded with a nylon web containing aluminum oxide particles or an abrasive cloth. The residue is then wiped with a solvent such as methyl ethyl ketone on a lint/grease-free cloth followed by a dry-cloth pickup. Cleaning is continued until the cloth shows no residue. It is recognized that this cleaning technique does not yield an environmentally durable bond, but line maintenance schedule opportunities preclude more sophisticated methods. Immediately upon completion of abrasive cleaning, the dent is filled with a room-temperature curing epoxy adhesive. The adhesive may be rigid (epoxy polyamine) or flexible (epoxy polyamide), depending on whether the surface to be filled is rigid or flexible. Subsurface air bubbles are eliminated during application and the adhesive is formed to the surrounding surfaces using a parting film cover within its work life. Heat can be used to accelerate the adhesive's cure. After curing, rough spots are sanded and the repair is covered with a

chromate-containing primer and a polyurethane paint topcoat. This
provides a certain amount of environmental durability.

Puncture and Gouge Repairs

These repairs utilize fast-curing epoxy adhesives with the ability
to withstand temperature excursions of −65 to +160°F without adhesion
problems. Epoxy polyamine formulations have been found to be satisfactory. Begin by assessing the damage and make sure that the adjacent area is not delaminated. Clean up the puncture or gouge by
trimming or routing/grinding and determine the repair diameter. Flush
the damaged area with a nonflammable solvent such as 1,1,1-trichloro-
ethane and dry using a hair dryer. Depending on the repair diameter,
follow the instructions given in Fig. 12. Solvent-clean and then install

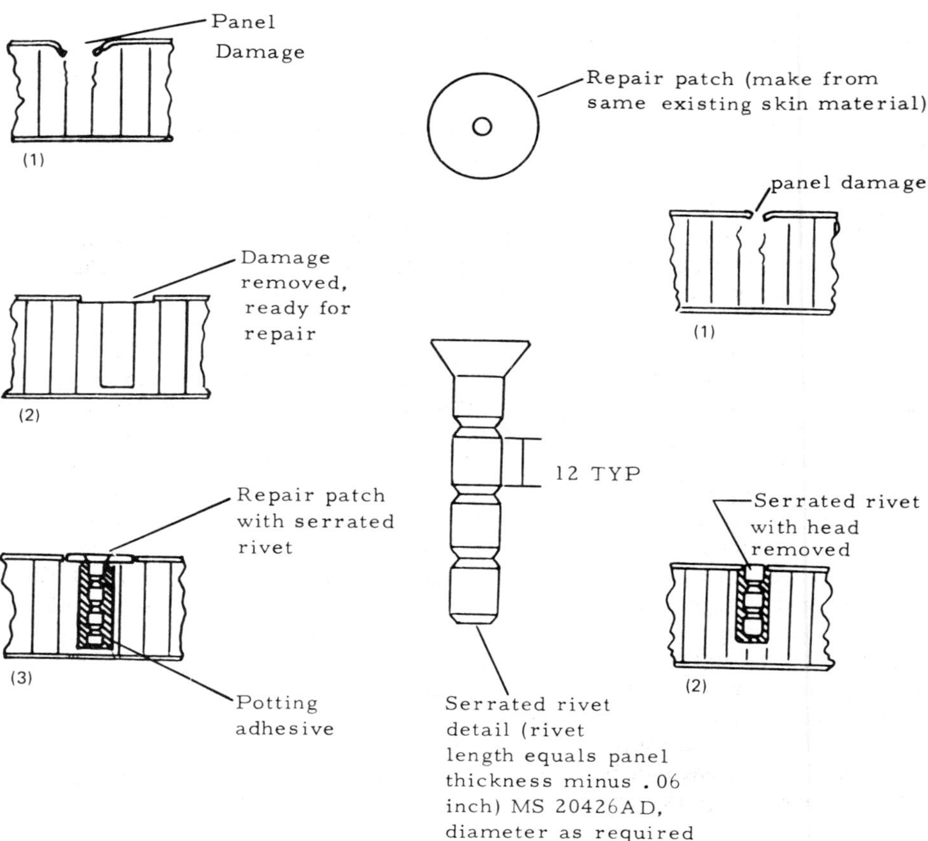

Figure 12   Rivet plug repair for punctures and gouges.

a serrated rivet and/or washer using a rigid, room-temperature curing epoxy adhesive. Make sure that the repair is flush with the panel surface. Repairs exposed to temperatures higher than 160°F can be potted with a room-temperature cure/heat postcure system to provide elevated-temperature strength. Cover the repair with a chromated primer and compatible polyurethane paint top coat.

Skin Damage or Delamination—No Honeycomb Damage

Damage to honeycomb panels can be removed by first sanding off all paint and primer in the damaged area. A minimum-size cutout to the skin is then required to remove the damage. This cutout can be circular or oblong. The edges of the resulting hole should be burr-free and approximately 90° to the surface. Any entrapped water can then be removed by heating the open core cells, being careful not to heat sealed parts of the panel above 180°F (for 250°F cured original structure). Adjacent parts of the panel should also be checked for water entrapment. In nonperforated honeycomb panels, each cell must be ventilated to remove water. Rather than drill a hole in the skin covering each cell, it is faster to remove the skin, dry the affected area, and replace the skin. After completing the cutout, vacuum the dust and chip residue. Clean the skin as described above for the dent repair and flush the honeycomb with nonflammable solvent. After drying, the Boeing Company specifies priming with a Buna N type of coating (MIL-S-4383) followed by impregnating 120 and/or 181 (1581/7781)-type fiberglass with an epoxy polyamide resin system. This system, although slow to cure, is resistant to hydraulic fluid and yields a structural repair. Additional environmental resistance is obtained by covering the repair with polyurethane paint over a compatible primer. The repair procedure follows:

1. Apply a thin coating of primer and allow to dry before the resin overcoat is applied.
2. The gauge of aluminum skin requiring repair determines the number of fiberglass plies to be used. Thickness to 0.020 in. requires two plies of 1581-style cloth, to 0.032 in. requires three plies, and to 0.040 in. requires five plies. If 120-style cloth is desired double the number of plies. Cut the number of fiberglass layers required. Each ply must be 1/2 in. larger in diameter or edge distance than the ply below. The final ply must overlap the edge of the repair by at least 1 1/2 in. (refer to Fig. 13).
3. Mix resin and catalyst. The weight of the mixture should approximate the weight of the dry fiberglass cloth.
4. Apply a thin coat of resin to the repair area (over the primed surface).
5. Place the first ply of fiberglass in position and work out the entrapped air. Stipple with a brush until the resin comes to the surface. Apply another layer of mixed resin and the next

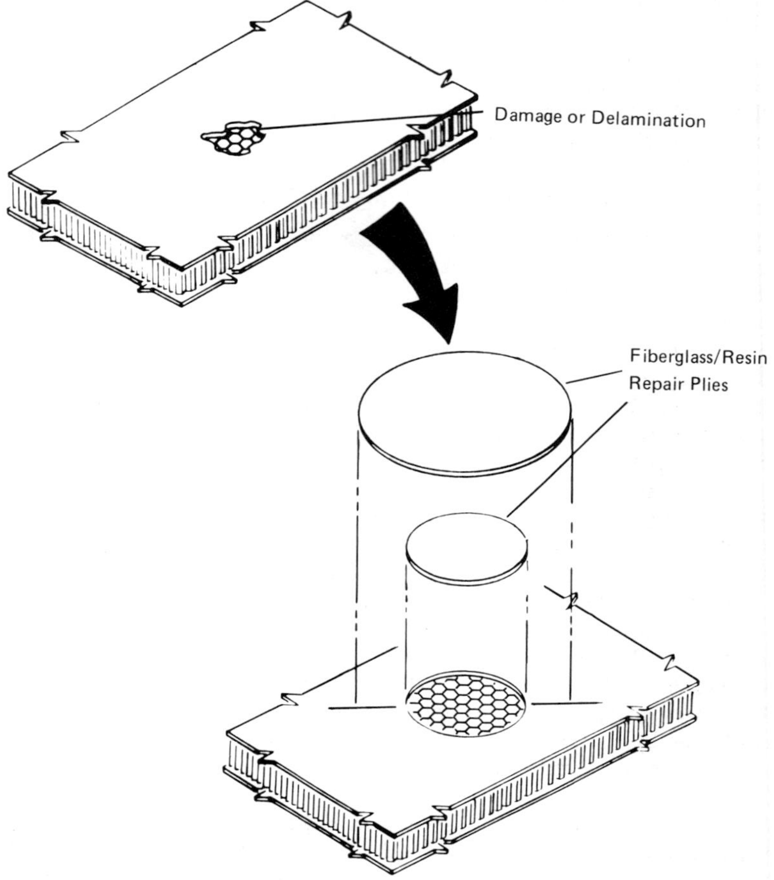

Figure 13   Schematic of Fiberglass-resin repair technique (no core damage).

ply. Repeat this operation until all plies have been applied.
Fair the resin out about 1/2 in. beyond the last ply to provide environmental protection for the primer. Complete the
repair within the resin's work life (pot life).

6.   Cover the repair with a parting film and squeegee to work out
excess air, being careful not to remove too much resin and
cause a "resin-starved" laminate. Wipe off resin outside the
parting film area.

7.   Allow the repair to cure. Vacuum may be used to increase the
strength of the repair. Typically, vacuum-bagged laminates

are approximately 20% stronger than those cured under contact
pressure (refer to Fig. 14). Vacuum can also be used to hold
the repair in position during cure, especially on under-wing or
vertical lay-ups. Heat in the form of lamps, blanket, or hot
air can be used to accelerate curing. Do not exceed a temperature which is 50 to 70°F below the panel's original cure temperature.

8. After the cure completion, remove the parting film, lightly sand
   the surface, and cover with a polyurethane primer and paint
   system.

There is a technique called the "preimpregnated cloth method" which
most manufacturers recommend as an alternative to conventional wet
lay-up techniques. Its utility is shown when working on vertical or
overhead positions.

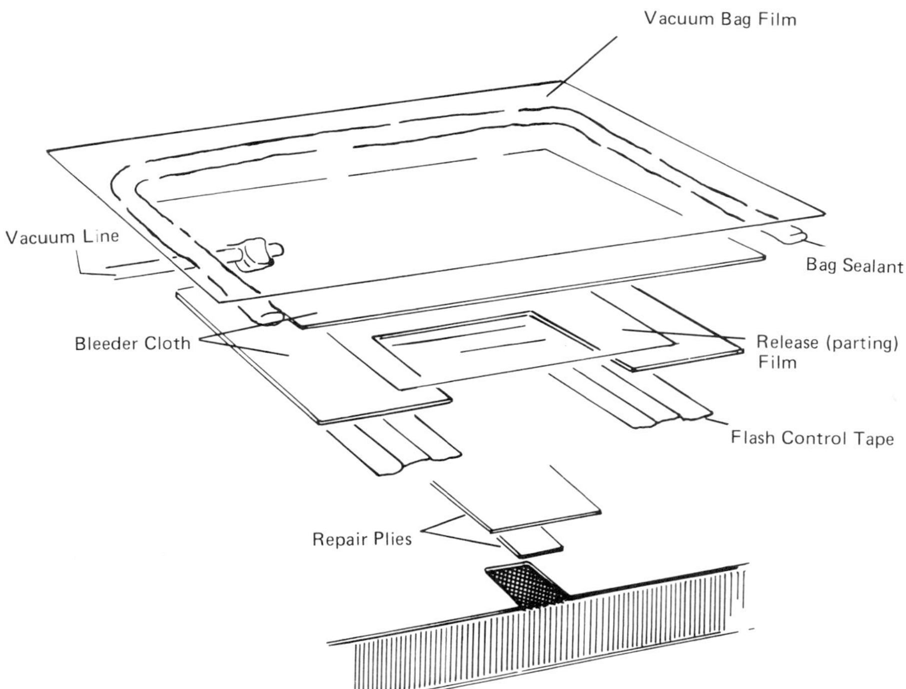

Figure 14 Application of pressure (vacuum) during cure for skin ply
replacement.

The procedure follows:

1. Precut the desired number of fiberglass plies to the required
    sizes.
2. Cut some parting film (hard material) at least 3 in. larger
    than the largest ply.
3. Mix the required amount of resin (approximately equal to the
    weight of fiberglass cloth).
4. Tape one piece of parting film to a flat surface. Spread a
    layer of mixed resin over the film.
5. Place one or two fiberglass plies on top of the parting film
    and resin.
6. Lay the other piece of parting film on top of the fiberglass
    plies and work the mixed resin into the cloth using a
    squeegee or roller. Be sure to wet the entire area of fiber-
    glass.
7. Peel back the top release film and cover the resin/cloth mix
    with another layer of resin. Add an additional one or two
    dry plies of fiberglass over the existing plies.
8. Replace the parting film and work the mixed resin into the
    fiberglass cloth.
9. Continue these steps until the required number of plies are
    used.
10. Cut the preimpregnated patch to the required outside size
    and apply a thin coating of resin to the previously prepared
    repair area.
11. Remove the inside layer of parting film and apply the patch
    over the damaged area. Remove and replace the remaining
    parting film with a larger piece and apply it over the repair.
12. Allow the patch to cure and finish the repair by covering
    with the appropriate primer and paint system.

Skin and Honeycomb Core Damage—Wet Lay-Up, Potted Core Technique

As in the previous skin damage repair sequence, proceed to cut
out the damage and prepare the area for bonding. Remove all loose
and damaged honeycomb but leave any well-bonded material (see Fig.
15). Flush with solvent and allow the solvent to evaporate before
adding potting compound. Potting compounds are selected for their
low density, stiffness, and compatibility for the honeycomb panel's
service temperatures. These repairs are usually small since the potting adds weight (and mass) to the panel, which could be a problem
in vibration areas and on balanced assemblies.

Mix the potting compound and apply it to fill the damaged area
as shown in Fig. 15. Allow the potting compound to cure and

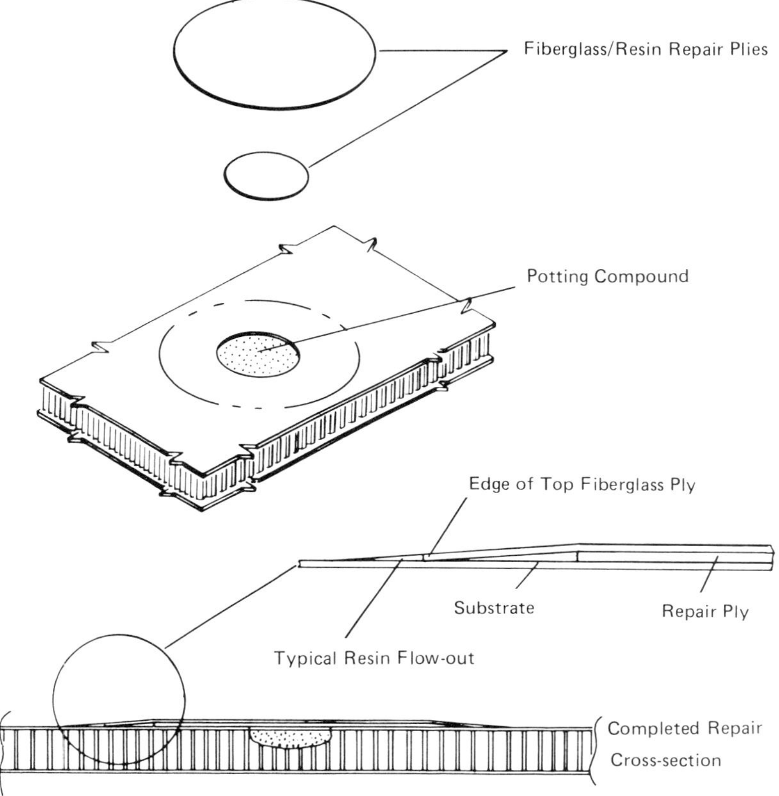

Figure 15   Potted core repair with fiberglass-resin overlay.

accelerate with heat if required. Sand the surface smooth and apply
the wet lay-up resin/fiberglass repair to one surface as discussed
previously.

Skin and Core Damage—Wet Lay-Up,
Honeycomb Core Replacement

Refer to Fig. 16 for a repair cross section. Cut out the damaged
honeycomb flush with the inside skin, being careful not to damage the
surface. A router and hole saw with stop is usually used for this operation. The damaged core is cut around the perimeter and then routed
from the inside adhesive surface. Before cleaning, make a core plug as
illustrated (Fig. 16) from the same material as the original honeycomb
panel. Then clean the core plug, core cutout area, and the area to be

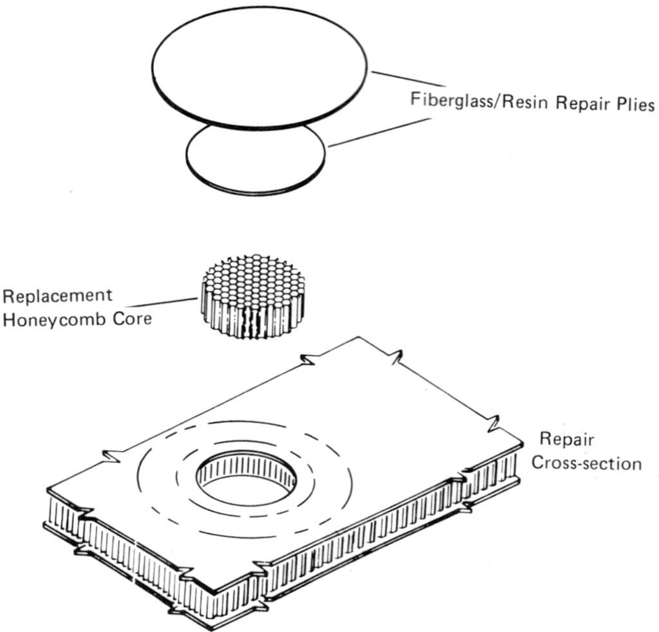

Figure 16   Honeycomb core plug repair with fiberglass-resin overlay.

repaired as described previously. Pot in the honeycomb core using potting compound material on the sides and bottom of the insert. Allow to cure and then proceed with the wet lay-up repair (assuming that the core was cut and potted to the correct thickness). To increase the repair strength, vacuum pressure can be used during cure of the potted honeycomb and fiberglass overlay (refer to Figs. 14 and 17).

Skin and Core Damage—Both Surfaces—Aerodynamic
Flush Patch (repair access from both sides of the
damaged panel)

Cut out damaged skin and honeycomb to gain access to the panel from the interior or internal skin surface. See Fig. 18 for repair cutout configuration. Clean all areas to be bonded and then tape a piece of parting film over the hole in the external skin. Proceed to lay up the required number of resin-fiberglass plies to provide for the skin repair (based on skin thickness). The first ply must mate with the hole in the skin and the last ply should just fit inside the honeycomb core cutout area (see Fig. 19). Using a potting compound whose curing

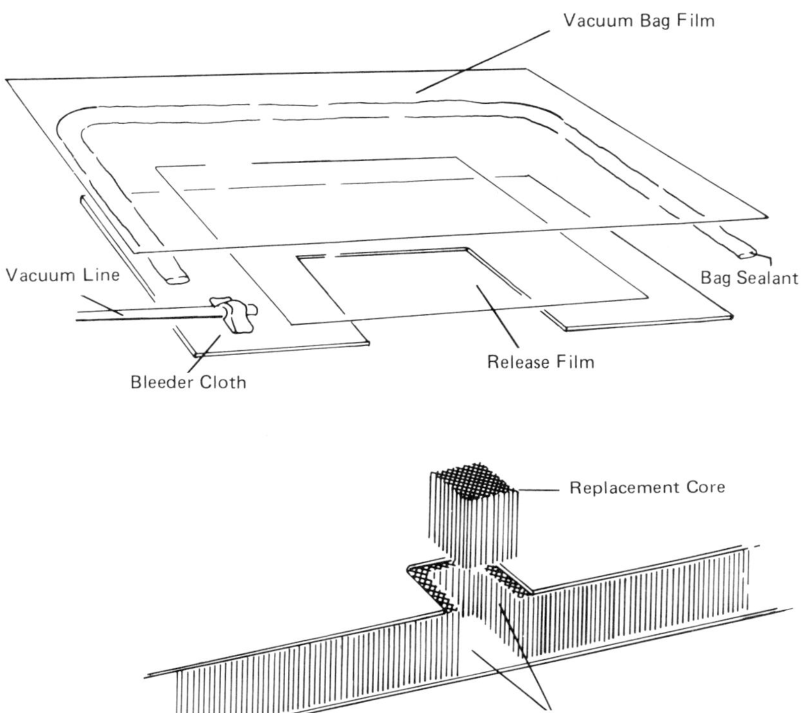

Figure 17   Application of pressure (vacuum) during cure for core replacement.

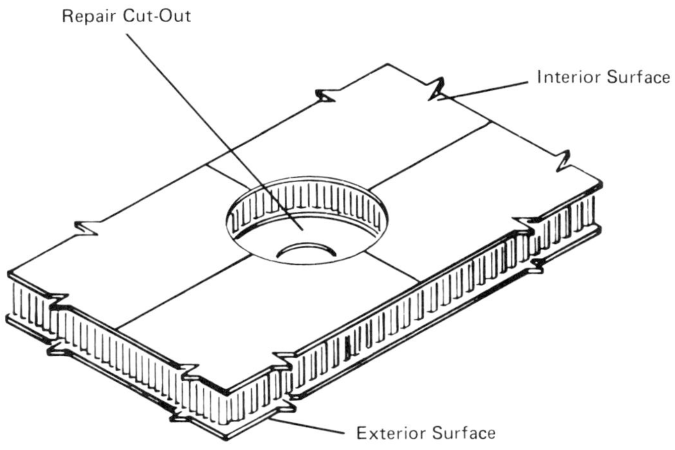

Figure 18   Repair cutout configuration for aerodynamic flush patch (surfaces reversed for clarity).

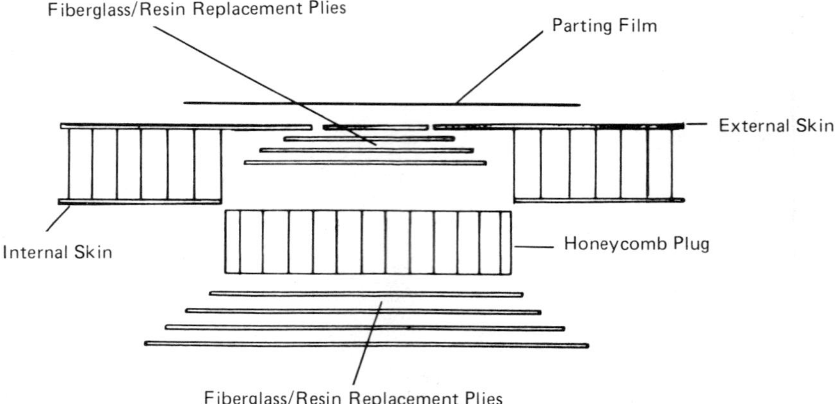

Figure 19   Honeycomb plug repair with fiberglass-resin overlay; aerodynamically flush.

agent is compatible with the wet lay-up resin, insert the core plug and allow the external skin repair and potting to cure. Mill the core plug flush with the internal skin when the potting and fiberglass lay-up have cured and flush out the residue with solvent. When the solvent has evaporated, lay up the required number of internal skin plies. Finish the repair after the resin has cured.

Should repair access be available from only one side of the panel, this type of repair can still be accomplished. The repair would start with the internal skin side, and the cutout configuration would be the same as shown in Fig. 19 (substitute internal skin for external skin). The wet lay-up resin-fiberglass patch should be precured and bonded to the inner skin. After a light sanding and cleaning, the repair would then proceed as described previously, but the external surface patch would not be flush aerodynamically.

Skin and Core Damage—Both Surfaces—Nonaerodynamic
Patch (access from both surfaces)

Begin the repair by removing the entire damaged area and clean as in previous procedures (refer to Fig. 20). Repair either side using a fiberglass-resin lay-up with the number of plies calculated from the original skin thickness. After curing, insert the core plug and pot around the perimeter and to the new skin. As soon as the potting compound has cured, mill the honeycomb flush with the skin and apply an additional fiberglass-resin lay-up. Finish as required after curing.

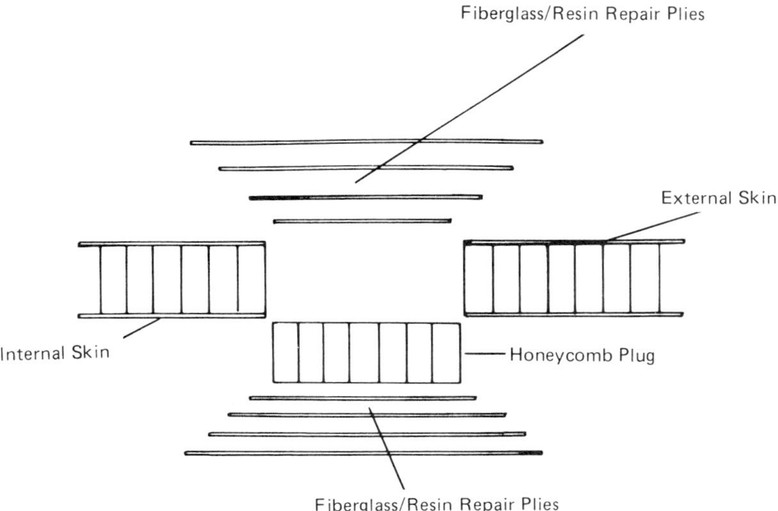

Figure 20 Honeycomb core plug repair with fiberglass-resin overlay; nonaerodynamic.

Trailing-Edge Damage Repair

Minor damage to the trailing edge of wedge-shaped structure can be repaired using a dent filler. Any scratches or irregular areas should be dressed out with a file or similar device. Provided that there is no delamination between the skins and honeycomb, thoroughly clean and then fill in the surface with a dent-filling compound (refer to Fig. 21). If the repair is visually objectionable, cover it with a thin-gauge aluminum wrap around skin which can be bonded over the trailing edge with a MIL-S-8802-type sealant. If special environmental protection is desired (e.g., Skydrol), apply an edge coating of silicone or a similar product.

Major damage to the trailing edge is repaired as shown in Fig. 22. Obvious areas of delamination should be cut out as shown. Depending on the size of the damage, the area is then cleaned and filled with potting compound or a new piece of honeycomb is cut to fit and potted. A fiberglass-resin wet lay-up repair (number of plies depending on skin thickness) to both surfaces completes the operation. Alternatively, if vacuum pressure is available to hold a trailing-edge wraparound doubler in position, the repaired area can be covered with a thin-gauge aluminum sheet bonded with MIL-S-8802 sealant.

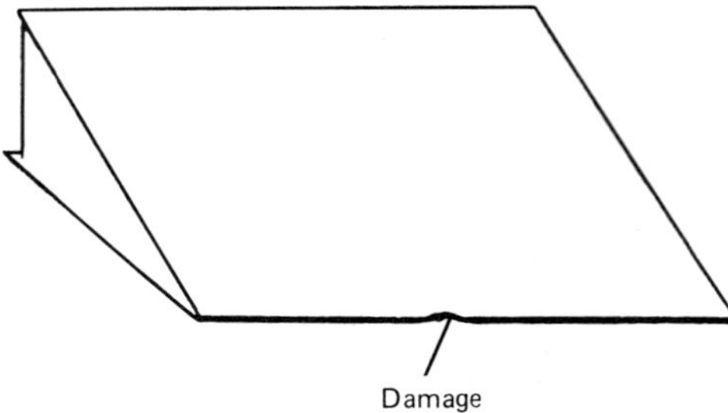

Dress out damage and fill. Cover with
fiberglass/resin overlay or metal wrap-around

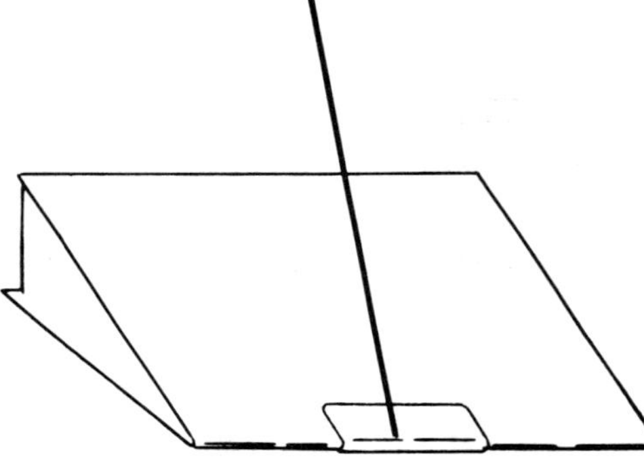

Figure 21   Minor trailing-edge damage repair.

Skin and Core Damage—Prefabricated Plug Repair—Nonaerodynamic

This type of repair can be accomplished in an expeditious manner
provided that repair inserts are available. Several different-diameter
plugs can be prefabricated using the same skin construction, honeycomb core, and adhesive as the original part. Standard repair cutout
sizes will then interface with these prefabricated plugs. See Fig. 23
for further details. The skin thickness of the repair plug will be
equivalent to that of the original panel, so it is important to determine
in advance if the repair will be accomplished from internal surface or
the external surface. If damage occurs through both skins, an insert
is bonded to the repair plug, which fills in the skin contour. This
configuration is similar to the fiberglass wet lay-up repair insert. In
any event, the damage must be dressed out and the area to be repaired
thoroughly cleaned. Prior to this, determine the location of the holes
for serrated rivet installation and line-drill both the repair plug and

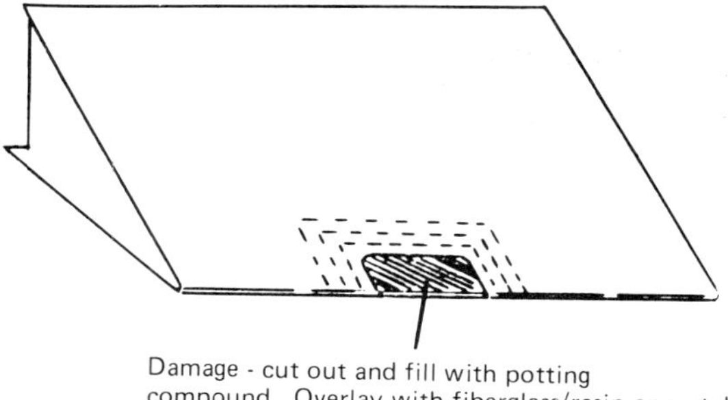

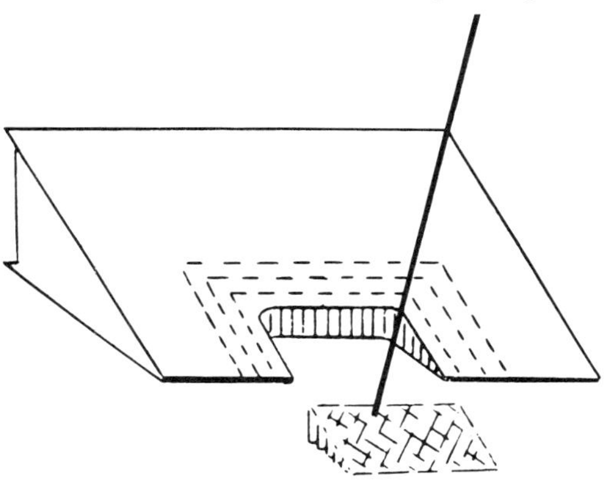

Figure 22   Major trailing-edge damage repair.

the panel. Using an L-shaped tool, open up the honeycomb inside the
panel hole to provide some space for the rivet potting compound. Then,
as shown in Fig. 23, glue the repair plug in the cutout with a room-
temperature or elevated-temperature curing structural epoxy adhesive.
If using a high-temperature adhesive, try to stay about 50°F below the
panel's original cure temperature and keep the panel and repair under
pressure while curing. At the same time that the plug is being installed
and cured, install the serrated rivets as described previously. After
the repair cures, fillet the edge with an environmental sealant and
cover the plug and panel with an environmentally durable primer and
paint system. This will reduce the incidence of corrosion in the repair
area. Note that the repair may be circular or oblong, with size limitations to be determined by the aircraft manufacturer.

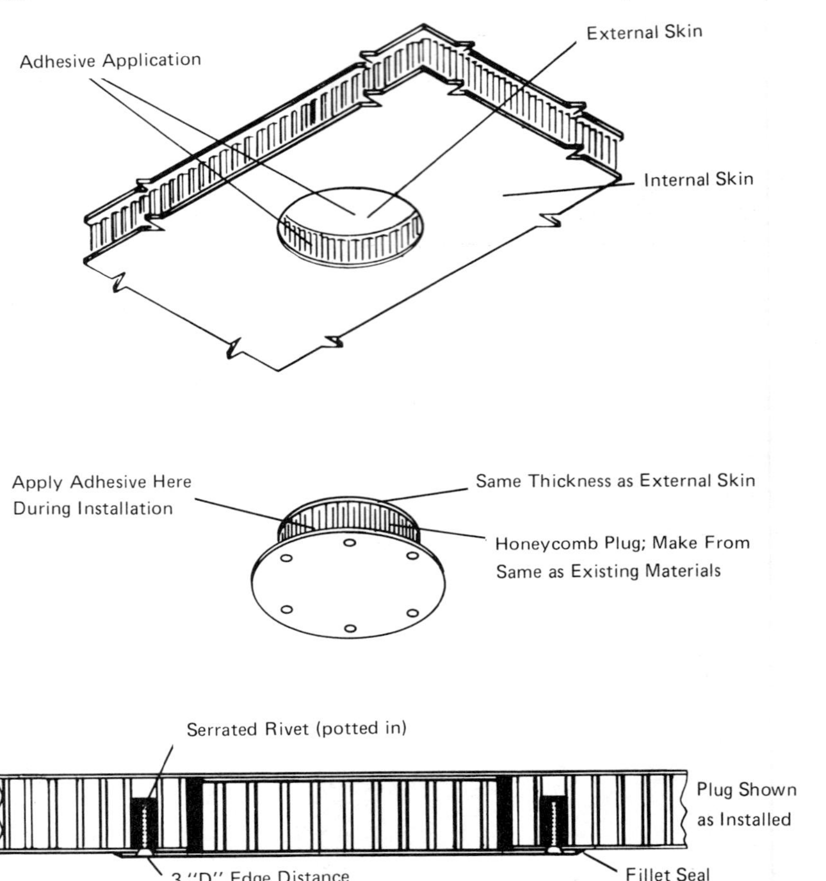

Figure 23   Honeycomb core/aluminum facing plug repair.

Skin and Core Damage—Prefabricated Plug—Aerodynamically Flush

This repair is similar to the plug configuration described previously and starts with removal of the damaged skin and honeycomb within allowable size limitations. The repair patch plug should be manufactured from the same materials as the original panel, with both skins the thickness of the damaged surface. After cleaning, the patch is bonded into the damaged panel cutout as described previously. The depth of the cutout must be controlled to mate with the thickness of the repair plug (see Fig. 24). After repair, the patch plug must fit flush with the existing skin. To complete the repair, seal around the perimeter with

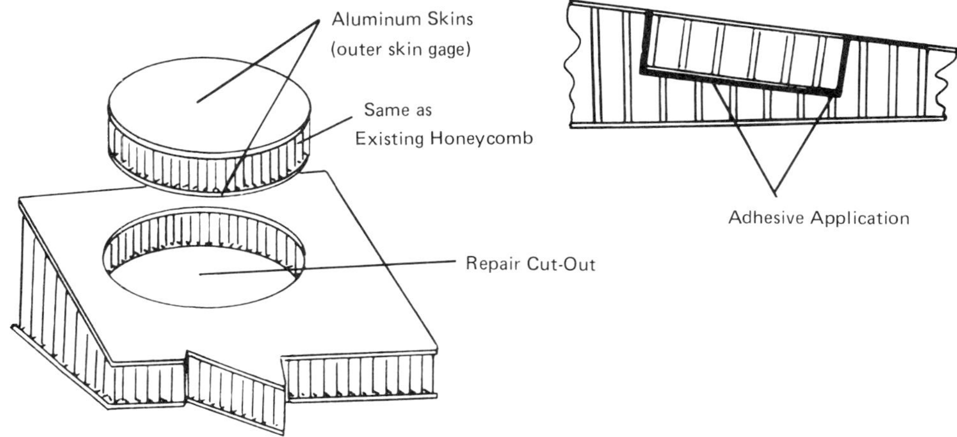

Figure 24   Honeycomb core/aluminum facing plug repair; aerodynamically flush.

an environmental sealant such as MIL-S-8802 and paint with polyurethane paint and a compatible primer.

C. Life-Limited Repairs

Life-limited repairs can be completed in 1 or 2 hr provided that preplanning and heat curing are accomplished. They must be inspected periodically for delamination, corrosion, or other bond degradation. Evidence of such degradation will require replacing the repair, replacing the assembly (time permitting), or accomplishing a permanent autoclave-type repair. Maintaining the environmental seal is the secret for a long-lived interim or life-limited repair. This seal should be replaced when deteriorated, as the repair is sensitive to moisture. Life-limited repairs such as those that follow should be confined to single skin areas of honeycomb panels, and the location of the repair should not interfere with sliding, moving, or stationary adjacent surfaces.

To accomplish this repair, have available several thicknesses of perforated 2024-T3 sheet coated with corrosion-inhibiting adhesive primer. Repair doublers are perforated with 1/8-in. holes on 2-in. centers and contain a minimum corner radius of 1/2 in. The preprimed doublers are packaged so that only a solvent wipe is required prior to bonding. An original thickness of 0.012 to 0.020 in. requires a .025. in. doubler, 0.021 to 0.040 in. requires a 0.050-in. doubler, and 0.040 to 0.063 in. requires a 0.080-in. doubler thickness. Doubler overlap distances as a function of damaged skin thickness are shown in Fig. 25.

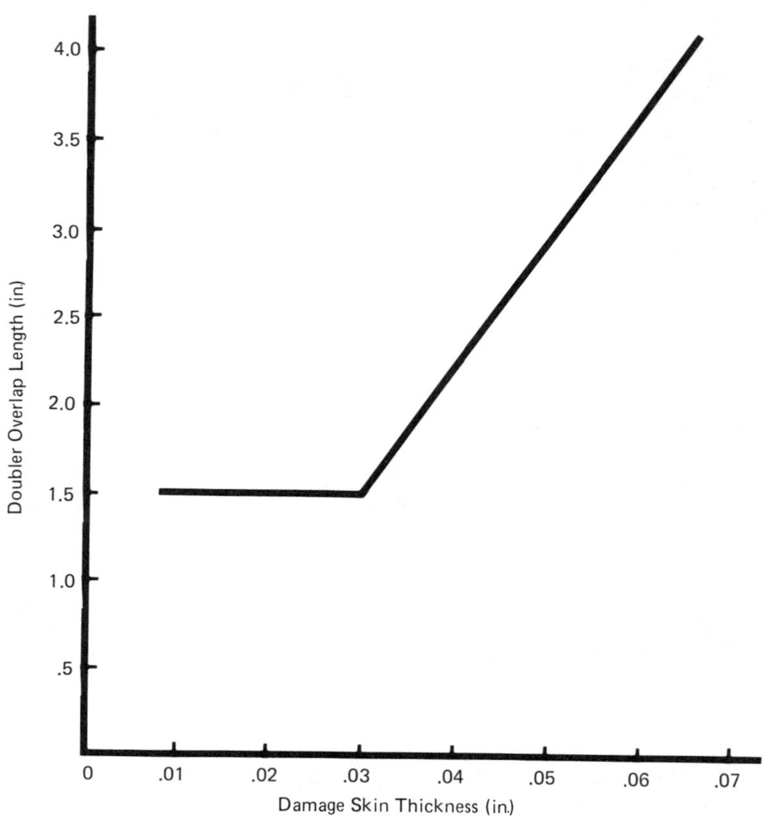

Figure 25   Overlap distance for aluminum doublers. When overlap exceeds 2.5 in. perforated doubler or mating face sheet.

First, evaluate the damage and measure the damaged skin thickness. Select and cut to size the proper preprimed, perforated repair doubler. Assuming that only one skin and the honeycomb are damaged, proceed as specified below. Refer to Fig. 26 for repair configuration.

1. Stop-drill skin cracks or tears and, within the perimeter of the repair doubler, perforate the dented skins with a pattern similar to that on the repair doubler. Be careful not to pene- trate the undamaged skin.
2. Mark the location of the repair doubler on the damaged surface. Drill holes through the doubler and skin on 1 1/2-in. spacings, 1/2 in. inside the outer perimeter, to fit repair rivets. Remove

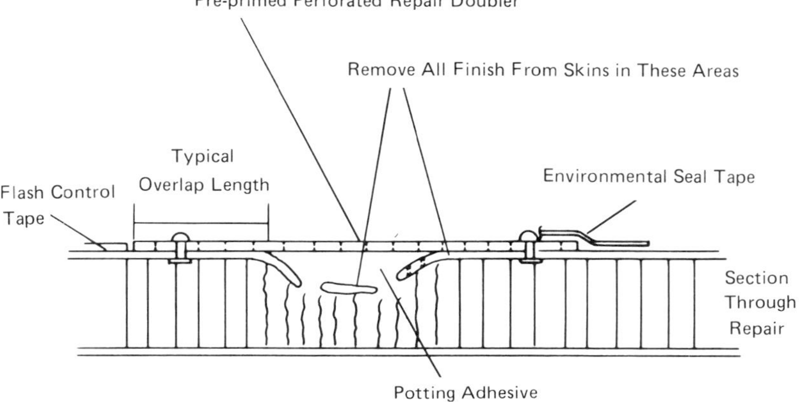

Figure 26   Fast heat-curing interim (life-limited) repair to aluminum honeycomb structures; damage limited to outside skin and honeycomb.

any protective finish in the doubler overlap area on the damaged surface.

3.   Vacuum any residue from the damage; clean the area with abrasive and solvent-wipe.

4.   Solvent-wipe the repair doubler with a clean solvent and purified cheesecloth.

5.   Mix the repair adhesive (a lightweight fast-curing epoxy is recommended) and force or inject the mixed material into cracks and holes. Fill the core and level the adhesive to the surface, coat the repair doubler, and place it in position on the damaged surface.

6.   Blind-rivet the doubler to the damaged surface within the adhesive's pot life.

7.   Apply a vacuum bag over the repair using a parting film in contact with the adhesive and heat-cure the patch. Maintain a repair cure temperature at least 50°F below the panel's original cure temperature.

8.   After cure, remove vacuum bag materials and determine that the repair contains no delamination. Cover with metallized polyester tape to act as an environmental seal.

For damage to both skins, repair from the outside surface as shown in Fig. 27. If the inside skin is cracked but not punctured, stop-drill and bond an aluminum doubler over the stop-drill hole. If the skin is punctured, cut out the damage (minimum size) and measure the skin thickness. Prepare a nonperforated doubler to fit inside the

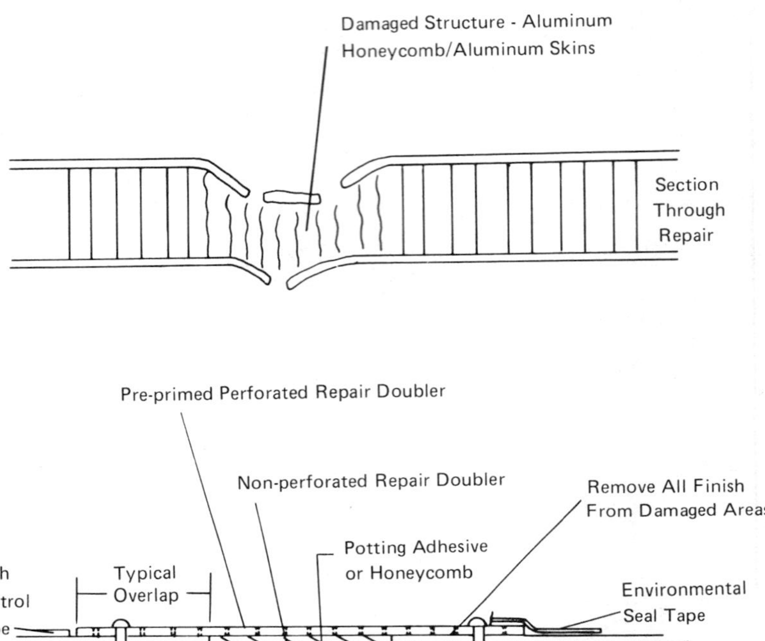

Figure 27   Fast heat-curing interim (life-limited) repair to aluminum honeycomb structure; damage to both skins, repair accomplished from outside.

inner skin. Deburr all edges. Drill the doubler and skin for rivet installation as described previously. Clean surfaces for bonding, mix adhesive, and apply adhesive to inner skin cutout and doubler area. Apply rivets within the adhesive's pot life. Immediately clean the doubler inside skin and proceed with the single skin repair described above. If honeycomb of the proper density and thickness is available, it should be used in place of the potting adhesive to reduce the weight of the repair.

D. Partial Repairs Using Heating Blanket and Vacuum Bag Technique

These repairs are accomplished using a 250°F or 350°F curing structural film adhesive which has been previously tested to ensure that it retains its strength and density after a vacuum pressure/heat exposure. The maximum allowable amount of applied vacuum during cure requires investigation to assure nonremoval of adhesive film constituents under heating conditions. The optimum heat-up rate is another parameter that must be determined in order to minimize adhesive porosity. Since portions of the vacuum pressure/heat-cured bond have a tendency to become slightly porous, proper sealing plus seal maintenance is required for long-lasting repairs. All perforations in the repair doubler, all adhesive flash, edges exposed by machining or trimming, and any fasteners through the bond should be sealed with polysulfide sealant and/or fluid-resisting primer and polyurethane paint. A typical partial repair configuration is shown in Fig. 28.

Two kinds of perforation patterns are used for vacuum cure repairs and the patterns depend on the size of the repair. Areas of metal-to-metal overlap for repairs up to 25 in.^{2} require perforating the inner skin surface to allow air escape during the bonding cycle. Drill 1/8-in. holes on 1.5-in. centers in the overlap area. Use at least two staggered rows of holes and recall that the proper doubler overlap distance as a function of thickness is shown in Fig. 25. All holes should be a minimum of 1/2 in. from the edge of the overlap. Partial repairs of greater than 25 in.^{2} should be manufactured from perforated metal for best results. Otherwise, the edges of the repair tend to pinch off when heat and vacuum are applied and the repair ovalizes. This results in poor bond strength in the center of the repair. A

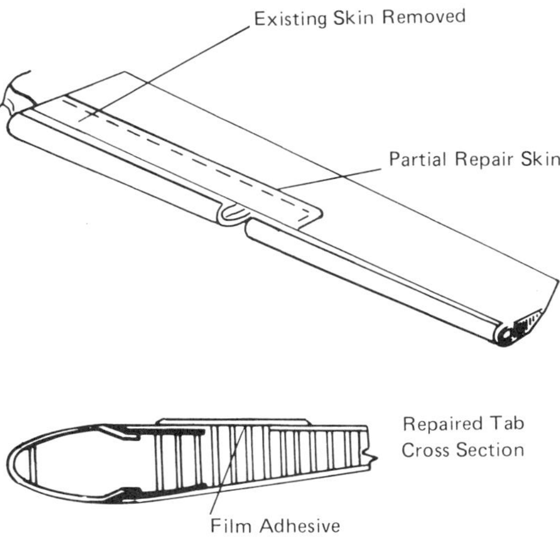

Figure 28 Honeycomb tab; typical partial repair configuration.

perforation pattern of 1/8-in. holes on 1-in. straight centers, 1.5 per square inch, is adequate and can be purchased commercially. As mentioned previously, different-size patches can be premanufactured by processing with a nonsilicated alkaline cleaner, sodium dichromate/ sulfuric acid etch, phosphoric acid anodizing, and then for 250°F curing adhesives, a corrosion-inhibiting adhesive primer.

Silicone heating blankets are fairly effective as a means of curing film adhesive. However, heat can also be obtained from cartridge heaters encased in a form-fitting aluminum block, radiant heat from heat lamps, a hot-air blower or gas heater, and so on, provided that an autoclave is unavailable. A temperature controller and thermocouples should be utilized to ensure that the structure does not overheat and that the requirements for heat-up rate and maximum cure temperature are maintained.

An important requirement for vacuum/heat repairs is that all parts must be prefitted and formed to ensure that the surfaces have good contact over the entire bonded area. All surfaces and edges must be free of burrs and other surface imperfections. This allows adhesive

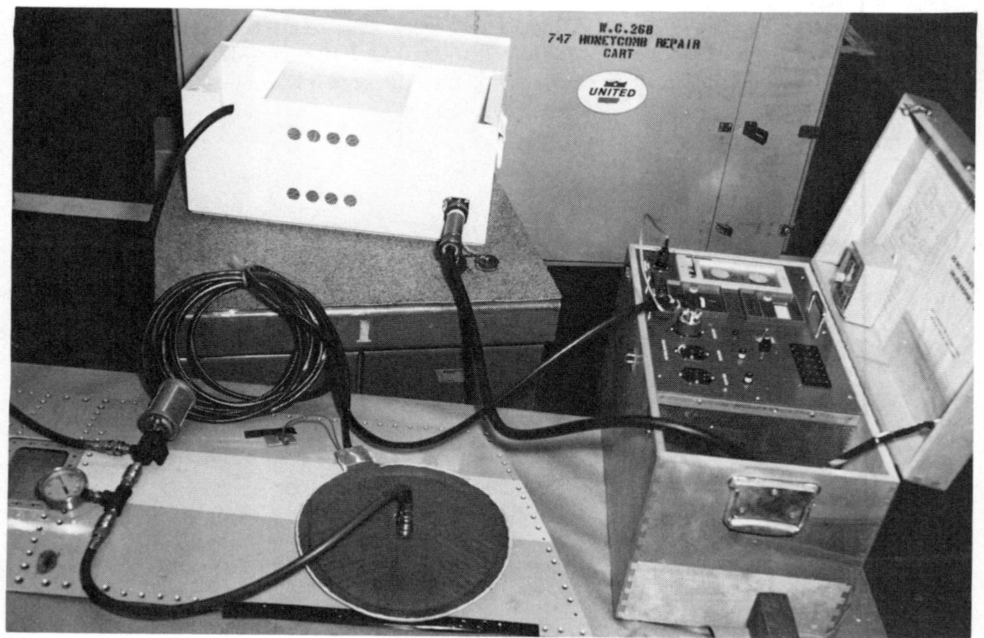

Figure 29   Vacuum heating blanket and instrumentation for curing the repair. (Courtesy United Airlines.)

to flow out of each perforation during cure, which indicates the presence of proper pressure on the repair. At present, vacuum/heat repairs are limited to a maximum of 280 in.^{2}. A silicone heating blanket with a power density of 5 W/in.^{2} is used. If a vacuum pump is unavailable, a vacuum aspirator can be commercially purchased which uses shop air pressure. This instrument is shown in Fig. 29.

One way of accomplishing a vacuum bag/heating blanket repair is to use a blanket which is its own vacuum bag. Blankets like this are expensive and usually limited to the smaller repairs. This type of repair will be illustrated keeping in mind that other styles of heating blankets and assembled vacuum bags (such as those shown in Figs. 14 and 17) can be used. The partial repair proceeds as follows:

1. Assess the damage and clean up the area as described previously. A typical puncture is shown in Fig. 30. Figure 31 shows the damage after cleanup.

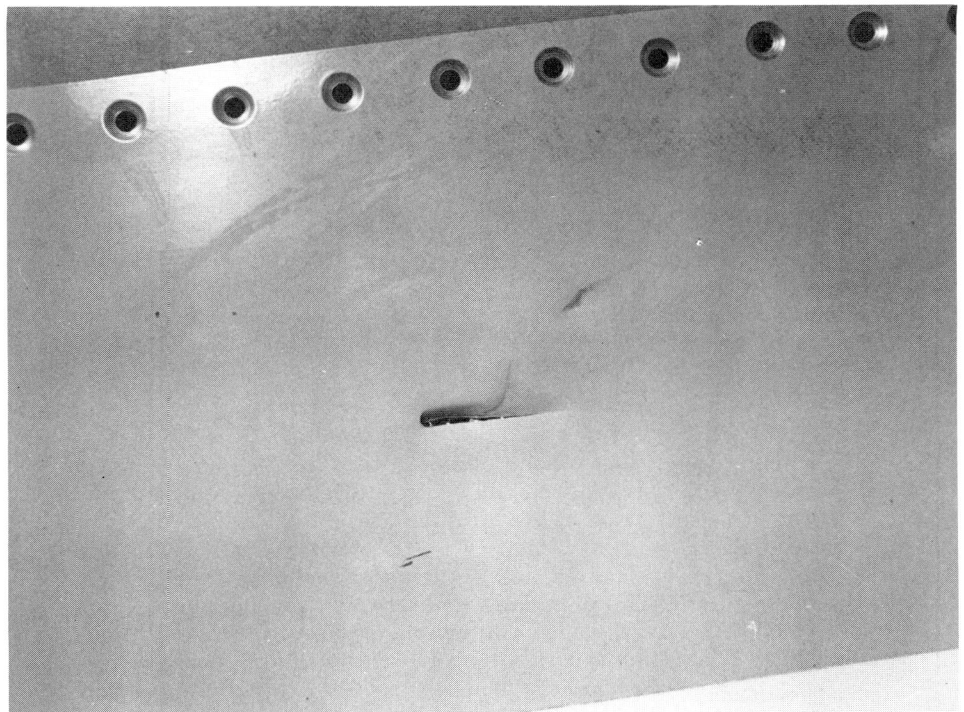

Figure 30 Punctured honeycomb surface. (Courtesy United Airlines.)

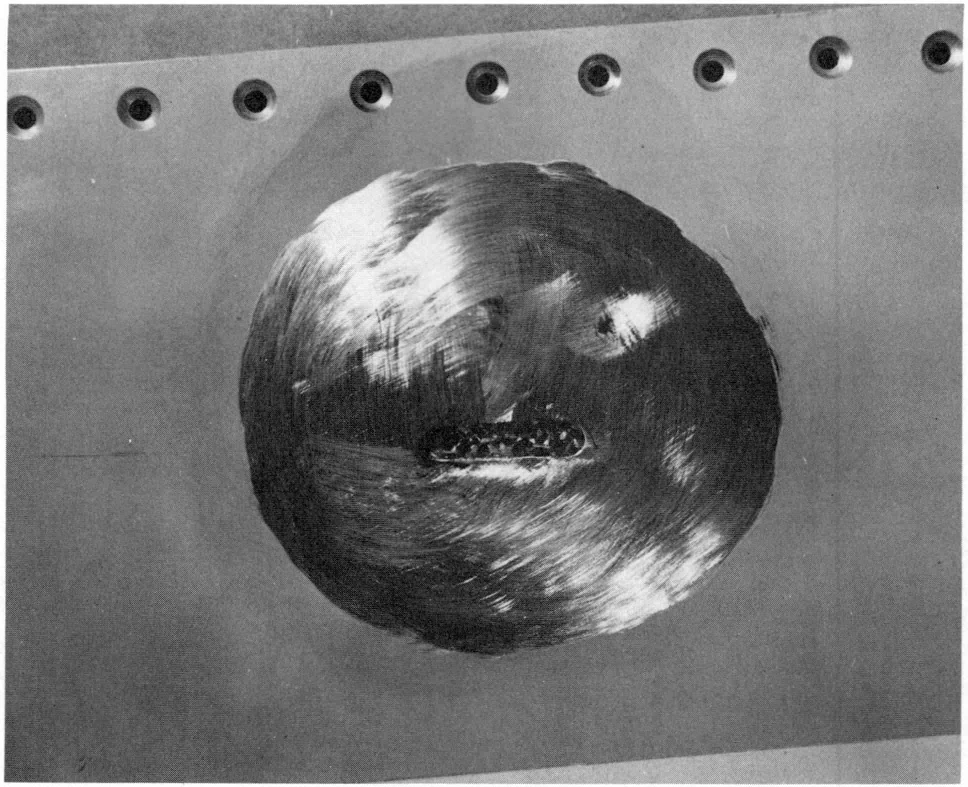

Figure 31   Same honeycomb surface as Fig. 30; cleaned and ready for repair. (Courtesy United Airlines.)

2.   Surface preparation on the damaged structure could consist of abrasion and solvent cleaning, followed by acid paste etching, being careful not to trap acid during application and rinsing. Aluminum foil tape is usually adequate for this purpose. The repair patch can be cleaned similarly or could be given an environmentally durable treatment described previously. Any finish in the damaged area is removed prior to final cleaning.
3.   Potting is applied and smoothed out prior to adhesive application. A heat-setting, expansion-type material that cures at the same temperature as the film adhesive can be used (see Fig. 32).

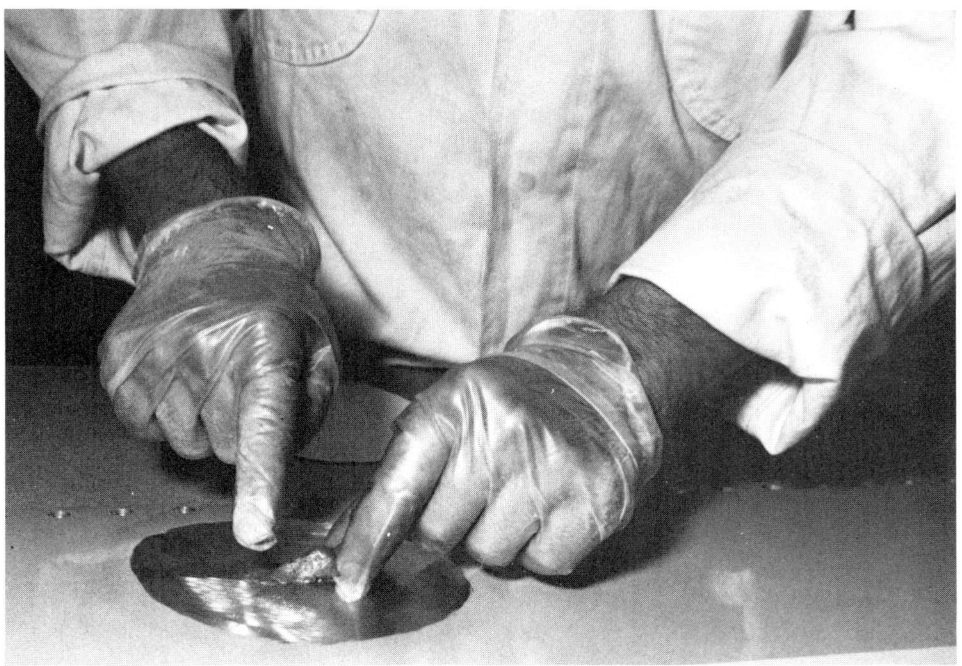

Figure 32 Potting/stabilizing the honeycomb with a foaming adhesive. (Courtesy United Airlines.)

4. Adhesive is applied to the repair patch (Fig. 33) and the patch is secured in position with metal bond flash control tape (see Fig. 34). This type of tape strips clean (without a residue) from the metal surface after cure. Since vacuum pressure is insufficient to form thicker skin contours, any joggle or overlapping repair must be preformed and a layer of adhesive approximately equal to the thickness of the existing skin must be placed in the joggle location. This will prevent void formation after adhesive curing with resulting "oil-canning" in service.

5. The vacuum heating blanket and instrumentation needed to pressurize and cure the repair are shown in Fig. 29. The instrumentation consists of a temperature indicator and controller, thermocouple connections, isolation transformer with ground-fault indicator, vacuum aspirator, and vacuum gage with associated hoses. The blanket itself contains a built-in

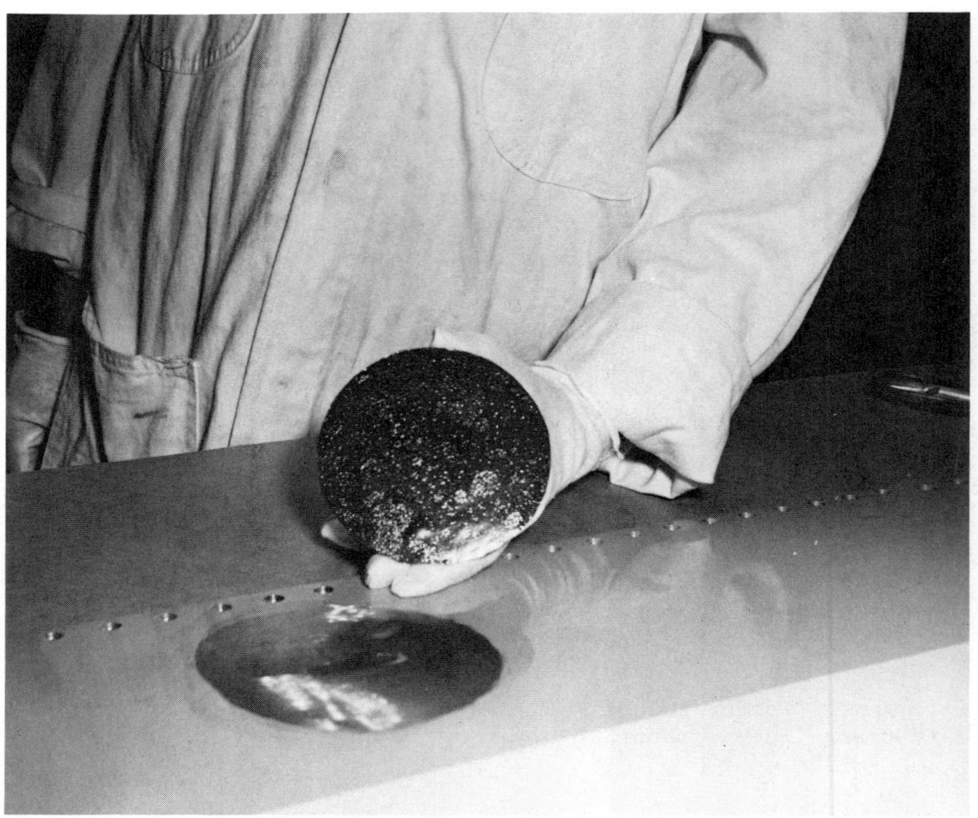

Figure 33   Film adhesive applied to the repair doubler.   (Courtesy United Airlines.)

thermocouple, a composite rubber/metal caul plate (to prevent mark-off) and a bleed pattern which eliminates the need for separate bleeder cloth.

a.   The thermocouple is located adjacent to the repair area and is connected to a jack plug as shown. A piece of release film is placed over the repair. The vacuum aspirator is connected to an 85-psi shop air line, and a vacuum hose is connected to the heating blanket.

b.   The control thermocouple within the blanket is located adjacent to the caul plate and the heating blanket is placed over the repair area. Hand pressure on the edge seal is required to obtain a vacuum. Independent thermocouples are used to determine the exact temperature of

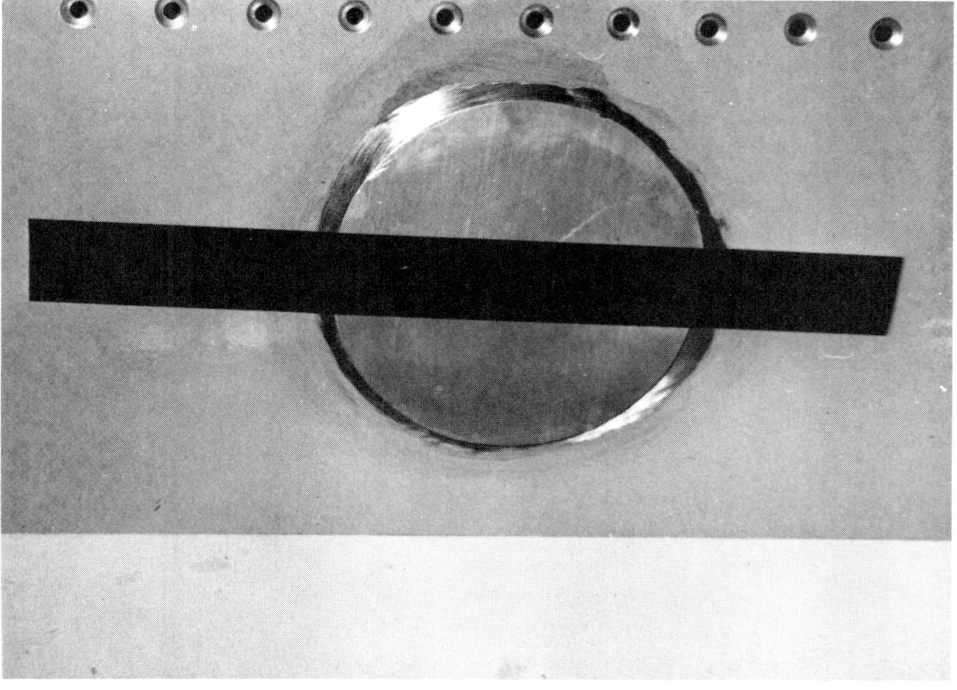

Figure 34   Repair doubler secured in position with flash control tape. (Courtesy United Airlines.)

the repair, since the control thermocouple is located near the vacuum bag heaters and may be affected by them.

c.   The blanket is connected to the electrical supply box which is then attached to the ground-fault indicator and isolation transformer. The isolation transformer is then connected to line voltage.

6.   Since the blanket is under vacuum, the next step is to raise the repair temperature incrementally to the curing range of the adhesive. The temperature controller is used for this purpose and increments usually do not exceed 3°F/min. The repair is then held at cure temperature for the required time.

7.   Upon cure completion, power is disconnected and vacuum is held on the blanket until the temperature falls below 125°F. Vacuum is then disconnected, the release film is removed over the repair, and any excess resin is peeled from the silicone blanket. The completed repair is shown in Fig. 35.

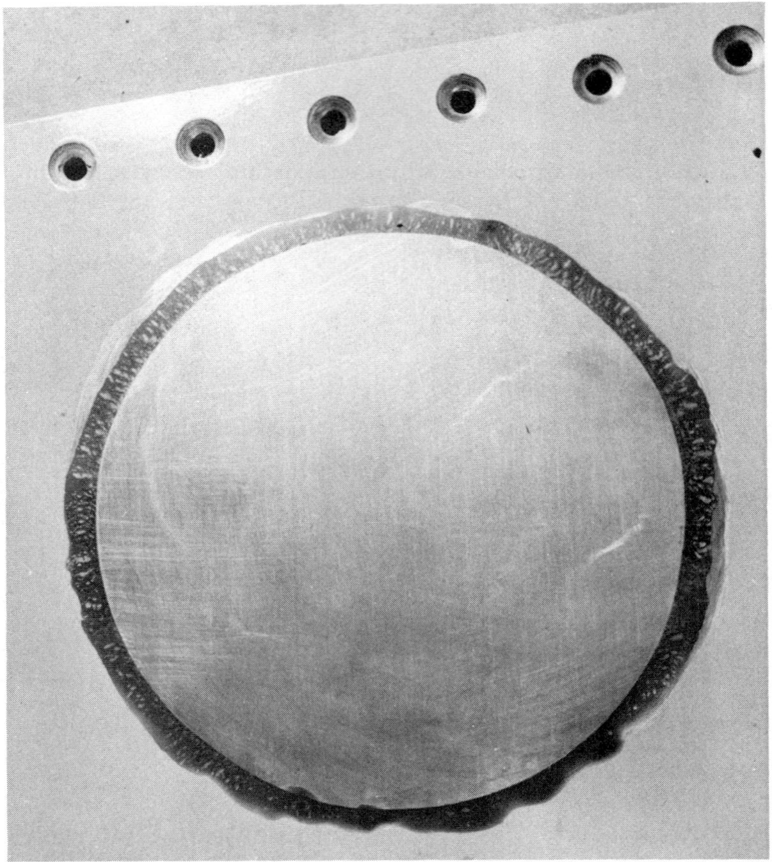

Figure 35   Completed repair, ready for final priming and painting.
(Courtesy United Airlines.)

As mentioned previously, a separate flat heating blanket (commercially available) placed inside a vacuum bag (Fig. 14) can be used in place of an integral vacuum bag/heating blanket. Placement depends on the surface contour and desired heating rate. For example, more heat must be used to penetrate the vacuum bag and vacuum insulating layer with the blanket outside the bag, hence heat-up rates are slower. However, it is not good practice to force the heating blanket, using vacuum pressure, against external fittings or over tight radii. This will damage the blanket and possibly cause an electrical malfunction.

E. Autoclave Repair Techniques

The autoclave is used for repair purposes when the amount of damage
to an adhesive-bonded part is extensive enough to require partial or
complete rebuilding. After years of flight service, many bonded structures contain a significant amount of corrosion and partial repairs are
no longer justified. The autoclave with a supporting tool backing up
the structure to be repaired, is used to provide the pressure-tempera-
ture environment for adhesive curing. Alternatively, as described in
Ref. 5, mechanical damage of the aircraft during production necessitates a permanent repair. If honeycomb or metal bonded structure
can be removed from the aircraft, it is processed through the autoclave as described below. If autoclave support tools are unavailable,
evacuated plastic bags filled with dry sand are sometimes used. Small
"shot" particles or balls are also used for the same purpose when
placed within a frame. If repairs must be accomplished on the aircraft,
pressure is sometimes applied through the use of removable fasteners,
secured through holes in perforated doublers and the original structure. Springs are used to exert curing pressure. See Ref. 5 for
further details. Since autoclave cure techniques have been discussed
earlier in the book, we will not elaborate on details except to show how
they affect the repair process. Similar lack of discussion of surface
treatments prior to bonding will exist because of their discussion elsewhere.

The repair of adhesive-bonded structure is basically accomplished
as follows:

1. Prepare the aluminum surface.
2. Prime the skins, spar, end ribs, and other metal details,
    excluding honeycomb.
3. Heat-bake the primer to drive off solvents and in some
    cases cure the film.
4. Verify correct primer thickness.
5. Apply film adhesive to skin details, locating spar, end ribs,
    and honeycomb as required; apply core splice and spar
    bond adhesives. Assemble test specimens if required.
6. Vacuum bag assemblies.
7. Autoclave cure assemblies.
8. Repeat steps 5 through 7 if more than one cure cycle,
    shaving honeycomb and adding more details as required.
9. Accomplish quality control inspection of cured structure
    and test specimens.
10. Seal and paint assembly.

Surface preparation varies depending on the need of the service.
The optimum treatment is applied to details requiring full skin, partial

repairs, and rebuilt structure. This treatment consists of hot-alkaline
cleaning, hot-acid etching and phosphoric acid anodizing, followed by
application of corrosion-inhibiting primer. Partial skin, spar, rib,
and honeycomb repairs for which limited life is anticipated need only
be alkaline cleaned and acid etched. Priming with a corrosion-inhibiting material is optional. Details that cannot be immersed in cleaning
and etching tanks are etched with acid paste or anodized with nontank
phosphoric acid. If the old adhesive is securely bonded with no evidence of corrosion or other deterioration, sanding and feathering to
the surrounding area is required. The adhesive is then solvent-wiped
until clean. Aluminum honeycomb core is liquid/vapor-degreased or
washed in clean solvent.

Other pertinent observations that relate to adhesive application
and cure as they apply to repair processes are as follows:

1. Avoid buildup of several layers of foaming film adhesives
    (core splice, spar bond) in areas where their heat of cure
    will not be distributed by metal details. The expanding ma-
    terial can also crush honeycomb if the "fit" is too tight.
2. Autoclave cure pressure should, in part, be determined by
    the condition of the old core when accomplishing partial re-
    pairs. Adjust the cure pressure downward if the honeycomb
    is deteriorated.
3. Particularly when curing 250°F adhesive-bonded assemblies
    with more than one cure cycle, the length of the first cycle
    may be reduced to 30 to 35 min since a later heat cycle will
    fully cure the assembly.

Figures 36 to 61 show the steps required to rebuild a typical aircraft spoiler. This illustrates the philosophy of repairing parts using
the latest environmentally durable processes for adhesive-bonded surfaces. The concept is to salvage as much of the old structure as possible. This means selecting a point in the service life of the part
where the spar, end ribs, and fittings can be saved. Beyond this
point in time, corrosion will necessitate the purchase of new structure.
Experience more or less determines the most optimum repair condition
without sacrificing some of the part's service life.

1. Repair begins by indicating the damaged or delaminated areas
    and deciding whether a partial repair or a full rebuild will be
    called for.
2. Dry ice is used to assist in peeling off the upper skin. A
    chisel starts the operation.
3. A "can opener" and some hand labor complete the skin peel
    process.
4. Shown are areas of clad dissolution and subsequent disbond.

Figure 36   Inspection indicates delaminated area. (Courtesy United Airlines.)

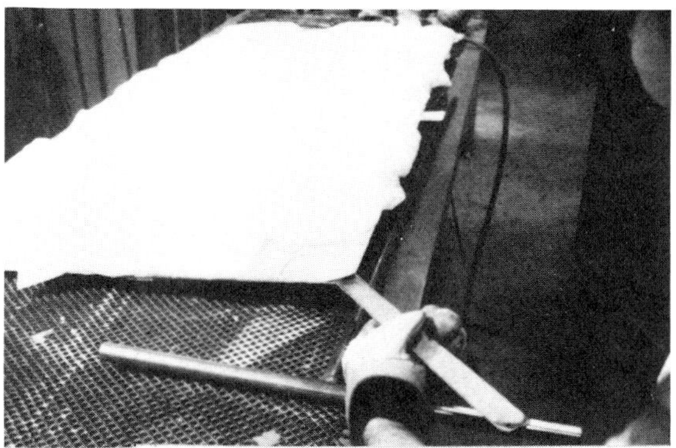

Figure 37   Removal of the upper skin. (Courtesy United Airlines.)

Figure 38   Upper skin removal with "can opener." (Courtesy United Airlines.)

Figure 39   Alclad dissolution and disbond.   (Courtesy United Airlines.)

Figure 40   Making a new skin.   (Courtesy United Airlines.)

Figure 41   Removal of the old, deteriorated structure.   (Courtesy United Airlines.)

Figure 42 Hot-alkaline cleaning immersion. (Courtesy United Airlines.)

Figure 43 Acid etch immersion after rinsing. (Courtesy United Airlines.)

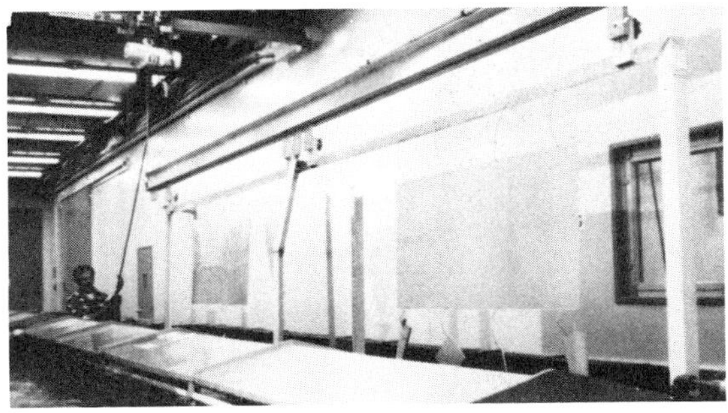

Figure 44 Water spray rinse after acid etch. (Courtesy United Airlines.)

Figure 45   Phosphoric acid anodizing (Boeing proprietary process).
(Courtesy United Airlines.)

Figure 46   Water spray rinse after anodizing.   (Courtesy United
Airlines.)

Figure 47   Air drying. (Courtesy United Airlines.)

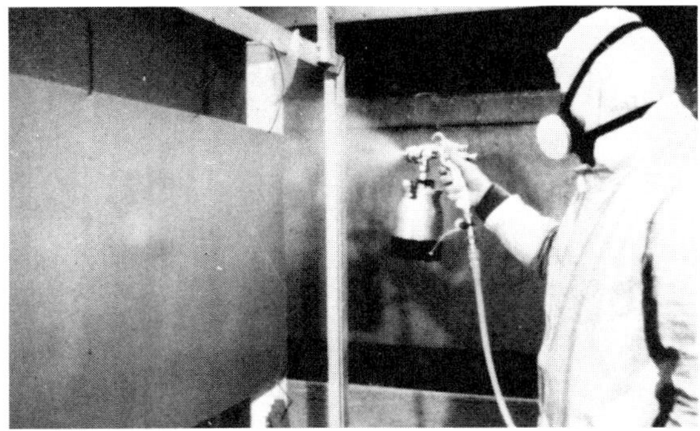

Figure 48   Applying corrosion-inhibiting primer.   (Courtesy United Airlines.)

Figure 49   Curing corrosion-inhibiting primer.   (Courtesy United Airlines.)

Figure 50   Assembly of spar to skin with film adhesive.   (Courtesy United Airlines.)

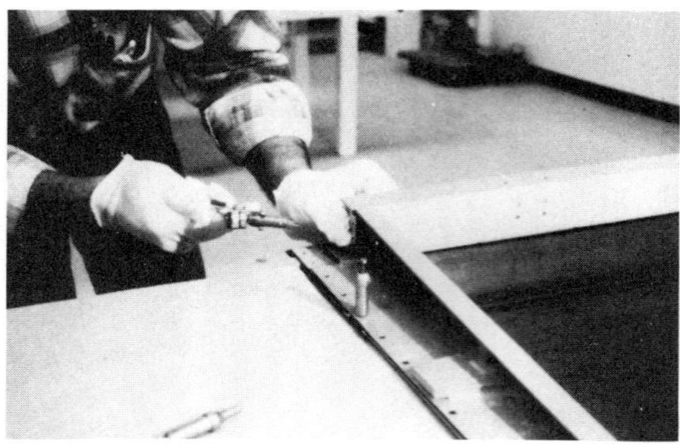

Figure 51 End rib and clip replacement. (Courtesy United Airlines.)

Figure 52 Placement of corrosion-inhibiting honeycomb. (Courtesy United Airlines.)

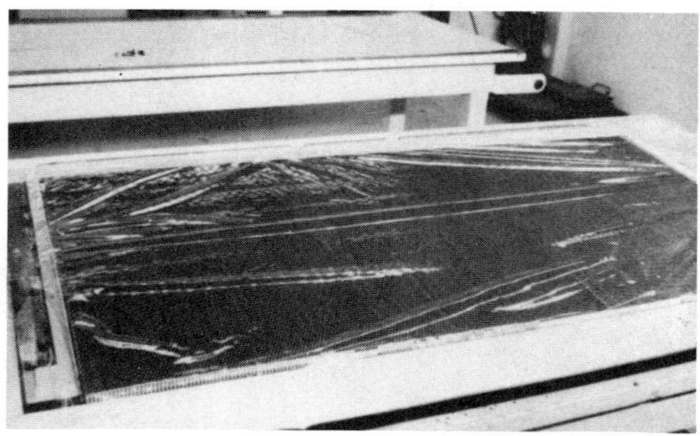

Figure 53 Release film placement during vacuum bagging. (Courtesy United Airlines.)

Figure 54   Ready for autoclave cure.   (Courtesy United Airlines.)

Figure 55   Shaping the honeycomb to final configuration.   (Courtesy United Airlines.)

Figure 56   Application of adhesive to lower surface skin and doublers. (Courtesy United Airlines.)

Figure 57 Positioning doublers with flash control tape. (Courtesy United Airlines.)

Figure 58 Bagging and supporting the assembly before final cure. (Courtesy United Airlines.)

Figure 59 Autoclave curing. (Courtesy United Airlines.)

Figure 60   Final assembly.   (Courtesy United Airlines.)

Figure 61   Finishing the structure.   (Courtesy United Airlines.)

5. A new upper skin is then outlined and sheared.
6. The old corroded structure is cut away.
7. After an immersion stripping process which removes finishes, sealant, adhesive, and adhesive foam, the spar and end ribs, together with skins and doublers, are racked and immersed in hot alkaline cleaning solution.
8. After spray rinsing, the rack is immersed in a hot sodium dichromate/sulfuric acid etch solution.
9. Figure 44 shows spray rinsing in demineralized water after exposure to hot acid.
10. Parts and rack are then immersed in a phosphoric acid anodizing solution.
11. Spray rinsing follows the anodizing operation.
12. The rack is then air dried.

13. 250°F curing adhesives require the application of corrosion-
    inhibiting adhesive primer to the freshly anodized surface;
    other systems may require a compatible primer. These
    protect the anodized surface from damage.
14. The corrosion-inhibiting primer is then cured. Other
    primers may require only partial cure and solvent removal.
15. The part details are disassembled from the handling rack
    and adhesive lay-up begins with the spar-to-upper skin
    assembly. The adhesive has already been inspected.
16. End ribs and clips are then added.
17. Corrosion-inhibiting honeycomb is located inside the rib/
    skin/spar assembly and trimmed to fit.
18. Release film is placed over the uncured assembly and test
    specimens. Bleeder cloth is added and a vacuum bag is
    located over the subassembly.
19. Figure 54 shows the vacuum-bagged spoiler assembly ready
    for the first-stage autoclave cure.
20. After the adhesive cure cycle, the vacuum bag and release
    materials are removed and the honeycomb is contoured to
    final shape.
21. The lower honeycomb surface skins and doublers are covered
    with adhesive and positioned over the contoured honeycomb.
22. Doublers are fixed in position with flash control tape.
23. Release film is placed over the assembly, bleeder cloth is
    added, and another vacuum bag is fabricated.
24. The surface undergoes a second autoclave cure.
25. The vacuum bag is removed and some fittings and seals are
    added. This is the final assembly.
26. The spoiler is covered with a polyurethane filaform corrosion-
    resisting primer and paint system.

Similar processes apply to other aircraft parts, such as flaps,
aileron tabs, rudder tabs, and flap vanes—in fact, any bonded
assembly.

REFERENCES

1. Marceau, J. A. and Y. Moji, A Wedge Test for Evaluating Adhe-
   sive Bond Surface Durability, *Adhes. Age*, Oct. 1977.
2. Marceau, J. A. and J. C. McMillan, Exploratory Development on
   Durability of Adhesive Bonded Joints, U.S. Air Force Materials
   Lab. Tech. Rep. AFML-TR-76-172, Wright-Patterson AFB, Ohio
   Oct. 1976.
3. McMillan, J. C., J. T. Quinlivan, and R. A. Davis, Phosphoric
   Acid Anodizing Aluminum for Structural Adhesive Bonding,
   SAMPE Q., Apr. 1976.

4. Douglas Aircraft Company, Primary Adhesively Bonded Structures Technology (PABST), Tech. Bull. 2, U.S. Air Force Contract F 33615-75-C-3016, Long Beach Calif., May 1975.
5. Horton, R. E., W. M. Scardino, and H. Croop, Adhesive Bonded Aerospace Structures Standardized Repair Handbook, U.S. Air Force Flight Dynamics Lab., Contract AF 33615-73-C-5171, Wright-Patterson AFB, Ohio, 1973.
6. Locke, M.C., R. E. Horton, and J. E. McCarty, Anodize Optimization and Adhesive Evaluations for Repair Applications, U.S. Air Force Materials Lab. Tech. Rep. AFML-TR-78-104, Wright—Patterson AFB, Ohio, July 1978.
7. Douglas Aircraft Company, Primary Adhesively Bonded Structure Technology (PABST), U.S. Air Force Contract F 33615-75-C-3016, 1975.

Index